AF062373

Cardiovascular Disease 2

Cellular and Molecular Mechanisms,
Prevention, and Treatment

GWUMC Department of Biochemistry
Annual Spring Symposia
Series Editors:
Allan L. Goldstein, Ajit Kumar, and J. Martyn Bailey
The George Washington University Medical Center

Recent volumes in this series:

ADVANCES IN MOLECULAR BIOLOGY AND TARGETED TREATMENT FOR AIDS
Edited by Ajit Kumar

BIOLOGY OF CELLULAR TRANSDUCING SIGNALS
Edited by Jack Y. Vanderhoek

BIOMEDICAL ADVANCES IN AGING
Edited by Allan L. Goldstein

CARDIOVASCULAR DISEASE
Molecular and Cellular Mechanisms, Prevention, and Treatment
Edited by Linda L. Gallo

CARDIOVASCULAR DISEASE 2
Cellular and Molecular Mechanisms, Prevention, and Treatment
Edited by Linda L. Gallo

CELL CALCIUM METABOLISM
Physiology, Biochemistry, Pharmacology, and Clinical Implications
Edited by Gary Fiskum

THE CELL CYCLE
Regulators, Targets, and Clinical Applications
Edited by Valerie W. Hu

GROWTH FACTORS, PEPTIDES, AND RECEPTORS
Edited by Terry W. Moody

NEURAL AND ENDOCRINE PEPTIDES AND RECEPTORS
Edited by Terry W. Moody

PROSTAGLANDINS, LEUKOTRIENES, AND LIPOXINS
Biochemistry, Mechanism of Action, and Clinical Applications
Edited by J. Martyn Bailey

PROSTAGLANDINS, LEUKOTRIENES, LIPOXINS, AND PAF
Mechanism of Action, Molecular Biology, and Clinical Applications
Edited by J. Martyn Bailey

Cardiovascular Disease 2

Cellular and Molecular Mechanisms, Prevention, and Treatment

Edited by

Linda L. Gallo

The George Washington University Medical Center
Washington, D.C.

Springer Science+Business Media, LLC

Library of Congress Cataloging-in-Publication Data

On file

Proceedings of the Fourteenth Washington International Spring Symposium at The George Washington University, held June 6–10, 1994, in Washington, D.C.

ISBN 978-1-4613-5805-3 ISBN 978-1-4615-1959-1 (eBook)
DOI 10.1007/978-1-4615-1959-1

Originally published by Plenum Press in 1995
Softcover reprint of the hardcover 1st edition 1995

10 9 8 7 6 5 4 3 2 1

PREFACE

The Fourteenth Washington International Spring Symposium, held in Washington, D.C., in June 1994, brought together over 400 leading scientists from 21 countries to review and update research on cardiovascular disease. This group satisfied the symposium goals of formulating a more comprehensive and integrated picture of the events contributing to atherosclerosis and of exploring modified gene expression as an approach to understanding the causes of atherosclerosis and providing clues to the prevention and treatment. This volume contains most of the papers presented at the eight plenary sessions together with selected contributions from the special sessions. The multidisciplinary nature of the chapters and their authors should stimulate the interests of biochemists, cell and molecular biologists, pathologists, pharmacologists, epidemiologists, nutritionists, and clinicians.

The volume is divided into eight sections which reflect the focus of the plenary sessions. Part 1 focuses on the pathophysiology of atherosclerotic plaques and predicts that the nature of the fibrous cap of atheroma determines plaque disruption and clinical events.

Papers in Part II deal with atherogenic lipoproteins. The introductory paper reviews the current view of the role of plasma lipoproteins in atherosclerosis. With respect to the newer members on the list, [oxidized LDL, Lp(a)] evidence is provided that suggests the involvement of one major gene in the development of oxidized LDL lipids, the expression of inflammatory genes, and the development of aortic fatty streaks. Further, we are alerted to a new functional polymorphism in Lp(a), specifically its lysine binding property, which should be considered when assessing the role of Lp(a) in atherosclerotic cardiovascular disease. Other papers in this section discuss apolipoprotein E isoforms and their relationship to LDL cholesterol levels and coronary artery disease and expand upon the basis for small dense LDL (hyperapo B) atherogenicity.

Papers in Part III are concerned with antiatherogenic lipoproteins. A complex, high affinity HDL receptor pathway is described which identifies candidate receptor proteins and involves the action of protein kinase C. Evidence in support of this signalling pathway is provided by studies in Tangier's fibroblasts in which there is a defect in HDL_3-mediated cholesterol efflux associated with reduced protein kinase C activation. Other papers in this section deal with the effectiveness of extracellular acceptors for promoting cholesterol efflux in vitro and correlate acceptor types found in human serum with longevity and coronary artery disease.

In Parts IV, V, and VI, attention is focused on the vessel wall as an integrator of pathophysiologic stimuli which play a role in the etiology of cardiovascular disease. Part IV deals with signal transduction pathways which regulate the expression of genes involved in vascular growth and inflammatory responses. The identity of FGF-1-inducible genes in connection with smooth muscle cell hyperplasia, redox-sensitive expression of adhesion molecules by endothelial cells, potentiation of PDGF gene expression and mechanisms of PDGF-induced cellular proliferation, the regulatory role of VEGF in endothelial cell proliferation, and gene expression associated with the developmental transition of smooth muscle cells from a proliferative to a quiescent phenotype are discussed.

Part V concentrates on adhesion pathobiology with a discussion of genetic deficiency syndromes which have provided important insights into the molecular basis and the biology of leukocyte adhesion. Results are presented which document the central role of the selectin-family of adhesion receptors in governing the migration patterns of different leukocyte classes. Further, the identity of endothelial ligands for the selectins and the implications for antiinflammatory therapeutics are discussed. Finally, the possible involvement of P-selectin in thrombogenesis and the domains of P-selectin and the P-selectin ligand that mediate adhesion are described.

Part VI is devoted to thrombosis and fibrinolysis with an emphasis on proteases, protease inhibitors, receptors, and thrombolytic therapies. Papers focusing on fibrinolysis in this section reveal a correlation between PAI-1 gene expression and the severity of atherosclerosis and demonstrate that monocytes produce PAI-2 when stimulated with oxidatively modified LDL, events which may imbalance the fibrinolytic system. The structure, function, and regulation of the urokinase plasminogen activator receptor are described and LRP is discussed in light of its dual ability to regulate plasma and cell surface proteinase levels and to play a role in the lipoprotein lipase promoted clearance of remnant lipoproteins. Turning to thrombosis, papers describe the molecular biology of its initiation, the structure and function of the thrombin receptor and the behavior of prothrombin at the de-endotheliased aorta.

Modified gene expression as an approach to provide greater understanding of the mechanisms, causes, and treatments for atherosclerosis is the subject of papers in Part VII. The progression from genetic dyslipoproteinemias to their molecular understanding to gene therapy unfolds. The advantages gained by simultaneous studies in mice and humans in the analysis of genetic factors contributing to atherosclerosis are described. Specific use of the techniques of gene transfer and disruption to modify the LDL and LRP receptors, "knockout" the apo C_1 gene, to overexpress apo B and apo (a) in mice, and to introduce growth factors into the artery wall are described and the conclusions drawn from these manipulations discussed. The last paper in this section describes the use of a PCR-based subtraction library to isolate novel genes regulated by dietary cholesterol.

In Part VIII, entitled Atherosclerosis Prevention and Public Policy, the important policy issues of who to screen, when to screen, who to treat and when to treat are debated.

Linda L. Gallo

Washington. D.C.

CONTENTS

PART I - ATHEROGENESIS

1. Lipid Accumulation and Plaque Disruption: Processes Triggering Clinical Instability in Coronary Disease - An Overview 1
B. Greg Brown, Xue Q. Zhao, Drew Poulin, and John J. Albers

2. Expression of 92 kDa Gelatinase in Human Atherosclerotic Lesions Following Recent Plaque Rupture 11
David L. Brown, Margaret S. Hibbs, Marianne Kearney, Carrie Loushin, Eric J. Topol, and Jeffrey M. Isner

3. Proteinases and Restenosis: Matrix Metalloproteinases and Their Inhibitor and Activator 19
Suresh C. Tyagi, Larry Meyer, Richard A. Schmaltz, H.K. Reddy, and Donald J. Voelker

PART II - ATHEROGENIC LIPOPROTEINS

4. Current Concepts of the Plasma Lipoproteins and their Role in Atherosclerosis - An Overview 31
H. Bryan Brewer, Jr.

5. Homeostasis of Lipid Oxidation in the Artery Wall 41
Alan M. Fogelman, Judith A. Berliner, Mahamad Navab, Ali Andalibi, Feng Liao, Linda L. Demer, Mary C. Territo, and Aldons J. Lusis

6. Molecular Basis for the Lysine Binding Polymorphism of Lipoprotein (a) 45
Angelo M. Scanu, Ditta Pfaffinger, Olga Klezovitch, and Celina Edelstein

7. Triglyceride-rich Lipoprotein Metabolism and Diabetes 49
George Steiner and Gary F. Lewis

8. Apolipoprotein E: Paradoxes Abound 57
William R. Hazzard, Deborah Applebaum-Bowden, and James G. Terry

9. The Role of Second Messenger Pathways in the Pathophysiology of Hyperapo-B and Premature Coronary Artery Disease 67
Peter O. Kwiterovich, Jr. and Mahnaz Motevalli

PART III - ANTI-ATHEROGENESIS/CHOLESTEROL REMOVAL MECHANISMS

10. The Role of HDL Receptors in Removal of Cellular Cholesterol 75
John F. Oram, Armando J. Mendez, Gordon A. Francis, and Edwin L. Bierman

11. The Defect in HDL_3 Mediated Efflux of Newly Synthesized Cholesterol is Associated with Impaired Activation of Protein Kinase C in Tangier Fibroblasts 79
Gerd Schmitz, Gerhard Rogler, Wolfgang Drobnik, Barbara Trumbach, Christoph Moellers, and Karl J. Lackner

12. Cholesterol Efflux from Cells in Culture, Reconstituted Particles, and Whole Serum 89
G.H. Rothblat, P. Yancey, W.S. Davidson, V. Atger, S. Lund-Katz, W.J. Johnson, M. de la Llera Moya, and M.C. Phillips

13. Apo A-1 Containing Particles and Atherosclerosis 97
Jean-Charles Fruchart, Garciella Castro, and Patrick Duriez

PART IV - SIGNAL TRANSDUCTION IN VASCULAR PROLIFERATION

14. Vascular Endothelium: An Integrator of Pathophysiologic Stimuli in Cardiovascular Disease - An Overview 105
Michael A. Gimbrone, Jr.

15. Identification of FGF-1-Inducible Genes by Differential Display 109
Jeffrey A. Winkles, Patrick J. Donohue, Debbie K.W. Hsu, Yan Guo, Gregory F. Alberts, and Kimberly A. Peifley

16. Antioxidants and Endothelial Expression of VCAM-1: A Molecular Paradigm for Atherosclerosis 121
Russell M. Medford

17. Mechanisms of Potentiation of PDGF in Atherosclerosis 129
Xin-Hua Lin, Zhao-Yi Wang, Hyeong Reh Kim, and Thomas F. Deuel

18. The Regulation of Normal and Pathological Angiogenesis by Vascular Endothelial Factor
Napoleone Ferrara 133

19. Characterization of Developmentally Associated Changes in Rabbit Vascular Smooth Muscle Cells 145
David K.M. Han and Gene Liau

PART V - ADHESION PATHOBIOLOGY

20. Molecular Basis and Pathologic Consequences of Neutrophil Adherence to Endothelium 153
John M. Harlan, Robert K. Winn, Sam R. Sharar, and Amos Etzioni

21. The Distribution of Adhesion Molecules in Normal and Atherosclerotic Arteries and Aortas 159
Dinah V. Parums

22. L-Selectin Regulation of Lymphocyte Homing and Leukocyte Rolling and Migration 173
Thomas F. Tedder, Anjun Chen, and Pablo Engel

23. L-Selectin, A Lectin-Like Receptor Involved in Normal Lymphocyte Recirculation and Inflammatory Leukocyte Trafficking 185
Steven D. Rosen

24. New Perspectives in P-Selectin Biology 191
Barbara C. Furie and Bruce Furie

25. Rat P-Selectin Mediates Neutrophil-Platelet Interactions via Two Sites (23-60, 76-90) Located on Its Lectin Domain 199
Elza Chignier, Marie-Hélène Sparagano, Lilian McGregor, Annie Thillier, Dorothée Pellecchia, Marié-Pierre Reck, and John McGregor

PART VI - THROMBOSIS/FIBRINOLYSIS

26. Regulation of Vascular Fibrinolysis by Type 1 Plasminogen Activator Inhibitor (PAI-1) 205
D. Seiffert, B.E. van Aken, and D.J. Loskutoff

27. Local Increase in PAI-2 on Stimulation of Monocytes with Modified LDL 211
Helen M. Ritchie, Alec Jamieson, and Nuala A. Booth

28. Structure, Function, and Regulation of the Urokinase Receptor 217
Francesco Blasi

29. Role of the LDL Receptor-Related Protein in Proteinase and Lipoprotein Catabolism 223
Dudley K. Strickland, Suzanne E. Williams, Maria Z. Kounnas, W. Scott Argraves, Ituro Inoue, Jean-Marc Lalouel, and David A. Chappell

30. Molecular Biology of Tissue Factor 235
Thomas A. Edgington and Wolfram Ruf

31. Thrombin Receptor: Structure and Function 243
Kenji Ishii, Ji Chen, Maki Ishii, Thien-Kahi H. Vu, Robert E. Gerszten, Tania Nanevicz, Ling Wang, and Shaun R. Couglin

32. Antiplatelet Effects of Direct Acting Thrombin Inhibitors and Platelet GPIIb/IIIa Antagonists: Comparative Analysis 255
Shaker A. Mousa and Thomas M. Reilly

33. Biosynthesis of Docosanoids by Human Platelet: Cardiovascular Properties 269
John W. Karanian, Hee Yong Kim, and Norman Salem, Jr.

34. The Hemostatic Response to Arterial Injury in vivo: Behavior of Prothrombin at the De-endotheliased Aorta Wall 279
M.W.C. Hatton, S.M.R. Southward, S.D. Serebrin, M. Kulczycky, and M.A. Blajchman

35. Plasma Coagulation Factors in Emergency Room Patients with Acute Chest Pain and Subsequent Hospitalization: Myocardial Infarction, Coronary Artery Disease, and Hypertension 291
C.F. Saladino, V. Misra, N. Sathish, R. Fox, S.E. Feffer, and E.A. Jonas

PART VII - MODIFIED GENE EXPRESSION/CLUES TO CAUSE, PREVENTION, AND TREATMENT OF ATHEROSCLEROSIS

36. Genetic Factors Contributing to Atherosclerosis: From Humans to Mice and Back Again 299
Craig H. Warden and Aldons J. Lusis

37. Genetic Manipulation of Lipoprotein Receptors: Implications for Lipid Metabolism and Atherosclerosis 307
Thomas E. Willnow, Shun Ishibashi, and Joachim Herz

38. Genetic Models of Vascular Disease 313
Elizabeth G. Nabel, David Gordon, Diane P. Carr, Takeshi Ohno, Zhiyong Yang, Hong San, and Gary J. Nabel

39. Generation and Characterization of Apolipoprotein C_1-Deficient Mice 317
Marten Hofker, Janine van Ree, Walter J.A.A. van den Broek, Jan M.A. van Deursen, Hans van der Boom, Rune Frants, Bé Wieringa, and Louis Havekes

40. Studies of Apolipoprotein B and Lipoprotein(a) in Transgenic Mice 323
Edward M. Rubin

41. Isolation of Novel Genes Regulated by Dietary Cholesterol by a PCR-Based Subtraction Library 327
Alan T. Remaley, U. Kurt Schumacher, H. Bryan Brewer Jr., and Jeffrey M. Hoeg

PART VIII - ATHEROSCLEROSIS PREVENTION AND PUBLIC POLICY

42. Cholesterol and Mortality: What Can Meta-Analysis Tell Us?. 333
David J. Gordon

43. Screening for High Blood Cholesterol: A Risky Enterprise 341
Steven B. Hulley

44. Cholesterol Lowering, Low Cholesterol, and Mortality 347
John C. LaRosa

45. Bavarian Cholesterol Screening Project (BCSP) 353
Peter Schwandt, Werner O. Richter, and Andreas C. Sönnichsen

Index 359

LIPID ACCUMULATION AND PLAQUE DISRUPTION: PROCESSES TRIGGERING CLINICAL INSTABILITY IN CORONARY DISEASE

B. Greg Brown, Xue Qiao Zhao,
Drew Poulin, and John J. Albers

From the Department of Medicine
Cardiology Division, University of Washington
School of Medicine, Seattle, WA

This review focuses on the interactions among arterial lipid content, disruptive changes in plaque structure, and subsequent clinical events. The concept that there are acute "triggering" risk factors for plaque disruption has been described in detail elsewhere (1, 2, 3) and we will summarize the salient aspects of this approach to understanding coronary artery disease.

Lipid Lowering and Atherosclerosis Progression: A number of published trials involving mostly symptomatic patients have examined the impact of a variety of different lipid-lowering interventions on coronary atherosclerosis progression and regression, as assessed from the angiogram. Those include NHLBI Type II (4) (resin), CLAS (5) (resin and niacin), POSCH (6) (partial ilial bypass), Lifestyle (7) (vegetarian diet and meditation), FATS (8) (resin plus niacin or lovastatin), UC-SCOR (9) (resin, niacin ± lovastatin), STARS (10) (diet ± resin), SCRIP (11) (exercise, lipid and B.P. drugs), CCAIT (12) (lovastatin), MARS (13) (lovastatin), and Heidelberg (diet and exercise). The highly significant consensus of these trials is that about one-half of all patients randomized to the **control** regimen undergo **progression**, while only onefourth of them do so if given **therapy**. Similarly, less than 10% of **control** patients experience regression, while about one-fourth do so with therapy. On average, percent stenosis in control patients worsens by about 3% stenosis (S) while improving by about 1.0 - 1.5%S with therapy.

Lipid Lowering and Prevention of Clinical Events: The landmark Lipid Research Clinics-Coronary Primary Prevention Trial (14) established that significant reduction of clinical coronary events, occurred in association with a 9% reduction in total cholesterol, relative to the dietary control, and a 13% reduction in LDL cholesterol accomplished with

Cardiovascular Disease 2
Edited by L.L. Gallo, Plenum Press, New York, 1995

diet and cholestyramine. Importantly, the magnitude of cardiovascular benefit correlated in a subgroup analysis with the degree of total and of LDL cholesterol reduction (14). The Helsinki Heart Trial (15) also achieved a significant reduction in total cardiac events, but not mortality. And the 15-year follow-up of the Coronary Drug project showed highly significant 11% reductions in cardiac and all-cause mortality only in the niacin-treated group (16).

Additional evidence that clinical events are decreased by lipid-lowering therapy is found in the angiographic trials listed above, which usually report 30-to-90% fewer events among treated patients. Indeed, the amount of clinical risk reduction seems out of keeping with the average 1-2%S regression in lesion severity and with the fact that only about 12% of all intensively treated lesions actually regress. How can regression of a small number of lesions result in a large reduction in the frequency of clinical events? To understand how, we must understand the series of pathobiological events in the plaque that turn a stable quiescent lesion into an unstable culprit lesion precipitating a clinical event.

Pathological Processes: This section focuses briefly on several clinically important aspects of plaque biology: lipid accumulation in the foam cells and core region, plaque fissures their adverse consequences and their healing, and vasoconstrictor tone.

LDL and more recently Lp(a) have been demonstrated in the intimal extracellular space, the cholesterol content of which has been shown to originate from plasma LDLc (17, 18). Lipid may also accumulate in the intima in subendothelial monocyte-derived macrophages (19, 20). Such "foam cell" formation is thought to occur by unregulated scavenger receptor uptake of oxidized LDL (21) and possibly of Lp(a) (22). Foam cells are abundant in precursor fatty streak lesions (23), and in the shoulders, cap, and basilar neovascular complex of advanced plaques (24). Lipid may enter the core region of the fibrous plaque by transmural flux (24) of its more mobile forms (lipoprotein particles, droplets, and vesicles (25, 26); or it may be deposited there during foam cell necrosis (23). In the core region, lipids coalesce into lower energy phases dictated by the local cholesterol, phospholipid and cholesteryl ester concentrations (27). Droplet and vesicular forms of the latter and cholesterol monohydrate crystals are the dominant core lipids (24, 27). While transmural flux of small perifibrous lipid droplets has been thought to initiate core lipid accumulation in the earliest human aortic lesions (25, 28), the contribution of foam cell necrosis to the continued accumulation of core lipid in the larger mature fibrous plaques remains to be determined. This question is important because of the possible therapeutic role of antioxidants (21), which, by preventing LDL oxidation, may act to prevent foam cell formation and, ultimately, core lipid accumulation.

As described below and illustrated in Figure 1, the fissuring of plaques is now recognized as the key event triggering abrupt arterial occlusion and ischemia. Also, "silent" fissuring can occur in the absence of clinical symptoms (3, 29, 30), suggesting another mechanism of plaque growth. Mural thrombus, or that formed at sites of intra-plaque hemorrhage can undergo a fibrous transformation due to ingrowth and organization by smooth muscle cells, thus expanding the plaque connective tissue mass. Evidence supporting this proposed mechanism of fibrogenesis is detailed elsewhere (3, 29, 31).

Increased vasoconstrictor tone worsens arterial narrowing and thus contributes to progressive obstruction. Atherosclerosis effects vascular tone by interfering with the

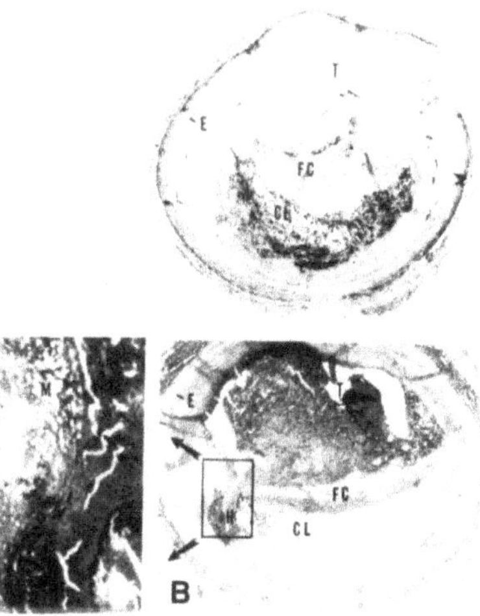

Figure 1. A) Histologic section through a structurally stable coronary plaque in a patient with vasospastic angina. Morphologic features include: E-Internal elastic lamina, **FC**--a thick fibrous cap composed largely of collagen and smooth muscle cells (**SMC**), **CL**--core lipid, here largely crystalline, and **T**--a small tag of thrombus. B) Section through a structurally unstable coronary plaque in a patient dying from myocardial infarction. The lumen, only moderately narrowed by the plaque, is acutely occluded by thrombus (**T**). There are many features in common with the section in A. In the unstable plaque, core lipid (some dislodged by sectioning artifact) comprises a much larger fraction of the plaque. The fibrous cap is much thinner than in A, is fissured (or vented) at its left shoulder, permitting **H**--a small pocket of hemorrhage in the plaque. This fissure, the associated hemorrhagic pocket, and the plaque shoulder, here rich in **M**--lipid-laden macrophages (round, bright spots), are shown at increased magnification in the inset. Also at higher magnification (not shown) the fibrous cap has few **SMC** but many **M.** Figure B reproduced from (54) with permission.

normal function of the endogenous vasodilator, EDRF, which is nitric oxide or an analog (32-37). This appears to account for the apparently paradoxical epicardial coronary vasoconstrictor effects of isometric and aerobic exercise in patients with CAD (38, 39). Since the impairment of function is experimentally reversed by reducing dietary cholesterol (40) despite persistence of intimal thickening, and since vascular SMCresponsiveness to direct dilators is largely unaltered by atherosclerosis, it is felt that vasorelaxant dysfunction is due to a direct effect of the atherogenic state on the endothelial release of EDRF. The mechanism of impairment is unknown, but LDL cholesterol and, more specifically, oxidized LDL have been implicated (40-42).

Reversal of Pathologic Processes: Convincing evidence that atherosclerosis can regress with lipid-lowering has come from non-human primate studies (27, 43-45). When monkeys with atherosclerosis induced by cholesterol-feeding are changed to a vegetarian "regression" diet, plasma lipids fall quickly to normal (140 mg/dl) and arterial lipid and connective tissue accumulations partially regress over 20-40 months. Collagen does not decrease much from its peak value (-20%) but elastin (-50%) and cholesterol (-60%) do (27, 44) and there is a fibrous transformation of the myointimal cellular response (46). The more mobile forms of cholesterol, including lipoproteins and cholesteryl esters in foam cells and in extracellular droplets, have been shown to regress; but the cholesteryl monohydrate crystals of the core lipid region are resistant to mobilization (27, 46). Histological measurements show that plaque size is reduced (43, 45). Particularly important, intimal foam cells are seen to virtually disappear within 6 months, although depletion of core cholesteryl esters proceeds over a time-course of several years (27).

Determinants of Plaque Disruption: Acute ischemic syndromes are most commonly precipitated when mild or moderate coronary lesions become disruptively transformed into severely obstructive culprit lesions. Such disruption usually involves fissuring of the fibrous cap of the atheroma, often with intramural hemorrhage and mural or occlusive thrombus. The plaque at high risk for such fissuring has a large core lipid pool and a structurally weakened fibrous cap. The cap can be weakened by the exodus or death of its smooth muscle cells, by an accumulation of lipid-laden macrophages, or by proteolytic or mechanical degradation of its collagen. Several evolving insights have greatly altered our understanding of the precipitation of plaque events leading to acute coronary events.

First, **mild** and **moderate** coronary lesions (<70% stenosis) may abruptly progress to severe obstruction, with resulting unstable angina, myocardial infarction, or death. In fact, a majority of clinical events occur under these circumstances (47, 48). Among patients undergoing thrombolytic therapy for acute MI, the severity of the atherosclerotic stenosis underlying the thrombotic occlusion was measured at less than 50% diameter stenosis in one-third of cases, and between 50% and 60% stenosis in another third (47). From another perspective, when the lesion precipitating an MI has, by chance, been seen on a recent angiogram, its pre-infarct severity averages 50% stenosis, and it will not usually possess visible features indicating that it is destined to soon become occluded (47-50). Although a given severe (≥ 70%) lesion is more likely to progress or totally occlude than a given mild or moderate lesion, clinical events are more frequently precipitated by lesions initially of the less severe type because these are much more numerous in the patient's anatomy (51), and also because the majority of occlusions of severe stenoses occur without an event (52).

A second insight was originally brought into focus by Constantinides (53) but is receiving renewed attention (3, 29, 30, 54-57). It is that, for the great majority of ischemic coronary events, a "culprit" lesion can be identified with variations of the following morphologic features at histologic examination: a.) a fissure, tear, or vent in the fibrous cap overlying the core lipid pool, b.) mural thrombus adherent at the site of the fissure, c.) bleeding into the core lipid region, and d.) severe arterial obstruction secondary to the composite mass of expanded plaque and thrombus.

Angiographic examples of plaques that have become unstable and caused a clinical event are shown in Figure 2. One can imagine the pathogenesis of each of these arteriographic examples in terms of the histologic section in Figure 1B. Figure 2A shows a hemorrhagic pocket in the atheroma connected to the lumen by a narrow-necked fissure, or vent. In such cases, it has long been debated whether increased internal pressure in the plaque (due to bleeding or to an inflammatory abscess) has burst the fibrous cap into the lumen, or whether a primary fissure in the plaque permits bleeding into the core region. Figures 2B is almost certainly an example of hemorrhage into the plaque, via an upstream fissure from the lumen, with resultant expansion of the plaque and obstruction of flow when the fibrous cap is driven into the lumen. Figure 2C shows an angiographic finding commonly called "ulceration" of the plaque. It may have been formed by hemodynamic or proteolytic erosion of a thin fibrous cap to unroof the core lipid region, or by an eruptive venting of a hemorrhagic plaque.

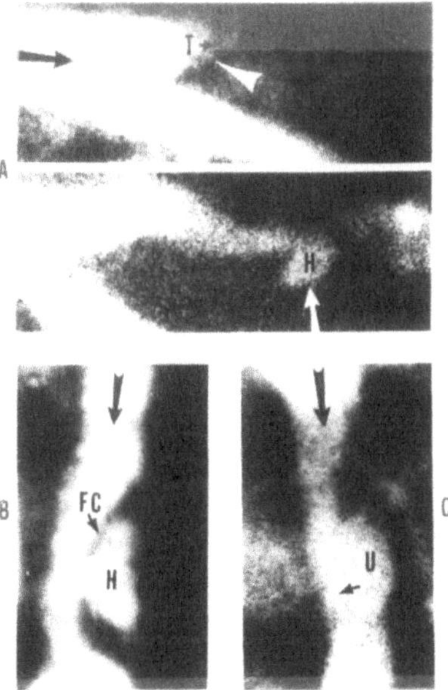

Figure 2. Highly magnified arteriographic images of structurally unstable plaques causing unstable angina or myocardial infarction. See text for descriptions.

A third insight is that there are aspects of plaque lipid composition that predict the risk of fissuring. Fissures are literally absent from the arterial intima if there are no atheroma. Among patients dying of non-cardiac causes, new fissures can be found in 9-17%, suggesting that not all fissures precipitate clinical events (30). The greater the core lipid content, the greater the likelihood of fissuring. Detailed histologic assessment of 86 infarct lesions confirmed these findings; in 83%, the intimal fissure extended from the lumen into an unstructured pool of extracellular lipid (55). Yet, in any given patient, only a small subgroup of plaques (perhaps one in eight) has a substantial core lipid accumulation. A fourth insight is that certain aspects of fibrous cap composition predict the risk of fissuring. The macrophage density in caps that fissure is greater than that in intact caps (53, 54). Fissuring occurs most commonly at the shoulder of an eccentric lipid-rich plaque (Figure 1B), a location of high macrophage density (55) and also of high circumferential stress when there is significant core lipid, according to computer models of repeated pulsatile distention of the diseased arterial cross-section (56, 57). Finally, the fibrous cap is thinned and weakened by the lack of smooth muscle cells and lysis of collagen. Cytotoxic agents, including macrophage secretory products and oxidatively modified LDL (58, 59) can transform a viable and structurally intact cap (Figure 1A) into one which is much more susceptible to fissuring (Figure 1B).

Prevention of Plaque Disruption: As described above, plaque fissuring is predicted by certain lipid-related aspects of plaque morphology including macrophage foam cell density, core lipid pool size, and possible cytotoxicity from oxidized LDL. Reduction of plasma LDL might be expected to reduce the likelihood of fissuring because of the experimentally demonstrated favorable effects of LDL reduction on the predictors described above (27, 43-66). As a consequence, the frequency of abrupt progression to clinical events should decline among patients in whom LDL has been therapeutically reduced. **Indeed, this has been the case.** Analysis of 13 coronary events among 146 FATS patients reveals that the events were associated with a culprit coronary lesion in the distribution of worsening ischemia which progressed significantly in severity from the baseline stenosis measurement to that at the time of the event (8). As seen in Figure 3, the "culprit" lesions causing the great majority of cardiac events (eight of nine) among the **conventionally** treated patients, arose from a pool of 414 lesions that were mild or moderate at baseline. By comparison, only one of 683 such lesions progressed to an event in the two **intensively** treated patient groups (p<.004, per patient or per lesion). However, (Figure 3) severe lesions did not appear to so benefit from lipid-lowering.

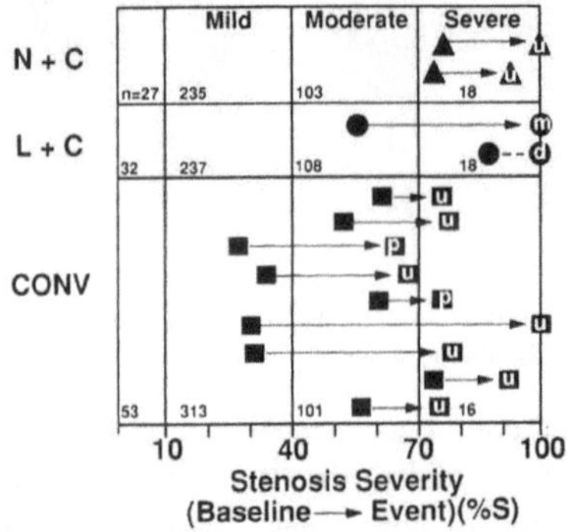

Figure 3. Chart showing culprit lesion changes associated with the thirteen coronary events as measured from 1316 lesions in 120 FATS patients. Among lesions exposed to intensive lipid-lowering therapy, only one of 683 mild or moderate lesions, at baseline, among 74 such patients progressed to a clinical event (CV death, MI, PICA or CABG, while 8 of 414 such lesions among 46 conventionally treated patients did so (per patient or per lesion, p < .004). By this standard, severe lesions did not appear to benefit from therapy. **N**--niacin, **C**--colestipol, **L**-lovastatin, **CONV**--conventional therapy, **U**--unstable angina event, **M**--myocardial infarction, **D**--death, **P**--progressive angina, % **S**--percent diameter stenosis. The number in each panel represents the number of lesions at risk, at baseline, in each subgroup.

SUMMARY

The consensus of evidence from angiographic trials demonstrates both coronary artery and clinical benefits from lowering of lipids, using a variety of regimens. The findings of reduced arterial disease progression and increased regression have been convincing but, at best, modest in their magnitude. In view of these modest arterial benefits, the associated reductions in cardiovascular events have been surprisingly great.

We believe the reduction in clinical events observed in these trials is best explained by the relationship of the lipid and foam cell content of the plaque to its likelihood of fissuring, and by the effects of lipid-lowering therapy on these "high risk" features of plaque morphology. The composite of data presented-here supports the hypothesis that lipid-lowering therapy selectively depletes (regresses) that relatively small but dangerous subgroup of fatty lesions containing a large lipid core and dense clusters of lipid-rich intimal macrophages. By doing so, these lesions are effectively stabilized and clinical event rate is accordingly decreased.

ACKNOWLEDGMENTS

The efforts of Robert Kelly in preparing this manuscript are greatly appreciated. This work was supported in part by NIH Grants RO1 HL 19451, PO1 HL 30086, and RO1 HL 42419 from the National Heart, Lung and Blood Institute and in part by a grant from the John L. Locke, Jr. Charitable Trust, Seattle, WA.

REFERENCES

1. Muller JE, Abela GS, Nesto RW, Tofler GH: Triggers, acute risk factors and vulnerable plaques: The lexicon of a new frontier. *JACC* 1994;23:809-13.
2. Brown BG, Zhao X Q. Sacco DE, Albers JJ: Lipid lowering and plaque regression: New insights into prevention of plaque disruption and clinical events in coronary disease. *Circulation* 1993:87:1781-91.
3. Fuster V, Badimon L, Badimon JJ, Chesebro JH: The pathogenesis of coronary artery disease and the acute coronary syndromes. *N Engl J Med* 1992;326:242-250; and 310-318.
4. Brensike JF, Levy RI, Kelsey SF, Passamani ER, Richardson JM, Loh IK, Friedewald W, Detre KM, Epstein SE: Effects of therapy with cholestyramine on progression of coronary atherosclerosis: Results of the NHLBI Type II coronary intervention study. *Circulation* 1984;69:313-324.
5. Blankenhorn DH, Nessim SA, Johnson RL, Sanmarco ME, Azen SP, Cachin-Hamphill L: Beneficial effects of colestipol-niacin therapy on coronary atherosclerosis and coronary venous bypass grafts. *JAMA* 1987;257:3233-3240.
6. Buchwald H, Varco RL, Matts JP, Long JM, Fitch LL, Campbell GS: Effect of partial ileal bypass on mortality and morbidity from coronary heart disease in patients with hypercholesterolemia-Report of the Program on Surgical Control of the Hyperlipidemias (POSCH). *N Engl J Med* 1990;323:946.
7. Ornish D, Brown SE, Scherwitz LW, Billings JH, Armstrong WT, Ports TA, McLanahan SM, Kirkeeide RL, Brand RJ, Bould KL: Can Lifestyle Changes Reverse Coronary Heart Disease? *Lancet* 1990;336:129-133.
8. Brown BG, Albers JJ, Fisher LD, Schaefer SM, Lin JT, Kaplan CK, Zhao XQ, Bisson BD, Fitzpatrick VF, Dodge HT: Regression of coronary artery disease as a result of intensive lipid-lowering therapy in men with high levels of apolipoprotein B. *N Engl J Med* 1990;323:1289-98.
9. Kane JP, Malloy MJ, Ports TA, Phillips NR, Diehl JC, Havel RJ: Regression of coronary atherosclerosis during treatment of familial hypercholesterolemia with combined drug regimens. *JAMA* 1990;264:3007.
10. Watts GF, Lewis B, Brunt JNH, Lewis ES, Coltart DJ, Smith LDR, Mann JI, Swan AV: Effects on coronary artery disease of lipid-lowering diet, or diet plus cholestyramine, in the St. Thomas'Atherosclerosis Regression Study (STARS). *The Lancet* 1992;339:563-569.
11. Haskell WL,Alderman E, Fain, JM, Maron DJ, Mackey SF, Superko HR, Williams PT, Johnstone IM, Champagne MA, Krauss RM, Farquhar JW: Effects of intensive multiple risk factor reduction on coronary atherosclerosis and clinical cardiac events in men and women with coronary atherosclerosis: The Stanford Coronary Risk Intervention Project (SCRIP)*Circulation* 1994;89:975 990.
12. (CCAIT) Waters D, Higginson L, Gladstone P, Kimball B, Lemay M, Boccuzzi JJ, Lesperanse J: Effects of monotherapy with an HMG-CoA reductase inhibitor on the progression of coronary atherosclerosis as assessed by serial quantitative arteriography: The Canadian Coronary Atherosclerosis Intervention Trial. *Circulation* 1994;89:959-968.
13. Blankenhorn DH, Azen SP, Kramsch DM, Mack WJ, Cashin-Hemphill L, eta l: Coronary angiographic changes with lovastatin therapy: The Monitored Atherosclerosis Regression Study (MARS). *Ann Intern Med.*
14. The Lipid Research Clinics Program: The Lipid Research Clinics Coronary Primary Prevention Trial Results: I. Reduction in incidence of coronary heart disease. *JAMA* 1984;251 351-364.
15. Manninen V, Elo MO, Frick MH, Happa K, Heinonen OP, Heinsalmi P, Helo P, Huttunen JK, Nikkila EA: Lipid alterations and decline in the incidence of coronary heart disease in the Helsinki Heart Study. *JAMA* 1988;260:641-51.
16. Canner PL, Berge KG, Wenger NK, Starmler J, Friedman L, Prineas RJ, Friedewald W: Fifteen year mortality in Coronary Drug Project patients: long-term benefit with niacin. *J Am Coll Cardio* 1986;8:1245-55.
17. Smith EB: The relationship between plasma and tissue lipids in human atherosclerosis. *Adv Lipid Res* 1974;12:1-49.

18. Rath M, Niendorf A, Reblin T, Dietel M, Krebber H-J, Beisiegel U: Detection and quantification of lipoprotein (a) in the arterial wall of 107 coronary bypass patients. *Arteriosclerosis* 1989;9579-592.
19. Gerrity RG: The role of monocyte in atherogenesis: I. Transition of blood-borne monocytes into foam cells in fatty lesions. *Am J Pathol* 1981;103:181-190.
20. Ross R: The pathogenesis of atherosclerosis-an update. N Engl J Med 1986;314:488-500.
21. Steinberg D, Parthasarathy S. Carew TE, Khoo JC, Witztum JL: Beyond cholesterol: modifications of low-density lipoprotein that increase its atherogenicity. *N Engl J Med* 1989;320:915-24.
22. Krempler F, Kostner GM, Roscher A, Bolzano K, & Sandhofer F: The interaction of human apo B containing lipoproteins with mouse peritoneal macrophages: A comparison of Lp(a) with LDL. *J Lipid Res* 1984;25, 283-287.
23. Stary HC: Evolution and progression of atherosclerosis in the coronary arteries of children and adults. In: Atherogenesis and Aging. Bates SR, Gangloff EC (Eds), Springer-Verlag, New York, 1987, p 56-63.
24. Guyton JR, Klemp KF: The lipid-rich core region of human atherosclerotic fibrous plaques. *Am J Pathol* 1989;134:705-717.
25. Guyton JR, Bocan TMA: Human aortic fibrolipid lesions. Progenitor lesions for fibrous plaques, exhibiting early formation of the cholesterol-rich core. *Am J Pathol* 1985;120:193-206.
26. Smith EB, Evans PH, Pownham MD: Lipid in the aortic intima: The correlation of morphological and chemical characteristics. *J. Atheroscler Res* 1967;7:171-186.
27. Small DM, Bond MG, Waugh D, Prack M, Sawyer JK: Physiochemical and histological changes in the arterial wall of nonhuman primates during progression and regression of atherosclerosis. *J Clin Invest* 1984;73:1590-1605.
28. Guyton JR, Bocan TM, Schifani TA: Quantitative ultrastructural analysis of perifibrous lipid and its association with elastin in nonatherosclerotic human aorta. *Arteriosclerosis* 1985;5:644-652.
29. Tracey RE, Devaney K, Kissling G: Characteristics of the plaque under a coronary thrombus. Virchows Arch Pathol Anat 1985;405:411-427.
30. Davies MJ, Krikler DM, Katz D: Atherosclerosis: inhibition or regression as therapeutic possibilities. *Brit Heart J* 1991;65:302-10.
31. Duguid JB: Thrombosis as a factor in the pathogenesis of aortic atherosclerosis. *J Pathol Bact* 1948;60:57-69.
32. McLenachan JM, Williams JK, Fish RD, Ganz P. Selwyn AP: Loss of flow-mediated endothelium dependent dilation occurs early in the development of atherosclerosis. *Circulation* 1991;84:1273-78.
33. Furchgott RF, Zawadzki JV: The obligatory role of endothelial cells in the relaxation of arterial smooth muscle by acetylcholine. *Nature* 1980;299:373-376.
34. Selke FW, Armstrong ML, Harrison DG: Endothelium-dependent vascular relaxation is abnormal in the coronary microcirculation of atherosclerotic primates. *Circulation* 1990;81:1568-93.
35. Moncada S, Palmer RM, Higgs EA: Nitric oxide physiology, pathophysiology, and pharmacology. *Pharmacol Rev* 1991;43:109-42.
36. Stamler JS, Simon DI, Osborne JA, Mullins ME, Jaraki O, Michel T, Singel DJ, Loscalzo J: S-nitrosylation of proteins with nitric oxide-synthesis and characterization of novel biologically active compounds. *Proc Natl Acad Sci* 89:444-448.
37. Chilian WM, Dellsperger KC, Layne SM, Eastham CL, Armstrong MA, Marcus ML, Heistad DD: Effects of atherosclerosis on the coronary micro-circulation. *Am J Physiol* 1990;258:H529-39.
38. Brown BG, Lee AB, Bolson EL, Dodge HT: Reflex constriction of significant coronary stenosis as a mechanism contributing to ischemic ventricular dysfunction during isometric exercise. *Circulation* 1984;70:18-24.
39. Hess OM, Bortone A, Eid K, Gage JE, Nonogi H, Grimm J, Krayenbuehl HP: Coronary vasomotor tone during static and dynamic exercise. *Eur Heart J* 1989, 10:suppl F:105-10.
40. Harrison DG, Armstrong ML, Freeman PC, Heistad DD: Restoration of endothelium-dependent relaxation by dietary treatment of atherosclerosis. *J Clin Invest* 1987;80 808-11.
41. Muegge A, Edwell JH, Peterson TE, Hofmeyer TG, Heistad DD, Harrison DD: Chronic treatment with polyethylene-glycolated superoxide dismutase partially restores endothelium-dependent vascular relaxations in cholesterol-fed rabbits. *Circ Res* 1991,69:1293-300.
42. Vita JA, Treasure CB, Nabel EG, McLenachan JM, Fish RD, Yeung AC, Veksktein VI, Selwyn AP, Ganz P: Coronary vasomotor response to acetyl-choline relates to risk factors for coronary artery disease. *Circulation* 1990;81:491-7.
43. Armstrong ML and Megan MB. Lipid depletion in atheromatous coronary arteries in rhesus monkeys after regression diets. *Circ Res* 1972; 30:675-6800.
44. Clarkson TB, Bond MG, Bullock BC, Marzetta CA: A study of atherosclerosis regression in Macaca mulatta. IV. Changes in coronary arteries from animals with atherosclerosis induced for 19 months

and then regressed for 24 or 48 months at plasma cholesterol concentrations of 300 or 200 mg/dl. *Exp Mol Pathol* 1981;34:345-68.

45. Armstrong MC, Megan MB: Arterial fibrous protein in cynomolygous monkeys after atherogenic and regression diets. *Circ Res* 1975;36:256-61.
46. Brown BG, Fry DL: The fate and fibrogenic potential of subintimal implants of crystalline lipid in the canine aorta. Quantitative histological and autoradiographic studies. *Circ Res* 1978:43;261-273.
47. Brown BG, Gallery CA, Badger RS, Kennedy JW, Mathey D, Bolson EL, Dodge HT: Incomplete lysis of thrombus in the moderate underlying atherosclerotic lesion during intracoronary infusion of streptokinase for acute myocardial infarction: Quantitative angiographic observations. *Circulation* 1986;73:653-661.
48. Ambrose JA, Tannenbaum MA, Alexopoulos D, Hiemdahl-Monsen CE, Leavy J. Weiss M, Borrico S. Gorlin R. Fuster V: Angiographic progression of coronary artery disease and the development of myocardial infarction. *J Am Coll Cardiol* 1988;12:56-62.
49. Little WC, Constantinescu M, Applegate RM, Kutcher MA, Burrows MT, Kahl FR, Santamore WP: Can coronary angiography predict the site of a subsequent myocardial infarction in patients with mild-to-moderate coronary artery disease? *Circulation* 1988;78:1157-1166.
50. Little WC: Angiographic assessment of the culprit coronary artery lesion before acute myocardial infarction. *Am J Cardiol* 1990;66:44G-47G.
51. Brown BG, Lin J-T, Kelsey S. Passamani ER, Levy RI, Dodge HT, Detre KM: Progression of coronary atherosclerosis in patients with probable familial hypercholesterolemia: Quantitative arteriographic assessment of patients in NHLBI Type II Study. *Arteriosclerosis* 1989,9:(Supplement I):I-81-I-90.
52. Webster MWI, Chesebro JH, Smith HC, Frye RL, Holmes DR, Reeder GS, Fuster V: Myocardial infarction and coronaty artery occlusion: A prospective 5-year angiographic study (abstract). *J Am Coll Cardiol* 1990;15 (Suppl A):218A.
53. Constantinides P: Plaque fissures in human coronary thrombosis. *J Athero Res* 1966;6I:1-17.
54. Constantinides P: Plaque hemorrhages, their genesis and their role in supra-plaque thrombosis and atherogenesis. In: Pathobiology of the Human Atherosclerotic Plaque. Glagov S, Newman WP, Schaffer SA, eds. Springer-Verlag, New York 1990, p 393-411.
55. Lendon CL, Davies MJ, Born GVR, Richardson PD: Atherosclerotic plaque caps are locally weakened when macrophage density is increased. *Atherosclerosis* 1991;87:87-90.
56. Richardson PD Davies MJ, Born GVR: Influence of plaque configuration and stress distribution on fissuring of coronary atherosclerotic plaques. *Lancet* 1989;3:94144.
57. Loree HM, Kam n RD, Strongfellow RG, Lee RT: Effects of fibrous cap thickness on peak circumferential stress in model atherosclerotic vessels. *Circ Res* 1992;71:850-858.
58. Hessler JR, Morel DW, Lewis LJ, Chisolm GM: Lipoprotein oxidation and lipoprotein-induced cytotoxicity. *Arteriosclerosis* 1983;3:215-222.
59 Yla-Herttuala S, Palinski W, Rosenfeld ME, Parthasarathy S, Carew TE, Butler S, Witztum JL, Steinberg D: Evidence for the presence of oxidatively modified low density lipoprotein in atherosclerotic lesions of rabbit and man. *J Clin Incest* 1989;84:1086-1095.

EXPRESSION OF 92 KDA GELATINASE IN HUMAN ATHEROSCLEROTIC LESIONS FOLLOWING RECENT PLAQUE RUPTURE

David L. Brown,[1] Margaret S. Hibbs,[2] Marianne Kearney,[3] Eric J. Topol[4] and Jeffrey M. Isner [3]

[1]Division of Cardiovascular Medicine
University of California, San Diego Medical Center
San Diego, CA 92103-8411
[2]Division of Rheumatology
Veterans Affairs Medical Center
Newington, CT 06111
[3]Division of Cardiovascular Medicine
St. Elizabeth's Hospital
Boston, MA 02135
[4]Department of Cardiology
Cleveland Clinic Foundation
Cleveland, OH 44195

INTRODUCTION

Coronary atherosclerosis frequently culminates with the development of the acute coronary ischemic syndromes, unstable angina, acute myocardial infarction and sudden cardiac death. These disparate syndromes occur along a clinical continuum linked by a common pathophysiologic event, intracoronary thrombosis (1-4). Thrombosis is initiated by rupture of atherosclerotic plaque and exposure of highly thrombogenic plaque constituents to coronary blood flow. The extent of the resulting thrombus and its location in the coronary circulation contribute to the different resultant clinical syndromes. Although, the proximate cause of plaque rupture is not known, pathological studies performed on patients dying suddenly of acute coronary thrombosis have documented a consistent relationship between sites of plaque rupture and the presence of intense macrophage infiltration (3,4).

Macrophages synthesize and secrete a diverse array of proteolytic enzymes capable of degradation of plaque constituents. One such family of enzymes, the matrix metalloproteinases, is capable of degrading all macromolecular constituents of the extracellular matrix . The mRNA for one member of this family, stromelysin, has been found in human atherosclerotic plaques (5). Another matrix metalloproteinase, the 92 kDa gelatinase is synthesized by macrophages in a highly regulated manner. It degrades collagen types IV, V and XI, which are found in atherosclerotic lesions and are resistant to degradation by the matrix metalloproteinases stromelysin and interstitial collagenase (6,7).

Cardiovascular Disease 2
Edited by L.L. Gallo, Plenum Press, New York, 1995

We sought to investigate whether excessive macrophage expression of the 92 kDa gelatinase may, by degradation of constituents of the extracellular matrix, result in loss of structural integrity of the plaque and contribute to plaque rupture. To test this hypothesis, we examined, by immunohistochemistry, atherectomy specimens obtained from patients with clinical evidence of recent plaque rupture (unstable angina) for expression of the 92 kDa gelatinase and compared the results to expression in normal internal mammary arteries and in atherectomy specimens from patients with coronary atherosclerosis but without clinical evidence of recent plaque rupture (stable angina).

METHODS

Patients

Atherectomy was performed on patients with symptomatic ischemic heart disease enrolled in the Coronary Angioplasty versus Excisional Atherectomy Trial (CAVEAT) (8). For this study, patients were considered to have unstable angina if they presented with a crescendo pattern of ischemic symptoms including symptoms at rest and electrocardiographic evidence of ischemia during symptoms. All of the 12 patients undergoing atherectomy in the CAVEAT trial who met this strict definition of unstable angina used for this study. Patients were considered to have stable angina if their symptoms were predictably exertional without any of the components of unstable angina. Ninety-two patients enrolled in CAVEAT underwent atherectomy for stable angina. Of this group of 92, 12 specimens were randomly chosen for inclusion in this study. Specimens of the left internal mammary artery were obtained surgically from 2 patients undergoing coronary artery bypass surgery.

Antibodies

Antibodies were prepared as previously described (9), either by immunizing rabbits with the 92 kDa form of neutrophil gelatinase obtained by preparative gel electrophoresis (IS3-70) or by immunizing rabbits with the purified native form of the proteinase (MH-1). Both antibodies have been shown to be monospecific by immunoblotting techniques(6) and gave identical results when immunostaining the tissues used for this study.

Tissue Preparation

Tissue specimens retrieved from the atherectomy device were prepared by placing them in 4% paraformaldehyde for 2 h followed by incubation in a 30% sucrose solution overnight. A portion of the specimen was post-fixed in 10% formalin for light microscopic analysis. Specimens were embedded in paraffin and stained for hematoxylin and eosin or elastic tissue trichrome.

Immunohistochemistry

Specimens were embedded in OCT (Miles Diagnostics, Elkhart, IN) and stored at $-70^{o}c$ prior to staining. In preparation for staining, specimens were cut, applied to slides and incubated in phosphate-buffered saline (PBS) for 5 min. Endogenous peroxidase activity was blocked by washing with 0.3% H_2O_2 for 5 min followed by 2 rinses in PBS. Slides were incubated with normal goat serum for 20 min. Primary antibody or negative control (normal rabbit serum), diluted 1:1000 in PBS/1% bovine serum albumin was applied for 45 min and followed by 2 5-min washes with PBS. Biotinylated anti-rabbit secondary antibody (Vector Laboratories, Burlingame, CA) was added for 45 min after which 2 5-min washes with PBS were performed. Slides were incubated in avidin-biotin complex (Vector Laboratories) for 30 min and then washed twice in PBS. Diaminobenzidine was applied to slides for 5 min followed by rinsing with distilled water. A 10 s counterstain with hematoxylin was performed. Staining of peripheral blood smears was used as a positive control.

Adjacent Section Immunostaining

Four frozen sections adjacent to those staining positively for 92 kDa gelatinase were cut, applied to slides and incubated in PBS for 5 min. Slides were then incubated in 0.3% H_2O_2 for 5 min. Non-specific protein binding was blocked by incubating slides in normal horse serum for 20 min. A different primary antibody, HAM56 for macrophages (Enzo Diagnostics, Farmingdale, NY), HHF35 for smooth muscle cells (Enzo Diagnostics) or DAKO-LCA (DAKO, Glostrup, Denmark) for lymphocytes or negative control (non-specific mouse IgG, Sigma Chemical Co., St. Louis, MO),was added to each of the 4 slides for 20 min. Slides were then rinsed in PBS. Secondary anti-mouse antibody (Signet Labs, Dedham, MA) was added for 20 min and followed by washing in PBS. Slides were then incubated with avidin-biotin complex (Signet Labs) for 20 min followed by a PBS wash. AEC (3 amino-9 ethyl carbazole) substrate was added to slides for 20 min. The slides were then washed with distilled water for 5 min and counterstained with hematoxylyn for 10 s. The identity of cells staining positively for 92 kDa gelatinase was determined by serially examining the 3 adjacent sections stained with cell-type specific antibodies for a positive immunoperoxidase reaction.

RESULTS

Expression of 92-kDa Gelatinase in Normal Internal Mammary Artery

Specimens of the internal mammary artery from 2 different patients undergoing bypass surgery were evaluated for expression of the 92 kDa gelatinase. Both specimens were histologically normal, without atherosclerosis. In neither artery was there any immunohistochemical evidence of 92 kDa gelatinase expression (Fig. 1).

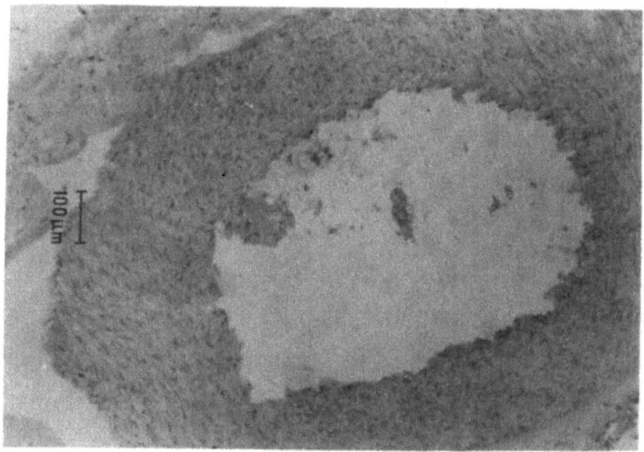

Figure 1. Negative immunoperoxidase staining of normal internal mammary artery specimen obtained at surgery. (25x magnification).

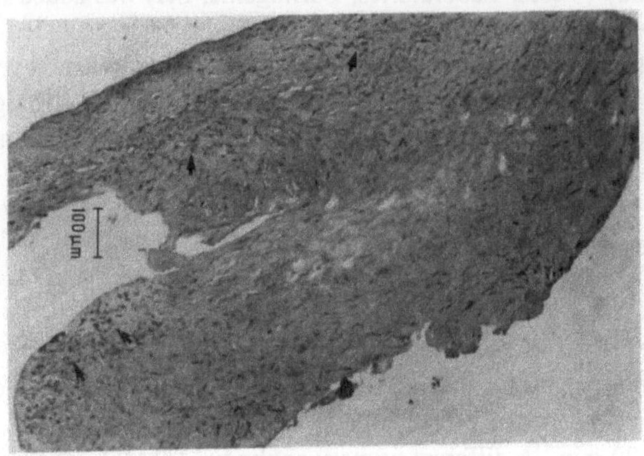

Figure 2. Atherectomy specimen obtained from a representative patient with unstable angina and positive immunoperoxidase reaction for 92 kDa gelatinase (25x magnification). Black arrows indicate positive intracellular staining.

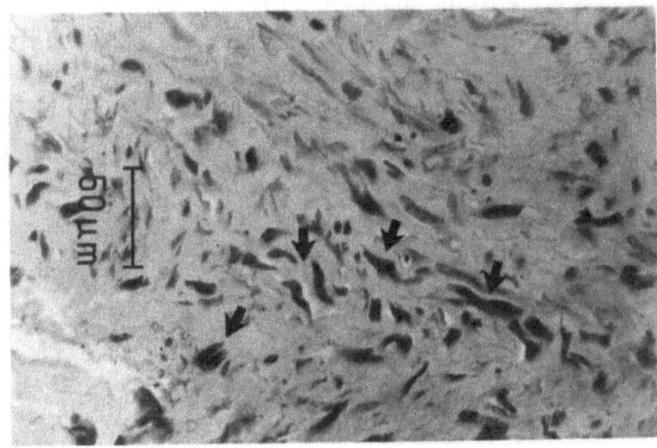

Figure 3. Atherectomy specimen obtained from same patient as in Fig. 2 with unstable angina and positive immunoperoxidase reaction for 92 kDa gelatinase (100x magnification). Black arrows indicate positive intracellular staining.

Expression of 92-kDa Gelatinase in Coronary Atherectomy Specimens from Patients with Unstable and Stable Angina

Of the coronary atherectomy specimens from the 12 patients with unstable angina, 10 (83%) stained positively for the 92 kDa gelatinase. Figures 2 and 3 illustrate representative positive specimens at low and high magnification from patients with unstable angina. Nine of 12 (75%) specimens from patients with stable angina stained positively for the 92 kDa gelatinase.

Extracellular vs. Intracellular Staining Pattern

The localization of positive staining was differed significantly between the unstable and stable angina specimens. All 10 of the positively stained specimens retrieved from patients with unstable angina demonstrated *intracellular* localization of the 92 kDa gelatinase. In three of these specimens positive staining for gelatinase was also observed in the extracellular space. In contrast, of the 9 positively stained specimens retrieved from patients with stable angina, only 3 displayed a pattern of intracellular staining. In the remaining 6 positively stained specimens, weak immunostaining was limited to the extracellular space.

Cellular Localization

Adjacent sections were successfully examined with antibodies specific for smooth muscle cells, macrophages and lymphocytes on 7 of the patients with unstable angina in whom positive intracellular staining was observed. There was insufficient quantity of specimen to perform adjacent section staining on 3 patients with positive intracellular staining. Positive immunostaining for the 92 kDa gelatinase was localized to macrophages in 7/7 specimens. Staining was also demonstrated in smooth muscle cells in 3/7 positive specimens. In 7/7 of these positive staining specimens 92 kDa gelatinase was identified in a small number of lymphocytes.

DISCUSSION

Acute coronary ischemia has been shown to be precipitated by atherosclerotic plaque rupture and subsequent intracoronary thrombosis (1-4). The biochemical factors predisposing to plaque rupture are poorly understood. The mRNA of stromelysin, one member of a family of proteases specific for constituents of the extracellular matrix, has been identified in atherosclerotic coronary arteries (5). In the current report, we provide evidence for the involvement of a second member of the metalloproteinase family, 92 kDa gelatinase in atherosclerotic coronary arterial lesions. In particular, the current findings are the first, to our knowledge, to: a) identify evidence for a metalloproteinase in atherosclerotic lesions at the *protein* level and b) relate the presence of a metalloproteinase to a specific clinical presentation of an acute ischemic coronary syndrome, namely unstable angina. This gelatinase was found in 83% of specimens obtained from the "culprit" coronary arteries of patients meeting a stringent definition of unstable angina. Macrophages constituted the major source of this metalloproteinase in all 7 positive sections examined by adjacent staining techniques. The intracellular localization of the 92 kDa gelatinase in all 10 positive specimens can be interpreted as evidence of active synthesis of this enzyme because macrophages do not store this metalloproteinase (10). In 3 of the 10 positive specimens 92 kDa gelatinase was also documented extracellularly.

The pattern of immunostaining of specimens from patients with stable angina was markedly different from that observed among patients with unstable angina. Only 3 of the 10 positive specimens in the stable angina group exhibited intracellular staining. The remaining 7 specimens demonstrated only weak extracellular staining. Thus, if active synthesis of enzyme, as indicated by intracellular staining, is considered to be pathophysiologically relevant in unstable angina, 83% of unstable angina specimens in this study were positive as compared to 25% of patients with stable angina.

There are several possible explanations for why all specimens from patients with unstable angina were not positive for 92 kDa gelatinase expression. One possible explanation is the sampling error inherent in the atherectomy procedure itself. The atherectomy catheter samples only an incomplete fraction of the atherosclerotic plaque. Thus, it is possible during the course of an atherectomy for unstable angina, that the actual site of plaque rupture was not excised. Furthermore unstable angina, by definition, is an episodic phenomenon. It is therefore possible that those patients with unstable angina in whom specimens were negatively stained were in a quiescent phase of the syndrome.

Evidence of plaque rupture has been found at autopsy in sections of coronary arteries of asymptomatic patients dying from non-cardiac diseases (11). Thus, silent plaque rupture resulting in non-occlusive thrombosis can occur in patients who do not develop the clinical hallmarks of any of the syndromes associated with plaque rupture. The 25% of patients in this study with clinically stable angina who nevertheless demonstrated intracellular staining for the 92 kDa gelatinase may have been in this category.

Alternatively, these patients had they not undergone atherectomy, might have soon progressed to frank plaque rupture and one of the acute coronary ischemic syndromes.

This study has documented only the presence of the 92 kDa gelatinase *protein* in coronary lesions. To achieve gelatinase *activity* at a specific site in the coronary artery requires both production of the protein in excess of its natural inhibitors and activation of the inactive precursor form of the protease. The presence of endogenous metalloproteinase inhibitors in coronary lesions of patients with acute ischemic syndromes was not investigated in this study but is worthy of future investigation. The antibodies to 92 kDa used in this study recognize both the active and inactive forms of the protease. Thus the presence of protein in these lesions does not necessarily prove the presence of gelatinase activity. Because the factors required for activation of 92 kDa gelatinase are incompletely understood, however, it is unknown whether constituents of the advanced atherosclerotic plaque will support enzyme activation.

In conclusion, the findings reported in this study provide evidence that active synthesis of 92 kDa gelatinase by macrophages is strongly associated with the clinical syndrome of unstable angina, possibly due to metalloproteinase-induced matrix degradation promoting plaque rupture. The factors which regulate macrophage metalloproteinase production within atherosclerotic lesions have not been identified. These findings however are potentially important from a fundamental standpoint because they suggest a pathogenic role for at least one of the metalloproteinases in the development of unstable angina. From a practical standpoint these findings raise the possibility that inhibitors of metalloproteinase activity-specifically of the 92 kDa gelatinase-might provide a novel form of therapy for stabilization of the atherosclerotic lesions of patients with coronary artery disease and prevention of acute ischemic syndromes.

REFERENCES

1. P. Constantinides. Plaque fissuring in human coronary thrombosis. *J Atheroscler Res* 6:1 (1966).

2. M. Friedman and G.J. Van de Bovenkamp. The pathogenesis of a coronary thrombus. *Am J Pathol* 48:19 (1966).

3. E. Falk. Plaque rupture with severe pre-existing stenosis precipitating coronary thrombosis: characteristics of coronary atherosclerotic plaques underlying fatal occlusive thrombi. *Br Heart J* 50:127 (1983).

4. M.J. Davies and A.C. Thomas. Plaque fissuring: the cause of acute myocardial infarction, sudden ischemic death, and crescendo angina. *Br Heart J* 53:363 (1985).

5. A.M. Henney, P.R. Wakeley, M.J. Davies, K. Foster, T. Hembry, G. Murphy, and S. Humphries. Localization of stromelysin gene expression in atherosclerotic plaques by in situ hybridization. *Proc Natl Acad Sci USA* 88:8154 (1991).

6. M.S. Hibbs, J.R. Hoidal, and A.H. Kang. Expression of a metalloproteinase that degrades native type V collagen and denatured collagens by cultured human alveolar macrophages. *J Clin Invest* 80:1644 (1987).

7. S.M. Wilhelm, I.E. Collier, B.L. Marmer, A.Z. Eisen, G. A. Grant, and G.I. Goldberg. SV-40-transformed human lung fibroblasts secrete a 92kDa type IV collagenase which is identical to that secreted by normal human macrophages. *J Biol Chem* 264:17213 (1989).

8. E.J. Topol, F. Leya, CA Pinkerton, PL Whitlow, B. Hofling, CA Simonton, R.R. Masden, PW Serruys, M.B. Leon, D.O. Williams, S. B. King III, D.B. Mark, J.M. Isner, D.R. Holmes, S.G. Ellis, K.L. Lee, G.P. Keeler, L.G. Berdan, T. Hinohara, and R.M. Califf for the CAVEAT Study Group. A comparison of directional atherectomy with coronary angioplasty in patients with coronary artery disease. *N Engl J Med* 329:221(1993).

9. M.S. Hibbs, K.A. Hasty, J.M. Seyer, A.H. Kang, and C.L. Mainardi. Biochemical and immunological characterization of the secreted forms of human neutrophil gelatinase. *J Biol Chem* 260:2493 (1985).
10. M.S. Hibbs. Expression of 92 kDa phagocyte gelatinase by inflammatory and connective tissue cells. *Matrix*. Supplement 1:51 (1992).
11. M. Davies, J. Bland, J. Hangartner, A. Angelini and A.C. Thomas. Factors influencing the presence or absence of acute coronary thrombi in sudden ischaemic death. *Eur Heart J* 10:203 (1989).

Key Words: Atherosclerosis Metalloproteinase Unstable angina Plaque Rupture Coronary Artery Disease

PROTEINASES AND RESTENOSIS: MATRIX METALLOPROTEINASE AND THEIR INHIBITOR AND ACTIVATOR

Suresh C. Tyagi, Larry Meyer, Richard A. Schmaltz[a],
Hanumanth K. Reddy, Donald J. Voelker,

Departments of Internal Medicine, Biochemistry, and Surgery[a],
The Dalton Cardiovascular Research Center,
University of Missouri-Health Sciences Center, and
The Harry S. Truman Veterans Administration Medical Center,
Columbia, MO 65212

PROTEINASES AND WOUND HEALING

Remodeling of the extracellular matrix (ECM) is the central part of normal tissue development and is involved in tumor growth and metastasis, angiogenesis, wound healing/repair, inflammation, embryonic development, post-partum involution of the uterus, bone and growth plate remodeling, ovulation, and the progression of certain diseases.[1,2] The process of connective tissue restructuring represents a balance between matrix production and its degradation. The breakdown of this highly organized extracellular environment is controlled, for the most part, by a gene family of matrix metalloproteinases (MMPs) which collectively can degrade virtually all components of connective tissues.[3,4]

At present, little is known about proteolytic system in normal or diseased vasculature. Post-translational factors can influence the activity and availability of tissue MMPs and their inhibitors (TIMPs) during vascular remodeling. While the synthesis of collagenase is regulated at the transcriptional level, the enzyme is secreted from the cells as an inactive zymogen and exists in the tissue in latent forms.[5] Therefore, the activation process of pro-collagenase in the extracellular milieu is an additional key step in the regulation of collagenolysis.

Pathophysiologic changes during restenosis suggest complex interactions of a myriad of biological processes, initiated by vessel injury and potentially dependent on the release of thrombogenic, vasoactive and mitogenic factors.[6] It has previously been proposed that there are three phases of vascular wound healing after tissue injury[7]: inflammation, granulation and matrix formation. During inflammation, fibronectin and plasminogen bind to biologically active substances and recruit inflammatory cells at the site of wound. From these cells growth factors are released which lead to granulation and cellular proliferation. Fibronectin/plasminogen in the ECM facilitate migration of endothelial cells from the wound

Cardiovascular Disease 2
Edited by L.L. Gallo, Plenum Press, New York, 1995

margin and fibroblasts and/or smooth muscle cells from adjacent tissue, where they proliferate and synthesize ECM. The third phase is matrix remodeling and deposition of proteoglycans and collagen, a step which involves proteinases, especially MMPs.

GROWTH AND CELLULAR RESPONSE OF RESTENOSIS

Cardiovascular disease accounts for considerable mortality and morbidity. The common scourge of cardiovascular system is atherosclerosis which induces functional and structural abnormalities in vessel wall. The changes include vasoconstriction, enhanced interaction of blood cells with the vessel wall, activation of coagulation mechanisms, and migration and proliferation of vascular smooth-muscle cells (SMC) and fibroblasts.[8-11] These vascular abnormalities play a significant role in pathogenesis of angina pectoris, myocardial infarction and myocardial cell loss and energy due to lack of oxygen metabolite.

Recently, clinical and experimental observations suggest that restenosis results from a complex interaction of biological processes, initiated by vessel injury and dependent on the release of thrombogenic, vasoactive, and mitogenic factors.[12-14] Restenosis is not a simple process. It is the result of numerous and diverse reparative processes each of which has its own time course of development (Figure 1).

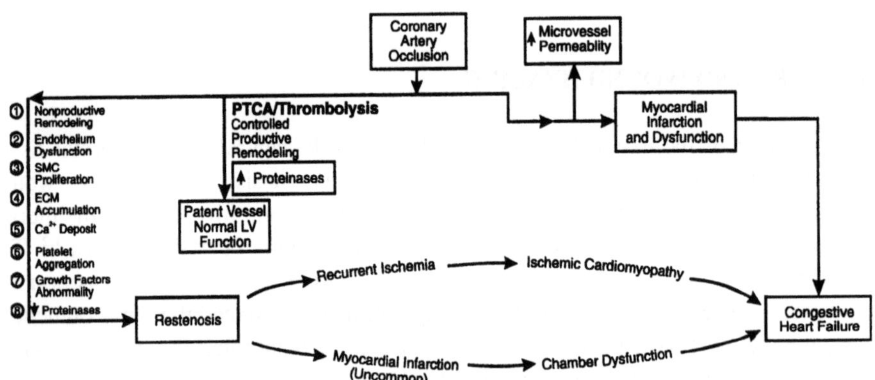

Figure 1. Possible mechanism of restenosis.

Early restenosis (within days of injury) most likely results from vasoconstriction and relaxation of the stretched vessel wall ("elastic recoil"). Later (within months) restenosis ensures the reparative process in which myointimal and fibrocellular proliferation is exaggerated and sufficiently severe to impede arterial blood flow.

We have been studying the role of elastinolysis and collagenolysis in arterial remodeling and disease.[15] Degradation of elastin[16] in the arterial intima by neutrophil elastase released during the inflammatory response, unstable angina pectoris and acute myocardial infarction[17] may trigger severe tissue damage. Neutrophil elastase can degrade elastin and myocardial TIMPs.[18] Hornebeck et al[19] showed that the content of aortic cross-linked elastin decreased with the degree of atherosclerosis, suggesting a role of elastinolytic enzymes in this disease. Several lines of evidence suggested that elastinolytic enzymes present in the human arterial wall are different from one another. Bellon et al[20] have shown that human arteries contain an elastinolytic enzyme which is immunologically different from the human pancreatic and neutrophil enzymes. Recently, Senior et al[21] have demonstrated that higher molecular weight collagenase/gelatinase (i. e. matrix metalloproteinase) can also degrade elastin in a gel matrix. Collectively these studies suggested involvement of a multifactorial proteolytic systems in arterial diseases, and the potentially important role of control of ECM synthesis and degradation in the pathophysiology of restenosis.

Intimal Fibrous Proliferation (IFP) Response

Platelets that adhere and aggregate at sites of endothelial disruption release their granular contents such as heparin-neutralizing factors and several mitogens, including platelet-derived growth factors, epidermal growth factor, transforming growth factor-β (TGF-β).[22] These factors release SMC from growth inhibition to their proliferation and migration from the media to the intima, and act as chemotactants for macrophases and neutrophils. The severity of this response to injury determines the degree of IFP. If excessive, it encroaches on the arterial lumen and compromises the flow.[23] In addition, platelet derived growth factor (PDGF), fibroblasts growth factor (FGF) and Insulin-like growth factor (IGF) (somatomedin C) are released from endothelial cells, macrophases, and SMC. These act synergistically with PDGF to induce intimal hyperplasia, and are capable of self-perpetuating their own mitogenic stimulation and SMC proliferation. Therefore, once SMC transform from a quiescent state (contractile phenotype) to a proliferative state (synthetic phenotype) and migrate to the intima, their continued activity may be self-perpetuated beyond the phase of platelet deposition.17,24,25

There is evidence that direct injury of the SMC itself can initiate the biological process leading to restenosis.[23] Endothelial cells synthesize heparin-like glycosaminoglycans, which are growth-inhibitory and maintain the underlying medial layer in a quiescent state. In addition, SMC also produce a growth-inhibitory heparin sulfate. When the endothelial layer is abraded or the SMC are disturbed by injury, the rapid proliferation of medial SMC may be due, in part, to loss of the growth-inhibitory factors.

In experimental animal models of restenosis, myointimal proliferation begins early after injury.[26] Medial SMC begins to proliferate within 24 to 48 h of injury. Approximately 4 days after injury SMC begin to migrate to the intima, where proliferation continues. This process of medial SMC proliferation, chemotaxis, and intimal proliferation is maximal by 1 week after injury. Nevertheless, intimal thickening continues for upto 8 weeks as ECM synthesis and turnover continues. In humans SMC proliferation has been identified by 17 days.[27]

PROTEINASES IN RESTENOSIS

Wound healing following mechanical revascularization of occluded coronary arteries, and the potential role of relative collagen synthesis and collagenase-mediated degradation may be important in understanding mechanism of restenosis. We have shown in normal myocardium the predominant presence of latent collagenases.[5] Furthermore, this latent collagenase can be activated by trypsin, plasmin, and organomercurials[18], and suggest an existence of an active and latent collagenolytic system in cardiovasculature. The thrombolytic proteinases can activate latent MMPs: plasmin can activate tissue procollagenase and urokinase plasminogen activator (uPA) can initiate proteolytic processing of the 72 kDa gelatinase.[28,29] It is not clear if the MMPs and their activity in restenosis are coordinated by their inhibitors and activators.

Proteinase activity and ECM proteins influence a variety of cellular events which are important developmentally and in the maintenance of homeostasis in normal vascular tissues. More recently, it has become clear that the ECM and proteinase activity may be involved in certain disease states.[30] The vascular ECM and ECM-derived peptides are in a constant state of flux and change as cells comprising the arterial wall modulate their behavior. These ECM related changes, in turn, may have a profound effect on the structural and functional properties of the tissue, as well as on the metabolic properties of the cell themselves. The major ECM components surrounding the cells in the arterial wall are collagens, elastic fibers, and proteoglycans.[31] In normal tissue, these macromolecules are distributed in the intimal, medial, and adventitial layers of the vascular wall in such a manner as to confer both structural integrity and viscoelasticity. We have shown that neutrophil elastase activity is regulated by elastin-derived peptides.[15]

The control of the composition of the ECM is critical to maintenance of the arterial lumen during disease states such as atherosclerosis. The long-term favorable outcome of mechanical revascularization of occlusive coronary artery disease with balloon angioplasty, directional coronary atherectomy, laser angioplasty, and other more recent techniques (e.g. rotablater, stents) is limited by restenosis.[32] Although the mechanism of restenosis are not well understood, studies have shown ECM accumulation[33,34] and cellular proliferation are operative in the wound healing response. The pathophysiology is poorly understood also, but most likely involves numerous growth factors cytokines, and products of coagulation.[35] Only recently has the potential role of the regional expression of matrix-degrading proteinases in human atherosclerotic plaques been implicated in restenosis.[36] Blood vessels undergo a marked intimal thickening in response to mechanical injury[37,38] and this process is believed to reflect one of the early stages in the development of the arteriosclerosis lesion. A number of studies have shown that part of this thickening is the consequence of the migration and proliferation of arterial muscle cells and accumulation of components of the ECM such as elastin, collagen and proteoglycans.[33, 34]

ECM and Proteinases

In a study of histopathological analysis on primary (untreated) and restenotic state reveals that in primary stenosis, 88% of the tissue recovered was atherosclerotic plaques, 9% was IFP with medial calcinosis, and 3% consisted of thrombus. In contrast, in restenotic tissue, 75% was IFP and 25% was classic atherosclerotic plaque (39). We compared MMPs activity with internal mammary artery tissue, and tissue derived from patients undergoing atherectomy for the first time for occlusive coronary artery disease. We have shown an increase in ECM production and a decrease in MMPs expression in restenotic tissue (Figure 2). This may suggest an imbalance of ECM synthesis and turnover during the cellular proliferative and intimal fibrous response after arterial injury.

The histopathologic characteristics of restenotic lesions differ from those of *de novo* arteriosclerotic lesions.[40] Our study confirms that quantitative differences exist in the composition of the ECM between tissue obtained from restenotic and primary atherosclerotic lesions. The greater predominance of collagen observed in restenotic lesions compared to native atherosclerotic lesions could be explained on the basis of increased production of ECM by vascular wall cells[41] or, alternatively by decreased turnover.

It is well documented that components of the ECM such as collagen, elastin and proteoglycans increase in blood vessels after mechanical injury. In a study using cDNA clones of collagen and elastin, collagen and elastin increased after arterial injury and correlated with an increase in mRNA levels coding for elastin and type I and III procollagen.[33] If at the same time enough MMPs are not expressed to account for increased synthesis of ECM, excessive deposition of ECM will occur that could impinge on the vascular lumen.

The increased synthesis and deposition of ECM components is a general response of a variety of cells to various forms of injury observed during wound healing and tissue repair.[42,43] Injury is often accompanied by cell proliferation, and at present it is unclear whether these ECM changes are directly related to the stimulation of cell division. A number of reports indicate that the synthesis of ECM components is increased when arterial cells are stimulated to divide.[44,45] However, it is also clear that ECM molecules also may be increased by other factors released at the wound sites, such as TGFβ, which does not stimulate cell proliferation.[46] These growth factors, in combination with proteoglycans, can also alter cell phenotypes.[47] This suggests that regulation of ECM deposition after arterial injury is a complex process.

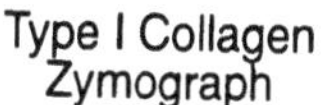

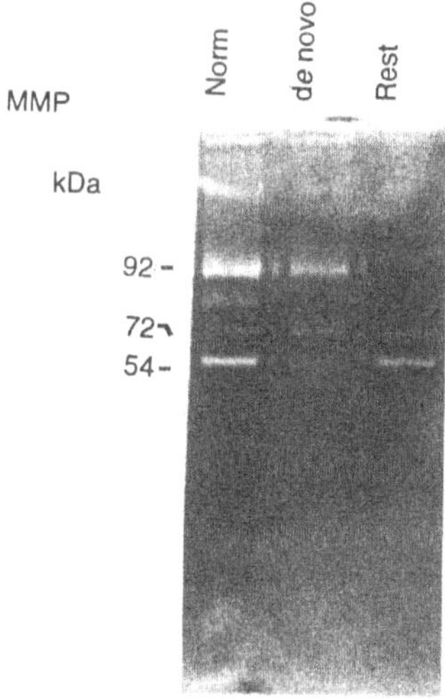

Figure 2. Expression of MMP activity in restenotic, *de* novo and normal tissue.

Currently, we do not know the type of collagen present in restenotic lesions. However, our study points out the possibility of an excess of type IV collagen. This is based upon the observation that we failed to find any type IV collagenase activity (i. e. 92-kDa gelatinase) in restenotic lesions (Figure 2). The data further suggest that the excessive production of collagen type IV may be the result of unopposed accumulation of material in the absence of type IV collagenase in patients who experience restenosis. Our quantitative pathological study does not differentiate among the different type of cells present in the resected lesion but only quantifies the cell numbers in the samples. We are currently attempting to understand the range, origins, and functions of intimal and lesion cells using specific antibodies to different cell types. Endothelial, smooth muscle, inflammatory cells and macrophages may be present in different amounts and stages of function.[48] However, some studies have predicted the presence of proliferative smooth muscle cell in restenotic lesions (49-51) without the critical evaluation for other cell types in primary and restenotic lesions.

Inhibitor and Activator of Proteinases. To increase our understanding of the potential role of metalloproteinases and their endogenous inhibitors and activators in remodeling of the vascular wall extracellular matrix following directional coronary atherectomy, we have shown collagenolytic activity in normal arterial tissue extracts, with the majority of the collagenolytic activity existing in the latent form. Only ~10% was in the active form, with the remaining in a pro or latent form. In tissue samples from restenotic lesions, we did not observe significant amounts of active or latent collagenases, as compared to samples from *de novo* atherosclerotic lesions. Storage of latent collagenases in the interstitium may provide one mechanism whereby activation could occur in the face of tissue injury, tissue remodeling and ECM degradation, especially since MMPs have been shown to degrade a variety of extracellular matrix components.[52] It has previously been shown that collagenases are secreted by cultured skin fibroblasts and are regulated post-transcriptionally by their TIMP.[53,54]

TIMPs function as endogenous inhibitors and regulators of both the latent and active forms of collagenase.[53,54] Additionally, TIMPs have been shown to demonstrate anti-mitogenic activity.[55] Thus, these inhibitors may play an important role in the control of MMPs post-translationally in tissue remodeling.[56] Increased or decreased inhibitor levels could lead to an imbalance in matrix degradation during injury response and other pathological conditions. We demonstrated that the TIMP levels in restenotic lesions were increased relative to both *de novo* atherosclerotic tissue and normal tissue (Figure 3).

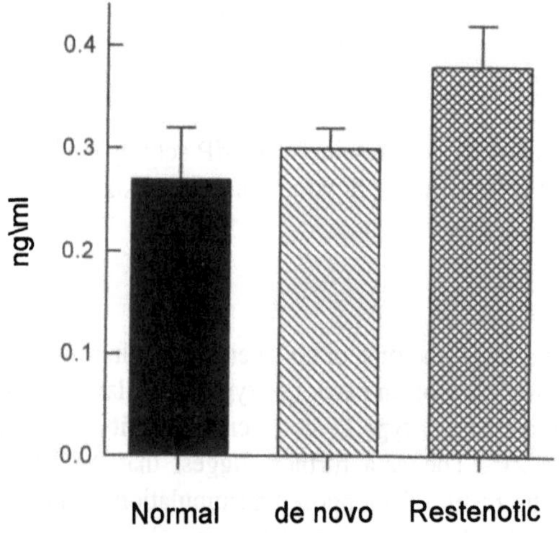

Figure 3. Expression of TIMP in restenotic, *de* novo and normal tissue

Thus, comparatively since there does not appear to be much active or latent collagenase in restenotic lesions, and since TIMPs are also a growth promoter, it is conceivable that endogenous inhibitors of metalloproteinases found in restenotic tissues may be important in the regulation of proliferation of arterial medial cells, pericytes, and fibroblasts that produce ECM components. In contrast, we failed to observe significant changes in the level of TIMP in *de novo* atherosclerotic tissue and normal arterial tissues.

The endogenous activators of MMPs have been characterized from culture media of rabbit synovial fibroblasts[57], rabbit uterine cervical fibroblasts[58] and bovine articular cartilage.[59] Although the modes of procollagenase activation by these activators have been shown to be variable[60,61], endogenous activators (i.e. tissue plasminogen activator (tPA) or uPA) have been shown to activate proMMPs in other tissues.[62] The presence of an endogenous procollagenase activator was previously proposed following experiments where latent collagenase in the culture medium of mouse bone explants was gradually activated after brief treatment of the medium with trypsin.[63] This observation was further substantiated by observations with other proteinases showing activation of procollagenase by stromelysin (MMP-3).[64] However, the mode of procollagenase activation by these activators was reported to be different in different tissues. Our studies showed the presence of tPA in tissues from restenosis lesions as well as from *de novo* atherosclerotic lesions. Tissue-PA may be involved in procollagenase activation in vasculature.

Our results also demonstrate that tissue MMPs and/or proMMPs present in vascular tissue may serve an important function in the lysis of an occlusive coronary thrombus by endogenous and exogenously administered lytic agents. The ability of a thrombolytic agent to dissolve an occlusive coronary artery thrombus is determined by several factors. After administration of a thrombolytic agent, the drug must be delivered to a thrombus and be associated with adequate substrate (plasminogen) and the appropriate metabolic environment for an enzymatic reaction (i. e. conversion of plasminogen to plasmin) to occur. The intrinsic composition (i. e. collagenous or non-collagenous) or ultrastructure of the thrombus itself also affects its lysability.

One potential mechanism for activation in intimal fibrous tissue may be through the tPA/plasmin system. Based on our results, a possible scheme of the activation of proMMPs in arteries can be postulated:

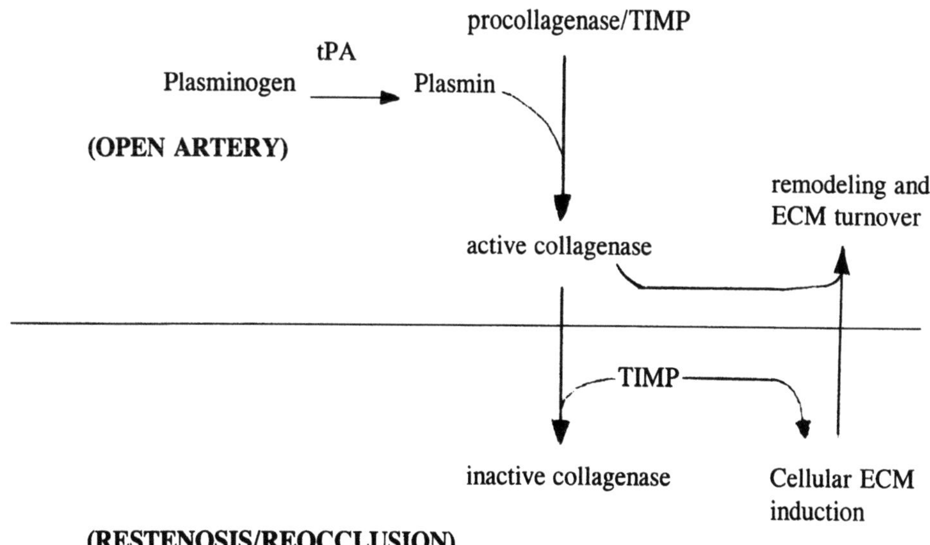

According to this scheme, if there is no sufficient active or pro/latent MMPs in a restenotic lesion, there will be abnormal tissue remodeling after arterial injury, leading to excess ECM deposition and compromise of the vessel lumen. At the same time if TIMP, any active MMP would be inhibited, along with the mitogenic action of TIMP to induce proliferation of fibroblasts that lay down ECM.

In all tissue samples that we studied, irrespective to their origin, we failed to observe changes in tPA levels. Our results suggest that in patients with repressed level of proMMPs, there is the possibility that tPA or streptokinase therapy may not be as efficacious. For example, clinical evidence suggests that not all coronary occlusions resolve following successful lysis by streptokinase and tPA. In a study[65,66] of 41 patients with myocardial infarction (MI) treated with lytic therapy, most survived except two in which lytic therapy was unsuccessful and the patients died. Our proposed model would predict the lack of adequate proMMPs expression at the site of collagenous thrombus in these two patients. The basis of this difference may be related to a genetic defect in arterial MMP expression.

SUMMARY

The results suggest that in arterial tissue from patients with angiographic restenosis there is an increased production of ECM collagen and a decrease in MMPs activity compared to both normal artery and atherosclerotic arterial samples from *de novo* patients undergoing an initial revascularization procedure of a significant coronary artery lesion. The expression of MMPs in restenotic tissue following mechanical revascularization is repressed, and may be associated with increased levels of TIMPs. However, endogenous activators of proMMPs do not change significantly in *de novo* atherosclerotic, restenotic, and normal vascular tissue. In conclusion, our data suggest the hypothesis that restenosis occurs, at least in part, secondary to decreased intimal turnover of collagen and decreased MMPs activity. Thus, the wound healing response characterized by increased collagen synthesis following mechanical revascularization may ultimately lead to restenosis if unopposed by MMPs degradation of excess collagen.

References

1. Liotta LA, Steeg PS, Stetler-Stevenson WG: Cancer metastasis and angiogenesis, an imbalance of positive and negative regulation, *Cell*, 64:327-36 (1991).
2. Matrisian LM: Metalloproteinases and their inhibitors in matrix remodeling, *Trends Genet*, 6:121-25 (1990).
3. Woessner JF, Jr.: Matrix metalloproteinases and thier inhibitors in connective tissue remodeling, *FASEB J*, 5:2145-54 (1991).
4. Murphy G, Hembry RM, Hughes CE, Fosang AJ, Hardinggham TE: Role and regulation of metalloproteinases in connective tissue turnover, *Biochem Soc Trans*, 8:812-15 (1990).
5. Tyagi SC, Matsubara L, Weber KT: Direct extraction and estimation of collagenase(s) activity by zymography in microquantitites of rat myocardium and uterus, *Clin Biochem*, 26:191-98 (1993).
6. Ip JH, Fuster V, Isreal D, Badimon L, Badimon J, Chesebro JH: The role of platelets, thrombin, and hyperplasia in restenosis after coronary angioplasty, *J Am Coll Cardiol*, 17:77B-88B, (1991).
7. Forrester JS, Fishbein M, Helfant R, Fagin J: A paradigm for restenosis based on cell biology: clues for the development of new preventive therapies, *J Am Coll Cardiol*, 17:758-69 (1991).

8. Luscher TF, Vanhoutte PM: The endothelium: modulator of cardiovascular function, CRC Press, Boca Raton, FL (1990).
9. Badimon JJ, Fuster V, Chesebro JH, Badimon L: Coronary atherosclerosis: a multifactorial disease, *Circulation*, 87:Supp II:II-3-II-16 (1993).
10. Ross R: The pathogenesis of atherosclerosis: a perspective for the 1990s, *Nature*, 362:801-09 (1993).
11. Luscher TF: The endothelium and cardiovascular disease-a complex relation, *N Eng J Med*, 330:1081-83 (1994).
12. Ip JH, Fuster V, Isreal D, Badimon L, Badimon J, Chesebro JH: The role of platelets, thrombin, and hyperplasia in restenosis after coronary angioplasty, *J Am Coll Cardiol*, 17:77B-88B, (1991).
13. Waller BF: Crackers, breakers, stretches, drillers, scrapers, shavers, burners, welders and melters-the future treatment of atherosclotic CAD? a clinical morphologic assessment, *J Am Coll Cardiol*, 13:969-87, (1989).
14. Waller BF, Pinkerton CA, Orr CM, Slack JD, Van Tassel JW, Peters T: Restenosis afterclinically successful coronary ballon angioplasty: a necropsy study of 20 pateints, *J Am Coll Cardiol*, 17:58B-70B, (1991).
15. Tyagi SC and Simon SR: Regulation of neutrophil elastase activity by elastin-derived peptide, *J Biol Chem*, 268:16513-18 (1993).
16. White JV, Haas K, Phillips S, Camerota AJ: Adventitial elastolysis is a primary event in aneurysm formation, *J Vasc Surg*, 17:371-80 (1993).
17. Dinerman JL, Mehta JL, Saldeen TGP, Emerson S, Wallin R, Davda R, Davidson A: Increased neutrophil elastase release in unstable angina rectoris and acute myocardial infarction, *J Am Coll Cadriol,* 15:1559-63, (1990).
18. Tyagi SC, Ratajaska A, Weber KT: Myocardial matrix metalloproteinases: localization and activation, *Mol Cell Biochem*, 126:49-59 (1993).
19. Hornebeck W, Adnet JJ, Robert L: Age dependent variation of elastin and elastase in aorta and human breast cancers, *Exp Gerontol*, 13:293-98 (1978).
20. Bellon G, Ooyama T, Hornebeck W, Robert L: Isolation and partial characterization of an elastase-type enzyme from human arterial wall by lima-bean trypsin inhibitor affinity chromatography, *Artery*, 7:290-302 (1980).
21. Senior RM, Griffin GL, Eliszar CJ, Shapiro SD, Goldberg GI, Welgus HG: Human 92- and 72- kilodalton type IV collagenases are elatases, *J Biol Chem*, 266:7870-75 (1991).
22. Ross R: The pathogenesis of atherosclerosis-an update, *N Engl J Med*, 314:488-500, (1986).
23. Liu MW, Roubin GS, King SB II: Restenosis after coronary angioplasty: potential biologic determinants and role of intimal hyperplasia, *Circulation*, 79:1374-87, (1989).
24. Libby P, Warner SJC, Salomon RN, Birinyi LK: Production of PDGF-like mitogen by SMC from human atheroma, *N Engl J Med*, 318:1493-98, (1988).
25. Walker LN, Bowen-Pope DF, Ross R, Reidy MA: Production of PDGF-like molecules by cultured arterial SMC accompanies proliferation after arterial injury, *Proc Natl Acada Sci* (USA), 83:7311-15, (1986).
26. Clowes AW, Clowes HM, Reidy MA: Kinetics of cellular proliferation after arterial injury III, endothelial and SM growth in chronically denuded vessels, *Lab Ivest*, 54:295-303, (1986).
27. Austin GE, Ratlif NB, Hollman J, Tabel S, Phillips DF: Intimal proliferation of SMC as an explanation for recurrent coronary artery stenosis after PTCA, *J Am Coll Cardiol*, 6:369-75, (1985).
28. Peuhkurinen KJ, Risteli L, Melkko JT, Linnaluoto M, Jounela A, Risteli J: Thrombolytic therapy with streptokinase stimulates collagen breakdown , *Circulation*, 83,1969-75 (1991).

29. Keski-Oja J, Lohi J, Tuuttila A, Tryggvason K, Vartio T: Proteolytic processing of the 72 kDa type IV collagenase by urokinase plasminogen activator, *J Exp Cell Res*, 202:471-76 (1992).
30. Hay E: Cell biology of the extracellular matrix, Plenum Press, New York, 1981.
31. Wight TN: Matrice extracellulaire et atherosclerose., Les Malaides De La Paroi Arterielle. Edited by J Camilleri, CL Berry, J Fiessinger, J. Bariety. Paris, Medicine-Sciences/Flammarion, pp 163-173, 1987.
32. Holmes DR, Vlietstra R, Smith H, et al.: Restenosis after coronary angioplasty (PTCA): a report from the PTCA registry of the national heart, lung and blood institute, *Am J Cardiol*, 53:77-81C (1984).
33. Boyd CD, Kniep AC, Pierce RA, Deak SB, Karboski C, Miller DC, Parker MI, Mackenzie JW, Rosenbloom J, Scott GE: Increased elastin mRMA levels associated with surgically induced intimal injury, *Connect Tissue Res*, 18:65-78 (1988).
34. Barnes MJ: Collagens in atherosclerosis. *Cell Relat Res*, 5:65-97 (1985).
35. Liu MW: Restenosis-Biological Aspects in *Complications of Coronary Angioplasty*, Edited by Black AJR, Anderson HV, Ellis SG: Vol:3, pp219-233, 1992.
36. Sukhova GK, Galis ZS, Lark MW, Lee RT, Libby P: Regional expression of matrix-degrading proteinases in human atherosclerotic plaques, *Circulation*, 88:I-174 (1993).
37. Clowes AW, Reidy MA, Clowes MM: Kinetics of cellular proliferation after arterial injury: I. smooth muscle growth in the absence of endothelial, *Lab Invest*, 49:327-33 (1983).
38. Clowes AW, Reidy MA, Clowes MM: Mechanism of stenosis after arterial injury, *Lab Invest*, 49:208-15 (1983).
39. Johnson DE, Hinohara T, Selmon MR, Braden LJ, Simpson JB: Primary peripheral arterial stenosis and restenosis excised by transluminal atherectomy: a histopathologic study, *J Am Coll Cardiol*, 15:419-25, (1990).
40. Garrats KN, Holmes DR, Bell MR, et al: Restenosis after directional coronary atherectomy: differences between primary atheromatous and restenotic lesions and influence of subintimal tissue resectio, *J Am Coll Cardiol*, 16:1665-71 (1990).
41. Forrester JS, Fishbein M, Helfant R, Fagin J: A paradigm for restenosis based on cell biology: clues for the development of new preventive therapies, *J Am Coll Cardiol*, 17:758-69 (1991).
42. Dvorak HF: Tumors: wounds that do not heal, Similarities between tumor stroma generation and wound healing, *N Engl J Med*, 315:1650-59 (1986).
43. Couchman JR, Hook M: Proteoglycans and wound repair, molecular and cellular biology of wound repair. Edited by RAF Clark, PM Henson, Plenum Press, New York, 1988, pp 437-470.
44. Hollman J, Thiel J, Schmidt A, Buddecke E: Increased activity of chrondrotin sulfate synthesizing enzymes during proliferation of arterial smooth muscle cells, *Exp Cell Res*, 167:484-94 (1986).
45. Wight TN, Potter-Perigo S, Aulinskas T: Proteoglycans and cell proliferation, *Am J Respir Dis*, 140:1132-35, (1989).
46. Chua CC, Chua BL, Zhao CZY, Krebs C, Diglio C, Perrin E: Effect of growth factors on collagen metabolism in cultured human heart fibroblasts, *Connect Tiss Res*, 26:271-81, (1991).
47. Flaumenhaft R, Rifkin DB: Extracellular matrix regulation of growth factor and protease activity, *Curr Opin Cell Biology*, 3:817-23 (1991).

48. Stary HC: Changes in the cells of atherosclerotic lesions as advanced lesions evolve in coronary arteries of children and young adults. In: Glagov S, Newman WP, Schoffer SA, eds. Pathobiology of the human artherosclerotic plaque, Springer-Verlag, New York, 93-106, 1990.
49. Pickering JG, Bacha PA, Weir L, Jekanowski J, Nichols JC, Isner JM: Prevention of smooth muscle cell outgrowth from human atherosclerotic plaque by a recombinant cytotoxin specific for the epidermal growth factor receptor, *J Clin Invest*, 91:724-29, (1993).
50. Simons M, Leclerc G, Safion RD, Isner JM, Weir L, Baim DS: Relation between activated smooth-muscle cells in coronary-artery lesions and restenosis anfter atherectomy, *N Engl J Med*, 328:608-13, (1993).
51. Leclerc G, Isner JM, Kearney M, Simons M, Safion RD, Baim DS, Weir L: Evidence implicating nonmuscle myosin in restenosis, *Circulation*, 85:543-53, (1992).
52. Dixit SN, Mainardi CL, Seyer JM, Kang AH: Covalent structure of collagen: amino acid sequence of alpha 2-CB5 of chick skin collagen containing the animal collagenase cleavage site, *Biochemistry*, 18:5416-22, (1979).
53. Goldberg GI, Marmer BL, Grant GA, Eisen AZ, Wilhelm S, He CS: Human 72-kDa type IV collagenase forms a complex with a tissue inhibitor of metalloproteinases designed TIMP-2, *Proc Natl Acad Sci*, 86:8207-11, (1989).
54. Gavrilovic J, Hembry RM, Reynolds JJ, Myrphy G: Tissue inhibitor of metalloproteinases (TIMP) regulates ectracellular type I collagen degradation by chondrocytes and endothelial cells, *J Cell Sci*, 87:357-62, (1987).
55. Moses MA, Langer R: A metalloproteinase inhibitor as an inhibitor of neovascularization, *J Cell Biochem*, 47:230-35, (1991).
56. Murphy G, Docherty AJP: In *The control of tissue damage* (Glauert AM, ed.,) pp 223-241, Elsevier Science Publishers, Amsterdam, 1990.
57. Vater CA, Nagase H, Harris ED Jr: Purification of an endogenous activator of procollagenase from rabbit synovial fibroblast culture medium, *J Biol Chem*, 258:9374-82, (1993).
58. Ishibashi M, Ito A, Sakyo K, Mori Y: Procollagenase activator produced by rabbit uterine cervical fibroblasts, *Biochem J* 241:527-34 (1987).
59. Treadwell BV, Neidel J, Pavia M, Towle CA, Trice ME, Mankin HJ: Purification and characterization of collagenase activator protein synthesized by articular cartilage, *Arch Biochem Biphys*, 251:715-23, (1986).
60. Murphy G, Cockett MI, Stephens PE, Smith BJ, Docherty AJP: Stromelysin is an activator of procollagenase. A study with natural and recombinant enzymes, *Biochem J*, 248:265-68, (1987).
61. Ito A, Nagase H: Evidence that human rheumatoid synovial matrix metalloproteinase 3 is an endogenous activator of procollagenase, *Arch Biochem Biophys*, 267:211-16, (1988).
62. Werb Z, Mainardi CL, Vater CA, Harris ED Jr: Endogenous activation of latent collagenase by rheumatoid synovial cells. Evidence for a role of plasminogen activator, *N Engl J Med* 296:1017-23, (1987).
63. Vaes G: Multiple steps in the activation of the inactive precursor of bone collagenase by trypsin, *FEBS Lett* 28:198-200, (1972).
64. Suzuki K, Enghild JJ, Morodomi T, Salvesen G, Nagase H: Mechanisms of activation of tissue procollagenase by matrix metalloproteinase 3 (stromelysin), *Biochemistry*, 29:10261-70, (1990).

65. Mathey DG, Kuck KH, Tilsner V, Krebber HJ, Bleifeld W: Non surgical coronary artery recanalization in acute transmural myocardial infarction, *Circulation*, 63:489-97, (1981).

66. Kennedy JW, Gensini GG, Timmis GC, Maynard C: Acute myocardial infarction treated with intracoronary streptokinase: a report of the society for cardiac angiography, *Am J Cardiol*, 55:871-77 (1985).

Key Words:

Extracellular matrix
Human vascular disease
Matrix metalloproteinase
Tissue inhibitor of metalloproteinase
Collagen
Elastin
Collagenase
Plasminogen activator
Restenosis
Atherosclerosis

CURRENT CONCEPTS OF THE PLASMA LIPOPROTEINS AND THEIR ROLE IN ATHEROSCLEROSIS

H. Bryan Brewer, Jr.

Molecular Disease Branch, National Heart, Lung, and Blood Institute
National Institutes of Health, Building 10, Room 7N115
10 Center Dr MSC 1666, Bethesda, MD 20892-1666

During the last two decades there have been major advances in our understanding of the role of the plasma lipoproteins, apolipoproteins, lipolytic enzymes, and lipoprotein receptors in cholesterol and lipoprotein metabolism. This new information has provided major insights into the role of cholesterol and lipoproteins in the pathogenesis of premature atherosclerosis. This report will briefly review our current concepts of lipoprotein metabolism and the role of the atherogenic and antiatherogenic plasma lipoproteins in the development of premature cardiovascular disease.

OVERVIEW OF HUMAN LIPOPROTEIN METABOLISM

The metabolism of the human plasma lipoproteins can be conceptually separated into three separate pathways. Two pathways involve lipoproteins containing apoB including chylomicrons and VLDL which are secreted from the intestine and liver, respectively. The third pathway involves HDL which contains the two major apolipoproteins, apoA-I and apoA-II. Schematic overviews of these three pathways for lipoprotein biosynthesis, transport, and catabolism are illustrated in Figure 1.

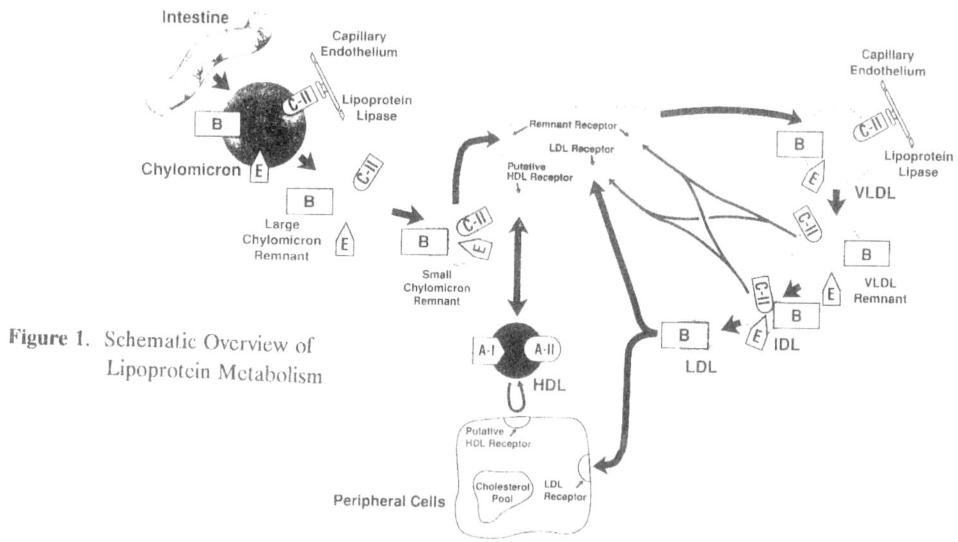

Figure 1. Schematic Overview of Lipoprotein Metabolism

Cardiovascular Disease 2
Edited by L.L. Gallo, Plenum Press, New York, 1995

Metabolic Cascades Of Lipoproteins Containing ApoB

The pathways involving the metabolism of lipoproteins containing apoB consist of two separate metabolic cascades (for general reviews see (1-4)). The first cascade of apoB containing lipoproteins involves triglyceride rich chylomicrons which transport dietary cholesterol and triglycerides from the intestine to peripheral tissues and ultimately the liver. Following secretion, chylomicrons acquire apoE and apoC-II which dissociate from HDL. ApoC-II activates lipoprotein lipase (LPL) which catalyses triglyceride hydrolysis and remodeling of triglyceride rich lipoproteins. Concomitant with triglyceride hydrolysis is the transfer of apolipoproteins as well as lipid constituents from chylomicrons to HDL. With lipopolysis chylomirons are converted to large and finally small chylomicron remnants. Chylomicron remnants are cleared from the plasma following the interaction with either the LDL receptors (5-8) or remmant receptors on the liver. The best current candidate for the putative remnant receptor is a glycosylated 600 kDa protein that has been designated the LDL receptor-related protein or LRP (9,10). ApoE is the primary ligand on chylomicron remnants for interaction with the liver lipoprotein receptors (11-13). The second apoB cascade involves triglyceride rich VLDL containing apoB secreted by the liver. ApoC-II and apoE are released from HDL and reassociate with the hepatogenous triglyceride rich VLDL secreted from the liver. ApoC-II activates LPL as outlined above and VLDL are serially converted to VLDL remnants, IDL, and finally LDL. ApoE and hepatic lipase, a second lipolytic enzyme, have been proposed to be required for the efficient conversion of IDL to LDL. During the metabolic conversion of VLDL to LDL approximately 50 percent of VLDL remnants and IDL are removed from the plasma by the liver. LDL, the final product of the VLDL lipoprotein cascade, interact with the LDL receptor present on the plasma membranes of liver as well as adrenal, and peripheral cells including fibroblasts and smooth muscle cells (5-8). The interaction of LDL with the LDL receptor initiates receptor mediated endocytosis and LDL catabolism. The two major apolipoproteins that serve as ligands for the LDL receptor are apolipoproteins B and E.

HDL Metabolism

A major role of HDL in lipoprotein metabolism is to transfer cholesterol from peripheral tissues to the liver. This hypothetical process termed reverse cholesterol transport involves the removal of excess cholesterol from peripheral cells by HDL (14,15). In this proposed pathway, HDL interacts with a putative HDL receptor that facilitates the transfer of intracellular cholesterol to the membrane for removal by HDL. HDL transports this cholesterol in plasma and delivers it to the liver via the putative HDL receptor for removal from the body by direct secretion into bile or following conversion to bile acids (16-22). An additional major pathway for the transport of cholesterol from peripheral cells to the liver is mediated by the cholesterol ester exchange protein (CETP) which exchanges cholesteryl esters and triglycerides between HDL and the apoB containing lipoproteins (19). Thus, cholesterol may be transported back to the liver either directly by HDL, or following exchange to lipoproteins in the apoB cascades (Figure 1).

MAJOR ATHEROGENIC AND ANTI-ATHEROGENIC PLASMA LIPOPROTEINS

Increased plasma levels of three different classes of lipoproteins including LDL, β-VLDL and Lp(a), and decreased levels of HDL have been associated with an increased risk of premature cardiovascular disease. A schematic overview of the interactions of the three major atherogenic lipoproteins and the anti-atherogenic HDL

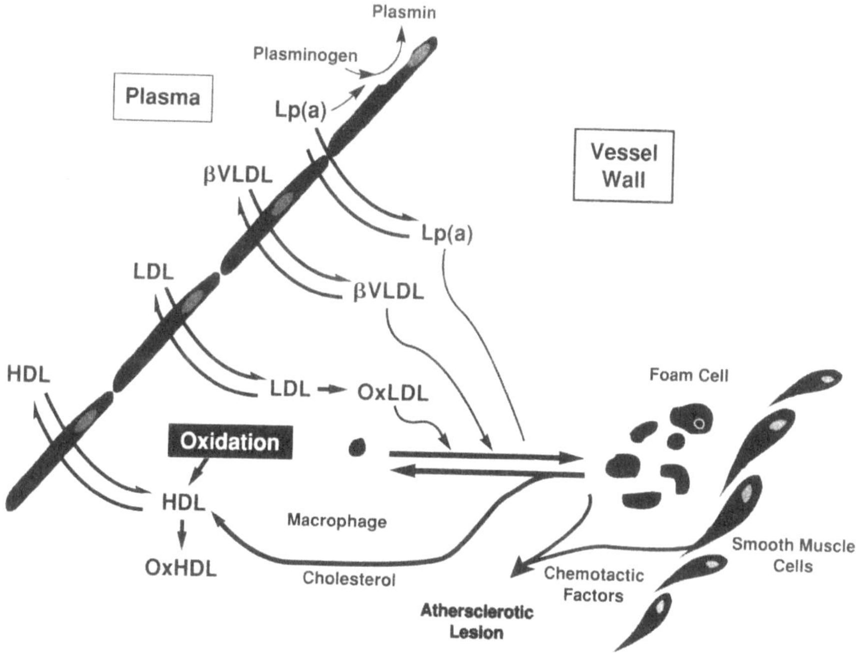

Figure 2. Role of the Atherogenic and Antiatherogenic Lipoproteins in the Development of the Atherosclerotic Lesion

LDL

in the development of the vascular plaque is summarized in Figure 2. In this conceptualization increased plasma levels of atherogenic lipoproteins accumulate in the vessel wall ultimately culminating in an increased uptake of cholesterol-rich lipoproteins by the macrophages resulting in the formation of foam cells which characterize the early atherosclerotic lesion. Smooth muscle cells may also take up lipoproteins and undergo conversion to foam cells.

The pathophysiological mechanisms involved in the development of atherosclerosis associated with increased plasma levels of LDL have recently been extensively studied (for reviews see (23-27). Native LDL is not readily taken up by the macrophage *in vitro* and incubation with LDL does not result in foam cell formation. An increasing body of data is consistent with the concept that oxidative modification of LDL is required for the uptake of LDL by macrophages with conversion to foam cells. Oxidative modification of LDL was observed following *in vitro* incubation with endothelial cells, smooth muscle cells, and macrophages, or following modification with malondialdehyde. Malondialdehyde is a byproduct of the metabolism of arachidonic acid in the biosynthesis of prostaglandins and is also formed during lipid peroxidation. Recent studies have also indicated that oxidized lipids within LDL may play an important role in the pathophysiology of the atherosclerotic lesion by stimulating the secretion of cytokinins and other factors which modulate endothelial cell function as well as facilitate the recruitment of plasma monocytes into the vessel wall.

Genetic dyslipoproteinemias characterized by an increased plasma level of LDL and premature cardiovascular disease include familial hypercholesterolemia, familial combined hyperlipoproteinemia, and polygenic hyperlipoproteinemia.

β-VLDL

Atherogenic remnants of triglyceride rich chylomicrons and VLDL which accumulate in type III hyperlipoproteinemia and in experimental animals fed diets high in cholesterol and saturated fat are designed **β-VLDL** (for reviews see (28-31). Efficient clearance of β-VLDL remnants requires apoE which functions as a ligand for binding to both remnant and LDL receptors facilitating clearance by the liver. An absence or structural mutations in apoE result in defective removal of remnants of triglyceride rich lipoproteins and the accumulation of plasma β-VLDL (11,12,28-31). The β-VLDL remnant lipoproteins have been proposed to be taken up by macrophages to form foam cells by either the LDL receptor or via a specific macrophage β-VLDL receptor pathway. Reported studies have not definitively established that **β-VLDL** must undergo oxidation prior to removal by the macrophage.

Lp(a)

Elevated plasma levels of Lp(a) have been reported to be an important independent risk factor for the development of premature cardiovascular disease. Plasma Lp(a) levels range from <1 to more than 100 mg/dl. Approximately 20 percent of the population have levels above 30 mg/dl which is associated with a two fold increase in the relative risk of premature cardiovascular disease. The mechanism(s) by which elevated plasma levels of Lp(a) increase the risk of premature heart disease remains to be established. Lp(a) may be taken up by macrophages resulting in cholesterol deposition and foam cell formation. Alternatively, it's atherogenic properties may be related to its role in increasing thrombosis. Lp(a) has been reported to interact with fibrin peptides, inhibit thrombolysis, and function as a competitive inhibitor of plasminogen for the plasminogen receptor present on endothelial cells (32,33)(Figure 2). The risk of cardiovascular disease associated with elevated plasma levels of Lp(a) may thus be due to either an increased risk of thrombosis or cellular uptake of cholesterol laden lipoproteins by macrophages or both.

HDL

HDL have been regarded as the principal anti-atherogenic lipoprotein in plasma (34-37). Based on the concept of reverse cholesterol transport (14,15), a reduced plasma levels of HDL is associated with a decreased efficiency in the transport of excess cholesterol from peripheral cells back to the liver and a greater potential risk of premature cardiovascular disease (Figure 2).

MOLECULAR HETEROGENEITY OF HDL

As reviewed above plasma HDL levels are inversely correlated with the development of premature cardiovascular disease. Because of the importance of HDL and the risk of cardiovascular disease the revised NCEP guidelines have recommended that individuals be screened with total cholesterol and HDL cholesterol. HDL screening will undoubtly identify a large number of individuals with low HDL cholesterol. However, recent studies have established that not all subjects with low HDL cholesterol levels are at increased risk for premature cardiovascular disease (38). Thus it will be a challenge for the physician to identify which individuals with low HDL cholesterol levels that are at increased risk for early heart disease.

In order to gain insight into the reason(s) that low plasma levels of HDL are not consistently associated with an increased risk of premature cardiovascular disease, studies have been initiated to evaluate the clinical importance of HDL heterogeneity. HDL contains several separate lipoprotein particles which now appear to have different functions in lipoprotein metabolism and reverse cholesterol transport (39-41). Several methods including electrophoresis, hydrated density, gradient gel electrophoresis, and affinity chromatography have been employed to separate and characterize the lipoprotein particles within HDL (40,42,43). The most effective current method available to classify lipoprotein particles in HDL is based on apolipoprotein composition (39). The two major apoA-I containing lipoprotein particles within HDL classified by apolipoprotein composition are lipoprotein containing only apoA-I or both apoA-I and apoA-II, designated LpA-I and LpA-I:A-II respectively. Two minor apoA-I containing lipoproteins include LpA-I:A-IV and LpA-I:E. The two major nascent HDL particles are codified as LpE and preβ-HDL. The C apolipoproteins and other minor apolipoproteins are also present on the major lipoprotein particles within HDL.

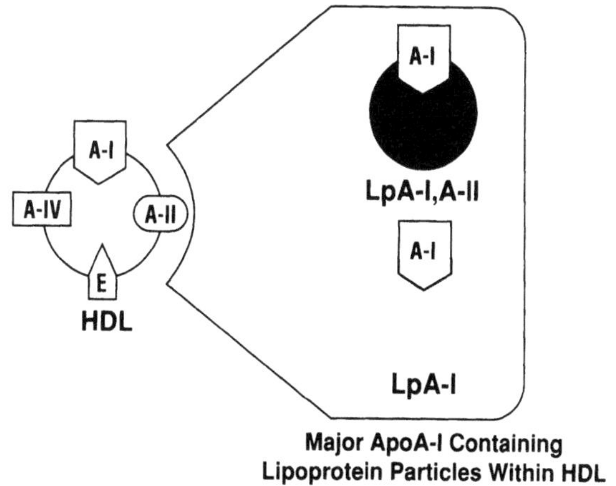

Figure 3. Major Lipoprotein Particles in High Density Lipoproteins

The relative importance of LpA-I and LpA-I:A-II in the protection against the development of premature cardiovascular disease have been recently addressed. LpA-I but not LpA-I:A-II has been reported to be inversely correlated with angiographically established coronary atherosclerosis (44). *In vitro* in cell culture studies have indicated that LpA-I but not LpA-I:A-II increased cholesterol efflux from OB1771 adipocytes (45). Analysis of LpA-I and LpA-I:A-II revealed that LpA-I:A-II not only was not effective in effluxing cholesterol from adipocytes but in fact inhibited the cholesterol efflux mediated by LpA-I (45). Thus, in cell culture studies only LpA-I was effective in facilitating the efflux of cellular cholesterol.

A separate approach to the evaluation of the antiatherogenic properties of LpA-I and LpA-I:A-II using transgenic mouse models. A comparison of the ability of LpA-I and LpA-I:A-II to protect against the development of diet induced atherosclerosis has been performed in transgenic mice on a high fat and cholesterol diet (46). Transgenic mice overexpressing human apoA-I and apoA-I + apoA-II were developed and the degree of atherosclerosis induced by a diet enriched in cholesterol and fat quantitated. LpA-I offered better protection than LpA-I:A-II against diet induced atherosclerosis in the transgenic mouse model.

Biochemical studies have also revealed different enzyme substrate specificity for LpA-I and LpA-I:A-II. In a comparison of the apoA-I containing lipoprotein particles, LpA-I:A-II was shown to be a better substrate for hepatic lipase than LpA-I (47). These results are consistent with the previous report that apoA-II increased the enzymic activity of hepatic lipase 1.5 fold in vitro (48). Thus, apoA-II may function as an important modulator of the enzymic activity of hepatic lipase in lipoprotein metabolism. Based on these result it is proposed that the LpA-I:A-II particles in HDL_2 are preferentially converted to particles within HDL_3.

Because of the emerging evidence that different lipoprotein particles in HDL may have a variable effect on the protection against atherosclerosis, a detailed series of metabolic studies were initiated to elucidate the kinetics of LpA-I and LpA-I:A-II (49). The kinetics of LpA-I and LpA-I:A-II have been analyzed in normolipidemic controls to gain insight into the potential metabolic differences between these two major apoA-I containing lipoprotein particles in man. LpA-I and LpA-I:A-II were separated from plasma or HDL by affinity chromatography utilizing antibodies to apoA-I and apoA-II. In these studies apoA-I and apoA-II were radiolabeled, incubated with plasma, and LpA-I and LpA-I:A-II isolated. In a separate approach isolated LpA-I and LpA-I:A-II were directly radiolabeled as lipoproteins. In the kinetic studies ^{125}I-LpA-I and ^{131}I-LpA-I:A-II were injected simultaneously in normal subjects, and the *in vivo* catabolism of LpA-I and LpA-I:A-II analyzed over 14 days. LpA-I particles were catabolized at a faster rate than LpA-I:A-II particles (49). Based on these results we have proposed that there are two parallel cascades of lipoprotein particles containing LpA-I and LpA-I:A-II within HDL_2 and HDL_3 (Figure 3). The results of the kinetic data support the view that LpA-I and LpA-I:A-II may have different metabolism and physiological functions in HDL metabolism. Clinical disorders associated with decreased levels of HDL can be divided into those that with no increase and dyslipoproteinemias with an increased risk of premature cardiovascular disease. An increased risk of vascular disease has been reported in familial hypoalphalipoproteinemia,thefamilialhypertriglyceridemia-hypoalpha-lipoproteinemia syndrome, apoA-I deficiency, and Tangier disease. No increased risk of premature heart disease is observed in lecithin cholesterol acyltransferase deficiency and selected kindreds with familial hypoalphalipoproteinemia (38). Of particular interest are this latter group of patients with familial hypo-alphalipoproteinemia and no increased risk of early heart disease. HDL metabolic studies have recently been completed in five kindreds with familial hypoalpha-lipoproteinemia with HDL cholesterol values ranging

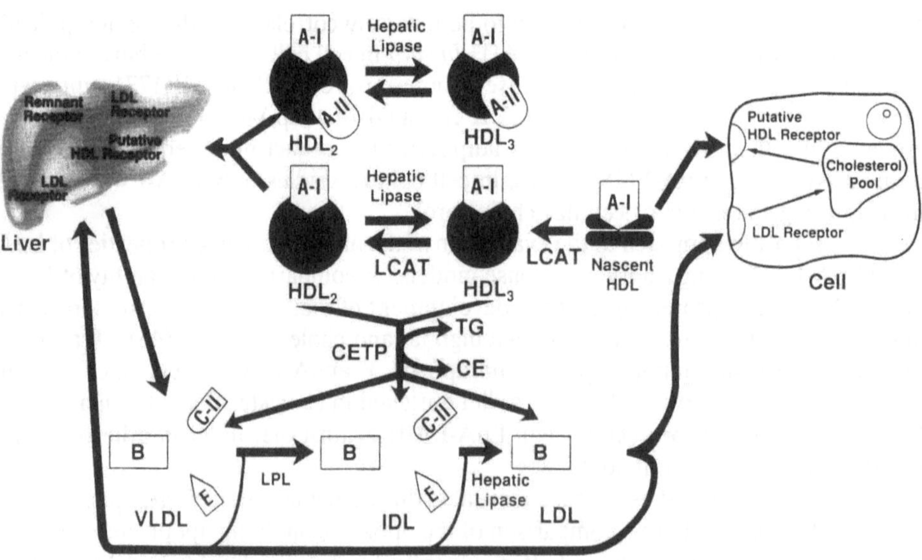

Figure 4. Schematic Overview of HDL Lipoprotein Metabolism

from 15 to 22 mg/dl. Clinical evaluation including ultrafast CT revealed no evidence of vascular disease. Radiolabeled LpA-I and LpA-I:A-II kinetic studies established that the decreased plasma levels of HDL were due to rapid catabolism of both LpA-I and LpA-I:A-II resulting in markedly reduced plasma levels of HDL. Thus despite very low HDL cholesterol levels there was no increased risk of premature heart disease.

The combined results from the analysis of the heterogeneity of HDL clearly establish that the separate lipoprotein particles in HDL have different metabolism as well as physiological functions. The evidence to date is consistent with LpA-I offering better protection against premature heart disease than LpA-I:A-II. In addition not all patients with low HDL cholesterol levels are at increased risk for premature heart disease. The challenge for the future will be to develop better tests of HDL function possibly including LpA-I and LpA-I:A-II levels to identify those individuals with low HDL levels with an increased risk that require drug treatment to raise their HDL levels.

REFERENCES

1. Brewer HB, Jr, Gregg RE, Hoeg JM, and Fojo SS. Apolipoproteins and lipoproteins in human plasma: an overview. Clin. Chem. 1988:34; pp.4-8.

2. Brewer HB, Jr, Gregg RE, and Hoeg JM. Apolipoproteins, lipoproteins, and atherosclerosis. In: Apolipoproteins, lipoproteins, and atherosclerosis.. Braunwald E ed. W.B. Saunders, Co., New York 1989:3; pp. 121-144.

3. Vega GL, Denke MA, and Grundy SM. Metabolic basis of primary hypercholesterolemia. Circulation. 1991:84; pp.118-128.

4. Schaefer EJ.Diagnosis and management of lipoprotein disorders. In: Diagnosis and management of lipoprotein disorders.. Rifkind BM ed. Marcel Dekker, Inc., New York 1991: pp. 17-52.

5. Sudhof TC, Goldstein JL, Brown MS, and Russell DW. The LDL receptor gene: a mosiac of exons shared with different proteins. Science. 1985:228; pp.815

6. Yamamoto T, Davis CG, Brown MS et al. The human LDL receptor: a cysteine-rich protein with multiple Alu sequences in its mRNA. Cell. 1984:39; pp.27-38.

7. Goldstein JL and Brown MS. The LDL receptor locus and the genetics of familial hypercholesterolemia. Annu. Rev. Genet. 1979:13; pp.259-89.

8. Goldstein JL, Brown MS, Anderson RG, Russell DW, and Schneider WJ. Receptor-mediated endocytosis: concepts emerging from the LDL receptor system. Annu. Rev. Cell Biol. 1985:1; pp.1-39.

9. Herz J, Hamann U, Rogne S, Myklebos O, Gausepohl H, and Stanley KK. Surface location and high affinity for a calcium of a 500 kDa liver membrane protein closely related to the LDL receptor suggest a physiological role as a lipoprotein receptor. EMBO J. 1988:7; pp.4119-4127.

10. Strickland DK, Ashcom JD, Williams S, Burgess WH, Migliorini M, and Argraves WS. Sequence identity between alpha2-Macroglobulin receptor and low density

lipoprotein receptor-related protein suggests that this molecule is a multifunctional receptor. J. Biol. Chem. 1990:265; pp.17401-17404.

11. Mahley RW, Innerarity TL, Rall SC, Jr., and Weisgraber KH. Plasma lipoproteins: apolipoprotein structure and function. J. Lipid Res. 1984:25; pp.1277-1294.

12. Davignon J, Gregg RE, and Sing CF. Apolipoprotein E polymorphism and atherosclerosis. Arteriosclerosis. 1988:8; pp.1-21.

13. Gregg RE and Brewer HB, Jr. The role of apolipoprotein E and lipoprotein receptors in modulating the in vivo metabolism of apolipoprotein B-containing lipoproteins in humans. Clin. Chem. 1988:34; pp.28-32.

14. Glomset JA, Janssen ET, Kennedy R, and Dobbins J. Role of plasma lecithin:cholesterol acyltransferase in the metabolism of high density lipoproteins. J. Lipid Res. 1966:7; pp.638-48.

15. Glomset JA. The plasma lecithins:cholesterol acyltransferase reaction. J. Lipid Res. 1968:9; pp.155-167.

16. Oram JF, Brinton EA, and Bierman EL. Regulation of high density lipoprotein receptor activity in cultured human skin fibroblasts and human arterial smooth muscle cells. J. Clin. Invest. 1983:72; pp.1611-1621.

17. Suzuki N, Fidge N, Nestel P, and Yin J. Interaction of serum lipoproteins with the intestine. Evidence for specific high density lipoprotein-binding sites on isolated rat intestinal mucosal cells. J. Lipid Res. 1983:24; pp.253-264.

18. Schmitz G, Niemann R, Brennhausen B, Krause R, and Assmann G. Regulation of high density lipoprotein receptors in cultured macrophages: role of acyl-CoA:cholesterol acyltransferase. EMBO. J. 1985:4; pp.2773-29.

19. Barbaras R, Puchois P, Grimaldi P, Barkia A, Fruchart JC, and Ailhaud G. Relationship in adipose cells between the presence of receptor sites for high density lipoproteins and the promotion of reverse cholesterol transport. Biochem. Biophys. Res. Commun. 1987:149; pp.545-554.

20. McKnight GL, Reasoner J, Gilbert T et al. Cloning and expression of a cellular high density lipoprotein-binding protein that is up-regulated by cholesterol loading of cells. J. Biol. Chem. 1992:267; pp.12131-12141.

21. Theret N, Delbart C, Aguie G, Fruchart JC, Vassaux G, and Ailhaud G. Cholesterol efflux from adipose cells is coupled to diacylglycerol production and protein kinase C activation. Biochem. Biophys. Res. Commun. 1990:173; pp.1361-1368.

22. Mendez AJ, Oram JF, and Bierman EL. Protein kinase C as a mediator of high density lipoprotein receptor-dependent efflux of intracellular cholesterol. J. Biol. Chem. 1991:266; pp.10104-10111.

23. Tall AR. Plasma lipid transfer proteins. J. Lipid Res. 1986:27; pp.361-367.

24. Steinberg D. Lipoproteins and atherosclerosis. A look back and a look ahead.. Arteriosclerosis. 1983:3; pp.283-301.

25. Steinberg D. Antioxidants and atherosclerosis: A current assessment. Circulation. 1991:84; pp.1420-1425.

26. Van Lenten BJ and Fogelman AM. Processing of lipoproteins in human monocyte-macrophages. J. Lipid Res. 1990:31; pp.1455-1466.

27. Haberland ME and Fogelman AM. The role of altered lipoproteins in the pathogenesis of atherosclerosis. Am. Heart J. 1987:113; pp.573-57.

28. Brewer HB, Jr, Zech LA, Gregg RE, Schwartz D, and Schaefer EJ. Type III hyperlipoproteinemia: diagnosis, molecular defects, pathology, and treatment. Ann. Intern. Med. 1983:98; pp.623-640.

29. Mahley RW. Dietary, fat, cholesterol, and accelerated atherosclerosis.. Atherosclerosis Rev. 1979:5; pp.1-34.

30. Gregg RE, Zech LA, Schaefer EJ, and Brewer HB, Jr. Type III hyperlipoproteinemia: defective metabolism of an abnormal apolipoprotein E. Science. 1981:211; pp.584-586.

31. Havel RJ. Familial dysbetalipoproteinemia. New aspects of pathogenesis and diagnosis. Med. Clin. North. Am. 1982:66; pp.441-454.

32. Loscalzo J. Lipoprotein(a). A unique risk factor for atherothrombotic disease. Arteriosclerosis 1990:10; pp.672-679.

33. Miles LA and Plow EF. Lp(a): an interloper in the fibrinolytic system. Thromb. Haemost. 1990:63; pp.331-335.

34. Miller GJ and Miller NE. Plasma-high-density-lipoprotein concentration and development of ischaemic heart-disease. Lancet. 1975:1; pp.16-19.

35. Gordon T, Castelli WP, Hjortland MC, Kannel WB, and Dawber TR. High density lipoprotein as a protective factor against coronary heart disease. The Framingham study. Am. J. Med. 1977:63; pp.707-714.

36. Miller NE, Thelle DS, Forde OH, and Mjos OD. The Tromso heart-study: high-density lipoproteins and coronary heart-disease: a prospective case-control study. Lancet. 1977:1; pp.965-968.

37. Gordon DJ and Rifkind BM. High-density lipoprotein--the clinical implications of recent studies. N. Engl. J. Med. 1989:321; pp.1311-1316.

38. Rader DJ, Ikewaki K, Duverger N et al. Very low high-density lipoproteins without coronary atherosclerosis. Lancet. 1993:342; pp.1455-1458.

39. Alaupovic P. Conceptual development of the classification systems of plasma lipoproteins. Protides of the biological fluids. Proc of 19th Colloquium. 1972:; pp.9-19.

40. Kostner G and Alaupovic P. Studies of the composition and structure of plasma lipoproteins. Separation and quantification of the lipoprotein families occurring in the high density lipoproteins of human plasma. Biochem. 1972:11; pp.3419-3428.

41. Osborne JC, Jr. and Brewer HB, Jr. The plasma lipoproteins. Adv. Protein. Chem. 1977:31; pp.253-337.

42. Nestruck AC, Niedmann PD, Wieland H, and Seidel D. Chromatofocusing of human high density lipoproteins and isolation of lipoproteins A and A-I. Biochim. Biophys. Acta. 1983:753; pp.65-73.

43. Cheung MC and Albers JJ. Characterization of lipoprotein particles isolated by immunoaffinity chromatography. Particles containing A-I and A-II and particles containing A-I but no A-II. J. Biol. Chem. 1984:259; pp.12201-12209.

44. Puchois P, Kandoussi A, Fievet P et al. Apolipoprotein A-I containing lipoproteins in coronary artery disease. Atherosclerosis 1987:68; pp.35-40.

45. Barbaras R, Puchois P, Fruchart JC, and Ailhaud G. Cholesterol efflux from cultured adipose cells is mediated by LpAI particles but not by LpAI:AII particles. Biochem. Biophys. Res. Commun. 1987:142; pp.63-69.

46. Schultz JR, Verstuyft JG, Gong EL, Nichols AV, and Rubin EM. ApoAI and apoAI + apoAII trangenic mice: a comparison of atherosclerotic susceptibility.. Circulation. 1992:86; pp.I-472

47. Mowri H-O, Patsch W, Smith LC, Gotto AM, Jr, and Patsch JR. Different reactivities of HDL2 subfractions with hepatic lipase. Circulation. 1990:82; p.558

48. Jahn CE, Osborne JC, Jr., Schaefer EJ, and Brewer HB, Jr. Activation of the enzymic activity of hepatic lipase by apolipoprotein A-II. Characterization of a major component of high density lipoprotein as the activating plasma component in vitro. Eur. J. Biochem. 1983:131; pp.25-29.

49. Rader DJ, Castro G, Zech LA, Fruchart JC, and Brewer HB, Jr. In vivo metabolism of apolipoprotein A-I on high density lipoprotein particles LpA-I and LpA-I, A-II. J. Lipid Res. 1991:32; pp.1849-1859.

KEYWORDS

Atherosclerosis
Apolipoproteins
Dyslipoproteinemias

HOMEOSTASIS OF LIPID OXIDATION IN THE ARTERY WALL

Alan M. Fogelman, Judith A. Berliner, Mahamad Navab, Ali Andalibi, Feng Liao, Linda L. Demer, Mary C. Territo, and Aldons J. Lusis.

Department of Medicine, Division of Cardiology, University of California Los Angeles, Los Angeles, CA 90024-1679

The subendothelial space is a complex network of collagen fibers and connecting fibrils.[1] Within two hours of injecting low density lipoprotein (LDL) into the femoral vein of a rabbit, the LDL was found trapped in the three dimensional matrix network.[2] The trapped LDL appeared to bind to the collagen fibers near the junction of connecting fibrils.[3] The bound LDL was no longer in solution and appeared to be trapped in a microenvironment that excluded water soluble antioxidants such as vitamin C.[4] The LDL lipids were then exposed to oxidative products released by the surrounding artery wall cells and oxidation of a mild degree occurred such that the resulting LDL was so minimally modified that it's bouyant density, R_f, and receptor recognition were not different from native LDL.[5] However, this mildly oxidized LDL, when applied to endothelial cells in culture induced the endothelial cells to bind monocytes but not neutrophils and to produce a chemotactic factor that induced monocyte migration.[5] All of the biologic activity in this mildly oxidized LDL was contained in the polar lipids.[5] The mildly oxidized LDL induced endothelial cells to express the gene and protein for the potent macrophage differentiation factor M-CSF.[6] The mildly oxidized LDL also induced endothelial cells and smooth muscle cells to express the gene and protein for the potent monocyte chemoattractant - monocyte chemotactic protein-1 (MCP-1).[7] Injection of the mildly oxidized LDL into mice produced an increase in blood M-CSF and expression of JE, the mouse homologue of the MCP-1 gene.[8] LDL given to co-cultures of human artery wall cells was sequestered in a microenvironment protected from aqueous antioxidants and, when recovered from the co-cultures, induced fresh co-cultures to express the gene and protein for MCP-1. As a result, when human monocytes were added to the co-cultures exposed to the artery wall cell- modified LDL, they migrated into the subendothelial space of the model system in response to the MCP-1 gradient.[4] All of the chemotactic activity of the co-cultures was blocked with neutralizing antibody to MCP-1. Pretreatment of LDL with antioxidants inhibited the ability of the co-cultures to generate the mildly oxidized LDL. Moreover, pretreatment of the artery wall cells with antioxidants before addition of the LDL prevented the generation of the mildly oxidized biologically active LDL.[4] However, once the LDL had been mildly oxidized by the artery wall co-culture, adding antioxidants did not prevent MCP-1 induction. Similarly, HDL blocked the oxidation of LDL by the co-culture but once the LDL was oxidized by the co-culture, adding HDL did not block MCP-1 induction.[4] Circulating monocytes expressed little MCP-1 mRNA but when placed in culture, they demonstrated a time dependent increase in message and protein that was also very density dependent, suggesting that as the monocytes migrated into the artery wall at sites of monocyte accumulation there were interactions that increased MCP-1 production and amplified the monocyte migration.[9] The induction of increased mRNA levels for JE, the mouse homologue of MCP-1, and other early response genes that were induced by the mildly oxidized LDL were found to be due to both increased transcription and increased mRNA stability.[10] The anti-inflammatory compound, leumedin, inhibited the production of biologically active mildly oxidized LDL by the co-culture without inhibiting conjugated diene formation.[11] In the co-culture, the interaction of the monocytes with the aortic wall cells resulted in a marked increase in connexin43 message and increased fibronectin production caused by IL-1 and IL-6.[12] The monocyte binding molecule induced on the surface of endothelial cells by mildly oxidized LDL was demonstrated to be different from all known monocyte binding molecules.[13] The induction of monocyte binding but not neutrophil binding by the mildly oxidized LDL was shown to be secondary to an elevation of cAMP levels induced by the mildly oxidized LDL and was associated with an activation of the transcription factor NFkB and a decrease in mRNA and protein expression of ELAM-1.[14] Inbred mice placed on an atherogenic diet accumulated similar amounts of lipid in their livers whether or not they were susceptible to develop aortic fatty streaks. However, the inbred mice that were susceptible to develop aortic fatty streaks demonstrated higher levels of conjugated dienes in their liver lipids than the mice that were resistant to the development of aortic fatty streaks.[15] Associated with this higher level of conjugated dienes was an activation of the transcription factor NFkB and the induction of inflammatory genes in the livers of the mice susceptible to develop aortic fatty streaks, but not in the mice resistant to the development of aortic fatty streaks.[15] Recombinant inbred strains derived from the parental strains were studied and indicated that the activation of NFkB and induction of inflammatory genes in the livers of the mice on an atherogenic diet cosegregated with the levels of conjugated dienes in hepatic lipids and with the development of aortic fatty streaks.[16] These results strongly suggest that at least one major gene is involved in both the development of oxidized lipids, the expression of inflammatory genes, and the development of aortic fatty streaks.

REFERENCES

1. J.S. Frank and A.M. Fogelman, The ultrastructure of the intima in WHHL and cholesterol-fed rabbit aortas prepared by ultra-rapid freezing and freeze-etching, *J Lipid Res* 30:967 (1989).

2. P.F.E.M. Nievelstein, A.M. Fogelman, G. Mottino, and J.S. Frank, Lipid accumulation in rabbit aortic intima two hours after bolus infusion of low density lipoprotein: A deep-etch and immuno-localization study for ultra-rapidly frozen tissue, *Arteriosclerosis and Thrombosis* 11:1795 (1994).

3. P. Nievelstein-Post, G. Mottino, A.M. Fogelman, and J.S. Frank, An ultrastructural study of lipoprotein accumulation in cardiac valves of the rabbit, *Arteriosclerosis and Thrombosis* 14:1151 (1994).

4. M. Navab, S.S. Imes, G.P. Hough, S.Y. Hama, L.A. Ross, R.W. Bork, A.J. Valente, J.A. Berliner, D.C. Drinkwater, H. Laks, and A.M. Fogelman, Monocyte transmigration induced by modification of low density lipoprotein in cocultures of human aortic wall cells is due to induction of monocyte chemotactic protein 1 synthesis and is abolished by high density lipoprotein, *J Clin Invest* 88:2039 (1991).

5. J.A. Berliner, M.C. Territo, A. Sevanian, S. Ramin, J.A. Kim, B. Bamshad, M. Esterson, and A.M. Fogelman, Minimally modified LDL stimulates monocyte endothelial interactions, *J Clin Invest* 85:1260 (1990).

6. T.B. Rajavashisth, A. Andalibi, M.C. Territo, J.A. Berliner, M. Navab, A.M. Fogelman, and A.J. Lusis, Modified low density lipoproteins induce endothelial cell expression of granulocyte and macrophage colony stimulating factors, *Nature* 344:254 (1990).

7. S.D. Cushing, J.A. Berliner, A.J. Valente, M.C. Territo, M. Navab, F. Parhami, R. Gerrity, C.J. Schwartz, and A.M. Fogelman, Minimally modified low density lipoprotein induces monocyte chemotactic protein (MCP-1) in human endothelial and smooth muscle cells, *Proc Natl Acad Sci* USA 87:5134-5138, 1990.

8. F. Liao, J.A. Berliner, M. Mehrabian, M. Navab, L.L. Demer, A.J. Lusis, and A.M. Fogelman, Minimally modified low density lipoprotein is biologically active in vivo in mice, *J Clin Invest* 87:2253 (1991).

9. S.D. Cushing and A.M. Fogelman, Monocytes may amplify their recruitment into inflammatory lesions by inducing monocyte chemotactic protein (MCP-1), *Arteriosclerosis and Thrombosis* 12:78 (1992).

10. R.W. Bork, K.L. Svenson, M. Mehrabian, A.J. Lusis, A.M. Fogelman, and P.A. Edwards, Mechanisms controlling competence gene expression in murine fibroblasts stimulated with minimally modified low density lipoprotein, *Arteriosclerosis and Thrombosis* 12:800 (1992).

11. M. Navab, S.Y. Hama, B.J. Van Lenten, D.C. Drinkwater, H. Laks, and A.M. Fogelman, A new antiinflammatory compound, leumedin, inhibits modification of low density lipoprotein, and the resulting monocyte transmigration into the subendothelial space of co-cultures of human aortic wall cells, *J Clin Invest* 91:1225 (1993).

12. M. Navab, F. Liao, G.P. Hough, L.A. Ross, B.J. Van Lenten, T.B. Rajavashisth, A.J. Lusis, H. Laks, D.C. Drinkwater, and A.M. Fogelman, Interaction of monocytes with co-culture of human aortic wall cells involves interleukins 1 and 6 with marked increases in connexin43 message, *J Clin Invest* 87:1763 (1991).

13. J.A. Kim, M.C. Territo, E. Wayner, T.M. Carlos, F. Parhami, C.W. Smith, M.E. Haberland, A.M. Fogelman, and J.A. Berliner, Partial characterization of the leukocyte binding molecule(s) on endothelial cells induced by minimally oxidized LDL, *Arteriosclerosis and Thrombosis* 14:427 (1994).

14. F. Parhami, Z. Fang, A.M. Fogelman, A. Andalibi, and J.A. Berliner, Minimally modified low density lipoprotein (MM-LDL)-induced inflammatory responses in endothelial cells are mediated by cAMP, *J Clin Invest* 92:471 (1993).

15. F. Liao, A. Andalibi, F.C. deBeer, A.M. Fogelman, and A.J. Lusis. Genetic control of inflammatory gene induction and transcription factor NFkB activation in response to an atherogenic diet in mice, *J Clin Invest* 91:2572 (1993).

16. F. Liao, A. Andalibi, J-H. Qiao, H. Allayee, A.M. Fogelman, and A.J. Lusis, Genetic evidence for a common pathway mediating oxidative stress, inflammatory gene induction and aortic fatty streak formation in mice, *J Clin Invest* In Press (1994).

MOLECULAR BASIS FOR THE LYSINE BINDING POLYMORPHISM OF LIPOPROTEIN(a)

Angelo M. Scanu, Ditta Pfaffinger, Olga Klezovitch, Celina Edelstein

Department of Medicine, Biochemistry and Molecular Biology and
Committee on Genetics,
University of Chicago
Chicago, IL 60637

Lp(a) and plasminogen share lysine binding property[1]. For plasminogen this binding function is localized in kringles 1 and 4, whereas for apo(a) the kringle (or kringles) responsible for this function has not been unequivocally identified[2]. However, there is evidence to suggest that kringle 4-37 contains the necessary structural elements for lysine binding[3,4]. Experimental support for this prediction has come from studies in rhesus monkeys showing that their apo(a) kringle 4-37 has a mutation associated with a lysine binding deficient Lp(a)[5]. This mutation is located in the nonpolar trough of the lysine binding pocket that comprises 7 amino acids (two anionic: asp55, asp57; two cationic: arg35, arg71 and three nonpolar: trp62, trp72 and phe64). According to the crystallographic studies on plasminogen kringle 4 by Tulinsky[6], trp72 plays a key role in lysine binding. Since lysine binding is related to fibrin binding and to the antifibrinolytic potential of Lp(a), a mutation at this level may render this lipoprotein comparatively less thrombogenic[5].

Prompted by the observations in the rhesus monkey we explored the possibility that a kringle 4-37-dependent lysine binding polymorphism may also occur in man. By using the technique previously described[5], we screened the plasma of 100 subjects selected from the Lipid and Hypertension Clinics of the University of Chicago for lysine binding. We found two unrelated subjects who exhibited a single apo(a) allele of approximately 120 kb, a single high molecular weight apo(a) isoform above the position of apoB100. The Lp(a) was defective in lysine binding and present in the plasma at concentrations below 1 mg/dl in terms of protein. Sequence analysis of the DNA fragment coding for the lysine binding pocket of apo(a) kringle 4-37, amplified by a technique previously developed in this laboratory[7], revealed a replacement of arg for trp in position 72 [8]. It is worth noting that the two subjects had neither a personal nor a familial history of atherosclerotic cardiovascular disease, ASCVD. They were both followed in the Clinic because of mild forms of hyperlipidemia.

In the course of these studies we have also observed a human apo(a) kringle 4-37 mutation, met 66→ thr, that occurred in about 50% of the subjects studied. This mutation, which was outside the lysine binding pocket, did not affect the lysine binding function of Lp(a).

The above results indicate that human apo(a) is mutable and that kringle 4-37 plays a dominant role in the lysine binding function of apo(a) via its lysine binding pocket domain. The results also show that sequence changes in this region can lead to changes in lysine binding function and from the thrombogenic viewpoint probably to a "benign" form Lp(a). The trp72→ arg mutation, probably reflecting homozygosity, was present in 2% of the subjects studied, although this frequency may not reflect that of the general population.

Cardiovascular Disease 2
Edited by L.L. Gallo, Plenum Press, New York, 1995

Subjects with a double apo(a) allele (about 80%) and two apo(a) size isoforms exhibited a significant inter-individual variability in Lp(a) lysine binding function. The mechanism of this variability is under investigation by taking into account that, besides genetic determinants, exogenous factors may influence the lysine binding of Lp(a), such as reductive or oxidative events.

The subjects who were lysine binding defective had also low plasma levels of Lp(a) suggesting a possible relationship between the two observations. For instance, one may speculate that lysine binding may be an important determinant in the initial non-covalent association step between apo(a) and apoB100. If this hypothesis were to be correct, the apo(a) of subjects with Lp(a) lysine binding deficiency, would fail to associate with apo B100 and be rapidly cleared from the circulation probably after hydrolysis by proteolytic enzymes in the plasma.

Table 1 HUMAN APO(a): KRINGLES 4-37 MUTATIONS

	Trp 72 → Arg	*Met 66 → Thr*
Frequency, %	1-2	40-50
Plasma Lp(a) levels	low	normal to high
Lysine binding	low to absent	within normal range
ASCVD	absent	present

From the above results summarized in Table 1 it is apparent that Lp(a) is functionally polymorphic in terms of lysine binding and that this polymorphism may affect the postulated thrombogenic potential of Lp(a). Thus, studies aimed at assessing the role of Lp(a) in ASCVD should take into account this functional polymorphism to complement the information derived from measurements of plasma Lp(a) concentrations and apo(a) size polymorphism[10].

ACKNOWLEDGMENTS

The original work by the Authors cited in this review was supported by NIH-NHLBI Program Project Grant 18577.

REFERENCES

1. A.M. Scanu, and G.M. Fless, Lp(a): Lipoprotein(a): Heterogeneity and biological relevance, *J. Clin. Invest.* 85:1709-1715 (1990).
2. A.M. Scanu, Structural basis for the presumptive athero-thrombogenic action of Lp(a). Facts and speculations, *Biochem.Pharm.* 46:1675-1680 (1993).
3. J.W. McLean, J.E. Tomlinson, W.J. Kuang, D.L. Eaton, E.Y. Chen, G.M. Fless, A.M. Scanu, and R.M. Lawn, cDNA sequence of human apolipoprotein(a) is homologous to plasminogen, *Nature* 330:132-137 (1987).
4. J., Jr. Guevara, J. Spurlino, A.Y. Jan, C-Y Yang, A. Tulinsky, B.V. Venkataram Prasad, J.W. Gaubatz, and J.D. Morrisett, Proposed mechanisms for binding of apo(a) kringle type 9 to apoB100 in human lipoprotein(a), *Biophys. J.* 64:686-700 (1993).
5. A.M. Scanu, L.A. Miles, G.M. Fless, D. Pfaffinger, J. Eisenbart, E. Jackson, J.L. Hoover-Plow, T. Brunck and E.F. Plow, Rhesus monkey Lp(a) binds to lysine Sepharose and U937 monocytoid cells less efficiently than human Lp(a). Evidence for the dominant role of Kringle 4-37, *J. Clin. Invest.* 91:283-291 (1993).
6. A.M. Mulichak, A. Tulinsky, and K.G. Ravichandran, Crystal and molecular structure of human plasminogen kringle 4 refined at 1.9-Å resolution, *Biochemistry* 30:10576-10580 (1991).

7. D. Pfaffinger, J. McLean and A.M. Scanu, Amplification of human apo(a) kringle 4-37 from blood lymphocytes DNA, *BBActa,* 1225:107-109 (1993).
8. A.M. Scanu, D. Pfaffinger, J.C. Lee, and J. Hinman, A single point mutation (Trp72-Arg) in human apo(a) kringle 4-37 associated with a lysine binding defect in Lp(a), *BBActa* in press (1994).
9. A.M. Scanu, D. Pfaffinger, O. Klezovitch and C. Edelstein, Genetically-determined polymorphism of kingle 4-37 of human apolipoprotein(a), *Circulation* (abstract) in press (1994).
10. G. Utermann, The mysteries of lipoprotein(a), *Science* 246:904-910 (1989).

Key words

Lp(a) polymorphism; lysine binding; apo(a) kringles; mutations.

TRIGLYCERIDE-RICH LIPOPROTEIN METABOLISM AND DIABETES

George Steiner and Gary F. Lewis

WHO Collaborating Center for the Study of Atherosclerosis in Diabetes,
The Toronto Hospital (General Division),
Toronto, Ontario, Canada. M5G 2C4

INSULINEMIA IN DIABETES

The most frequent form of hyperlipidemia in diabetes is hypertriglyceridemia[1]. Therefore, this paper will focus on triglyceride-rich lipoprotein metabolism. The major metabolic effects of diabetes stem from abnormalities in insulin. Hence, the chapter will deal with the effect of insulin on triglyceride-rich lipoprotein metabolism.

In those situations in which diabetes results from injury (surgical, immunologic, chemical etc) to the pancreatic B-cell and in which exogenous insulin is not administered, diabetes is associated with an absolute deficiency of insulin. However, in the majority of patients with diabetes, the plasma levels of insulin are actually greater than, or equivalent to those in the nondiabetic population. In those individuals insulin resistance, the pancreas attempts to compensate by secreting more insulin. This insulin is secreted into the hepatic portal circulation and then partially cleared by the liver. Thus, the hyperinsulinemia that is pancreatic, or endogenous in origin is even greater in the hepatic circulation than it is in the peripheral circulation. In contrast to endogenous hyperinsulinemia, those who are treated with insulin have hyperinsulinemia in which the insulin levels in the hepatic portal and the peripheral circulation have equally elevated[2]. Hence, in the majority of those with diabetes, the hyperglycemia does not merely reflect a deficiency of insulin. Rather, it reflects a deficiency of insulin's effect on glucose metabolism.

In light of the above considerations, a discussion of plasma triglyceride metabolism in diabetes should be divided into an examination both of the insulin deficiency and of hyperinsulinemia. It should be recognized that changes occurring in association with hyperinsulinemia may or may not be a direct consequence of insulin itself. In order to emphasize that the changes in triglyceride-rich lipoprotein metabolism may result either from insulin of from some of the many hormonal and metabolic alterations that accompany hyperinsulinemia, the chapter will refer to the hyperinsulinemic state. depending on the source of insulin, the hyperinsulinemic state can be subdivided into the endogenously and exogenously hyperinsulinemic states. Also, depending on the duration of exposure to insulin, these may be further subdivided into those states in which the hyperinsulinemia is chronic (for example long

Cardiovascular Disease 2
Edited by L.L. Gallo, Plenum Press, New York, 1995

term experimental insulin administration or long term compensation for insulin resistance in conditions such as obesity) and those in which it is acute (for example in the immediate period after an injection or infusion of insulin).

PLASMA TRIGLYCERIDE METABOLISM IN INSULIN DEFICIENCY

Severe insulin deficient in humans (i.e. diabetic ketoacidosis) is often accompanied by severe hypertriglyceridemia. However, both for ethical reasons and because these individuals are not in standardized metabolic conditions it is difficult to examine triglyceride-rich lipoprotein metabolism in depth in humans under these conditions. Therefore, some years ago we turned to diabetic dogs in order to explore the reasons for this hypertriglyceridemia[3]. The dogs were made diabetic by pancreatectomy. Appropriate controls were studied in order to exclude the effects of surgical stress itself.

Changes were noted in both the production and removal of triglyceride-rich lipoproteins in these dogs. Within the first 12 to 16 hours after insulin deprivation the plasma levels of free fatty acids, a substrate for triglyceride production, increased and the activity of lipoprotein lipase fell. Both the increase in plasma free fatty acids and the decrease in lipoprotein lipase were necessary for the development of the hypertriglyceridemia. Interestingly, after 48hrs of insulin deficiency the dogs' plasma triglyceride levels begin to decline. This is inspite of a continued elevation of free fatty acid levels and a reduction of lipoprotein lipase activity. Although kinetic studies were not done, these data imply that triglyceride-rich lipoprotein production declines after 48hrs of insulin deficiency to a rate that is less than the rate of triglyceride-rich lipoprotein removal. They also suggest that the effects of insulin on triglyceride-rich lipoprotein production are time dependent.

PLASMA TRIGLYCERIDE METABOLISM IN CHRONIC HYPERINSULINEMIA - *IN VIVO STUDIES*

A positive relation between the plasma levels of insulin and triglyceride has long been recognized[4]. Insulin is known to increase the activity of lipoprotein lipase measured in post-heparin plasma and in biopsies of adipose tissue[5]. Thus, it can increase the rate of removal of triglyceride-rich lipoproteins from the circulation. Hence, the raised plasma concentration of triglyceride-rich lipoproteins in the hyperinsulinemic state implies that the rate of production of these lipoproteins also increases, and does so even more that does the rate of their removal.

In an attempt to understand the mechanisms responsible for this, plasma triglyceride-rich lipoprotein production has been studied in humans and in animals in the hyperinsulinemic state. These in vivo studies have generally shown that in the chronically hyperinsulinemic state (as seen in obese humans, glucorticoid treated humans or rats receiving insulin over two weeks), triglyceride-rich lipoprotein production is increased[6-8]. In rats that were given insulin chronically this increase was seen despite a reduction in plasma free fatty acid levels. This suggested that the increase in triglyceride production in chronic hyperinsulinemia was not dependent on the supply of free fatty acids from the plasma and that another source, perhaps *de novo* lipogenesis, might provide the free fatty acids from which the triglycerides are made. The increase in triglyceride production in chronically hyperinsulinemic rats supplemented with fructose was greater than that observed in chronically hyperinsulinemic rats supplemented with an equal amount of glucose[9]. A very large proportion of fructose, in contrast to glucose, is first metabolized in the liver. This raised the possibility that fructose served as a more efficient source of substrate or energy for triglyceride synthesis in the chronically hyperinsulinemic state.

PLASMA TRIGLYCERIDE METABOLISM IN ACUTE HYPERINSULINEMIA - *IN VIVO STUDIES*

In contrast to the chronically hyperinsulinemic state, plasma triglyceride concentrations fall promptly in response to the acute administration of insulin[10]. In part this reflects the insulin-induced increase in lipoprotein lipase activity, referred to earlier. In addition it may reflect a change in the rate of triglyceride-rich lipoprotein production.

Triglyceride-rich lipoprotein production decreases in response to the acute administration of insulin (acute hyperinsulinemia) in both humans[10,11] and experimental animals[12]. This response to insulin may be blunted, but not obliterated in obesity[10], probably because of the associated insulin resistance. The reduction in triglyceride-rich lipoprotein production during acute hyperinsulinemia is accompanied by a reduction in plasma free fatty acid concentrations. We have studied triglyceride-rich lipoprotein production when this insulin-induced decline in free fatty acid was prevented by infusing oleate complexed to albumin into rats[12] or Intralipid and heparin into humans[13] . In the rat studies, acute hyperinsulinemia increased triglyceride-rich lipoprotein production when the fall in free fatty acid concentration was prevented. In the human studies, the acute hyperinsulinemia-associated declined in triglyceride-rich lipoprotein production was significantly blunted. Both of these studies suggested that in acute hyperinsulinemia free fatty acid suppression plays an important role in the reduced triglyceride-rich lipoprotein production seen during acute hyperinsulinemia. The human studies suggested that some other effect, possibly a direct insulin action on the liver, might also play a role in the
reduced triglyceride-rich lipoprotein production.

Combining the information from the studies of triglyceride-rich lipoprotein production in the chronically hyperinsulinemic, and the acutely hyperinsulinemic states raises the following possibility with respect to the source of the fatty acids in the triglyceride-rich lipoprotein triglyceride. In the acutely hyperinsulinemic state the plasma free fatty acid levels may be important and *de novo* lipogenesis may play little role in determining the rate of triglyceride-rich lipoprotein triglyceride production. By contrast, in the rat studies, in the chronically hyperinsulinemic state, plasma free fatty acid levels fall and yet triglyceride-rich lipoprotein triglyceride production increases. This suggests that in chronic hyperinsulinemia *de novo* lipogenesis may play an important role in the production of triglyceride-rich lipoproteins. Again, as in the studies of insulin deficiency, this would raise the possibility that the effects of insulin on *in vivo* hepatic triglyceride-rich lipoprotein production are dependent on the time period over which insulin levels are altered.

EFFECTS OF HYPERINSULINEMIA ON TRIGLYCERIDE-RICH LIPOPROTEIN PRODUCTION - *IN VIVO* vs *IN VITRO*

For the most part, the results of studies performed in isolated hepatocytes or HepG2 cells are consistent with the *in vivo* data discussed above. Several groups have found that adding insulin directly to cultures of normal rat hepatocytes or HepG2 cells, for periods up to approximately 16hrs, inhibits the secretion of VLDL[14-20]. This inhibition may be attenuated if the cells are incubated in the presence of oleate[20], or if they are exposed to insulin for longer periods (48 to 72hrs)[16]. Hepatocytes isolated from chronically hyperinsulinemic rats were shown to produce VLDL at a rate greater than that seen with cells from normal rats[21].

We and others have noted that adding insulin directly to the medium perfusing isolated rat livers increases the rate at which they secrete triglyceride-rich lipoproteins[22-26]. Our studies showed this stimulation when the perfusate had erythrocytes in order to permit adequate oxygen supply to the organ, and when an adequate supply of free fatty acid was maintained[25]. We also found that the livers from chronically hyperinsulinemic rats had a greater rate of triglyceride-rich lipoprotein secretion than did livers from normal rats[25]. Furthermore, although the livers from the chronically hyperinsulinemic animals were resistant to insulin, they still responded to sufficiently high insulin concentrations with an increase in triglyceride-rich lipoprotein secretion.

SUMMARY

These data point out the complexity of hyperinsulinemia's effects on hepatic triglyceride-rich lipoprotein production. The effects may be a direct response to insulin, may be mediated by other hormonal or metabolic effects of insulin or may be modulated by a variety of coexisting influences. The availability of free fatty acids has already been shown to play an important role, at least in the acute response to insulin. In the whole animal the availability of this substrate is greatly influenced by the action of insulin at an extrahepatic site, adipose tissue.

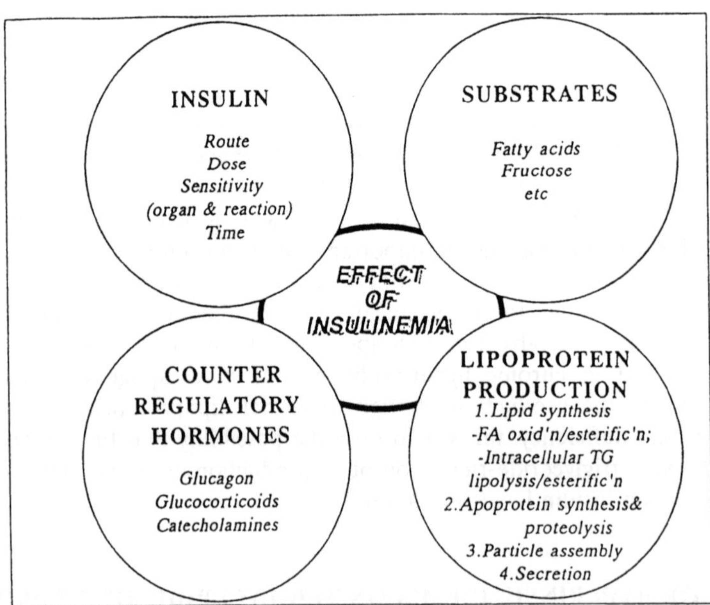

Figure 1. A schematic representation of the multiplicity of factors that might influence the effects of hyperinsulinemia or hypoinsulinemia on the production of triglyceride-rich lipoproteins by the liver.

Theoretically the effects of the hyperinsulinemic state may, at least in part, be due to the actions of counter-regulatory hormones. However, the fact that the chronically hyperinsulinemic rats were normoglycemic and that measured levels of glucagon were normal[27] suggest that this did not play a major role.

The stimulatory effect of adding insulin directly to the medium perfusing normal livers suggests that the hormone has a direct effect to enhance triglyceride-rich lipoprotein production. A similar conclusion would be suggested from the stimulation of triglyceride-rich lipoprotein production seen with 48 to 72 hr exposure of hepatocytes to insulin. By contrast, the inhibitory effect of short-term exposure of hepatocytes to insulin, while also suggesting a direct action of the hormone, would suggest that it has the opposite effect. Clearly these studies all raise the possibility that the duration of exposure to altered levels of insulin may be important in determining the type of response that will occur. The differences between the responses of whole livers and isolated cell preparations needs to be resolved, but may reflect the importance of the cellular architecture of the whole organ and/or the manner in which oxygen and substrates are delivered to the liver's cells. There are also a large number of intracellular metabolic processes that can influence the ultimate secretion of the mature triglyceride-rich lipoprotein. These include fatty acid synthesis, fatty acid oxidation vs esterification, hepatic storage and lipolysis of triglyceride, apoB synthesis, intracellular proteolysis of apoB, packaging of the triglyceride-rich lipoprotein lipids and proteins, and secretion of the triglyceride-rich lipoprotein. Insulin already has been shown to influence many of these, and in future may be found to influence others. Some new facets of the multiple effects of hyperinsulinemia can be learned from each of the models in which the effects of insulin on triglyceride-rich lipoprotein production have been studied.

ACKNOWLEDGMENTS

This chapter was prepared under the aegis of, and reports data from work supported by The Medical Research Council of Canada, The Heart and Stroke Foundation of Ontario and The Canadian Diabetes Association.

REFERENCES

1. G. Steiner, Atherosclerosis, the major complication of diabetes. *in* "Comparison of Type I and Type II Diabetes", M. Vranic , C.H. Hollenberg, and G. Steiner, eds., Plenum Publishing, New York (1985).
2. E.A. Nikkila . High density lipoproteins in diabetes. Diabetes 30(suppl2):82(1981).
3. G. Steiner, M.E. Poapst, and J.K. Davidson, Production of chylomycron-like lipoproteins from endogenous lipid by the intestine and liver of dogs. Diabetes 24:263 (1975).
4. J.M. Olefksy, J.W. Farquhar, and G.M. Reaven, Reappraisal of the role of insulin in hypertriglyceridemia. Am J Med. 57:551(1974).
5. E.A. Nikkila , J.K. Huttunen , and C. Ehnholm, Postheparin plasma lipoprotein lipase and hepatic lipase in diabetes mellitus. Relationship to plasma triglyceride metabolism. Diabetes 26:11(1977).
6. D.A. Streja, E.B. Marliss, and G. Steiner, The effects of prolonged fasting on plasma triglyceride kinetics in man. Metabolism. 26:505(1977).
7. D.C. Cattran, G. Steiner, D.R. Wilson, and S.S.A. Fenton, Hyperlipidemia after renal transplantation: natural history and pathophysiology. Ann Int Med. 91:554(1979).

8. G. Steiner, F.J. Haynes, G. Yoshino, and M.Vranic, Hyperinsulinemia and in vivo very-low-density lipoprotein-triglyceride kinetics. Am. J. Physiol. 246:E187(1982)
9. T. Kazumi, M. Vranic, and G. Steiner, Triglyceride kinetics: effects of dietary glucose, sucrose, or fructose alone or with hyperinsulinemia. Am. J. Physiol. 250:E325(1986)
10. G.F. Lewis, K.D. Uffelman, L.W. Szeto, and G. Steiner, Effects of acute hyperinsulinemia on VLDL triglyceride and VLDL apo B production in normal weight and obese individuals. Diabetes 42:833(1993).
11. G.F. Lewis, B. Zinman, K.D. Uffelman, L.W. Szeto, B. Weller, and G. Steiner, Very low density lipoprotein (VLDL) production is decreased to a similar extent by acute portal versus peripheral venous insulin delivery in humans. Am. J. Physiol. (in press)
12. K. Ferguson, J. Mamo, and G. Steiner, (in preparation)
13. G.F. Lewis, K.D. Uffelman, L.W. Szeto, B. Weller, and G. Steiner, (in preparation)
14. P.N. Durrington, R.S. Newton, D.B. Weinstein, and D. Steinberg, Effects of insulin and glucose on very-low-density lipoprotein triglyceride secretion by cultured rat hepatocytes. J. Clin. Invest. 70:63(1982).
15. W. Patsch, S. Franz, and G. Schonfeld, Role of insulin in lipoprotein secretion by cultured rat hepatocytes. J. Clin. Invest. 71:1161(1983).
16. Bartlett S.M., and Gibbons G.F. Short- and longer-term regulation of very-low-density lipoprotein secretion by insulin, dexamethasone and lipogenic subsrates in cultured hepatocytes. A biphasic effect of insulin. Biochem J. 249:37-43, 1988.
17. J.M. Duerden , S.M. Bartlett, and G.F. Gibbons, Long term maintenance of high rates of very-low-density lipoprotein secretion in hepatocyte cultures. A model for studying the direct effects of insulin and insulin deficiency in vitro. Biochem J. 263:937(1989).
18. C.E. Sparks, J.D. Sparks, M. Bolognino, A. Salhanick, P.S. Strumph, and J.M. Amatruda, Insulin effects on apolipoprotein B lipoprotein synthesis and secretion by primary cultures of rat hepatocytes. Metabolism 35:1128(1986).
19. A.C. Beynen, H.P. Haagsman, L.M.G. Van Golde, and M.J.H. Geelen, The effects of insulin and glucagon on the release of triacylglycerols by isolated rat hepatocytes are mere reflections of the hormonal effects on the rate of triacylglycerol synthesis. Biochim et Biophys Acta. 665:1(1981).
20. N.Dashti, and G.Wolfbauer, Secretion of lipids, apolipoproteins, and lipoproteins by human hepatoma cell line, HepG2: effects of oleic acid and insulin. J. Lipid Res. 28:423(1987).
21. A.I. Salhanick, M.L. Deichman, J.M. Amatruda, Chronic in vivo and acute in vitro effects of insulin on apolipoprotein B synthesis and secretion in rat hepatocytes. Diabetes 41(suppl1):17A(1992).
22. W.F. Woodside, and M. Heimberg, Effects of antiinsulin serum, insulin and glucose on output of triglycerides and on ketogenesis by the perfused rat liver. J. Biol. Chem. 251: 13(1976).
23. M.E. Laker, and P.A. Mayes, Investigations into the direct effects of insulin on hepatic ketogenesis, lipoprotein secretion and pyruvate dehydrogenase activity. Biochem. Biophys. Acta. 795:427(1984).
24. D.L. Topping , and P.A. Mayes, Insulin and non-esterified fatty acids. Acute regulators of lipogenesis in perfused rat liver. Biochem. J. 204:433(1982).
25. M. Raman, and G. Steiner, Effect of insulin on VLDL - triglyceride secretion and glucose production in the perfused rat liver. Diabetes 39(Suppl 1):45A(1990)
26. D.L. Topping, G.B. Storer, and R.P. Trimble, Effects of flow rate and insulin on triacylglycerol secretion by perfused rat liver. Am. J. Physiol. 255:E306(1988).

27. L.P. Brubaker, T. Kazumi, T. Hirano, M. Vranic, and G. Steiner, Failure of chronic hyperinsulinemia to suppress pancreatic glucagon in vivo in the rat. Can. J. Physiol. Pharmacol. 69:437,(1991)

APOLIPOPROTEIN E: PARADOXES ABOUND

William R. Hazzard, Deborah Applebaum-Bowden
James G. Terry, B.S.[2]

[1]Head, Section of Endocrinology
[2]Section of Endocrinology
Bowman Gray School of Medicine of Wake Forest University
Medical Center Boulevard
Winston-Salem, NC 27157-1052

Apolipoprotein E was so dubbed two decades ago when apolipoproteins were being named in alphabetical sequence as their properties were detailed (apo E was first characterized as the "arginine-rich" peptide). Subsequently its properties have been virtually completely described[1,2]: it is a peptide composed of 299 amino acids with a molecular weight of 34,000 daltons. It contains 22 amino acid repeats, forming amphipathic helices. It is synthesized primarily in the liver (though it is virtually ubiquitous in tissues, recently receiving special notoriety because of its distribution in the central nervous system in the areas of injury, and, specifically, its concentration in the plaques of the brains of patients with Alzheimer's disease). It is secreted in glycosylated form and is a constituent of all plasma lipoprotein classes (chylomicrons, very low density lipoproteins [VLDL], intermediate density lipoproteins [IDL], and high density lipoproteins [HDL]) except low density lipoproteins (LDL) (though traces of apo E may be found in more buoyant LDL particles, especially in animal species other than the human). Like apo B, it serves as a ligand for the E/B (LDL) receptor, and its binding domain has been characterized in detail. It has also been reported to constitute the principal apolipoprotein ligand for the apo E ("chylomicron") receptor in the liver, which has otherwise been characterized as the lipoprotein receptor protein (LRP). The gene that codes for apo E has been sequenced, cloned, and localized on chromosome 19 at the C 19q13.1 locus, closely linked to the apo C-I and apo C-II genes and more remotely linked to the gene for the LDL receptor.

Cardiovascular Disease 2
Edited by L.L. Gallo, Plenum Press, New York, 1995

The distribution of apo E among lipoprotein classes in human plasma varies as a function of the relative concentrations of triglyceride (TG) - rich lipoproteins (VLDL and IDL) vs. HDL (more in VLDL and less in HDL in hypertriglyceridemic states).[3-5] Furthermore, while total plasma apo E levels do not change following oral fat ingestion, during alimentary lipemia a portion of HDL-apo E migrates to the surface of plasma chylomicrons (after their entry to the plasma compartment from the thoracic duct). Its activity vis-a-vis the apo E receptor (LRP) is such that its addition together with cholesterol enriched beta-VLDL to media containing hepatocytes increases LRP expression (in contrast to the effect of apo C, especially C-III, which decreases LRP expression).[6,7] It is hypothesized that in the course of chylomicron remnant removal (predominantly above S_f400 and virtually complete by S_f60) apo E facilitates hepatic uptake of such cholesterol-rich remnants, serving in turn to down-regulate HMG CoA reductase (and hepatic cholesterol synthesis) and the LDL receptor.[1] As such apo E plays a central role in the mechanism whereby VLDL-IDL-LDL cholesterol concentrations are reciprocally (albeit imperfectly) regulated by dietary cholesterol intake in the human. Apo E (relative to C-III) also appears to be a key factor regulating hepatic uptake of endogenous TG-rich lipoproteins[1]: at the larger, more buoyant, TG-rich end of the S_f400-12 spectrum, the concentration of the C-apoproteins (and notably C-III) is highest relative to apo E, and hepatic uptake is minimal; as TG is progressively hydrolyzed (principally via lipoprotein lipase [LPL] - apo C-II at the upper end of the cascade and hepatic triglyceride lipase [HTGL] at the lower end), the apo C's are lost to HDL, apo E percent concentration increases, and the E/C-III ratio rises dramatically. This facilitates hepatic uptake of E-containing VLDL and IDL remnants via the B/E receptor (for which apo E is a more powerful ligand than apo B). Particles lacking apo E proceed to the S_f12-0 range of LDL, where they may be removed by the LDL receptor as a function of their apo B content. Particles not removed from plasma via these high affinity, receptor-mediated pathways may be removed via low affinity, non-regulated, "scavenger" mechanisms, notably by macrophages that can become cholesterol-laden and incorporated into arterial wall atherosclerotic plaques. The latter outcome seems to be favored when lipoprotein remnant or LDL plasma transit time is prolonged and/or LDL is modified as by oxidation of its phospholipid moieties. As with chylomicron remnant uptake, efficient, apo-E facilitated hepatic uptake of endogenous VLDL-IDL lipoprotein remnants (ca. S_f100-12) serves to downregulate the LDL receptor and HMG CoA reductase (and endogenous cholesterol synthesis) as a function of intra-hepatic unesterified cholesterol pool size. This pool appears to be tightly regulated in the human. An excess of hepatic unesterified cholesterol is prevented via five coordinately regulated mechanisms: the aforementioned downregulations of endogenous hepatic cholesterol synthesis via HMG CoA reductase and of the LDL receptor, enhanced cholesterol esterification (to cholesterol oleate) via hepatic acylcholesterol acyltransference (ACAT), secretion of hepatic cholesterol into nascent VLDL, and biliary excretion (directly or after conversion to bile acids via 7-OH hydroxylase). Cholesterol delivered to the liver at the VLDL-IDL stage in this cascade process represents a mixture of that remaining following VLDL-TG lipolysis and that delivered from HDL via the cholesterol ester transfer protein (CETP). Studies in this laboratory have identified a zone of narrow density range (and S_f distribution) within IDL characterized by a dramatic per particle increase in cholesterol content co-incident with a reciprocal decrease in TG (termed the "lipid transfer zone").[8] Missing from this comprehensive scheme of the role of apo E in the regulation of human lipoprotein metabolism is one additional potential point of control: intestinal cholesterol absorption, which has been reported to be modulated by apo E (see below).

THE ROLES OF APOLIPOPROTEIN E ISOFORMS UNIQUE TO THE REGULATION OF HUMAN LIPOPROTEIN METABOLISM

Thus far reported only in the human, apolipoprotein E exists in three common polymorphisms determined by three alleles at a single locus.[1,2,9] These vary in composition and charge by arginine-cysteine substitutions at amino acids #112, 145, and 158: a cys for arg substitution at 112 confers the lowest isoelectric point (apo E_2), a cys for arg at 145 results in an isoform of intermediate charge (apo E_3), and an arginine at all three positions confers the highest net charge (apo E_4) (other, rarer variants are also being reported in increasing numbers, but only a few appear to have physiological significance, and those will not be discussed further here[1]). The phenotype of a given individual is co-dominantly inherited in autosomal fashion.[9] Thus there are six common phenotypes which vary relatively broadly among populations. In general, however, the commonest gene, ϵ_3, is most prevalent and hence the "wild type", while ϵ_4 is of intermediate frequency and ϵ_2 is of slightly lower prevalence. In a United States population of 168 subjects[9] studied these alleles were distributed as ca. 9% ϵ_2, 77% ϵ_3, and 14% ϵ_4, giving rise to six phenotypes, with frequencies as follows (1): $E_{2/2}$ 2.4%, $E_{3/2}$ 10.1%, $E_{4/2}$ 4.2%, $E_{3/3}$ 61.3%, $E_{4/3}$ 20.8%, and $E_{4/4}$ 1.1%. That these various apo E phenotypes exert significant metabolic impact is suggested by their associated mean total plasma and, especially, LDL cholesterol concentrations, which are inversely related to "phenotype score" ($E_{2/2}$ having the lowest and $E_{4/4}$ having the highest associated mean levels[1]). As such the gene coding for apo E phenotype represents the most potent genetic determinant of population plasma cholesterol concentrations reported to date.[3] Not surprisingly, the gradient of LDL cholesterol levels determined by apo E phenotype is paralleled by the gradient in associated coronary disease prevalence, be it clinically overt, disclosed at angiography, or in persons with silent ischemia detected by treadmill testing and thallium-scintigraphy.[3]

The charge characteristics of the three principal apo E isoforms appear to relate importantly to its distribution among lipoproteins and metabolic functions. For example, treatment of apo E_2 with cysteamine causes a redistribution of apo E from HDL (where much is complexed with A-II) to VLDL.[3] These charge differentials also appear to affect profoundly the binding affinity of the apo E ligand: apo E_2 is poorly bound to the B/E receptor (displacing ^{131}I-LDL with low efficiency), as opposed to the high affinity of DMPC complexes containing either E_3 or E_4 for the B/E receptor in fibroblast systems. Once again, treatment of apo E_2 with cysteamine largely overcomes this poor binding affinity. A similar relative affinity for the apo E/LRP/"chylomicron remnant" receptor appears to prevail: E_2 is bound little if at all to the LRP when added to test systems together with beta-VLDL, while E_3 and E_4 promote expression of the LRP in such systems.

THE PARADOXES OF APO E

Synthesis of the above information produces a paradox that demands explanation: if apo E_2 is recognized poorly if at all by both the chylomicron and B/E receptors, why do individuals of the $E_{2/2}$ phenotype have the lowest LDL concentrations? A potential resolution to this paradox has been offered by hypothesizing that diminished chylomicron and VLDL remnant hepatic cholesterol uptake upregulates the LDL receptor, leading to enhanced LDL clearance via the apo B ligand. If such were the case, one would also predict upregulation of HMG CoA reductase in those of the $E_{2/2}$ phenotype. As such this hypothesis is consistent with the report of highest endogenous hepatic cholesterol biosynthesis in those of the apo $E_{2/2}$ phenotype, intermediate in those of $E_{3/3}$, and lowest in those with $E_{4/4}$ or $E_{4/3}$ phenotype.[10] (However, such compensatory enhanced cholesterol biosynthesis would have to be incomplete; otherwise there would be no net reduction in LDL levels.) Also consistent with this schema has been the reported direct correlation between apo E phenotype "score" and the LDL apo B concentration ($E_{2/2}<_{2/3}<_{2/4}=_{3/3}<_{3/4}<_{4/4}$) and its inverse correlation with the fractional catabolic rate (FCR) of LDL. Also consistent with this hypothesis is the inverse relationship between the plasma concentrations of apo E and apo B across apo E phenotype scores: apo E levels are highest and apo B levels lowest in those with the apo $E_{2/2}$ phenotype[1]; at the opposite extreme, those with apo $E_{4/4}$ have the lowest total plasma apo E and the highest apo B concentrations (these distributions are coupled with the predicted distributions of VLDL and LDL: those of $E_{2/2}$ phenotype have the highest VLDL-TG and the lowest LDL-cholesterol levels, while those of $E_{4/4}$ demonstrate the opposite pattern). However, this rationale would not be sufficient to explain all observations unless an additional mechanism were postulated, one that either reduced chylomicron remnant cholesterol flux or removed chylomicron remnant cholesterol via an alternative pathway that did not increase the hepatic unesterified cholesterol concentration. One proposed such explanation has been an inverse correlation between apo E phenotype score and fractional dietary intestinal absorption[11,12,13], those of the $E_{2/2}$ phenotype having been reported to exhibit the lowest and those of $E_{4/4}$ phenotype the highest percentage intestinal cholesterol absorption. But could such an explanation, supported by preliminary but not widely confirmed data, explain the entire gradient of LDL cholesterol concentrations across the gradient of apo E phenotype "scores"? Thus it has been deduced that the apo E phenotype must also be associated with enhanced hepatic cholesterol excretion, specifically via the biliary tract as cholesterol and/or after conversion to bile acids.[13] However, to date only preliminary evidence in support of this hypothesis has been reported.[10,13] So must one search for yet another alternative? Perhaps enhanced chylomicron (and VLDL) remnant removal via an extrahepatic mechanism? Via alternative hepatic mechanism(s)? Driven by further enhanced biliary cholesterol/bile acid excretion?

Yet another series of observations must also be reconciled with this schema, and here a second paradox emerges. Whereas the vast majority of persons of the apo $E_{2/2}$ phenotype are among those with the lowest plasma cholesterol concentrations, a subset (ca. 1% of those with the $E_{2/2}$ phenotype or 1 in 10,000 of the population at large) have among the very highest cholesterol concentrations.[1] These are individuals with the so-called type III hyperlipidemia (otherwise named dysbetalipoproteinemia, broad beta disease, or, the term coined Gofman in his original description, xanthoma tuberosum). Such persons are at extraordinary risk to premature atherosclerosis of both coronary and peripheral arteries. They have severe elevations of VLDL and, especially, IDL, both unusually enriched in cholesterol and apo E, and of slow, beta electrophoretic migration, while at the same time their LDL levels are depressed. Thus, the co-inheritance of an additional genetic cause of hyperlipoproteinemia and/or additional influences of diet and hormones appear to convert those at lowest risk of hypercholesterolemia and associated atherosclerosis to among those at highest risk. In our experience this has almost invariably represented the coincidence of the apo $E_{2/2}$ phenotype with familial combined hyperlipidemia (FCHL, probably equivalent to hyperapobetalipoproteinemia, with uncertain and no doubt variable overlap with familial dyslipidemic hypertension). This association was most evident in the largest kindred with type III hyperlipidemia reported to date.[14] The pedigree of this family was equally overrepresented by those with type III hyperlipidemia and those with other abnormal lipoprotein phenotypes, specifically with the various combinations of elevated VLDL, LDL or both (i.e., meeting the criteria for FCHL), type III resulting when the $E_{2/2}$ phenotype was present in a hyperlipidemic member and FCHL (in a II or IV lipoprotein pattern) when hyperlipidemia occurred in a member without the $E_{2/2}$ phenotype. The fundamental basis of FCHL remains conjectural, but apo B overproduction (and secretion as VLDL) with resulting high plasma apo B concentrations appears to be a metabolic hallmark of the disorder. But why would high apo B flux convert those with the lowest LDL levels (non-hyperlipidemic persons of the $E_{2/2}$ phenotype) to those with the highest VLDL-cholesterol levels (those with type III hyperlipidemia)? Of note, both normolipidemic and hyperlipidemic (type III) individuals of the $E_{2/2}$ phenotype have depressed LDL levels. Thus, while a metabolic explanation for this second paradox remains obscure, it is evident that a pathway of direct VLDL-IDL clearance without throughput to LDL must be hypothesized. This putative pathway must selectively recognize the E_2 ligand and be of high efficiency yet limited capacity, resulting in the accumulation of abnormal, cholesterol-enriched, atherogenic beta-VLDL when it becomes saturated in circumstances of high VLDL apo-B flux, as in FCHL.

CAN A PARADOX RESOLVE A PARADOX?

And here a third paradox emerges. However, this one may offer promise of resolving the metabolic puzzle of apo E. This is the paradoxical hypolipemic response of persons with type III hyperlipidemia to estrogen.[15] It is well established that estrogen (usually given as hormone replacement therapy to post-menopausal women) increases VLDL-TG and apo B synthesis and secretion. Thus we approached estrogen replacement requested by a woman with type III with great caution, hospitalizing her on a metabolic ward on a fat-free diet for a trial of such therapy to avoid the risk of chylomicronemic pancreatitis that might occur should the severely increased hypertriglyceridemia that we predicted have ensued. However, instead she demonstrated a paradoxical hypolipidemic response, with profound reductions in TG, cholesterol, and beta-VLDL. This response, subsequently confirmed repeatedly in our and other laboratories, suggests that estrogen selectively promotes removal of VLDL by a pathway not requiring E_2 and which may indeed preferably recognize E_2 (and predictably > E_3 > E_4) or at least represent the preferred removal pathway of remnants lacking E_3 or E_4.

Beyond the description of the profound, paradoxical hypolipidemic effect of estrogen in apo $E_{2/2}$ subjects with type III hyperlipidemia (who had co-inherited FCHL) and the qualitatively similar (but quantitatively less profound) response of those with FCHL (and other apo E phenotypes than $E_{2/2}$) to estrogen replacement (unpublished) is the observation by Applebaum-Bowden et al. of a major reduction in total plasma apo E concentrations in postmenopausal, normolipidemic women treated with ethinyl estradiol (1 mcg/kg/day) in a controlled clinical investigation during constant dietary intake.[16] This disclosed a ca. 35% reduction in plasma apo E concentrations, the only intervention described to date to demonstrate such a reduction. Moreover, pilot studies of the distribution of apo E (measured by radioimmunoassay) in lipoproteins of one subject before and during estrogen replacement (Figure 1) demonstrated a shift in distribution of apo E from HDL to VLDL-IDL (while TG levels and VLDL apo B concentrations also increased). Thus it appears likely that estrogen may affect the affinity of VLDL for apo E and perhaps thereby enhance clearance of VLDL-IDL via the enriched apo E content of such lipoproteins. This hypothesis is consistent with the important work of Walsh et al.[17] demonstrating that estrogen replacement in postmenopausal women is associated simultaneously with both increased synthesis and secretion of large VLDL and also enhanced clearance of those

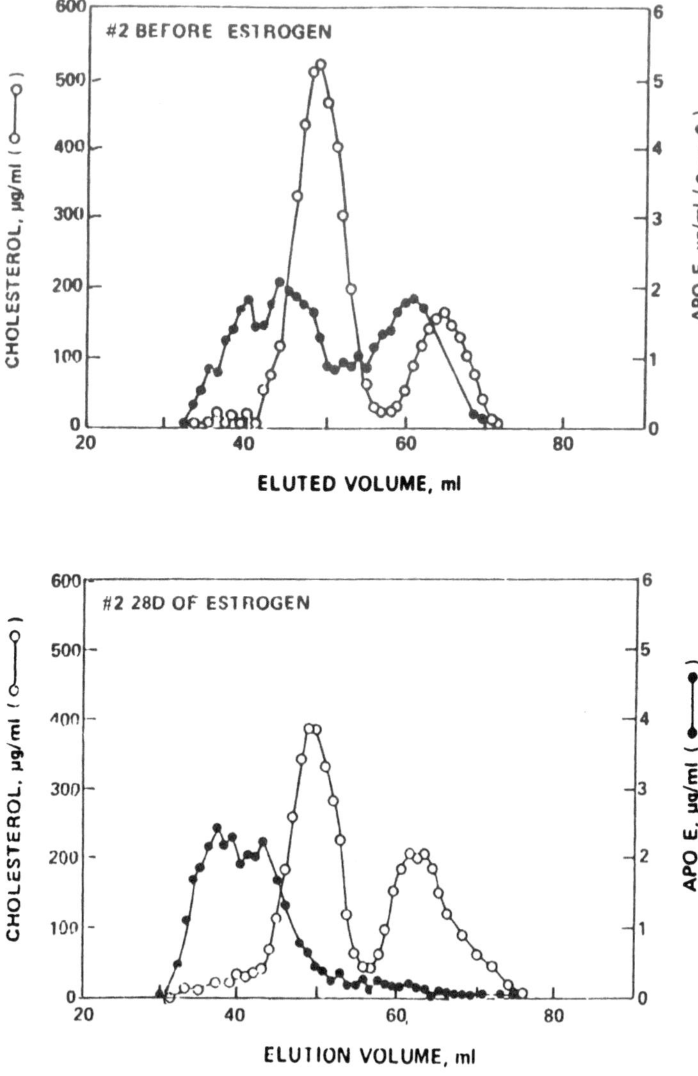

Figure 1: Apolipoprotein E distribution across VLDL, IDL, LDL, and HDL (with increasing eluted volume, respectively) in a postmenopausal woman before and during supplementation with ethinyl estradiol (1 µg/kg/d), demonstrating the selective reduction in HDL apo-E with estrogen.

endogenous particles (presumably by the liver), with diminished conversion to denser VLDL (S_f60-20), IDL and LDL. Thus apo E (of unspecified phenotype in the Walsh et al. studies) concentrated in these VLDL may increase their uptake and prevent their further progress down the catabolic cascade through IDL and LDL. In this sense estrogen may exaggerate the effects of apo E phenotype (and vice versa), which have demonstrated reduced conversion of VLDL remnants to LDL in subjects of apo $E_{2/2}$ than $E_{3/3}$, which in turn show less conversion than in subjects of $E_{4/4}$ status.[18] Other unpublished studies from our laboratory (as well as those reported by others[17]) have demonstrated enhanced LDL clearance in subjects treated with supplemental exogenous estrogen. Thus it may well be that estrogen enhances LDL receptor activity primarily by promoting hepatic uptake of both chylomicron remnants[19] and larger VLDL when they contain E_3 or E_4. However, as noted above, such uptake must not result in an expanded hepatic unesterified cholesterol pool but if anything should have the opposite effect (perhaps via enhanced cholesterol secretion directly or after conversion to bile acids) so that LDL receptor activity is also upregulated. Estrogen also profoundly decreases HTGL activity[20]; this may explain the TG-rich nature of IDL and, indeed, LDL (and HDL) in women treated with estrogen, and, furthermore, the reduced reverse cholesterol transport evident across the "lipid transfer zone" in women treated with supplemental hormone estrogen replacement therapy.[8]

Thus one must deduce that estrogen facilitates removal of chylomicron remnants and VLDL cholesterol as well as IDL and LDL cholesterol via multiple mechanisms, several of which are dependent upon recognition of apo E by the liver and, by extrapolation, enhanced hepatic throughput and biliary excretion of that cholesterol thus removed (otherwise LDL receptors would have been turned off by the enhanced cholesterol uptake, as would endogenous hepatic biosynthesis). However, this remains a speculation and must be tested in direct metabolic experiments.

Finally, as noted above, one must also deduce facilitated recognition of apo $E_{2/2}$ chylomicron and VLDL remnants via a mechanism (as yet to be delineated) not absolutely dependent on E_3 or E_4 and perhaps relatively favorable to the E_2 ligand. That estrogen (and most likely fibric acid derivatives) should facilitate such uptake may be hypothesized because of the extraordinary vulnerability of apo $E_{2/2}$ subjects to hepatic apo B overproduction, notably when FCHL is co-inherited, and their extraordinary response to these agents (without, obviously, a change in their apo E phenotype). But what is this pathway? Perhaps remarkably, even speculation as to its existence has not previously appeared in the literature.

Thus significant enigmatic components remain as to the regulatory role of apo E in human lipoprotein metabolism. Perhaps most exciting, however, has been the apparent relevance of apo E (and associated cellular cholesterol transport?) to structure and function in the central nervous system. Apo E is secreted by macrophages[1], and such secretion may be critical in the regulation of cholesterol flux and associated cellular processes in micro environments. One of these of special interest is the central nervous system, specifically in those at risk to Alzheimer's disease. Following the startling report by Roses and Strittmatter recently several laboratories have reported an increased prevalence of Alzheimer's disease in those with the ϵ_4 allele (and they have suggested reduced vulnerability in those of apo E_2 and E_3 phenotypes) (reviewed in [21]). Pursuing this now widely confirmed observation at the cellular level, investigators have recently reported profound alterations in the morphology of the neuronal response to damage depending on the form of apo E in the medium of such cells grown in culture[22]: apo E_4 (co-incubated with rabbit beta-VLDL as a source of cholesterol) causes neuronal sprouting in a constrained pattern, whereas apo E_3 promotes longer, less entangled dendritic growth. Therefore present research is focusing upon extra-vascular and specifically neural tissue responses to apo E mediation of the cellular reparative response to injurious stimuli, apo E phenotype in the human appearing to play a major, differential role in the nature and outcome of that response and in turn one's vulnerability to one of the most devastating diseases of old age.

REFERENCES

1. Mahley, R.W. and Rall, S.C. Type III hyperlipoproteinemia (dysbetalipoproteinemia): Role of apo E in normal and abnormal lipoprotein metabolism. In: Metabolism: Basis of Inherited Disease. 6th edition. 1989;1:1195-1213.
2. Mahley, R.W. Apo E: Cholesterol transport protein with expanding role in cell biology. Science 1988;240:622-30.
3. Weisgraber, K.H. Apo E distribution among human plasma lipoproteins: role of cysteine-arginine interchange at residue 112. J Lipid Res 1990;31:1503-11.
4. Steinmetz, A. et al. Differential distribution of apo E isoforms in human plasma LPs. Arterio 1989;9:405-11.
5. Borghini, I. et al. Distribution of apo E between free and A-II complexed forms in VLDL and HDL: functional implications. Biochimic et Biophysica Acta 1991; 1083:139-146.
6. Kowal, R.C. et al. LDL receptor-related protein mediates uptake of CE derived from apo E-enriched lipoproteins. Proc Natl Acad Sci USA 1989;86:5810-14.
7. Kowal, R.C. et al. Opposing effects of apolipoproteins E and C on lipoprotein binding to low density lipoprotein receptor-reltaed protein. J Biol Chem 1990; 265(18):10771-79.
8. Phair, R.D., Hazzard, W.R., Applebaum-Bowden, D. Theoretical consequences of hypotheses concerning reverse cholesterol transport. In: High density lipoproteins and Atherosclerosis II. N.E. Miller, ed. Elsevier Science Publishers. Amsterdam, The Netherlands. 1989;313-320.
9. Breslow, J.L. Apolipoprotein genetic variation and human disease. Physiol Rev 1988;68(1):106-108.
10. Miettinen, T.A., Gylling, H., Vanhanen, H., and Ollus, A. Cholesterol absorption, elimination, and synthesis related to LDL kinetics during varying fat intake in men with different apoprotein E phenotypes. Arteriosclerosis and Thrombosis 1992;12:1044-1052.

11. Kesaniemi, Y.A., and Miettinen, T.A. Metabolic epidemiology of plasma cholesterol. Ann of Clin Res 1988;20:26-31.

12. Kesaniemi, Y.A., Ehnholm, C., and Miettinen, T.A. Intestinal cholesterol absorption efficiency in man is related to apoprotein E phenotype. J Clin Invest 1987;80:578-581.

13. Miettinen, T.A. Impact of apo E phenotype on the regulation of cholesterol metabolism. Ann Med 1991;23:181-186.

14. Hazzard, W.R., Warnick, G.R., Utermann, G., Albers, J.J. Genetic transmission of isoapolipoprotein E phenotypes in a large kindred: relationship to dysbetalipoproteinemia and hyperlipidemia. Metabolism 1981;30:79-88.

15. Kushwaha, R.S., Hazzard, W.R., Gagne ,C., Chait, A., Albers, J.J. Type III hyperlipoproteinemia: paradoxical hypolipidemic response to estrogen. Ann Intern Med 1977;87:517-525.

16. Applebaum-Bowden, D., McLean, P., Steinmetz, A., Fontana, D., Matthys, C., Warnick, G.R., Cheung, M., Albers, J.J., Hazzard, W.R. Lipoprotein, apolipoprotein, and lipolytic enzyme changes following estrogen administration in postmenopausal women. J Lipid Res 1989;30:1895-1906.

17. Walsh, B.W., Schiff, I., Rosner, B., Greenberg, L., Raumkas, V., Sacks, F.M. Effects of postmenopausal estrogen replacement on the concentrations and metabolism of plasma lipoproteins. N Eng J Med 1991;325:1196-1204.

18. Demant, T. et al. Influence of apo E polymorphism on apo B-100 metabolism in normolipidemic subjects. J Clin Invest 1991;88:1490-1501.

19. Jäckle, S., Rinninger, F., Greeve, J., Greten, H., and Windler, E. Regulation of the hepatic removal of chylomicron remnants and ß-very low density lipoproteins in the rat. Jour Lipid Res 1992;33:419-429.

20. Applebaum, D.M., Goldberg, A.P., Pykalisto, O.J., Brunzell, J.D., Hazzard, W.R. Effect of estrogen on post-heparin lipolytic activity: selective decline in hepatic triglyceride lipase. J Clin Invest 1977;59:601-608.

21. Meeting Briefs: Neuroscientists reach a critical mass in Washington. Science 1993;262:1210.

22. Nathan, B.P., Bellosta, S., Sanan, D.A., Weisgraber, K.H., Mahley, R.W., Pitas, R.E. Differential effects of apolipoproteins E3 and E4 on neuronal growth in vitro. Science 1994;264:850-852.

THE ROLE OF SECOND MESSENGER PATHWAYS IN THE PATHOPHYSIOLOGY OF HYPERAPOB AND PREMATURE CORONARY ARTERY DISEASE

Peter O. Kwiterovich, Jr. and Mahnaz Motevalli

Lipid Research Atherosclerosis Unit
Departments of Pediatrics and Medicine
Johns Hopkins University Medical School
600 N. Wolfe Street / CMSC 604
Baltimore, Maryland 21287-3654

The development of coronary artery disease (CAD) results from the interplay of a variety of genetic and environmental factors. A number of risk factors for CAD have been described including elevated plasma levels of total cholesterol and its major carrier, low density lipoprotein (LDL), a low level of high density lipoprotein (HDL), hypertension, cigarette smoking, obesity, diabetes mellitus, physical inactivity and a diet high in total fat, saturated fat and cholesterol (1). Whether a high level of plasma triglyceride and its major carrier, very low density lipoprotein (VLDL) is an independent risk factor for CAD in controversial, but many patients with CAD have hypertriglyceridemia (2). However, all of these risk factors in aggregate account for only about 50 percent of the risk for developing CAD (3).

Disorders of lipid metabolism, namely, familial hypercholesterolemia (FH), familial combined hyperlipidemia (FCHL) and familial hypertriglyceridemia (FHTG) have been found in about 20 percent of families with premature CAD (4). However, the genetic and biochemical bases of these lipid disorders has only been elucidated in FH, namely, the molecular defects in the LDL receptor described by Brown, Goldstein and co-workers (5). Thus, the fundamental defect in premature CAD has been elucidated in only about five percent of the families.

There has been an interest in whether the plasma apolipoproteins may provide additional information about risk of CAD, namely, the major apolipoprotein of VLDL and LDL, apoB, and the major apolipoprotein of HDL, namely, apoA1. In many but not all studies, apoB appears to predict CAD better than LDL cholesterol and apoA1 better than HDL cholesterol (6). In premature CAD, apoB appears the best lipid risk factor discriminator in women, while apoA1 appears the best lipid discriminator in men (6). The distribution of plasma apoB in women with a premature CAD appears bimodal while it is markedly skewed to the right in men, suggesting the influence of a major gene (6).

In 1980, Sniderman, Kwiterovich and co-workers (7) found that a number of patients with angiographically documented CAD had an elevated plasma LDL apoB level

Cardiovascular Disease 2
Edited by L.L. Gallo, Plenum Press, New York, 1995

($\geq$ 120 mg/dl), in the presence of normal or borderline high LDL cholesterol levels (130-160 mg/dl). This disproportionate increase in the LDL apoB level, compared with the LDL cholesterol level, was termed hyperapobetalipoproteinemia (hyperapoB). The phenotype of hyperapoB was subsequently shown to be due to an increased number of small, dense LDL particles, that were depleted in core cholesteryl ester and relatively enriched in apoB, providing a low ratio of LDL cholesterol to apoB (i.e. < 1.2) in plasma (8). Patients with hyperapoB may be normotrigly-ceridemic or hypertriglyceridemic, depending upon whether the increased number of small, dense LDL particles is accompanied by a normal VLDL or an elevated VLDL level (2,8,9). About one-third of patients with premature CAD were found to have hyperapoB (2).

The increased numbers of small, dense LDL particles in hyperapoB results from the overproduction of VLDL apoB in liver (10) (Fig. 1). With increased VLDL overproduction, there appears to be an increased transfer of core triglyceride from VLDL to LDL in exchange for cholesteryl ester from LDL to VLDL (11). The triglyceride in the cholesteryl ester - depleted LDL is hydrolyzed by lipoprotein lipase (LPL), producing a small, dense LDL particle. Both in vivo (10) and in vitro (12) studies indicate that the LDL receptor activity is normal in hyperapoB. However, the removal of small, dense LDL is slower than that of normal sized LDL in both normal and hyperapoB patients (10). Teng et al (13), using a monoclonal antibody to that portion of apoB recognized by the LDL receptor, showed that the immunochemical reactivity of this antibody to small, dense LDL was less than that of normal LDL, indicating that apoB polypeptide was oriented differently on the surface of small, dense LDL, potentially interfering with its affinity for the LDL receptor. In addition to the overproduction of VLDL particles in hyperapoB, there appears to be a delayed clearance of postprandial triglyceride-enriched lipoproteins in this disorder (Fig. 1) (14).

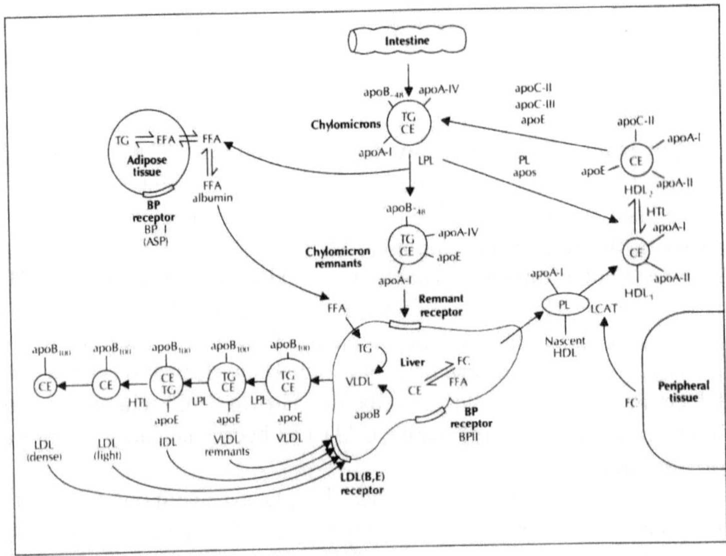

Figure 1. Schematic diagram of hypothetical pathways that may be involved in the pathogenesis of hyperapoB and FCHL. TG - triglyceride; FFA - free fatty acids; CE - cholesteryl esters; PL - phospholipids; LPL - lipoprotein lipase; LCAT - lecithin cholesterol acyl transferase; HTL - hepatic triglyceride lipase; BP - serum basic proteins; ASP - acylation stimulatory protein. Reproduced with permission (15).

Patients with FCHL also have been shown to have small, dense LDL particles and increased VLDL overproduction (reviewed in 15). Thus, hyperapoB and FCHL appear to be "metabolic cousins". The phenotype of hyperapoB is present in some families with FCHL. Patients with hyperapoB may also have hypertension, obesity, diabetes and a low level of HDL cholesterol (2). Thus, there also appears to be a overlap of hyperapoB with familial dyslipidemic hypertension, LDL subclass pattern B and syndrome X (15). HyperapoB is often familial and was found in one-third of the children of patients with premature CAD and hyperapoB (16). HyperapoB may be due to a influence of a major gene but the precise genetic basis for hyperapoB, FCHL and related syndromes has not been elucidated (15). Nine candidate genes have been proposed for hyperapoB and FCHL (15) (Table 1). Recent evidence indicates that the APOB gene (15) and the lipoprotein lipase gene (17), are not common causes of hyperapoB and FCHL. The possible roles of the other genes have recently been reviewed in detail (15); the focus here will be upon studies related to the serum basic proteins (BP) and the putative BP receptor (Fig. 1).

Table 1. Candidate genes for fundamental defects in hyperapoB and FCHL.

1. APOB
2. Trans DNA binding protein for APOB.
3. Lipoprotein lipase gene.
4. APOC-III
5. APOA-IV
6. ATHS on chromosome 19.
7. INSR gene for insulin receptor.
8. Gene for putative "basic protein receptor".
9. Gene(s) for serum basic proteins.

Modified after reference 15.

Abnormalities in removal of postprandial triglyceride-enriched lipoproteins in hyperapoB patients appeared, in preliminary studies, to be accompanied by an abnormal increase in the levels of postprandial free fatty acids (11). This lead was followed by in vitro studies in which the incorporation of $[^{14}C]$ oleate and $[^{3}H]$ palmitate into triglycerides in fibroblasts and adipocytes from hyperapoB patients were found to be decreased by about 50 percent compared with normal cells (15). Since these experiments were performed in the presence of lipoprotein-deficient serum, a partially purified plasma fraction that promoted acylation stimulatory activity was next isolated from human serum by Cianflone and co-workers (18). Cianflone et al (19) then isolated a small basic protein (called acylation stimulatory protein, or ASP) from normal human serum. ASP was shown to approximately double the incorporation of oleate into triglyceride in normal cultured human fibroblasts, an effect that was decreased by half in hyperapoB fibroblasts (20). Cianflone and co-workers (20) also found evidence for deficient binding of ^{125}I ASP to hyperapoB fibroblasts, which was accompanied by a proportional decrease in the stimulation of triglyceride formation. The apparent binding constant (K_d) was normal, but the specific binding of ASP was decreased by about 50 percent, indicating that the mutant cells manifested half the normal number of ASP binding sites. More recently, evidence has been put forward that ASP has homology to C_{3a} desarg (21). Adipsin, an enzyme in adipocytes involved in the proteolytic cleavage of the third component of complement, leads to the generation of C_{3a} desarg. Thus, this biochemical pathway system may be of importance in the pathophysiology of hyperapoB.

Work from our laboratory has focused on the isolation, characterization and pathophysiologic effects of three serum basic proteins (BP), isolated from normal human serum and termed BP I (Mr 14,000, pI 9.10), BP II (Mr 27,500, pI 8.48), and BP III (Mr 55,000, pI 8.73) (22). These basic proteins differ significantly in their amino acid compositions (22,23). BP I appears to be the same protein as ASP (22). BP II and BP III appear different from ASP and other lipid carrier proteins (22,23). BP I, BP II and BP III all stimulate the incorporation of ^{14}C oleate in lipid esters in normal fibroblasts, an effect that was time and concentration dependent (22,23).

We have found that BP I doubled the mass of cell triglyceride in normal human fibroblasts, and that there is 50 percent decrease in such stimulation of triglyceride synthesis by BP I in hyperapoB fibroblasts (24). In distinct contrast, BP II stimulated abnormally the formation of cholesteryl ester mass (an average of six fold) in hyperapoB cells. No abnormalities in hyperapoB fibroblasts were observed with BP III.
Both our work and that from the Montreal laboratory indicated that the effect(s) of the basic proteins appeared to involve a high affinity mechanism (18-24). We next asked the question whether the effects of BP I and BP II may be modulated via second messenger pathways.

We first examined whether the effects of BP I and BP II were blocked by H-7 (1-C5-isoquinoline-sulfonyl)-2-methyl piperozine dihydrochloride), an inhibitor of protein kinase C. In normal fibroblasts, the stimulatory effect of BP I upon [^{14}C] oleate incorporation into triglyceride was decreased significantly by H-7 in a concentration dependent fashion (Fig. 2, top) (24). Similar observations were made in the hyperapoB cells. For BP II, we found in normal cells that H-7 inhibited the incorporation of ^{14}C oleic acid into cell cholesteryl esters. The abnormal stimulatory effect of BP II on cholesteryl ester formation in hyperapoB cells was actually stimulated at the lower concentration of H-7, but inhibited at higher concentrations, suggesting that protein kinase C was also involved in modulating the effect of BP II (Fig. 2, bottom) (24).

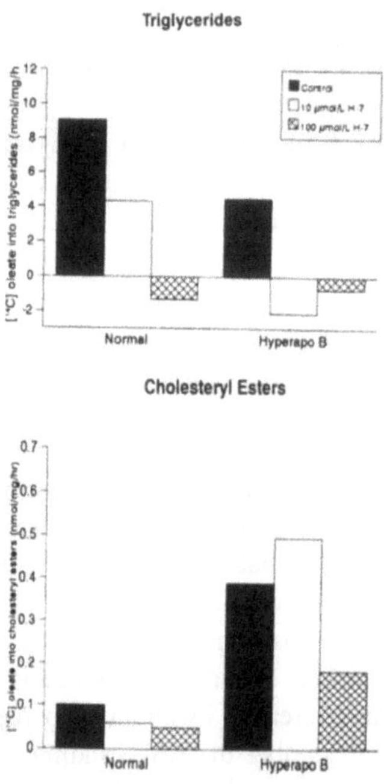

Figure 2. Effect of BP I (top) and BP II (bottom) on the stimulation of the rate of incorporation of ^{14}C oleate into cell triglyceride and cholesteryl esters, respectively, in the absence (black bar) or presence of 10 umol/L (white bar) or 100 umol/L (cross-hatched bar) of H-7, an inhibitor of protein kinase C (24). Normal fibroblasts are on the left, and hyperapoB fibroblasts on the right. Reproduced with permission (24).

We next stimulated protein kinase C, using C:8, an analogue of diacylglycerol, which activates protein kinase C (Fig. 3). When normal cells were incubated in F-12 without basic proteins, C:8 stimulated [^{14}C] oleate incorporation into triglyceride approximately five fold over baseline; BP I doubled this effect in both control and C:8 treated normal cells (Fig. 3, top) (24). In the hyperapoB cells, the addition of C:8 to F-12 medium alone stimulated triglyceride formation seven fold; when BP I was added to C:8

in the hyperapoB cells, there was much less additional stimulatory effect of BP I than occurred in normal cells (Fig. 3, top) (24). These results together suggested that the intracellular component of the phosphatidylinositol pathway is normal in hyperapoB cells, and confirm further their cellular defect in the normal response to BP I.

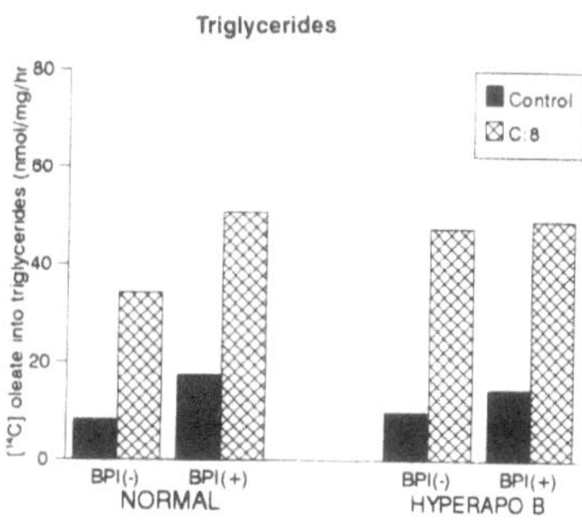

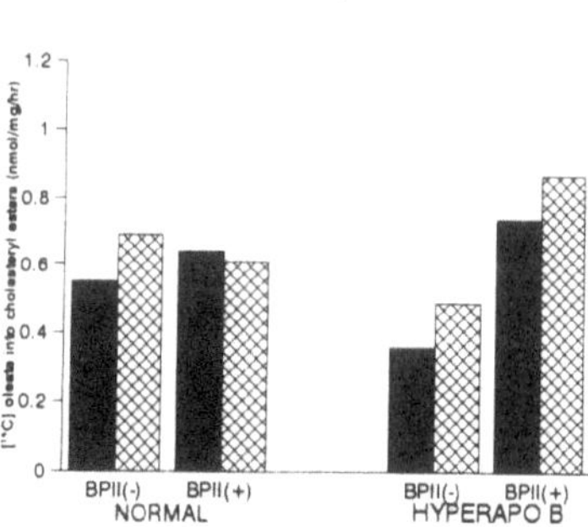

Figure 3. Effect of C:8, an analogue of diacylglycerol, in the absence (-) or presence (+) of BP I (top) and BP II (bottom), on the stimulation of the rate of incorporation of ^{14}C oleate into cell triglyceride and cholesteryl esters, respectively (24). Normal fibroblasts are on the left and hyperapoB fibroblasts are on the right. Reproduced with permission (24).

In normal cells, BP II stimulated [^{14}C] oleate incorporation into cholesteryl esters a small amount; the addition of C:8 to either F-12 or BP II containing medium did not stimulate cholesteryl ester formation (Fig. 3, bottom) (24). In hyperapoB cells, the greater stimulation of cholesteryl ester formation with BP II was accentuated markedly by C:8, suggesting that the abnormal response of hyperapoB cells to BP II may be modulated by this pathway.

Genistein, a highly specific inhibitor of protein tyrosine kinase (25), was used to determine if the effects of the serum basic proteins (BP I, BP II) were mediated through tyrosine phosphorylation, and if there was a deficiency in such a process in hyperapoB fibroblasts. Cells from 8 hyperapoB subjects (6 unrelated kindreds) and 6 normals were incubated for 24 hr in lipid-free medium, at which time medium containing oleate:albumin alone (control), or BP I or BP II (6 μg/ml medium) ± genistein (25 μg/ml medium) were added for 6 hr. The effect of genistein was time (6 hr max) and concentration (25 μg/ml medium, nadir) dependent (26). The mass of lipid (μg/mg cell protein) in the control ± genistein was subtracted from that with BP ± genistein.

The stimulation of the mass of triglyceride (mean±SEM) in normal cells (60.0±6.8)with BP I was significantly reduced in hyperapoB cells (29.1±6.0) (p=.02); genistein inhibited the effect of BP I on mass of triglyceride (34.2±4.8) (p=.008) in normal cells, but did not affect the mass of triglyceride in hyperapoB cells (33.3±3.3) (26). BP II abnormally stimulated the mass of unesterified cholesterol (16.6±2.3) (p=.003) in hyperapoB cells, an effect that was inhibited by genistein (1.87±1.5) (p=.006)

(26). In distinct contrast, genistein + BP II markedly stimulated the mass of cholesteryl esters (from 2.8±1.3 to 28.3±4.9) ($p=.007$) in normal cells, but inhibited the mass of cholesteryl esters in hyperapoB cells (from 6.8±2.5 without genistein to -0.2±2.4 with genistein) ($p=.07$). This "crossover" effect of genistein + BP II produced about a 30 fold difference in mass of cholesteryl esters between normal and hyperapoB cells. Tyrosine kinase phosphorylation mediates the effect of BP I and BP II in normal cells, and appears deficient in hyperapoB cells.

SUMMARY

The normal effect of BP I and BP II appears to require tyrosine kinase phosphorylation; however, it is not known whether this action is at a transmembrane tyrosine kinase receptor, or a non-receptor, membrane associated tyrosine kinase molecule, or at another post-receptor site. The PKC second messenger pathway appears to be involved in the action of BP I and BP II, but is normal in hyperapoB cells. The fundamental defect in hyperapoB cells may reside in a cell surface molecule, perhaps a transmembrane tyrosine kinase receptor. Upon activation, such a tyrosine kinase receptor may also phosphorylate other proteins, such as phospholipase C (PLC), γ-1, one of several PLC isoforms that cleaves phosphatidyl inositol 4,5 biphosphate (PIP_2) to second messengers, diacylglycerol and inositol triphosphate which in turn stimulate protein kinase C (27). Definitive answers to these questions await the identification and isolation of the putative receptor for the serum basic proteins.

ACKNOWLEDGMENTS

This work was supported by the following grants from the National Institutes of Health 1 P50 HL47212-03 (Specialized Center of Research in Arteriosclerosis), HL31497, General Clinical Research Center Program, RR-52, RR-35 and CLINFO. We thank Pauline Gugliotta for preparation of this manuscript.

REFERENCES

1. Expert Panel on Detection, Evaluation and Treatment of High Blood Cholesterol in Adults. Summary of the Second Report of the National Cholesterol Education Program (NCEP) Expert Panel on Detection, Evaluation, and Treatment of High Blood Cholesterol in Adults (Adult Treatment Panel II). JAMA 269:3015-3023, 1993.
2. Kwiterovich PO Jr, Coresh J, Bachorik PS: Prevalence of Hyperapobetalipoprotein-emia and Other Lipoprotein Phenotypes in Men (< 50 years) and Women (< 60 years) with Coronary Artery Disease. Amer. J. Cardiol. 71:631-639, 1993.
3. Wilson PWF, Castelli WP, Kannel WB: Coronary risk prediction in adults (the Framingham Heart Study). Am. J. Cardiol. 59:91G-4G, 1987.
4. Goldstein JL, Schrott HG, Hazzard WR, Bierman EL, Motulsky AG: Hyperlipidemia in coronary heart disease II. Genetic analysis of lipid levels in 176 families and delineation of a new inherited disorder, combined hyperlipidemia. J. Clin. Invest. 52:1544-1568, 1973.
5. Hobbs HH, Brown MS, Goldstein JL: Molecular Genetics of the LDL Receptor Gene in Familial Hypercholesterolemia. Hum. Mutation 1:445-466,1992.
6. Kwiterovich PO, Coresh J, Derby C, Smith HH, Bachorik PS and Pearson TA. Comparison of the plasma levels of apolipoproteins B and A-1, and other risk factors in men and women with premature coronary artery disease. Am. J. Cardiol. 69:1015-1021, 1992.
7. Sniderman A, Shapiro S, Marpole D, Malcolm I, Skinner B, Kwiterovich PO Jr. The association of coronary atherosclerosis and hyperapobetalipoproteinemia (increased protein but normal cholesterol content in human plasma low density lipoprotein). Proc. Natl. Acad. Sci. U.S.A. 97:604-608, 1980.

8. Teng B, Thompson GR, Sniderman AD, Forte TM, Krauss RM, Kwiterovich Jr PO: Composition and distribution of low density lipoprotein fractions in hyperapobetalipoproteinemia. Normolipidemia and familial hypercholesterolemia. Proc. Natl. Acad. Sci. U.S.A. 80:6662-6666,1983.
9. Sniderman A, Wolfson C, Teng B, Franklin F, Bachorik P, Kwiterovich PO Jr: Association of hyperapobetalipoproteinemia with endogenous hypertriglyceridemia and atherosclerosis. Ann. Intern. Med. 97:833-839, 1982.
10. Teng B, Sniderman AD, Soutar AK, Thompson GR: Metabolic basis of hyperapobetalipoproteinemia. Turnover of apolipoprotein B in low density lipoprotein and its precursors and subfractions compared with normal and familial hypercholesterolemia. J. Clin. Invest. 77:663-672, 1986.
11. Sniderman A, Kwiterovich PO Jr: Hyperapobetalipoproteinemia and LDL and HDL_2 Heterogeneity. Proceedings from a workshop on Lipoprotein Heterogeneity. NIH Publication No. 87-2646. US Department of Health and Human Services: Washington, DC; 293-304, 1987.
12. Ladias JAA, Kwiterovich PO Jr, Smith HH, Miller M, Bachorik PS, Forte T, Lusis AJ, Antonarakis SE: Apolipoprotein B-100 Hopkins ($Arginine_{4019}$$\rightarrow$Tryptophan): A new apolipoprotein B-100 variant in a family with premature atherosclerosis and hyperapobetalipoproteinemia. JAMA 262:1980-1988, 1989.
13. Teng B, Sniderman AD, Krauss RM, Kwiterovich Jr PO, Milne RW, Marcel YL: Modulation of apolipoprotein B antigenic determinants in human low density lipoprotein subclasses. J. Biol. Chem. 260:5067-5072, 1985.
14. Genest J. Sniderman AD, Cianflone K, Teng B, Wacholder S, Marcel Y, Kwiterovich PO Jr: Hyperapobetalipoproteinemia: Plasma lipoprotein responses to oral fat load. Arterioscler. Thromb. 6:297-304, 1986.
15. Kwiterovich PO Jr: Genetics and molecular biology of familial combined hyperlipidemia. Current Opin. Lipid. 4:133-143, 1993.
16. Sniderman AD, Teng B, Genest J, Cianflone K, Wacholder S, Kwiterovich Jr PO: Familial aggregation and early expression of hyperapobetalipoproteinemia . Am. J. Cardiol. 55:291-5, 1985.
17. Nevin DN, Brunzell TD, Deeb SS. The LPL gene in individuals with familial combined hyperlipidemia and decreased LPL activity. Arterioscler. Thromb. 14:869-873, 1994.
18. Cianflone K, Kwiterovich Jr PO, Walsh M, Forse A, Rodriguez MA, Sniderman AD: Stimulation of fatty acid uptake and triglyceride synthesis in human cultured skin fibroblasts and adipocytes by a serum protein. Biochem. Biophys. Res. Commun. 144:94-100, 1987.
19. Cianflone KM, Sniderman AD, Walsh MJ, Vu HT, Gagnon J, Rodriguez MA: Purification and characterization of acylation stimulating protein. J. Biol. Chem. 264:426-430, 1989.
20. Cianflone KM< Maslowska MH, Sniderman AD: Impaired response of fibroblasts from patients with hyperapobetalipoproteinemia to acylation stimulating protein. J. Clin. Invest. 85:722-730, 1990.
21. Baldo A, Sniderman AD, St-Luce S, Avramoglu RK, Maslowska M, Hoang B, Monge JC, Bell A, Mulay S, Cianflone K: The adipsin-acylation stimulating protein system and regulation of intracellular triglyceride synthesis. J. Clin. Invest. 92:1543-1547, 1993.
22. Kwiterovich PO Jr, Motevalli M, Miller M: Acylation stimulatory activity in hyperapobetalipoproteinemic fibroblasts: Enhanced cholesterol esterification with another serum basic protein, BP II. Proc. Natl. Acad. Sci. U.S.A. 87:8980-8984, 1990.
23. Kwiterovich PO Jr, Motevalli M, Miller M, Bachorik PS, Kafonek SD, Chatterjee S, Beaty T and Virgil D. Further insights into the pathophysiology of hyperapobetalipoproteinemia: Role of basic proteins I, II, III. Clin. Chem. 37(3):317-326, 1991.
24. Kwiterovich PO Jr, Motevalli M, Miller M: The effect of three serum basic proteins on the mass of lipids in normal and hyperapoB fibroblasts. Arterioscler Thromb. 14:1-7, 1994.
25. Akiyama T, Ishida J, Nakagawa S, Ogawara H, Watanabe S. Itah N. Shibuya M, and Fukami Y. Genistein, a specific inhibitor of tyrosine protein kinases. J Biol. Chem. 262:5592-5595, 1987.
26. Kwiterovich PO and Motevalli M. Inhibition of tyrosine kinase phosphorylation differentially alters the effects of serum basic proteins in normal and hyperapoB fibroblasts. Circulation in press, 1994 (Abst.).
27. Koch CA, Anderson P, Moran MF, Ellis C, Parson T. SH_2 and SH_3 domains. Elements that control interactions of cytoplasmic signaling proteins. Science 252:668-674, 1992.

THE ROLE OF HDL RECEPTORS IN REMOVAL OF CELLULAR CHOLESTEROL

John F. Oram, Armando J. Mendez,
Gordon A. Francis, and Edwin L. Bierman

Department of Medicine
University of Washington
Seattle, WA 98195

The ability of HDL to protect against atherogenesis may be related to its role in "reverse cholesterol transport", a pathway by which cholesterol is transported from peripheral cells to the liver for excretion from the body. The first step of this pathway, the removal of cholesterol from cells, may be mediated by binding of HDL to a specific cell-surface receptor. Through this receptor interaction, HDL may continually deplete peripheral cells of excess cholesterol, thereby preventing the buildup of cholesterol in artery wall cells that is an early feature of the developing atherosclerotic lesion.

Cellular Cholesterol Excretory Pathways

Cholesterol is essential for cell growth and viability, largely because of its role as a structural component of membranes. Accumulation of excess cellular cholesterol, however, can be detrimental to cells and promote atherogenesis in the artery wall. Thus, cells have developed several regulated pathways to maintain their cholesterol content within narrow limits.[1] When cells ingest more cholesterol than is required for membrane synthesis, the excess cholesterol is either converted to cholesteryl esters and stored as lipid droplets in the cytoplasm or it is excreted from cells to HDL particles.

At least two independent processes contribute to cholesterol excretion from cells to HDL. Plasma membrane cholesterol can desorb from the cell surface, diffuse through the aqueous layer surrounding the cell, and be sequestered by the phospholipid layer on the surface of HDL particles.[2] The second process involves the direct interaction of HDL apolipoproteins with cell-surface binding sites.[3,4] This interaction facilitates excretion of select pools of excess cholesterol that are accessible to esterification by the enzyme acyl CoA:cholesteryl acyltransferase (ACAT). This latter process appears to have an active component involving intracellular signals that modulate cholesterol trafficking between intracellular compartments and the plasma membrane.[5,6]

Cardiovascular Disease 2
Edited by L.L. Gallo, Plenum Press, New York, 1995

The HDL Receptor Hypothesis

The findings that HDL actively stimulates excretion of select pools of cellular cholesterol raise the possibility that HDL apolipoproteins are interacting with cell-surface receptors. Cell culture studies have shown that HDL interacts with specific, high-affinity binding sites on the cell surface that are upregulated in response to cholesterol loading of cells, consistent with the idea that these sites function as receptors that rid cells of excess cholesterol. The interaction of HDL apolipoproteins with these high-affinity binding sites elicits signals involving formation of diacylglycerol and activation of protein kinase C.[5,6] These signals stimulate translocation of excess cholesterol from cellular pools that are accessible to ACAT to cell-surface domains that are accessible to HDL apolipoproteins (Figure 1). The apolipoprotein particles readily pick up the translocated cholesterol and remove it from the cell. Thus, the HDL receptor functions to communicate to cells that appropriate sterol acceptors are present at the cell surface, and cholesterol is diverted from intracellular storage sites into an excretory pathway.

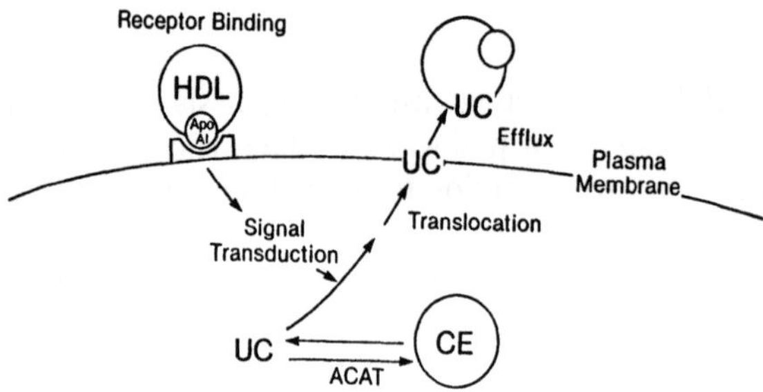

Figure 1. Model for the cellular HDL receptor pathway. Abbreviations: Apo AI, apolipoprotein AI; UC, unesterified cholesterol; CE, cholesteryl esters.

The relative activity of the HDL receptor pathway is highly dependent upon the growth state and cholesterol status of cells. When cells are proliferating, they require cholesterol for continual synthesis of membranes, and the number of HDL receptors on the cell surface is relatively low.[7,8] When cells are growth-arrested, such as highly differentiated cells, less cholesterol is required for membrane synthesis and the number of HDL receptors per cell is higher than that for proliferating cells. When these growth-arrested cells are overloaded with cholesterol, there is a marked increase in the number of cell-surface receptors.[7-11] Thus, the HDL receptor pathway is regulated in response to the cell's need to excrete excess cholesterol.

The "Microdomain" HDL Receptor Model

Although the physical and metabolic properties of HDL receptors are unknown, a plausible model is that the receptors are components of plasma membrane microdomains with a highly ordered composition of protein, cholesterol, and phospholipids. These domains may be ports of entry and exit of cholesterol transported between the plasma membrane and intracellular sites, and they may function as "sterolstats" that tightly regulate the overall membrane cholesterol content. Thus, when cells are in a sterol-deficient state, demand for cholesterol as a plasma membrane structural component may deplete cholesterol from these domains, eliciting signals to mobilize more intracellular cholesterol for replenishment (Figure 2, bottom right). This mobilization may relieve feedback repression of cholesterol biosynthetic enzymes and the LDL receptor, increasing production and delivery of more cholesterol. As cells are overloaded with cholesterol, the cholesterol content of the regulatory microdomains may increase, altering cholesterol trafficking so as to divert excess cholesterol into intracellular compartments for storage as esters (Figure 2, bottom left). These microdomains may also contain specialized proteins, including receptors for HDL apolipoproteins. With cholesterol loading of cells, these receptors may increase in number or activity and promote reversible binding of HDL apolipoproteins to these regions, allowing apolipoprotein particles to remove cholesterol from the microdomains and transport it from the cell (Figure 2, top). This localized depletion of cholesterol may stimulate translocation of more intracellular cholesterol to these sites. Such a process would continually deplete cells of cholesterol until the intracellular cholesterol content decreased to levels below that required to induce HDL receptors, at which time the cholesterol excretion rate would return to basal levels.

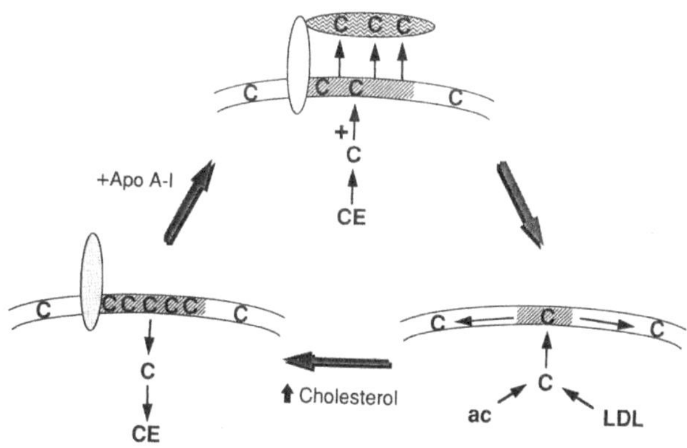

Figure 2. The microdomain model for the HDL receptor pathway. Abbreviations: C, unesterified cholesterol; CE, cholesteryl esters.

The actual molecules responsible for binding apolipoproteins have not been identified conclusively, although several candidate proteins have been reported. One of these proteins, termed HBP, has been cloned and partially characterized.[12] It has several features predicted for an HDL receptor, including localization to the plasma membrane, induction by cholesterol loading of cells, and promotion of HDL binding to the cell surface when overexpressed in cells. The structure of HBP, however, does not conform to that of any known plasma membrane receptor, in that it lacks a classic signal sequence and membrane spanning domain. Instead, HBP appears to bind to membranes through its multiple amphipathic helices. Signaling molecules responsible for mediating the effects of apolipoproteins on cholesterol trafficking also have not been identified. It is possible that they represent G proteins, tyrosine kinases, or other signaling proteins that are localized to regulatory microdomains of the plasma membrane. The activities of these proteins may be regulated in response to receptor binding of apolipoproteins or to changes in microdomain lipid composition.

Clinical Implications

It is becoming increasingly evident that HDL can stimulate excretion of cholesterol from cells by a complex receptor-mediated pathway involving multiple proteins. Candidates for some of these proteins are just now being identified and characterized. Mutations in these proteins can impair the ability of HDL to clear excess cholesterol from cells of the artery wall, leading to early development of heart disease. Identification of these mutations could help define the precise genotype that underlies some of the forms of heart disease that are frequently associated with low plasma levels of HDL. The HDL receptor pathway also has important implications as a target for drug therapy. A receptor agonist, for example, may enhance the rate of clearance of cholesterol from peripheral cells and reduce the deposition of sterol in the artery wall. Thus, a more complete understanding of the cellular mechanism involved in cholesterol excretion may suggest therapeutic approaches for the treatment of some forms of heart disease.

REFERENCES

1. M.S. Brown and J.L. Goldstein, A receptor-mediated pathway for cholesterol homeostasis, *Science* 232:34 (1986).
2. M.C. Phillips, W.J. Johnson, and G.H. Rothblat, Mechanisms and consequences of cellular cholesterol exchange and transfer, *Biochim. Biophys. Acta* 906:223 (1987).
3. E.A. Brinton, J.F. Oram, C.-H. Chen, J.J. Albers, and E.L. Bierman, Binding of high density lipoprotein to cultured fibroblasts after chemical alteration of apoprotein amino acid residues, *J. Biol. Chem.* 261:495 (1986).
4. J.F. Oram, A.J. Mendez, J.P. Slotte, and T.F. Johnson, High density lipoprotein apolipoproteins mediate removal of sterol from intracellular pools but not from plasma membranes of cholesterol-loaded fibroblasts, *Arterioscler. Thromb.* 11:403 (1991).
5. A.J. Mendez, J.F. Oram, and E.L. Bierman, Protein kinase C as a mediator of high density lipoprotein receptor-dependent efflux of intracellular cholesterol, *J. Biol. Chem.* 266:10104 (1991).
6. N. Theret, C. Delbart, G. Aguie, J.D. Fruchart, G. Vassaux, and G. Ailhaud, Cholesterol efflux from adipose cells in coupled to diacylglycerol production and protein kinase C activation, *Biochem. Biophys. Res. Commun.* 173:1361 (1990).
7. M.J. Oppenheimer, J.F. Oram, and E.L. Bierman, Downregulation of high density lipoprotein receptor activity of cultured fibroblasts by platelet-derived growth factor, *Arteriosclerosis* 7:325 (1987).
8. M.J. Oppenheimer, J.F. Oram, and E.L. Bierman, Up-regulation of high density lipoprotein receptor activity by gamma-interferon associated with inhibition of cell proliferation, *J. Biol. Chem.* 263:19318 (1988).
9. J.F. Oram, E.A. Brinton, and E.L. Bierman, Regulation of HDL receptor activity in cultured human skin fibroblasts and human arterial smooth muscle cells, *J. Clin. Invest.* 72:1611 (1983).
10. E.A. Brinton, R. Kenagy, J.F. Oram, and E.L. Bierman, Regulation of high density lipoprotein binding activity of aortic endothelial cells by treatment with acetylated low density lipoprotein, *Arteriosclerosis* 5:329 (1985).
11. G. Schmitz, R. Niemann, B. Brennhausen, R.M. Krauss, and G. Assmann, Regulation of high density lipoprotein receptors in cultured macrophages: role of acyl CoA:cholesterol acyltransferase, *EMBO J.* 4:2773 (1985).
12. G.L. McKnight, J. Reasoner, T. Gilbert, K.O. Sundquist, B.M. Hokland, P.A. McKernan, J. Champagne, C.J. Johnson, M.C. Bailey, R. Holly, P.J. O'Hara, and J.F. Oram, Cloning and expression of a cellular high density lipoprotein-binding protein that is up-regulated by cholesterol loading of cells, *J. Biol. Chem.* 267:12131 (1992).

THE DEFECT IN HDL_3 MEDIATED EFFLUX OF NEWLY SYNTHESIZED CHOLESTEROL IS ASSOCIATED WITH IMPAIRED ACTIVATION OF PROTEIN KINASE C IN TANGIER FIBROBLASTS

Gerd Schmitz, Gerhard Rogler, Wolfgang Drobnik, Barbara Trümbach, Christoph Moellers, Karl J. Lackner

Institute for Clinical Chemistry and Laboratory Medicine, University of Regensburg, Federal Republic of Germany

INTRODUCTION

The removal of excess cholesterol from peripheral cells is the first step in reverse cholesterol transport and many studies demonstrated a net removal of cellular cholesterol in response to HDL_3 [1,2]. This function has been suggested to account for the inverse correlation between coronary heart disease risk and plasma levels of HDL-cholesterol[3]. In addition to unspecific cholesterol desorption from the plasma membrane[4], depletion of cholesterol may be achieved by binding of HDL apolipoproteins to specific surface receptors and subsequent induction of cholesterol transport from internal stores to the plasma membrane [5,6].

Tangier disease is a rare, autosomal recessive disorder of cellular lipid and lipoprotein metabolism, characterized by severe reduction of serum high density lipoproteins (HDL) and cholesteryl ester deposition in various tissues. The reduced levels of HDL, apoA-I (< 1% of normal) and apoA-II (5-10% of normal) in Tangier disease are due to rapid catabolism of HDL and its apolipoproteins, while synthetic rates are within the normal range[7,8,9]. The increased catabolism of HDL is most likely caused by a defect in cellular lipid transport. Tangier mononuclear phagocytes (MNPs) were demonstrated to have a defect in HDL_3 retroendocytosis by our group[10]. Also fibroblasts show morphologic abnormalities of the Golgi apparatus and abnormalities in the intracellular traffic of lipoproteins and lipids, indicating that the cellular defect of the disease is expressed in both cell types[11].

These findings led us to study HDL_3 induced transport of cellular cholesterol and HDL_3 mediated release into the medium in cultured skin fibroblasts from Tangier patients. Since HDL_3 is not internalized appreciably by fibroblasts, they must release cholesterol from the cell membrane into the medium. In this study we chow a reduced efflux of cellular cholesterol from Tangier fibroblasts. Studies in adipose cells[12] and fibroblasts[13] indicate that HDL_3 induced cholesterol efflux of newly synthesized cholesterol depends on the activation of PKC. Therefore, also HDL_3 induced signal transduction cascades were analyzed in normal and Tangier fibroblasts.

Cardiovascular Disease 2
Edited by L.L. Gallo, Plenum Press, New York, 1995

METHODS

Patients Cutaneous fibroblasts were obtained from two patients homozygous for Tangier disease: Patient 1 (E.G.), a 60-year-old woman, (triglycerides 2,94 - 4, 89 mmol/l; cholesterol 2,02 - 2,67 mmol/l); patient 2 (J.S.), a 57-year-old man, brother of patient 1 (triglycerides 1,58 - 2,24 mmol/l; cholesterol 1,16 - 1,50 mmol/l). Four lines of control fibroblasts (G.M., T.L., N.F., R.W.) were cultured from the cutis of normolipidemic individuals who underwent abdominal surgery.

Cell Culture Fibroblasts were cultured according to standard conditions in DMEM supplemented with 10% fetal calf serum (FCS) in a humidified 5% CO_2 atmosphere at 37°C.

Determination of Cholesterol Efflux from Fibroblasts De novo synthesized cholesterol was labelled with [^{14}C]-mevalonolactone, cell membrane cholesterol by incorporation of [^{14}C]-cholesterol and the pathway of LDL derived cholesterol was investigated by reconstitution of delipidized LDL with [^{3}H]-cholesteryl linoleate. After reaching confluence, fibroblasts were incubated with DMEM containing 10% LPDS for 48h to deplete the cells of cholesterol. Thereafter, cells were incubated for 3 hrs at 15°C with either 0,5 µCi/ml [^{14}C]-cholesterol to label membrane cholesterol, or 2,0 µCi/ml [^{14}C]-mevalonolactone to label newly synthesized cholesterol. To label lysosomal cholesterol, cells were incubated with reconstituted LDL (0.29 µCi/ml containing 52 nmol cholesteryl linoleate) for 3 hrs at 37°C. Cells from three dishes were harvested to determine cholesterol synthesis or cholesterol incorporation. The other cells were incubated in DMEM containing 1% BSA supplemented with increasing concentrations of HDL_3 or phospholipid micelles as indicated. Aliquots of the medium were taken at the time points specified. Radioactivity in the medium was determined by liquid scintillation counting. Specific HDL_3-mediated efflux is defined as the difference between efflux in the presence of HDL_3 and 1% BSA minus the efflux in the presence of 1% BSA only. Activation of PKC was achieved by incubation of cells in the presence of 10^{-5} M of the membrane permeable 1,2-dioctanoylglycerol (1,2-DOG) [13].

Determination of Cellular Lipids Lipid extractions were performed according to the method of Bligh and Dyer[14]. Cellular lipids were separated by high performance thin layer chromatography (HPTLC).

Determination of intracellular calcium concentration Confluent quiescent fibroblasts were trypsinated and collected. Cells were loaded with 1 µM fura-2 pentaacetoxymethyl ester at 37°C in HEPES-buffer at a density of 10^6 cells/ml. Fluorescence was measured at 20°C while stirring in a Hitachi F-2000 spectrofluorometer (Raitingen, Germany) at excitation wavelengths of 340 and 380 nm and at an emission wavelength of 505nm.

Cell fractionation for protein kinase C assay Fibroblasts in DMEM/BSA were exposed to specific stimuli as indicated in the figure legend. After a 5 min incubation cells were chilled on ice, washed and then scraped from the flasks into ice-cold sonication buffer (20 nM Tris-HCl, pH 7.4, 0.5 mM EDTA Na_2, 0.5 mM EGTA, 0.25 M sucrose, 50 µg/ml leupeptin, 5 µg/ml antipain, 5 µg/ml aprotinin, 50 µg/ml phenyl-methylsulfonyl fluoride, 10 mM 2-β-mercaptoethanol). After centrifugation at 1000 x g for 3 min, cell pellets were sonicated and centrifuged at 128.000 x g for 20 min. The supernatant, representing the cytosolic fraction, was withdrawn and the pellet, representing the membrane fraction, was suspended in sonication buffer and sonicated for 2 x 30 sec.

Western blot analysis Equal amounts (50 µg) from cytosolic and membrane proteins from fibroblasts were resolved by sodium dodecylsulfate-polyacrylamide gel electrophoresis (SDS-PAGE) (8% gels) [28]. Proteins were transferred to nitrocellulose membranes and probed with specific antibodies for PKC-α (monoclonal), δ, ε, and ζ (polyclonal) using standard procedures.

Assay of inositolphosphates: Fibroblasts (in 6-well plates) were cultured in the presence of 9 µCi/ml [^{3}H]-myo-inositol during the quiescence period. Thereafter, unincorporated radioactivity was removed by washing twice with HBSS. Cultures were preincubated for 30 min with Hepes-buffered saline solution (HBSS) supplemented with 10 mM LiCl prior to addition of stimuli. Stimulation was terminated by aspiration of solution and subsequent addition of 0.2 volumes of ice-cold 20 % perchloric acid and one volume of HBSS/10 mM LiCl to cell layers. Then 25 µg phytic acid hydrolysate dissolved in 4 µl H_2O was added to each well, and after 20 min at 4°C proteins were removed by centrifugation at 2000 g for 20 min at 4°C. The supernatants were titrated with cold 10 N KOH to pH 7.5 and kept on ice. The precipitated $KClO_4$ was removed by centrifugation, and supernatants were diluted with H_2O to a volume of 10 ml. Samples were loaded onto 1 ml Dowex AG1X8 columns and inositol phosphate were resolved by elution with the following solutions: free inositol with distilled water, inositol monophosphate ($InsP_1$) with 0.2 M ammonium formate / 0.1M formic acid, inositol bisphosphate ($InsP_2$) with 0.4 M ammonium formate/0.1 M formic acid and inositol trisphosphate ($InsP_3$) with 0.8 M ammonium formate/0.1 M formic acid. A sample of each fraction was taken for scintillation counting.

Diacylglycerol Determination: In order to measure the biphasic ("early" and "late") generation of diacylglycerol, confluent, quiescent fibroblasts (in 6-well plates) were exposed to stimuli for 30 and 300 seconds. Stimulations were terminated by aspiration of medium and addition of 1 volume of ice-cold methanol containing 0.25 % HCl (v/v). Cells were scraped from the dish and the suspension was mixed with 2 volumes of ice-cold chloroform. For complete lipid extraction this mixture was kept for 1 h at 4°C before removal of proteins by centrifugation. $CaCl_2$ (0.05 mM 1/5th volume) was added to the supernatant and the resultant two phases were separated by centrifugation at 2,000 x g for 10 min at 4°C. The upper phase was carefully removed and the chloroform phase was evaporated to dryness under a stream of nitrogen. Diacylglycerol was measured by radioenzymatic conversion of diacylglycerol to [^{32}P]phosphatidic acid and subsequent separation by thin layer chromatography.

RESULTS

Efflux of Newly Synthesized Sterol from Fibroblasts.

To analyze efflux of newly synthesized sterols, cells were labeled with radioactive mevalonolactone. Since differences in sterol synthesis between control and Tangier fibroblasts might lead to differences in sterol efflux, de novo cholesterol synthesis was determined after incubation for 24 hrs with [^{14}C]-mevalonolactone (0,5 µCi/ml). Control fibroblasts contained 6680 $\pm$ 280 dpm newly synthesized cholesterol/mg cell protein which was not significantly different from Tangier fibroblasts with 7270 $\pm$ 1400 (S.D.) dpm/mg cell protein. This indicates that uptake of mevalonolactone and synthesis of cholesterol are similar in both cell types.

Control fibroblasts labeled at 15°C for 3 hrs with [^{14}C]-mevalonolactone as precursor of endogenous sterol synthesis showed an increased sterol efflux at 37°C with increasing HDL_3 concentrations in the medium (fig 1). The average specific HDL_3-mediated efflux from control fibroblasts calculated as the increase over efflux in the presence of 1% BSA only, ranged from 3% and 18% for HDL_3 concentrations between 10µg/ml and 100µg/ml. Most of the HDL_3-specific efflux occurred within the first four hours of incubation (fig 1). In Tangier fibroblasts, the efflux of sterol with 1% BSA was not significantly different from control fibroblasts. However, there was almost no specific HDL_3-mediated efflux of sterol (0-2.5%) during the whole incubation time in both patient's cell lines (fig. 1). With the lowest concentration of 10µg/ml HDL_3 the difference between control and Tangier fibroblasts was only significant for the 12 and 24 hr time points, due to the minor HDL_3-mediated efflux from control cells with this concentration. With the higher HDL_3 concentrations the differences were significant.

If the reduced efflux of newly synthesized sterol from Tangier fibroblasts to HDL_3 was related to the ability of HDL_3 to induce cholesterol translocation to the cell membrane, efflux to a potent nonspecific cholesterol acceptor should be similar in Tangier and control fibroblasts. Therefore, efflux to protein free phosphatidylcholine/sodium taurocholate micelles, a known potent cholesterol acceptor which does not stimulate cholesterol translocation, on the efflux of newly synthesized sterol was determined. PC-micelles were used in a concentration of 100 µg/ml PC. Both in Tangier and control cells, cholesterol efflux induced by PC-micelles was significantly higher than efflux induced by 1% BSA. There was no difference between the two cell lines. In Tangier cells, efflux of newly synthesized sterol to PC-micelles was higher than the HDL_3-induced efflux. These data suggest that the passive removal of cholesterol from the cell membrane, which is mainly dependent on the chemical composition and affinity to cholesterol of the acceptor is normal.

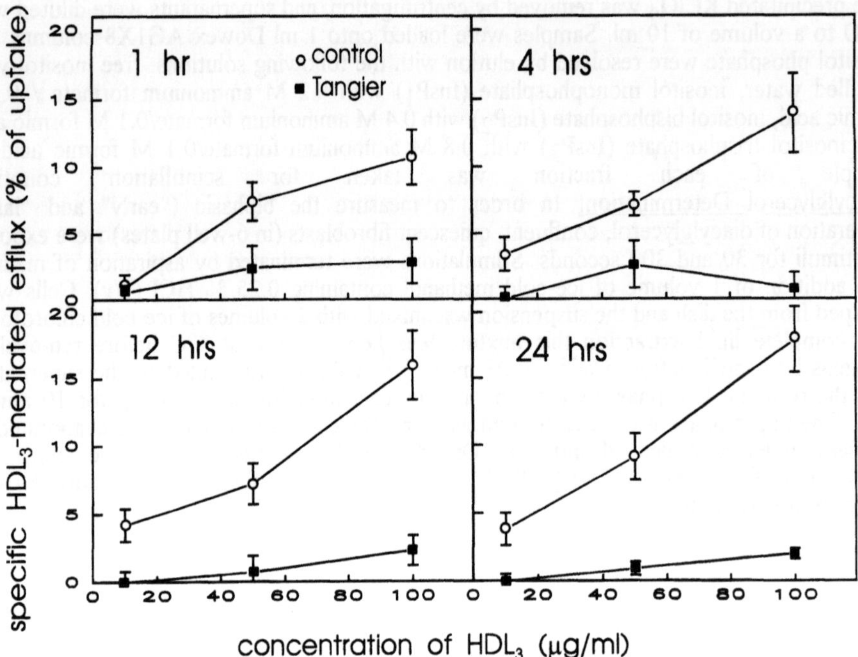

Figure 1 Specific HDL_3-mediated efflux of newly synthesized cholesterol from Tangier and control fibroblasts. Cells were prelabeled with [^{14}C]-mevalonolactone for 3 hrs at 15°C to avoid intracellular cholesterol transport and subsequently incubated in DMEM with 1% BSA or 1% BSA plus 10µg/ml, 50µg/ml, or 100µg/ml HDL_3 for 24 hrs. Efflux is measured as radioactivity in the medium per mg cell protein. In Tangier fibroblasts specific HDL_3-mediated efflux of newly synthesized cholesterol is almost absent for all HDL_3 concentrations. Data represent means (± S.D.) from six experiments per cell line.

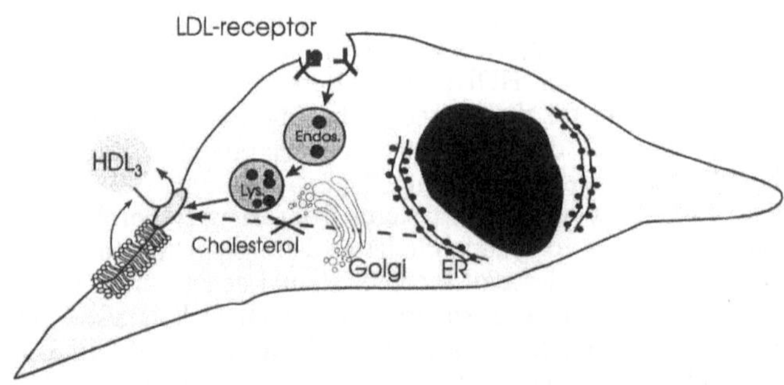

Figure 2 Schematic representation of the defect in the translocation of newly synthesized cholesterol from intracellular stores to the plasma membrane in Tangier fibroblasts. Whereas desorption of membrane cholesterol and LDL derived cholesterol are normal, the transport of newly synthesized cholesterol from ER to the cell membrane appears to be impaired.

Effect of PKC stimulation on efflux of newly synthesized cholesterol. PKC was activated by addition of 10^{-5} M 1,2-DOG to the incubation medium, and the effect on sterol efflux to BSA and HDL_3 was determined. As expected, there was a large increase of BSA-mediated sterol efflux after PKC stimulation in control cells (N.F.), indicating that PKC stimulation leads to an increased availability of cholesterol for membrane desorption. There was only a small further increase in HDL_3-mediated sterol efflux , which is compatible with the concept, that HDL partially activates PKC. In Tangier fibroblasts (J.S.), there was a similar increase in BSA-mediated sterol efflux as in control cells. In addition, PKC stimulation considerably increased HDL_3-mediated efflux of newly synthesized sterols. In fact, HDL_3-mediated sterol efflux after PKC stimulation was similar to control cells.

Redistribution of protein kinase C in Tangier and normal fibroblasts: Fibroblasts (both normal and Tangier) express substantial amounts of PKC α, ε and ζ and trace amounts of PKCδ, while PKC β_1,β_2, γ and η were undetectable. Thus we investigated the ability of HDL_3, LDL and PMA to activate PKC α, ε and ζ in Tangier and normal fibroblasts. Activation of PKC was determined by evaluating translocation from cytosolic to cell membrane fractions. In unstimulated Tangier fibroblasts 30.4 ± 10% of PKCα was recovered in the membrane fraction, compared to only 15.1 ± 4% in normal fibroblasts. Unstimulated Tangier and normal fibroblasts did not differ with respect to membrane associated PKCε. HDL_3 (50 μg/ml, for 5 minutes) increased membrane associated PKCα by 2.8-fold in normal fibroblasts, but did not elicit PKCα translocation in Tangier fibroblasts. Likewise, HDL_3 increased (by 2.1-fold) the membrane bound fraction of PKCε in normal fibroblasts, but was without effect on PKCε in Tangier fibroblasts (fig 3). PMA (100 nM, 5 minutes), which was used as a positive control, promoted translocation of PKC α and PKCε in both cell types.

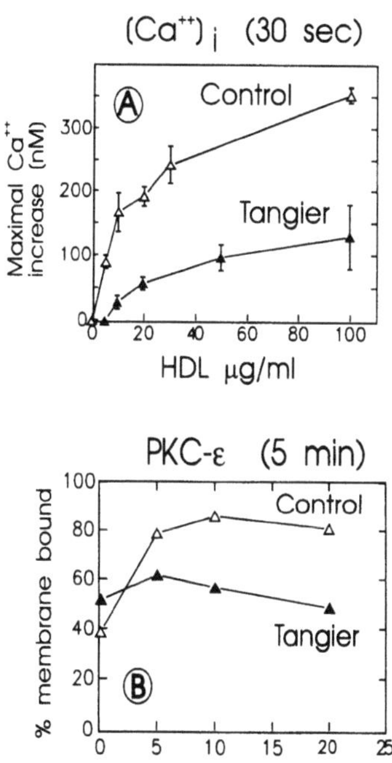

Figure 3 **A:** Dose response of HDL_3 and LDL induced $[Ca^{2+}]_i$ signals in Tangier and normal fibroblasts. Fura-2 loaded fibroblasts were stimulated with the indicated concentration of HDL_3. Each point shows the maximal $[Ca^{2+}]_i$ obtained after stimulation (reached within 30 seconds) and represents mean values ± SD of 9 independent experiments. **B:** Effect of HDL_3 on redistribution of protein kinase C_ε (PKC_ε) in Tangier and normal fibroblasts. Fibroblasts were stimulated with different HDL_3-concentrations for 5 minutes. The distribution of PKCε between membrane and cytosolic fractions was analysed by Western blotting techniques. Data for membrane associated PKC are given, and membrane bound PKCε is expressed relative to their respective total amounts (cytosolic plus membrane, 100%). Results represent mean values from three independent experiments.

$[Ca^{2+}]_i$ transients in Tangier and normal fibroblasts The ability of HDL_3 to cause $[Ca^{2+}]_i$ transients was examined in normal and Tangier fibroblasts. With both cells, peak calcium levels were reached about 30 seconds after addition of HDL_3 (50μg/ml) and returned to basal levels after about 200 seconds. However, the maximum $[Ca^{2+}]_i$ increase in Tangier fibroblasts (75 nM) was markedly lower than that in normal fibroblasts (210 nM). To elucidate this difference further the $[Ca^{2+}]_i$ response of Tangier and normal fibroblasts to various concentrations of HDL_3 was examined. As illustrated in figure 3, the maximum response in both cells was obtained at 100 μg/ml HDL_3, and even at this saturating concentration the maximum $[Ca^{2+}]_i$ increase in Tangier fibroblasts (136 ± 35 nM) was much lower than that in normal fibroblasts (315 ± 69 nM).

Formation of inositol phosphates in Tangier and normal fibroblasts Because the calcium experiments indicated a defective calcium release from internal stores, the formation of inositol phosphates in response to HDL_3 was examined. A 30 second incubation of normal fibroblasts with 50 μg/ml HDL_3 resulted in a 160 ± 15% increase of the $InsP_2$ content per mg cell protein relative to the $InsP_2$ content of unstimulated normal fibroblasts (100 ± 12%). At 60 seconds the $InsP_2$ (98 ± 2%) content returned to basal levels. Stimulation of Tangier fibroblasts with 50 μg/ml HDL_3 did not result in an increased formation of $InsP_2$ after either 30 seconds (96 ± 4%) nor after 60 seconds (73 ± 14%).

Formation of 1,2-diacylglycerol in Tangier and normal fibroblasts The lack of inositol phosphate formation in Tangier cells after HDL_3 -stimulation indicated a defective PI-PLC activation, and thus the formation of 1,2-DAG in response to HDL_3 was examined in Tangier and normal fibroblasts. In normal fibroblasts HDL_3 (50 μg/ml)-induced increases in 1,2-DAG after 30 and 300 seconds were 179 ± 13% and 136 ± 20 %, respectively. HDL_3 did not induce 1,2-DAG generation in Tangier fibroblasts after either 30 seconds or 300 seconds stimulation.

Membrane desorption and efflux of LDL cholesterol Desorption of cholesterol from the cell membrane did not differ between control and Tangier fibroblasts, independent whether the extracellular cholesterol acceptor was bovine serum albumin or HDL_3. In addition, there was no difference in the efflux of LDL-derived cholesterol between control and Tangier cells, providing further evidence for undisturbed membrane desorption as the last step in cholesterol release.

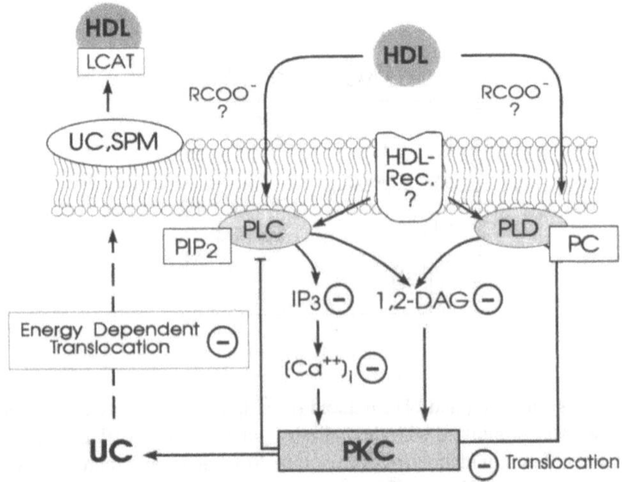

Figure 4 Schematic representation of the determined alterations in HDL_3 mediated signal transduction of Tangier fibroblasts. HDL_3 mediated activation of signal cascades was impaired compared to control fibroblasts. The production of IP_3, early 1,2-DAG and Ca^{2+} was reduced compared to normal fibroblasts leading to impaired translocation of PKC. Late 1,2 DAG-levels, probably generated by PLD were not different.

DISCUSSION

Analysis of cellular cholesterol traffic in normal cells has revealed that there are different transport mechanisms for cholesterol from cholesterol stores or cholesterol poor intracellular membranes, e.g. the endoplasmatic reticulum, to the cell membrane. These obviously depend on the origin of cholesterol (for review see 15). DeGrella and Simoni showed that when cells are pulsed with precursors of sterol synthesis, newly synthesized cholesterol is labeled within minutes[16]. Transport of newly synthesized cholesterol from the endoplasmatic reticulum is energy dependent and completely abolished by temperatures below 15°C. At 37°C, transport takes between 10 and 60 min.

In contrast to these findings, transport of cholesterol to the cell membrane taken up via the LDL receptor and the lysosomal route is not inhibited by energy poisons, indicating that it is not energy dependent. Lysosomal cholesterol appears somewhat faster in the cell membrane than newly synthesized cholesterol. However, transport time (2-40 min) is similar to newly synthesized cholesterol. Thus, there is evidence for two at least in part independent transport routes of cellular cholesterol to the cell membrane. Defects in either of these pathways might affect cholesterol homeostasis of the cell and reverse cholesterol transport.

In the present study, sterol transport was determined by measuring efflux to extracellular acceptors like HDL_3, BSA or PC-micelles. To this end cellular cholesterol pools were labeled in three different ways: [1] incorporation of labeled cholesterol into the cell membrane lipid pool by diffusion; [2] uptake of labeled cholesteryl esters by the LDL-receptor pathway via lysosomes; and [3] incorporation of labeled mevalonolactone into newly synthesized sterols.

Our data provide evidence that there is a defect in HDL_3-mediated efflux of newly synthesized sterol in Tangier fibroblasts, whereas desorption of cholesterol from the cell membrane and transport of lysosomal cholesterol are not disturbed. We found an almost complete absence of the typical increase in efflux of newly synthesized sterol induced by HDL_3 in Tangier fibroblasts. The most likely explanation for this observation is a defect in the transport of sterols from the endoplasmatic reticulum to the cell membrane. This could be either a defect in one or more steps in the transport process itself or a defect in the regulation of transport. Efflux of newly synthesized sterol to 1% BSA alone and to PC-micelles was similar in Tangier and normal fibroblasts. This observation suggests that the impaired efflux to HDL_3 is a specific phenomenon for this particle. It has been shown recently that HDL apolipoproteins induce sterol transport to the cell membrane for desorption by activating PKC[12,13]. This suggested to us that PKC activation or another signal induced by HDL leading to translocation of cellular cholesterol to the cell membrane was defective in Tangier fibroblasts. We could show that incubation of Tangier fibroblasts with HDL_3 does not lead to normal activation of PKC, as determined by analysis of translocation of PKC to the membrane. Therefore we studied the effects of pharmacologic PKC activation on HDL_3-mediated efflux of newly synthesized cholesterol. If PKC was activated, there was no difference in HDL_3-mediated efflux between control and Tangier fibroblasts. This is evidence that the genetic defect in Tangier disease leads to an inadequate stimulation of PKC by HDL_3, resulting in retention of cholesterol in cellular pools. Further experiments were performed to clarify the defect in signal transduction. We demonstrated that, compared to normal fibroblasts, the intracellular Ca^{2+} signaling response to HDL_3 in Tangier fibroblasts is greatly diminished. The concentration response profiles show that the impaired Ca^{2+} signal in Tangier fibroblasts is not due to a reduced sensitivity of Tangier fibroblasts for HDL_3, but to a diminished response capacity, which at maximum was only about 40% of that in normal fibroblasts.

The role of $InsP_3$, a product of PI-PLC activity, in promoting Ca^{2+} release from intracellular pools is well established[17]. Therefore we considered that the impaired calcium signal response to HDL_3 in Tangier fibroblasts might reflect an inability of HDL_3 to activate PI-PLC. Ca^{2+} release from intracellular pools in normal fibroblasts is in agreement with previously presented data, which demonstrated the ability of HDL_3 to activate PI-PLC. In accordance with the Ca^{2+} signaling data, bradykinin was found to promote accumulation of $InsP_2$ in both normal and Tangier fibroblasts, whereas an $InsP_2$ accumulation in response to HDL_3 was observed in normal fibroblasts, but not in Tangier fibroblasts. Our inability to detect significant concomitant increases in $InsP_3$ is probably due to its rapid metabolic conversion to $InsP_2$, a process thought to play a role in termination of $InsP_3$-induced calcium release from internal pools[17]. The inability of HDL_3 to promote either the accumulation of inositol phosphates or intracellular Ca^{2+} mobilization in Tangier fibroblasts therefore indicates an inability of HDL_3 to activate PI-PLC in these cells. Additional support for this conclusion is derived from data demonstrating a lack of effect of HDL_3 on 1,2-DAG generation, the second product of PI-PLC activity[17], in Tangier fibroblasts. To further investigate defective HDL_3-induced signal transduction in Tangier fibroblasts, we studied a cellular event downstream to 1,2-DAG generation, namely the activation of PKC. The activity of PKC was evaluated by its translocation from cytosolic to membrane fractions[18]. In contrast to the PKC translocation response to HDL_3 in normal fibroblasts, exposure of Tangier fibroblasts to HDL_3 did not increase the membrane association of either Ca^{2+}-dependent PKCα or Ca^{2+}-independent PKCε. These data are in accordance to the apparent lack of PI-PLC activation (i.e. no generation of the second messengers $InsP_3$ and 1,2-DAG) by HDL_3 in Tangier fibroblasts. We could conclusively demonstrate that HDL_3 are unable to activate PI-PLC in Tangier fibroblasts and therefore no PKC activation occurs in response to these stimuli. Since PKC activation plays a crucial role in mediating HDL_3 induced cholesterol efflux[12,13], we conclude that impaired HDL_3-induced signal transduction is responsible for reduced HDL_3 mediated efflux of newly synthesized cholesterol in Tangier fibroblasts. The ability of 1,2-DAG to normalize cholesterol efflux strongly supports this conclusion. Efflux of lysosomal cholesterol, which was not different in Tangier and normal fibroblasts, seems to be independent of HDL_3 induced signal transduction. However, the presented data support the concept that PKC isoforms are involved in the removal of cholesterol under physiological conditions.

Abbreviations: apo - apolipoprotein; BK - bradykinin; BSA - bovine serum albumin; $[Ca^{2+}]_i$ - intracellular calcium concentration; 1,2-DAG - 1,2-diacylglycerol; DMEM - Dulbecco's modified eagle medium; FCS - fetal calf serum; fMLP - f-Met-Leu-Phe; G-protein- guanine nucleotide binding protein; HDL_3 - high density lipoprotein 3; HPTLC - high performance thin layer chromatography; $InsP_1$ - inositol monophosphate; $InsP_2$ - inositol biphosphate; $InsP_3$ - 1,4,5-trisphosphate; LDL - low density lipoprotein; PA - phospatidic acid; PBS - phosphate buffered saline; PC - phosphatidylcholine; PC-PLC - phosphatidylcholine specific phospholipase C; PI-PLC - phosphatidyl-inositol-specific phospholipase C; PKC - protein kinase C; PMA - phorbol 12-myristate 13-acetate.

REFERENCES

1 Daniels,R.J.,L.S. Guertler, T.S. Parker, and D. Steinberg. Studies on the rate of efflux of cholesterol from cultured human skin fibroblasts. *J. Biol. Chem.* 256: 4978-4983 (1981).

2 Fielding, C.J. and P.E. Fielding. Evidence for a lipoprotein carrier in human plasma catalyzing sterol efflux from cultured fibroblasts and its relationship to lecithin:cholesterol acyltransferase. *Proc. Natl. Acad. Sci. USA.* 78: 3911-3914 (1981)

3 Miller, N.E.. Mechanisms and approaches to therapy. In: Miller, N.E. (ed): Atherosclerosis. Raven press, New York. 156-168 (1984)

4 Phillips, M.C., W.J. Johnson, and G.H. Rothblat. Mechanisms and consequences of cellular cholesterol exchange and transfer. *Biochim. Biophys. Acta* 777: 209-276. (1987)

5 Fielding, C.J. and P.E. Fielding. Cholesterol transport between cells and body fluids. Role of plasma lipoproteins and the plasma cholesterol esterification system. *Med. Clin. N. Am.* 6: 363-373. (1982)

6 Slotte, J.P., J.F. Oram, and E.L. Bierman. Binding of high density lipoproteins to cell receptors promotes translocation of cholesterol from intracellular membranes to the cell surface. *J. Biol. Chem.* 262: 12904-12907 (1987)

7 Glickman, R.M., P.H.R. Grenn, R.S. Lees, A. Tall. Apoprotein A-I synthesis in normal intestinal mucosa and in Tangier disease. *N. Engl. J. Med.* 299: 1424-1427 (1978)

8 Law, S.W., H.B. Brewer Jr.. Tangier disease: The complete amino acid sequence for proapo AI. *J. Biol. Chem.* 260: 12810-12814 (1985)

9 Assmann, G., G. Schmitz, H.B. Brewer Jr. Familial HDL deficiency: Tangier disease. In: Scriver C.R., Beaudet A.S., Sly W.S., Valle D. (eds): *The Metabolic Basis of Inherited Disease*, 6th ed. New York, McGraw-Hill Book co: 1267-1282 (1989)

10 Schmitz, G., H. Robenek, U. Lohmann, G. Assmann. Interaction of high density lipoproteins with cholesteryl ester laden macrophages: Biochemical and morphological characterization of cell surface binding, endocytosis and resecretion of high density lipoproteins by macrophages. *EMBO J.* 4: 613-622 (1985)

11 Robenek, H., G. Schmitz. Abnormal processing of Golgi elements and lysosomes in Tangier disease. *Arteriosclerosis and Thrombosis* 11: 1007-1020 (1991)

12 Theret, N., C. Delbart, G. Aguie, JC Fruchart, G. Vassaux, G. Ailhaud. Cholesterol efflux from adipose cells is coupled to diacylglycerol production and protein kinase C activation. *Biochem. Biophys. Res. Commun.* 173: 1361-1368 (1990)

13 Mendez, A.J., J.F. Oram, and E.L. Bierman. Protein kinase C as a mediator of high density lipoprotein receptor-dependent efflux of intracellular cholesterol. *J. Biol. Chem.* 266:10104-10111 (1991)

14 Bligh, E.G. and Dyer, W.J. A rapid method of total lipid extraction and purification. *Can. J. Biochem. Phys.* 37, 911-917 (1959)

15 Liscum, L. and Dahl, N.K. Intracellular cholesterol transport. *J. Lipid Res.* 33, 1239-1254 (1992)

16 DeGrella, R.F. and Simoni, R.D. Intracellular transport of cholesterol to the plasma membrane. *J. Biol. Chem.* 257, 14256-14262 (1982)

17 Berridge, M.J., and R.F. Irvine. Inositol phosphates and cell signalling. *Nature* 341: 197-205 (1989)

18 Hug, H., T.F. Sarre. Protein kinase C isoenzymes: divergence in signal transduction. *Biochem. J.* 291:329-343 (1993)

CHOLESTEROL EFFLUX FROM CELLS IN CULTURE: STUDIES WITH LIPID-FREE ACCEPTORS, RECONSTITUTED PARTICLES AND WHOLE SERUM

G. H. Rothblat, P. Yancey, W. S. Davidson, V. Atger*, S. Lund-Katz, W. J. Johnson, M. de la Llera Moya and M.C. Phillips

Medical College of Pennsylvania, Philadelphia. PA and *Hopital Broussais, Paris, France

INTRODUCTION

The movement of cholesterol from cells to acceptor lipoproteins is the first step in the process of reverse cholesterol transport. Both cellular factors and the characteristics of the acceptors modulate the rate at which the cholesterol molecules leave the cell and are picked up by the extracellular acceptors. To gain more information on the factors that influence the efficiency of various acceptors we have conducted a series of studies using acceptor particles of increasing complexity. The simplest system consisted of lipid-free apolipoprotein (apo) AI or synthetic peptides. Increasing complexity was produced when these peptides were reconstituted into disc-like structures composed of phospholipid and the different peptides. The last, and most complex, experimental cholesterol efflux system was one in which whole human serum was added to cells. In all of the studies we have quantitated the release of radiolabeled cholesterol from the cells. In the studies using lipid-free peptides or reconstituted particles the release of the labeled cholesterol reflects the net movement of cholesterol from cells to acceptors, since the acceptors were initially free of cholesterol and no significant cholesterol influx could occur. In the studies using whole serum, the quantitation of labeled cholesterol efflux does not predict the net change in cell cholesterol content since influx would be occurring from a variety of lipoproteins in the serum.

Cardiovascular Disease 2
Edited by L.L. Gallo, Plenum Press, New York, 1995

Efflux of cell lipids to lipid-free apo AI and synthetic peptides

In the present studies we examined the process of lipid efflux to lipid-free apo AI from mouse peritoneal macrophages and L-cells, and utilized synthetic peptides of defined structure to determine what structural properties of apo AI stimulate lipid efflux. Apo AI and the other exchangeable apoproteins of lipoproteins are composed of a varying number of 22-residue-long amino acid repeats [1]. These repetitive segments are amphipathic alpha-helical domains which mediate interaction of the protein with lipid [1]. Synthetic peptide 18A is an 18 amino acid peptide with an amino acid sequence which forms a single amphipathic alpha-helix. This helix is similar to the class A domains present in apolipoproteins [1,2]. There is no sequence homology between this synthetic peptide and naturally occurring apolipoproteins [1]. Blocked-18A ($Ac18ANH_2$) is the 18A molecule modified to contain an acetyl group at the N-terminal and an amide group at the C-terminal. This neutralizes the charges at the ends of the molecule, and stabilizes and lengthens the amphipathic helical segment giving this peptide higher lipid binding affinity than 18A [1,3]. Peptide 37pA is a dimer of two 18A molecules that are covalently joined by a proline residue so that there are two helical domains separated by the proline within the molecule [4].

The comparison of the molar acceptor concentrations (EC_{50}) which promoted half-maximal efflux indicated that the order of efficiency with which the acceptors stimulate cholesterol efflux from both cell types is apo AI > 37pA > $Ac18ANH_2$ > 18A. On a mass basis, the relative order of efficiency for the peptides in both cell types was similar to that observed when the data was expressed on a molar basis (apo AI ≈ 37pA > $Ac18ANH_2$ > 18A), except that 37pA and apo AI were equally effective. The order of efficiency with which the different acceptors stimulated phospholipid efflux was similar to that observed with cholesterol efflux in both cell types. However, in contrast to cholesterol release, saturation of phospholipid efflux was observed only when apo AI was used as an acceptor with both cell types. In addition, the amount of phospholipid release from L-cells and macrophages to the peptides was significantly greater than that to apo AI. The EC_{50} values for cholesterol and phospholipid efflux to apo AI were in good agreement for both cell types. In contrast, the EC_{50} values for phospholipid efflux from mouse macrophages and L-cells to the peptides were 2- to 5-fold higher than those determined for cholesterol efflux.

It is difficult to reconcile all of our observations on the basis of a single mechanism for lipid efflux to the apoprotein and peptides used in the present study. Rather, it is probable that different phenomena are operating, depending on factors such as incubation time, acceptor type and acceptor concentration. We propose that at low acceptor concentrations the primary mechanism is the transient and reversible interaction of the lipid-free apoprotein/peptide with the plasma membrane. The dissociation of the acceptor from the membrane, together with the newly acquired phospholipid, would produce a particle that was capable of promoting cholesterol efflux. The apoprotein/phospholipid complex formed by this process may resemble the preß-HDL particle that has been reported to be an efficient acceptor of cell cholesterol [5,6]. We also propose that alternative mechanisms for lipid release operate at higher concentrations of lipid-free acceptors, particularly the peptides. In this case the release of subcellular particles or vesicles may occur, the origin of which could be either the plasma membrane or intracellular membranes. The data suggest that these vesicles would be rich in phospholipid and depleted of cholesterol. Large vesicular structures would be expected to be relatively inefficient as cholesterol acceptors [7], and their presence in the incubation media may not elicit significant cholesterol efflux above that obtained with the small preß-HDL-like particle discussed above. The scenario presented above is only one of a number of different combinations of mechanisms that could be occurring upon the exposure of cells to lipid-free apoproteins.

Efflux to reconstituted particles

To further address the effects of the amphipathic α-helix structure on cellular free cholesterol efflux, the three class A helical peptides discussed and apo AI were complexed to dimyristoyl phosphatidylcholine (DMPC) to make discoidal complexes that were used as acceptors of cell cholesterol. The three peptides strongly mimic the lipid-binding characteristics of the amphipathic segments of apolipoproteins and form discoidal complexes with DMPC that are similar in diameter (11-12 nm) to those formed by human apo AI when reconstituted at a 2.5:1 (w:w) phospholipid to protein ratio. The abilities of these complexes to remove radiolabeled free cholesterol were compared in experiments using cultured mouse L-cell fibroblasts; efflux of free cholesterol from both the plasma membrane and the lysosomal pools was examined. All four discoidal complexes were equally efficient acceptors of cell membrane free cholesterol when compared at saturating acceptor concentrations of >200 µg DMPC/ml medium. However, at the same lipid concentration, protein-free, DMPC small unilamellar vesicles (SUV) were significantly less efficient. A 10-fold higher V_{max} for the apoprotein/peptide-containing acceptors was observed when compared to that produced by SUV and this difference is likely due to a reversible interaction of apoprotein or peptide with the plasma membrane that changed the lipid packing characteristics in such a way as to increase the rate of free cholesterol desorption from the cell surface. This interaction required amphipathic α-helical segments but it was not affected by the length, number or lipid-binding affinity of the helices. Furthermore, the efflux efficiency was not dependent on the amino acid sequence of the helical segments which suggests that this interaction is not mediated by a specific cell surface binding site.

The abilities of the peptide- or apo AI-containing particles to induce efflux of intracellularly-derived free cholesterol were also compared. To achieve this, the lysosomal pool of free cholesterol was labeled via reconstituted LDL (rLDL) that contained [^{3}H]cholesterol oleate. The cells had been previously exchange-labeled with [^{14}C]free cholesterol to allow for the concurrent measurement of plasma membrane free cholesterol efflux. At a concentration of 100 µg DMPC/ml, the 18A/DMPC particle appeared to be the most effective acceptor of cell membrane free cholesterol with the apoAI-containing particle being the least efficient; the 37pA/DMPC and Ac-18A-NH_2/DMPC particles exhibited intermediate efficiencies. Qualitatively similar results were observed for the efflux of lysosomally-derived [^{3}H]free cholesterol and [^{14}C]free cholesterol originating in the plasma membrane. These results demonstrate that the peptide- and apoAI-containing complexes have the same relative ability to remove plasma membrane-derived free cholesterol as lysosomally-derived free cholesterol.

Efflux to whole serum

A cell culture system was employed to test a large number of samples of human serum for their ability to stimulate the efflux of cell cholesterol [8]. The extent of efflux obtained with each specimen was correlated with the serum concentrations of lipids (i.e. cholesterol, triglycerides), apoproteins (i.e. apo B, apo AI, apo AII) and lipoprotein subfractions (i.e. HDL_2, HDL_3, LpAI, LpAI/AII). The serum samples used in these studies were obtained from the clinical chemistry laboratory at the Hopital Broussais in Paris, France. All samples were from males who had been selected by means of a cholesterol screening program conducted at their worksites.

No relationships were observed between cell cholesterol efflux and the concentrations of apo B, triglycerides or LDL in the test sera. However, statistically significant correlations were obtained between efflux and all serum parameters related to HDL. The highest correlation was obtained between efflux and the level of total HDL cholesterol ($r = 0.68$, $p = <0.0001$, $n = 113$). Strong, and similar, relationships were observed between efflux and the concentrations of both HDL_2 and HDL_3. A comparison between efflux and LpAI or LpAI/AII indicated that efflux more closely correlated with the LpAI ($r = 0.57$, $p = <0.0001$) than with the LpAI/AII ($r = 0.26$, $p = 0.002$). Although there was a very significant correlation between total HDL and efflux, there are many instances in which sera having similar HDL levels stimulated very different rates of cholesterol release. The quantification of HDL subfractions according to either their cholesterol or apoproteins contents results in a considerable overlap in HDL populations. Thus, based on partial correlation analysis LpAI and HDL_3 emerge as the stronger independent serum parameters influencing the fractional efflux of cholesterol from cells.

A goal of this investigation was to establish the extent to which human sera differed in efflux potential and to establish the correlations between cell cholesterol efflux and the different apoproteins and lipoproteins in the sera. There is no correlation between efflux and serum parameters associated with apo B-containing lipoproteins. Thus, although LDL can accept/exchange cholesterol with cells [9-13], the presence of apo B-containing lipoproteins does not appear to affect the rate of removal of cell cholesterol. The highest correlation to cholesterol efflux was obtained with total HDL cholesterol, followed closely by a number of other HDL-related parameters. These correlations suggest that there is not a single HDL subclass that is totally responsible for the efflux, but rather a number of particles contribute to the process of removal of cell cholesterol.

Summary

No single mechanism can explain the data that we have obtained in studies of cellular cholesterol efflux using a variety of different extracellular acceptors. Clearly the desorption of cholesterol from the plasma membrane and its diffusion through the aqueous phase is a fundamental process, and may be the primary process mediating the efflux that occurs to a protein-free acceptor such as a small unilamellar vesicle (for a review see [14]).

The presence of either apo AI or the synthetic peptides on a phospholipid-containing particle enhances the rate of removal of cholesterol from cells. We propose that the efflux that is occurring in the presence of apoprotein-containing acceptors is mediated, at least in part, by the interaction of the apoprotein with the cell membrane. The stimulation of efflux by the synthetic peptides that lack of sequence homology to any region of the apo AI sequence suggests that the particles do not interact with a specific cell surface site. Therefore, the observed increase in free cholesterol efflux to apoprotein-containing particles most likely results from a relatively non-specific interaction of the apoprotein with the plasma membrane. Such an interaction could conceivably occur in two ways. The first is that the apoprotein could remain lipoprotein-bound and still interact with the cell surface, perhaps at specific domains on the plasma membrane [15]. The second type of interaction involves the lipoprotein-unassociated protein. In this case the free protein could either promote free cholesterol and phospholipid efflux by itself , consistent with the data discussed above, or it could bind to the cell surface and modify the lipid packing characteristics of the membrane in such a way as to increase the desorption rate of cholesterol molecules [16]. It is probable that all of these proposed mechanisms can contribute to the efflux of cellular cholesterol, and the relative contribution of each mechanism to overall efflux will be determined by the combination of acceptor particles present in the extracellular environment.

ACKNOWLEDGEMENTS

This work was supported by Program Project Grant HL22633 and a Minority Investigator Research Supplement (MIRS) to this grant, NSF Research Planning Grant MCB-9308279 (MM), National Institutes of Health Training Grant HL07443, a pre-doctoral fellowship to W.S. Davidson from the American Heart Association, Southeastern Pennsylvania Affiliate, and NATO Collaboration Research Grant 930317 (VA). We wish to thank Drs. G. M. Anantharamaiah and J. P. Segrest for supplying the synthetic peptides used in these studies.

REFERENCES

1. Segrest JP, Jones MK, De Loof H, Brouillette CG, Venkatachalapathi YV, Anantharamaiah GM. The amphipathic helix in the exchangeable apolipoproteins: a review of secondary structure and function. *J Lipid Res.*1992;33:141-166.

2. Anantharamaiah GM, Jones JL, Brouillette CG, Schmidt CF, Chung BH, Hughes TA, Bhown AS, Segrest JP. Studies of synthetic peptide analogs of the amphipathic helix. *J Biol. Chem* .1985;260:10248-10255.

3. Venkatachalapathi YV, Phillips MC, Epand RM, Epand RF, Tytler EM, Segrest JP, Anantharamaiah GM. Effect of end group blockage on the properties of a class A amphipathic helical peptide. *Proteins* 1993;15:349-359.

4. Sokoloff L, Rothblat GH. Regulation of sterol synthesis in L-cells: steady-state and transitional responses. *Biochim. Biophys. Acta* 1972;280:171-181.

5. Castro GR, Fielding CJ. Early incorporation of cell-derived cholesterol into pre-B-migrating high-density lipoprotein. *Biochemistry* 1988;27:25-29.

6. Francone OL, Fielding CJ. Initial steps in reverse cholesterol transport: the role of short-lived cholesterol acceptors. *Eur. Heart J.* 1990;11:218-224.

7. Phillips MC, Johnson WJ, Rothblat GH. Mechanisms and consequences of cellular cholesterol exchange and transfer.. *Biochim. Biophys. Acta* 1987;906:223-276.

8. de la Llera Moya M, Atger V, Paul JL, Fournier N, Moatti N, Giral P, Friday KE, Rothblat GH. A cell culture system for screening human serum for ability to promote cellular cholesterol efflux: relationships between serum compnents and efflux, esterification and transfer. *Arterioscler. Thromb.* 1994;In Press:

9. Bates SR, Rothblat GH. Regulation of cellular sterol flux and synthesis by human serum lipoproteins. *Biochim. Biophys. Acta* 1974;360:39-55.

10. Johansson J, Carlson LA, Landou C, Hamsten A. High density lipoproteins and coronary atherosclerosis. A strong inverse relation with the largest particles is confined to normotriglyceridemic patients. *Arterioscler. Thromb.* 1991;11:174-182.

11. Lund-Katz S, Hammerschlag B, Phillips MC. Kinetics and mechanism of free cholesterol exchange between human serum high- and low-density lipoprotein. *Biochemistry* 1982;21:2964-2969.

12. Francone OL, Fielding CJ, Fielding PE. Distribution of cell-derived cholesterol among plasma lipoproteins: a comparison of three techniques. *J. Lipid Res* .1990;31:2195-2200.

13. Nakamura R, Ohta T, Ikeda Y, Matsuda I. LDL inhibits the mediation of cholesterol efflux from macrophage foam cells by apoA-I-containing lipoproteins. *Arterioscler. Thromb.* 1993;13:1307-1316.

14. Johnson WJ, Mahlberg FH, Rothblat GH, Phillips MC. Cholesterol transport between cells and high density lipoproteins. *Biochim. Biophys. Acta* .1991;1085:273-298.

15. Rothblat GH, Mahlberg FH, Johnson WJ, Phillips MC. Apolipoprotein, membrane cholesterol domains, and the regulation of cholesterol efflux. *J. Lipid Res* .1992;33:1091-1098.

16. Letizia JY, Phillips MC. Effects of apolipoproteins on the kinetics of cholesterol exchange. *Biochemistry* 1991;30:866-873.

APO A-I CONTAINING PARTICLES AND ATHEROSCLEROSIS

Jean-Charles Fruchart, Graciela Castro, and Patrick Duriez

SERLIA & Unité INSERM 325
Institut Pasteur
1, rue du Professeur Calmette
59019 Lille Cédex - France

INTRODUCTION

Epidemiological and clinical studies showing an association between decreased concentrations of high-density-lipoprotein (HDL) cholesterol and increased risk of premature coronary artery disease (CAD)[1,2] have generated interest in the mechanism through which HDL prevents atherosclerosis. The HDL have been historically defined as lipoproteins with densities between 1.063 and 1.20 g/ml[3]. Human HDL consists of a collection of particles differing in size, density and apolipoprotein content[4]. Over the years, ultracentrifugation and, subsequently, polyanion precipitation and gradient gel electrophoresis have been used to fractionate HDL into subclasses[5]. Recognition of the importance of the apolipoproteins (apo) not only in the formation and structural stability of lipoproteins but also in their metabolism has led to the separation of HDL into further subpopulations according to their apolipoprotein composition rather than their physicochemical properties.

On the basis of this classification system, it is now recognized that HDL contains at least two types of apo A-I containing lipoprotein particles that may have different metabolic, functional and clinical significance. One species contains both apo A-I and apo A-II (LpA-I:A-II) as main protein components, whereas in the other, apo A-II is absent (LpA-I).

Our purpose in this article is to describe the recent progress made in isolation, characterization, quantification and determination of the clinical significance of LpA-I and LpA-I:A-II.

ISOLATION AND COMPOSITION OF LpA-I AND LpA-I:A-II

LpA-I and LpA-I:A-II are currently purified from total plasma by sequential immunoaffinity chromatography[6]. Some authors do not find any difference in lipid composition between LpA-I and LpA-I:A-II[6] but others have claimed that the percentage of triglyceride and the cholesteryl ester/total cholesterol ratio are lower in LpA-I than in LpA-I:A-II[7]. The lipid/protein ratio appears to be higher in LpA-I than in LpA-I:A-II. The molar ratio of apo A-I to apo A-II in LpA-I:A-II is 1.5. Small quantities of apo A-IV, Cs, D, E are found in both fractions[8].

Cardiovascular Disease 2
Edited by L.L. Gallo, Plenum Press, New York, 1995

Of considerable significance is the finding that proteins stimulating reverse cholesterol transport (LCAT, CETP) and other proteins such as apo J are mainly present in LpA-I and not in LpA-I:A-II[9,10].

METABOLISM OF APO A-I CONTAINING PARTICLES

Our understanding of the metabolism of apo A-I containing particles is limited. Recently, it has been shown that although both particles are synthesized by the liver, LpA-I is produced only by the intestine[11]. The metabolic interrelation between the two subpopulations is not well established, but it appears that LpA-I is catabolized at a faster rate than LpA-I:A-II[12].

PHYSIOLOGICAL ROLE OF APO A-I CONTAINING PARTICLES

Cellular Studies

One of the key questions is whether HDL particles have different physiological role. To gain some insight into the mechanisms by which cholesterol movement takes place in peripheral cells, cultured adipose cells were used. Adipose tissue is the main organ of cholesterol storage in the body and contains mostly non esterified cholesterol. Moreover, rat adipocytes can accumulate and release, on feeding and fasting, respectively, large amounts of cholesterol, suggesting that these peripheral cells may represent a relaxed form of control of cholesterol homeostasis. This observation has been advantageous in the study of cholesterol efflux from cholesterol-preloaded adipose cells in culture, using Ob1771 adipose cells (a subclone of Ob17 cells established from the epididymal fat pad of the Ob/Ob mouse as a model of peripheral cells). In the presence of LDL, these cells can accumulate cholesterol via the LDL receptor pathway. After cholesterol preloading with LDL, long-term exposure to LpA-I particles promoted cholesterol efflux, whereas not efflux was observed in the presence of LpA-I:A-II[13]. The ligands that recognize the cell surface HDL binding sites have been identified as apo A-I, apo A-IV and apo A-II[14,15]. It has been proposed[15] that apo A-I and apo A-IV play the role of agonists and apo A-II that of antagonists of cholesterol efflux. It has been reported that HDL3 induces a protein kinase C dependent translocation of cholesterol from intracellular membrane to the cell surface in human fibroblasts or bovine endothelial cells[16]. We recently demonstrated[17] that cholesterol efflux from adipose cells is couples to diacylglycerol production and protein kinase C activation. The fact that the binding of apo A-I/liposomes, but not apo A-II liposomes, produces diacylglycerol strongly supports the role of apo A-II as an antagonist in the production of cholesterol efflux. However, it has been recently reported that both LpA-I and LpA-I:A-II demonstrate equal ability to promote efflux of cholesterol from several types of cells, such as fibroblasts, smooth muscle cells and Fu5AH[18,19].

CLINICAL SIGNIFICANCE OF LpA-I AND LpA-I:A-II MEASUREMENTS

Quantitative Determination of LpA-I and LpA-I:A-II

A number of methods for direct quantification of apo A-I containing lipoprotein particles in human plasma have been developed : immunoprecipitation[20], two phase electroimmmunoassay[21], and enzyme-linked differential antibody immunosorbent assay[22]. These method are well adapted for use in research laboratories but are time-consuming and inaccurate. The recent development of a differential electroimmunoassay allows the direct measurements of LpA-I[23]. By using a large excess of anti-A-II antibodies, LpA-I:A-II are retained in one peak and LpA-I migrates as a second peak. This new system can provide specific and reproducible determination of LpA-I in human plasma.

Longevity

Assuming that octogenarians who have survived periods of life during which the incidence of coronary artery disease is very high should have several protective factors, we have compared in octogenarians and younger control subjects (30 -50 years) the levels of apo A-I containing particles : LpA-I is significantly elevated and LpA-I:A-II clearly reduced in octogenarians[24].

Brewer et al.[11] recently investigated LpA-I and LpA-I:A-II in a kindred with hyperalphalipoproteinemia and decreased risk of coronary artery disease. The selective increase in LpA-I in the 60 year-old putative homozygote proband with a family history of longevity supports the concept that these particles may represent the "anti-atherogenic" fraction of HDL.

Coronary Artery Disease (CAD)

LpA-I but not LpA-I:A-II are decreased in normolipemic, angiographically documented CAD patients when compared to a group of asymptomatic subjects and a group of patients with arteriographically normal coronary arteries[25]. Nevertheless, in a recent similar study where the triglyceride levels were higher in the patients than in the controls, it was found that both LpA-I and LpA-I:A-II were reduced to a similar degree in patients with CAD[26].

A case control study of apo A-I containing particles has been performed in three populations of contrasting risk for CAD (ECTIM study)[27]. Male patients with myocardial infarction and controls were recruited in two French Centers (Strasbourg and Toulouse) and in Northern Ireland (Belfast). The standardized mortality rates in Belfast, Strasbourg and Toulouse were respectively 348, 102 and 78 per 100 000 for the test populations. LpA-I and LpA-I:A-II levels were lower in the patients than in the controls, but the level of LpA-I was statistically significantly different among the three populations : the LpA-I level was much lower in the Belfast population than in the French populations, in both controls and patients. The multivariate analyses suggest that LpA-I/HDL cholesterol ratio is very significant. Recently, we have observed that the level of LpA-I (but not LpA-I:A-II) is lower in children whose parents had a premature CAD than in a control group who had no familial history of CAD[28].

Primary and Secondary Dyslipoproteinemias

Some dyslipoproteinemias have specific profile of apo A-I particles. For instance, type III dyslipoproteinemia is characterized by a decrease of LpA-I and an increase of LpA-I:A-II[29]. The HDL decrease observed in patients with chronic renal failure who had to undergo hemodialysis is mainly due to a decrease in LpA-I:A-II[30]. Non-insulin dependent diabetes mellitus is characterized by a specific decrease in LpA-I[31].

Diet

Diet also modify LpA-I concentration. The effect of dietary polyunsaturated (P) : saturated (S) fat ratio on apo A-I containing particles has been investigated. While the total fat and cholesterol intake being kept constant, it was found that a high P/S diet (compared to a low P/S) leads to a decrease in LpA-I but not LpA-I:A-II[32].

Alcohol consumption increase HDL circulating levels. When the antiatherogenic role of HDL was proposed, some authors suggested a beneficial effect of chronic alcohol consumption on CAD. We have investigated in 344 men the relationship between LpA-I,

LpA-I:A-II and alcohol consumption[33]. As the alcohol intake rises, LpA-I:A-II levels increased whereas LpA-I levels decreased.

Studies with octogenarians and patients with CAD suggest that LpA-I is the main antiatherogenic particle, therefore, it seems unlikely that chronic alcohol consumption would have an antiatherogenic effect, at least through changes bases upon its action on LpA-I and LpA-I:A-II levels.

Effects of Drugs on Apo A-I Particles

Considering the preliminary results obtained in clinical and epidemiological studies, it is obviously interesting to study the effect of drug therapy on lipoprotein particles defined by their apolipoprotein composition.

Two main questions may be raised concerning the effects of drugs :

- do compounds with various mechanisms of action lead to different effects on HDL particles ?

- is there any relationship between the pharmacological modulation of a particular lipoprotein family and the change in cardiovascular morbidity and mortality ?

We now have some information to answer the former but further investigation is certainly needed to answer the latter.

Atmeh et al.[22] showed that LpA-I may be increased by the use of nicotinic acid, whereas probucol leads to a decrease in this particle. In contrast, nicotinic acid decreased LpA-I:A-II and probucol had no major effect on its concentration. We have shown that fenofibrate decreases LpA-I and increases LpA-I:A-II[34] whereas hydroxymethylglytaryl-CoA reductase inhibitors (simvastatin and pravastatin) have different effects. Simvastatin increased LpA-I (particularly when the baseline levels were low)[34] but had no effect on LpA-I:A-II. Pravastatin increased both LpA-I and LpA-I:A-II[35]. Cholestyramine, a bile acid sequestrant also increased these two types of particles. Because it has been suggested that LpA-I may represent the particle that is involved in cholesterol efflux from peripheral cells, we speculate that the increasing effect on LpA-I may potentiate the beneficial cardiovascular effect of low-density lipoprotein cholesterol reduction also seen with these compounds. Inversely, the decreasing effect obtained with fenofibrate might be considered as a potentially harmful effect. However, kinetic study are necessary to determine whether the increasing effect of HMG-CoA reductase inhibitors and cholestyramine is due to oversynthesis or undercatabolism and whether the decreasing effect of fenofibrate is due to undersynthesis or overcatabolism. The clinical importance of the effect of drugs on LpA-I may depend on these findings.

Studies in transgenic animals

Recent study[36] had shown that the serum from transgenic mice expressing human apo A-I are able to induce a cholesterol efflux than the serum from transgenic mice expressing both human apo A-I and apo A-II.

CONCLUSION

As peripheral cells are unable to degrade cholesterol, a pathway by which intact cholesterol molecules can be removed from the cells is essential for cholesterol homeostastis. The process whereby cholesterol is removed from peripheral cells and transported to the liver has been called reversed cholesterol transport. This removal of excess cholesterol is mediated by HDL. With different types of cultivated cells, we have demonstrated that a subpopulation of HDL, specifically, LpA-I free of apo A-II (LpA-I), mediates translocation of intracellular cholesterol to the plasma membrane and induces cholesterol efflux. Therefore, LpA-I is probably an important lipoprotein involved in the reverse cholesterol transport and an antiatherogenic lipoprotein particle candidate.

The introduction of new immunological methods of measurement of LpA-I and LpA-I:A-II levels in the blood allows to study large populations. Clinical and epidemiological studies highly suggest that LpA-I is the main antiatherogenic particle. This particle could be very accurate in order to predict the development of premature atherosclerosis. The molecular analysis of lipoprotein particles in terms of apolipoprotein content provide a new basis for the classification of dyslipoproteinemic states and the effects of diet and hypolipidemic drugs.

REFERENCES

1. T. Gordon, W.P. Castelli, M.C. Hjortland, W.B. Kannel, and T.R. Dawber, High density lipoprotein as a protective factor against coronary heart disease : the Framingham Study, *Am. J. Med.* 62:707 (1977).
2. N.E. Miller, O.H. Forte, D.S. Thelle, and O.D. Mjos, The Tromso heart study : high density lipoprotein and coronary heart disease : a prospective case control study, *Lancet* i:965 (1977).
3. R.J. Havel, H.A. Eden, and J.H. Bragdon, The distribution and chemical composition of ultracentrifugally separated lipoproteins in human serum, *J. Clin. Invest.* 34:1345 (1955).
4. P. Alaupovic, The physicochemical and immunological heterogeneity of human plasma high density lipoproteins, *in*: "Clinical and Metabolic Aspects of High-Density Lipoproteins", N.E. Miller, G.J. Miller, eds., Elsevier Science Publishers B.V., Amsterdam (1985).
5. J.C. Fruchart, S. Marcovina, and P. Puchois, Laboratory measurements of plasma lipids and lipoproteins, *in* : "Human Plasma Lipoprotein", J.C. Fruchart, J. Shepherd, eds., Walter de Gruyter, Berlin (1989).
6. M.C. Cheung, and J.J. Albers, Characterization of lipoprotein particles isolated by immunoaffinity chromatography : particles containing A-I and A-II and particles containing A-I but not A-II, *J. Biol. Chem.* 259:12201 (1984).
7. T. Ohta, S. Hattori, S. Nishiyama, and I. Matsuda, Studies on the lipid and apolipoprotein compositions of two species of apo A-I containing lipoproteins in normolipidemic males and females, *J. Lipid Res.* 29:721 (1988).
8. P. Alaupovic, E.D. Bekaert, and E. Koren, Isolation, characterization and quantitation of subclasses of plasma lipoproteins defined by their apolipoprotein composition, *Atherosclerosis Rev.* 20:179 (1990).
9. M.C. Cheung, A.C. Wolf, K.D. Lum, J.H. Tollefson, and J.J. Albers, Distribution and localization of lecithin : cholesteryl acyl-transferase and cholesteryl ester transfer activity in apo A-I containing lipoproteins, *J. Lipid Res.* 27:1135 (1986).
10. H.V. De Silva, W.D. Stuart, and C.R. Duvic, A 70-kDa apolipoprotein designated apo J is a marker for subclasses of human plasma high density lipoproteins, *J. Biol. Chem.* 265:13240 (1990).
11. H.B. Brewer, D. Rader, S. Fojo, and J.M. Hoeg, Frontiers in the analysis of HDL structure, function, and metabolism, *in*: "Disorders of HDL", L.A. Carlson, ed., Smith-Gordon and Company Ltd, London (1990).
12. D. Rader, J.R. Schaefer, M.R. Kindt, L.A. Zecht, J.C. Fruchart, and H.B. Brewer, Differential in vivo metabolism of HDL subclasses LpA-I and LpA-I:A-II in man, *Clin. Res.* 38:204A (1990).
13. A. Barkia, P. Puchois, N. Ghalim, G. Torpier, G. Ailhaud, and J.C. Fruchart, Differential role of apolipoprotein A-I containing particles in cholesterol efflux from adipose cells, *Atherosclerosis* 87:135 (1991).
14. A. Steinmetz, R. Barbaras, N. Ghalim, V. Clavey, J.C. Fruchart, and G. Ailhaud, Human apolipoprotein A-IV binds to apolipoprotein A-I/A-II receptor sites and promotes cholesterol efflux from adipose cells, *J. Biol. Chem.* 265:7859 (1990).
15. R. Barbaras, P. Puchois, J.C. Fruchrt, A. Pradines-Figueres, and G. Ailhaud, Purification of an apolipoprotein A binding protein from mouse adipose cells, *Biochem. J.* 269:767 (1990).

16. J.P. Slotte, J.F. Oram, and E.L. Bierman, Binding of high density lipoproteins to cell receptors promotes translocation of cholesterol from intracellular membranes to the cell surface, *J. Biol. Chem.* 262:12904 (1987).
17. N. Theret, C. Delbart, G. Aguie, J.C. Fruchart, G. Vassaux, and G. Ailhaud, Cholesterol efflux from adipose cells is coupled to diacylglycerol production and protein kinase C activation, *Biochem. Biophys. Res. Commun.* 173:1361 (1990).
18. S. Oikawa, A.J. Mendez, M.C. Cheung, J.F. Oram, and E.L. Bierman, Effect of apo A-I and apo A-I:A-II HDL particles in intracellular cholesterol efflux, *Circulation* (abstract) 84(suppl II):2711 (1991).
19. W.J. Johnson, E.P.C. Kilsdonk, A. Van Tol, M.C. Phillips, and G.H. Rothblat, Cholesterol efflux from cells to immunopurified subfractions of human high density lipoprotein : LpA-I and LpA-I:A-II, *J. Lipid Res.* 32:1992 (1991).
20. M.C. Cheung, and J.J. Albers, Distribution of high density lipoprotein particles with different apoprotein composition : particles with A-I and A-II and particles with A-I but not A-II, *J. Lipid Res.* 23:747 (1982).
21. R.F. Atmeh, J. Shepherd, and C.J. Packard, Subpopulations of apolipoprotein A-I in human high density lipoproteins. Their metabolisms properties and response to drug therapy, *Biochim. Biophys. Acta* 751:175 (1983).
22. E. Koren, P. Puchois, P. Alaupovic, J. Fesmire, A. Kandoussi, and J.C. Fruchart, Quantification of two types of apolipoprotein A-I containing lipoprotein particles in plasma by enzyme-linked differential antibody immunosorbent assay, *Clin. Chem.* 33:38 (1987).
23. H.J. Parra, H. Mezdour, N. Ghalim, J.M. Bard, and J.C. Fruchart, Differential electroimmunoassay of human LpA-I lipoprotein particles on ready-to-use plates, *Clin. Chem.* 36:1431 (1990).
24. G. Luc, J.M. Bard, S. Lussier-Cacan, S. Bouthillier, H.J. Parra, J.C. Fruchart, and J. Davignon, High-density lipoprotein particles in octogenarians, *Metabolism* 40:1238 (1991).
25. P. Puchois, A. Kandoussi, P. Fiévet, J.L. Fourrier, M. Bertrand, E. Koren, and J.C. Fruchart, Apolipoprotein A-I containing lipoproteins in coronary artery disease, *Atherosclerosis* 68:35 (1987).
26. J.J. Genest, J.M. Bard, J.C. Fruchart, J.M. Ordovas, P.F.W. Wilson, and G.J. Shaefer, Plasma apolipoproteins (a), A-I, A-II, B, E and C-III containing particles in men with premature coronary artery disease, *Atherosclerosis* 90:149 (1991).
27. F. Cambien, H.J. Parra, D. Arveiler, J.P. Cambou, A. Evans, A. Bingham, Lipoprotein particles in patients with myocardial infarction and controls, *Circulation* 82(Suppl III):348 (1990).
28. P. Amouyel, D. Isorez, J.M. Bard, M. Goldman, P. Lebel, G. Zylberberg, and J.C. Fruchart, Parenteral history of early myocardial infarction is associated with decreased levels of lipoparticle A-I in adolescents, *Arterioscl. Thromb.* 13:1640 (1193).
29. S. Lussier-Cacan, J.M. Bard, L. Boulet, A.C. Nestruck, A.M. Grother, J.C. Fruchart, and J.C. Fruchart, and J. Davignon, Lipoprotein composition changes induced by fenofibrate in dysbetalipoproteinemia type III, *Atherosclerosis* 78:167 (1989).
30. C. Cachera, A. Kandoussi, K. Equagoo, J.C. Fruchart, and A. Tacquet, Evaluation of apolipoprotein A-I containing particles in chronic renal failure patients undergoing hemodialysis, *Am. J. Nephrol.* 10:171 (1990).
31. J.C. Fruchart, Insulin resistance and lipoprotein abnormalities, *Diabete Metab.* 17:244 (1991).
32. F. Fumeron, L. Brigant, H.J. Parra, J.M. Bard, J.C. Fruchart, and M. Apfelbaum, Lowering of HDL_2 cholesterol and lipoprotein A-I particle levels by increasing the ratio of polyunsaturated to saturated fatty acids, *Am. J. Clin. Nutr.* 53:655 (1991).

33. P. Puchois, N. Ghalim, G. Zylberberg, P. Fiévet, C. Demarquilly, and J.C. Fruchart, Effect of alcohol intake on human apolipoprotein A-I containing lipoprotein subfractions, *Arch. Intern. Med.* 150:1638 (1990).
34. J.M. Bard, H.J. Parra, R. Camare, G. Luc, O. Ziegler, C. Dachet, E. Bruckert, P. Douste-Blazy, P. Drouin, B. Jacotot, J.L. De Gennes, U. Keller, and J.C. Fruchart, A multicenter comparison of the effects of simvastatin and fenofibrate therapy in severe primary hypercholesterolemia, with particular emphasis on lipoproteins defined by their apolipoprotein composition, *Metabolism* 4:498 (1992).
35. J.M. Bard, H.J. Parra, P. Douste-Blazy, and J.C. Fruchart, Effect of pravastatin, an HMG CoA reductase inhibitor, and cholestyramine, a bile acid sequestrant, on lipoprotein particles defined by their apolipoprotein composition, *Metabolism* 39:269 (1990).
36. J.C. Fruchart, C. De Geitère, B. Delfly, G.R. Castro - Unpublished results.

VASCULAR ENDOTHELIUM: AN INTEGRATOR OF PATHOPHYSIOLOGIC STIMULI IN CARDIOVASCULAR DISEASE

Michael A. Gimbrone, Jr.

Vascular Research Division,
Department of Pathology,
Brigham and Women's Hospital
Boston, Massachusetts 02115

Vascular endothelium comprises the continuous lining of the cardiovascular system, and, as such, forms a dynamic interface between circulating blood and all other tissues and organs. Therefore, it is strategically situated to monitor blood-borne and/or locally generated stimuli and to adaptively alter its functional state. This physiologic process typically proceeds without notice, contributing to normal homeostasis[1]. However, nonadaptive changes in the functional properties of the vascular endothelium, provoked by various pathologic stimuli, can result in localized, acute and chronic, alterations in the interactions of cellular and macromolecular components of circulating blood with the arterial wall. These include: altered permeability to plasma lipoproteins; increased cytokine and growth factor production; imbalances in procoagulant and fibrinolytic activities; and hyperadhesiveness for blood leukocytes. These manifestations, collectively termed "endothelial dysfunction", play an important role in the initiation, progression and clinical complications of vascular diseases[2].

Various types of pathophysiologically relevant stimuli of endothelial dysfunction have been identified, including immuno-regulatory cytokines such as tumor necrosis factor and interleukin-1, viral infection and transformation, bacterial toxins (especially Gram-negative endotoxin), oxidatively modified lipoproteins, and advanced glycosylation end-products. At a molecular level, perhaps the best studied paradigm is the "activation" of endothelial cells by inflammatory cytokines such as TNF and IL-1, and bacterial lipopolysaccharide (endotoxin)[3]. This process involves a coordinated sequence of events, initiated by cell surface receptor activation, that culminates in a pattern of gene expression that typically is characteristic for a particular cytokine or mixture of cytokines. At the level of the nucleus, a number of these stimuli appear to converge into final common pathways of transcriptional regulatory factors, such as the NFkB system[4].

In addition to these soluble cytokine stimuli, biochemical forces generated by the pulsatile flow of blood through the branched arterial tree, such as oscillating wall shear stresses, and cyclic stretching, can also influence a variety of endothelial functions[5]. Certain of these biomechanically induced effects appear to involve the modulation of endothelial gene expression at the transcriptional level. Recently, a "shear stress response

Cardiovascular Disease 2
Edited by L.L. Gallo, Plenum Press, New York, 1995

element (SSRE)" has been discovered in the platelet-derived growth factor (PDGF) B-chain gene that appears to be involved in these processes. This cis-acting transcriptional element has been shown by deletion analysis to be necessary for shear-induced activation of the PDGF-B chain gene, and also is present in the promoters of several other shear-inducible endothelial genes[6]. Interestingly, ICAM-1, an endothelial-leukocyte adhesion molecule that contains the SSRE in its promoter, shows selective upregulation in response to physiologic levels of laminar shear stress, compared with E-selectin and VCAM-1, that do not contain this regulatory element[7]. This differential pattern of induction of these endothelial-leukocyte adhesion molecules by a biomechanical force is in contrast to the coordinate pattern of induction typically observed with other activating stimuli. Experimental analysis of the transduction mechanisms that link externally applied forces to genetic regulatory events within the nucleus thus may provide new insights into the endothelial activation process.

Recent studies suggest that an early manifestation of endothelial dysfunction in the atherosclerotic process is the expression of inducible mononuclear leukocyte-selective adhesion molecules, such as VCAM-1 (termed the "ATHERO-ELAM", in the rabbit)[8]. This observation has several conceptual as well as practical implications. First, this provides an objective (and potentially quantitative) index of endothelial activation in a vascular disease process. Second, this inducible leukocyte adhesion molecule may be a potential target for "anti-adhesive" therapeutic interventions. Third, characterization of analogous changes in human atherosclerotic lesions may provide novel markers for the early stages of this complex disease process, potentially useful for both diagnosis and therapy. Clearly, the successful application of such diagnostic and therapeutic strategies will require better understanding of the basic mechanisms of endothelial activation in the context of human atherosclerotic vascular disease.

In summary, the endothelial lining of the cardiovascular system is a dynamically mutable interface which can exhibit a spectrum of adaptive changes. Various components of its vast repertoire of autacoids, growth factors, vasoactive, hemostatic and fibrinolytic substances often contribute to agonist/antagonist balances that have important implications for the function of the vascular lining, adjacent vascular cells and interacting blood constituents. By virtue of its unique anatomical position, the endothelium also plays an important role in the local transduction and integration of diverse biological stimuli, including circulating hormones and bacterial products, locally generated cytokines, and even biomechanical forces. Thus, in an important sense, the phenotype of an endothelial cell is a reflection of the local pathophysiologic milieu. As our knowledge of the stimuli and consequences of dysfunctional endothelial phenotypes increases, so will our working concepts of the pathogenesis of vascular diseases.

ACKNOWLEDGMENTS

Research in the author's laboratory has been supported primarily by grants from The National Heart, Lung and Blood Institute, and the American Heart Association and its Massachusetts Affiliate, and an unrestricted grant for cardiovascular research from the Bristol-Myers Squibb Research Institute. The author wishes to acknowledge his colleagues and collaborators in the experimental studies summarized here, especially Drs. Tucker Collins, Ramzi Cotran, Myron Cybulsky, C. Forbes Dewey Jr., F. William Luscinskas, Tobi Nagel, and Nitzan Resnick.

REFERENCES

1. M.A. Gimbrone, Jr. Vascular endothelium in health and disease, in: Molecular Cardiovascular Medicine", E. Haber, ed., Scientific American Medicine, New York (in press, 1994).
2. N. Simionescu and M. Simionescu, eds., "Endothelial Cell Dysfunctions". Plenum Press, New York (1992).
3. J. S. Pober and R. S. Cotran. Cytokines and endothelial cell biology. Physiol. Rev. 70:427 (1990).
4. T. Collins. Endothelial nuclear factor-kB and the initiation of the atherosclerotic lesion. Lab. Invest. 68:499 (1993).
5. P.F. Davies and S.C. Tripathi. Mechanical stress mechanisms and the cell: an endothelial paradigm (Mini Review). Circ. Res. 72:239 (1993).
6. N. Resnick, T. Collins, W. Atkinson, D.T. Bonthron, C.F. Dewey Jr. and M.A. Gimbrone, Jr. Platelet-derived growth factor B chain promoter contains a cis-acting fluid shear-stress-responsive element. Proc. Natl. Acad. USA, 90:4591 (1993).
7. T. Nagel, N. Resnick, W. Atkinson, C.F. Dewey, Jr., M.A. Gimbrone, Jr. Shear stress selectively upregulates ICAM-1 expression in cultured human vascular endothelial cells. J. Clin. Invest. (In press, 1994).
8. M. Cybulsky and M.A. Gimbrone, Jr. Endothelial expression of a mononuclear leukocyte adhesion molecule during atherogenesis. Science. 251:788-791 (1991).

IDENTIFICATION OF FGF-1-INDUCIBLE GENES BY DIFFERENTIAL DISPLAY

Jeffrey A. Winkles, Patrick J. Donohue, Debbie K.W. Hsu, Yan Guo, Gregory F. Alberts, and Kimberly A. Peifley

Department of Molecular Biology
Holland Laboratory
American Red Cross
Rockville, MD 20855

INTRODUCTION

Endothelial cells and smooth muscle cells, the most abundant cell types in the blood vessel wall, normally have a low replication rate in the adult animal. However, endothelial cell proliferation and concomitant neovascularization is associated with tumor growth and the pathogenesis of numerous angiogenesis-dependent diseases.[1] Also, accelerated smooth muscle cell replication plays a central role in atherogenesis,[2] vascular graft stenosis,[3] and restenosis of vessels following angioplasty or atherectomy.[4] Fibroblast growth factor (FGF)-1 and FGF-2, also commonly known as acidic and basic FGF, respectively, are two of the polypeptide mitogens that are likely to be important mediators of vascular cell growth *in vivo*. They are both potent angiogenic factors[5,6] and smooth muscle cell mitogens.[7,8] They are expressed by vessel wall cells[9,10] and by monocyte-derived macrophages within human atheroma.[9,10] FGF-2 has also been detected in human platelets[11] and T lymphocytes.[12] Direct evidence for the involvement of FGF-2 in rat balloon injury-induced medial smooth muscle cell proliferation was reported by Lindner and Reidy.[13] Therefore, it is possible that inhibition of FGF expression or action may be of therapeutic benefit in the treatment of human cardiovascular disease. One strategy that may prove effective for inhibition of FGF mitogenic activity is to interrupt the FGF intracellular signaling pathway. In this review, we describe our approach to identify proteins that are involved in FGF-1 mitogenic signal transduction and thus potential targets for therapeutic intervention.

FGF-1 MITOGENIC SIGNAL TRANSDUCTION

The majority of studies investigating the molecular mechanisms of FGF signaling have not used human vascular cells, but instead immortalized murine cell lines such as NIH 3T3 fibroblasts. These cells are relatively easy to maintain, express cell surface

Cardiovascular Disease 2
Edited by L.L. Gallo, Plenum Press, New York, 1995

FGF receptors, can be arrested in G_0 by serum-starvation, and display a strong mitogenic response to FGF-1 stimulation.[14] It is known that FGF-1 binding to responsive cells stimulates receptor autophosphorylation and the tyrosine phosphorylation of specific proteins, including phospholipase C-γ,[14,15] Shc,[16] raf-1,[17] ERK-1,[16] ERK-2,[16] and cortactin.[18] The phosphorylation-mediated activation of phospholipase C-γ promotes phosphatidylinositol hydrolysis, which in turn results in protein kinase C activation and Ca^{2+} mobilization. The role of these intracellular events in FGF signaling is unclear; two groups have reported that phosphatidylinositol turnover is not necessary for FGF-induced mitogenesis.[19,20]

Another cellular response to FGF-1 stimulation is the enhanced expression of specific genes. Growth-regulated genes are generally classified into one of two groups: immediate-early or delayed-early. Immediate-early genes are rapidly and transiently expressed following mitogenic stimulation.[21] Their transcriptional activation is independent of *de novo* protein synthesis; therefore, the cellular factors necessary for gene induction pre-exist in quiescent cells and need only to be modified following stimulation. Examples of FGF-1-inducible immediate-early genes in NIH 3T3 cells include those encoding c-Fos,[22] c-Jun,[22] c-Myc,[22] early growth response gene-1 (unpublished results) and thrombospondin-1 (unpublished results). The rapid induction of early growth response gene-1 mRNA is illustrated in Figure 1.

Delayed-early genes are first expressed a few hours after mitogen addition and transcript levels can remain elevated for relatively long periods of time.[21] These genes are not activated if protein synthesis is inhibited; indeed, there is evidence that at least some delayed-early genes are regulated by newly synthesized immediate-early transcription factors. For example, the addition of antisense *c-myb* and *c-myc* oligonucleotides inhibits cdc2 expression[23] and antisense *c-fos* oligonucleotides prevent transin gene activation.[24] The FGF-1-regulated (FR)-1 gene, which encodes an aldose reductase-related protein, is an example of an FGF-1-inducible delayed-early gene in NIH 3T3 cells.[25] We have noted that four additional FGF-1-inducible mRNAs, encoding phosphofructokinase,[26] ornithine decarboxylase (unpublished results), glyceraldehyde 3-phosphate dehydrogenase (Figure 1) and proliferin (Figure 1), are expressed with kinetics typical of delayed-early genes but the effect of protein synthesis inhibition has not yet been tested. It is presently unknown whether any of the FGF-1-inducible immediate-early or delayed-early proteins described above are required for FGF-1-stimulated cell proliferation.

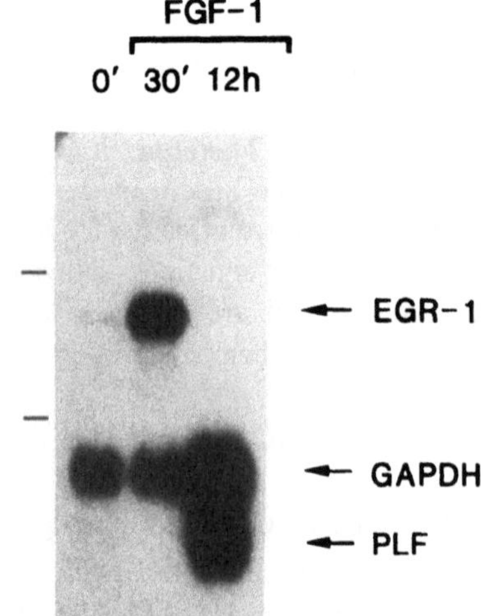

Figure 1. Effect of FGF-1 on early growth response gene-1 (Egr-1), glyceraldehyde 3-phosphate dehydrogenase (GAPDH) and proliferin (PLF) mRNA levels in NIH 3T3 fibroblasts. Serum-starved cells were either left untreated or treated with 10 ng/ml human recombinant FGF-1 and 5 U/ml heparin for 30 min or 12 h. RNA was isolated and equivalent amounts of each sample analyzed by Northern blot hybridization[25] using ^{32}P-labeled Egr-1, GAPDH or PLF cDNA probes. The upper and lower tick marks on the left side indicate the positions of 28S and 18S rRNA, respectively.

GROWTH FACTOR-RESPONSIVE GENES AND CELLULAR PROLIFERATION

Previous reports have indicated that some growth factor-inducible genes encode proteins that perform functions critical for cellular proliferation. In most of these studies, the expression or activity of a specific gene product was inhibited by either adding antisense oligonucleotides to cell culture media, transfecting cells with antisense expression constructs, or microinjecting specific monoclonal antibodies into the cell nucleus or cytoplasm. The effect of these treatments on cell cycle progression *in vitro* was then monitored. These studies have indicated that the mitogen-inducible proteins c-Fos,[27-29] c-Jun,[29,30] c-Myc,[31-33] c-Myb,[34] c-Ras,[35] p53,[36] and proliferating cell nuclear antigen,[37,38] are important for the proliferative response of mitogen-treated cells. In the context of vascular biology, recent studies applying the antisense oligonucleotide approach *in vivo* have indicated that c-Myc,[33] c-Myb[39] and the mitogen-inducible cdc2 protein kinase[40] are involved in the smooth muscle cell hyperplasia associated with balloon catheter-induced injury of the rat carotid artery.

IDENTIFICATION OF NOVEL FGF-1-INDUCIBLE GENES

In consideration of the results described above, it is reasonable to conclude that one approach that may prove successful for identifying proteins involved in FGF-stimulated mitogenesis is to isolate and characterize cDNA clones representing FGF-1-inducible mRNAs. Genes regulated by serum,[41-47] PDGF,[48] EGF,[49] IGF-1,[50] TGF-β1,[51] TNF-α,[52] IL-1,[53] IL-2[54,55] or CSF-1[56] have been successfully identified by subtracted cDNA probe hybridization or differential screening of cDNA libraries. However, these two strategies are technically challenging, laborious, require significant amounts of cellular mRNA, and may not identify genes encoding relatively rare transcripts. We have used an alternative method, based on the reverse transcription-polymerase chain reaction (RT-PCR) technique, to identify FGF-1-inducible genes.[26] As this work was in progress, similar approaches to identify differentially-expressed mRNA species were described by Liang and Pardee,[57] Welsh et al.,[58] and Bauer et al.,[59] and termed "differential display (DD)", "RNA fingerprinting using arbitrarily primed PCR (RAP)", or "differential display reverse transcription PCR (DDRT-PCR)", respectively.

In our experiments, RNA is isolated from quiescent or FGF-1-treated NIH 3T3 cells and converted into cDNA using random primers and reverse transcriptase. This cDNA is then used in PCR assays containing various combinations of degenerate oligonucleotide primers designed to recognize protein domains conserved in different classes of signaling proteins; for example, zinc finger, leucine zipper, protein tyrosine kinase and Src homology 2 domains. Amplification products are separated by electrophoresis on agarose gels and visualized by ethidium bromide staining. Two examples of differential display gels are shown in Figure 2. In these experiments, three cDNA fragments, of ~700-, ~230-, or ~790-basepairs (bands #1, #2, or #3, respectively), were specifically amplified in the reactions using RNA isolated from FGF-1-treated cells.

The cDNA fragments isolated by this approach can then be purified, re-amplified and subcloned into an appropriate plasmid vector. Typically, the cDNA inserts are then isolated, radiolabeled and used as probes in Northern blot hybridization experiments to confirm that the respective clone actually represents an FGF-1-inducible mRNA. Also, nucleotide sequence data is obtained and assessed for homology to sequences deposited in the various databases. Depending on the Northern blot and DNA sequencing results, a particular RT-PCR-derived cDNA may then be used as a probe to isolate larger cDNA clones from a murine cDNA library.

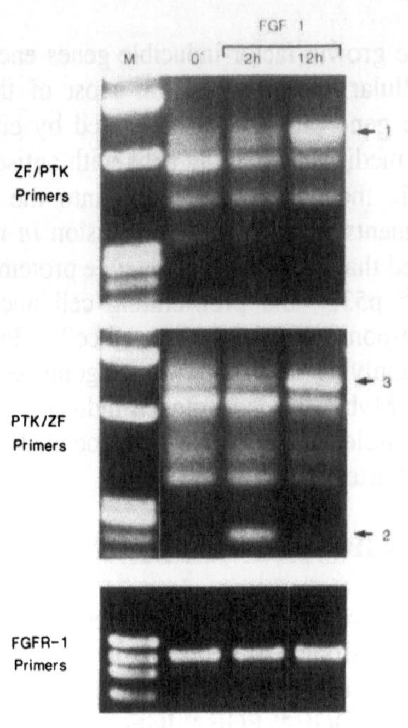

Figure 2. Identification of three FGF-1-inducible mRNAs by differential display. Serum-starved NIH 3T3 cells were either left untreated or treated with FGF-1 and heparin for 2 or 12 h. RNA was isolated, random-primed cDNA was synthesized, and PCR performed as described.[26] For PCR, either sense zinc finger (ZF)/antisense protein tyrosine kinase (PTK) (top panel), antisense PTK/sense ZF (middle panel), or sense and antisense FGF receptor-1 (bottom panel) oligonucleotide primers were used. PCR was performed with these latter primers to demonstrate the integrity of the three cDNA templates. Amplification products were separated by agarose gel electrophoresis and visualized by ethidium bromide staining. DNA size markers (M) are present in the first lane. The cDNA fragments representing the three differentially-expressed transcripts are indicated with arrows.

We have identified numerous distinct FGF-1-inducible genes using our differential display technique. Both immediate-early and delayed-early kinetics of mRNA expression have been observed. Three different temporal expression patterns are illustrated in Figure 3. FR-1 mRNA expression was detected in both quiescent and FGF-1-stimulated cells; however, FR-1 mRNA levels were elevated at 4, 8 and 12 h after growth factor addition. FR-1 mRNA induction does not occur if protein synthesis is inhibited using cycloheximide.[25] FR-2 mRNA was not detected in quiescent cells but was expressed in response to FGF-1 addition. The expression level was maximal at 1 h post-stimulation and then returned to near basal levels by 12 h. FR-2 mRNA induction can still occur in cycloheximide-treated cells (unpublished results). FR-3 transcripts were also only expressed in FGF-1-stimulated cells; in this case, maximal levels of mRNA expression were evident at 8 and 12 h post-stimulation.

Although our interest was to isolate genes encoding proteins with particular structural motifs, it is clear from the DNA sequence information obtained to date that many of the cDNA clones represent proteins devoid of these domains. Since many of the oligonucleotide primers were degenerate, a relatively low PCR annealing temperature was used. Consequently, the primers were able to anneal to cDNA sequence coding regions of modest homology. This is how the Src homology 2/zinc finger primer combination amplified fatty acid synthetase cDNA and the Src homology 2/leucine zipper primer combination amplified sarco(endo)plasmic reticulum Ca^{2+}-ATPase cDNA.[26] In addition, some primers hybridized to partially homologous sequences within 5'- or 3'-untranslated regions. This is how the zinc finger/ protein tyrosine kinase primer combination amplified the aldose reductase-related FR-1 cDNA.[25] Nevertheless, a subset of the genes identified to date do appear to encode proteins containing the targeted structural domains.

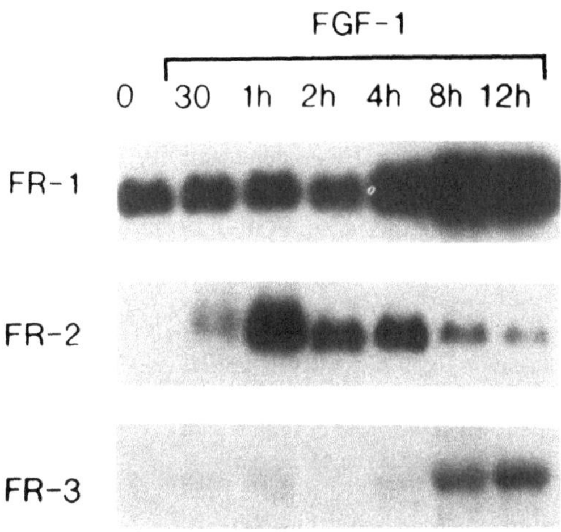

Figure 3. Effect of FGF-1 on FR-1, FR-2 and FR-3 mRNA levels in NIH 3T3 fibroblasts. Serum-starved cells were either left untreated or treated with FGF-1 and heparin for the indicated time periods. RNA was isolated and equivalent amounts of each sample analyzed by Northern blot hybridization as described.[25] The ^{32}P-labeled cDNA probes used are indicated on the left.

ARE THERE FGF-SPECIFIC CELLULAR RESPONSES?

In general, many immediate-early or delayed-early genes are activated to a similar extent when quiescent cells are stimulated with different growth-promoting agents. Most of the FGF-1-inducible genes we have examined to date are also expressed following FGF-2 or serum treatment of quiescent NIH 3T3 cells. Whole blood serum is a complex mixture containing plasma constituents as well as factors released from platelets during the process of coagulation. It contains numerous polypeptide mitogens, including PDGF, TGF-β, EGF and IGF-I but relatively low amounts of FGF-1[60] and FGF-2 (<10 pg/ml).[60,61] Interestingly, the FR-1 gene is differentially expressed in response to FGF or serum stimulation. This result is shown in Figure 4, where the temporal expression pattern of FR-1 mRNA is compared to that of FR-2 mRNA, a typical immediate-early transcript induced by serum as well as various purified polypeptide growth factors. It is evident from this experiment that serum treatment actually reduces the basal level of FR-1 mRNA expression detected in unstimulated cells.

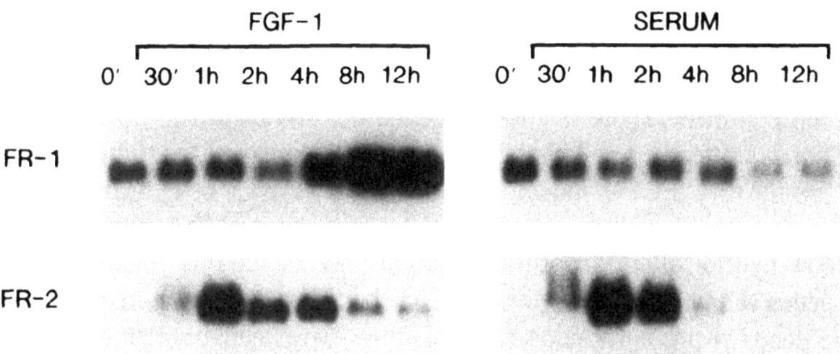

Figure 4. Effect of FGF-1 or serum treatment on FR-1 and FR-2 mRNA levels in NIH 3T3 fibroblasts. Serum-starved cells were either left untreated or treated with either 10 ng/ml FGF-1 and 5 U/ml heparin or 10% calf serum for the indicated time periods. RNA was isolated and equivalent amounts of each sample analyzed by Northern blot hybridization as described.[25] The ^{32}P-labeled cDNA probes used are indicated on the left.

An experiment was also performed to determine the effect of combined mitogenic stimulation with serum and FGF-1. As illustrated in Figure 5, FGF-1 could still elevate FR-1 mRNA levels even in the presence of 10% calf serum. This indicates that the positive effect on FR-1 gene activity by FGF-1 is dominant over the negative effect exerted by serum. Taken together, these results demonstrate that alternative growth promoters, in this case FGF-1 or serum, can induce distinct genomic responses in a common cell type.

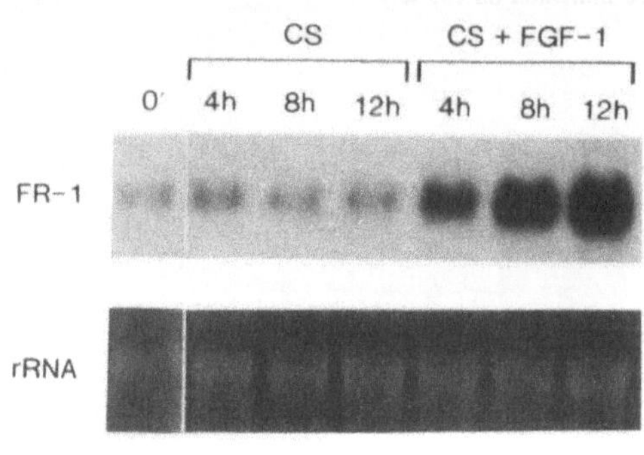

Figure 5. Effect of FGF-1 and serum treatment on FR-1 mRNA levels in NIH 3T3 fibroblasts. Serum-starved cells were either left untreated or treated with 10% calf serum (CS) in the absence or presence of 10 ng/ml FGF-1 and 5 U/ml heparin for the indicated time periods. RNA was isolated and equivalent amounts of each sample analyzed by Northern blot hybridization[25] using a ^{32}P-labeled FR-1 cDNA probe. The bottom panel is a photograph of the 28S rRNA band as visualized by ethidium bromide staining. This demonstrates equivalent amounts of RNA were present in each gel lane.

FGF-1 SIGNALING IN VASCULAR CELLS

It is likely that the molecular mechanism of FGF-1 mitogenic signal transduction is similar, but perhaps not identical, in different cell types. Thus, we reasoned that the NIH 3T3 cell system could be used to identify genes encoding proteins important for FGF-1 stimulation of vascular endothelial or smooth muscle cell growth. At the present time we do know that at least some of the genes we have identified are also activated in FGF-1-stimulated vascular cells. For example, FR-12 mRNA is expressed with similar kinetics in FGF-1-treated NIH 3T3 cells or rat aortic smooth muscle cells (Figure 6). Although it appears that the degree of induction is significantly less in the smooth muscle cells, this may reflect poor cross-hybridization between the murine cDNA clone and the rat transcripts. We conclude that at least some aspects of the FGF-1 mitogenic signaling program are shared between non-vascular and vascular cell types.

SUMMARY

A group of genes encoding proteins with diverse functions are transcriptionally activated following the addition of serum or purified polypeptide growth factors to quiescent cell cultures. There is good evidence that the expression of at least some of these proteins is critical for cell cycle progression. We have used a differential display technique to identify FGF-1-inducible genes in NIH 3T3 cells. In comparison to alternative approaches, such as differential screening of a cDNA library or subtracted cDNA probe hybridization, the differential display method is relatively simple and candidate genes can be quickly identified. The characterization of numerous novel FGF-1-inducible genes is presently underway. Future experiments will investigate whether the newly identified FGF-1-inducible proteins perform functions important for FGF-1-stimulated cell proliferation.

Figure 6. Effect of FGF-1 on FR-12 mRNA levels in NIH 3T3 fibroblasts or rat aortic smooth muscle cells. Serum-starved cells were either left untreated or treated with 10 ng/ml human recombinant FGF-1 and 5 U/ml heparin (fibroblasts) or 10 ng/ml bovine brain-derived FGF-1 (smooth muscle cells) for the indicated time periods. RNA was isolated and equivalent amounts of each sample analyzed by Northern blot hybridization[25] using a ^{32}P-labeled FR-12 cDNA probe. FR-12 mRNA expression in murine NIH 3T3 cells (panel A) or rat aortic smooth muscle cells (panel B) is shown. The upper and lower tick marks on the left side of each blot indicate the positions of 28S and 18S rRNA, respectively.

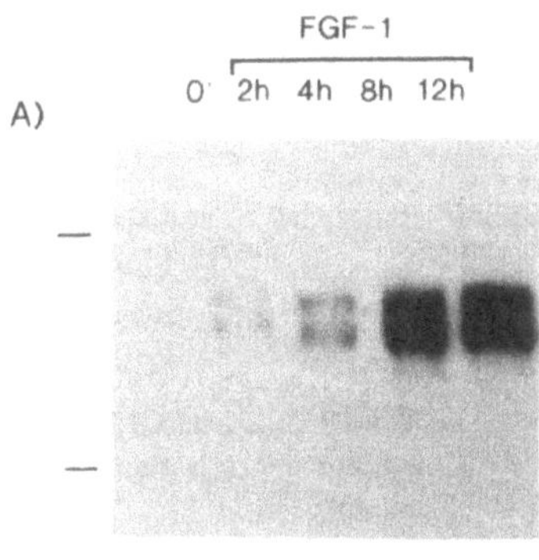

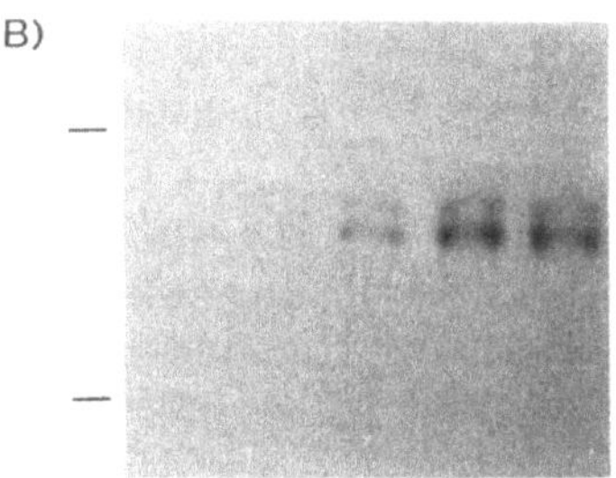

ACKNOWLEDGMENTS

This work was supported, in part, by National Institutes of Health grant HL39727 and a Grant-in-Aid from the American Heart Association (to J.W.). The authors are grateful to Dr. W. Burgess for providing the FGF-1 and also thank Ms. S. Appleby for DNA sequencing and Ms. C. Wawzinski for her help in the preparation of this manuscript.

REFERENCES

1. J. Folkman and Y. Shing, Angiogenesis, *J. Biol. Chem.* 267:10931 (1992).

2. R. Ross, The pathogenesis of atherosclerosis: A perspective for the 1990's, *Nature* 362:801 (1993).

3. A.W. Clowes, Intimal hyperplasia and graft failure, *Cardiovasc. Pathol.* 2:179S (1993).

4. J.N. Wilcox, Molecular biology: Insight into the causes and prevention of restenosis after arterial intervention, *Am. J. Cardiol.* 72:88E (1993).

5. L-Q. Pu, A.D. Sniderman, R. Brassard, K.J. Lachapelle, A.M. Graham, R. Lisbona, and J.F. Symes, Enhanced revascularization of the ischemic limb by angiogenic therapy, *Circulation* 88:208 (1993).

6. E.R. Edelman, M.A. Nugent, L.T. Smith, and M.J. Karnovsky, Basic fibroblast growth factor enhances the coupling of intimal hyperplasia and proliferation of vasa vasorum in injured rat arteries, *J. Clin. Invest.* 89:465 (1992).

7. S. Banai, M.T. Jaklitsch, W. Casscells, M. Shou, S. Shrivastav, R. Correa, S.E. Epstein, and E.F. Unger, Effects of acidic fibroblast growth factor on normal and ischemic myocardium, *Circ. Res.* 69:76 (1991).

8. V. Lindner, D.A. Lappi, A. Baird, R.A. Majack, and M.A. Reidy, Role of basic fibroblast growth factor in vascular lesion formation, *Circ. Res.* 68:106 (1991).

9. E. Brogi, J.A. Winkles, R. Underwood, S.K. Clinton, G.F. Alberts, and P. Libby, Distinct patterns of expression of fibroblast growth factors and their receptors in human atheroma and non-atherosclerotic arteries: Association of acidic FGF with plaque microvessels and macrophages, *J. Clin. Invest.* 92:2408 (1993).

10. S.E. Hughes, D. Crossman, and P.A. Hall, Expression of basic and acidic fibroblast growth factors and their receptor in normal and atherosclerotic human arteries, *Cardiovasc. Res.* 27:1214 (1993).

11. G. Brunner, H. Nguyen, J. Gabrilove, D.B. Rifkin, and E.L. Wilson, Basic fibroblast growth factor expression in human bone marrow and peripheral blood cells, *Blood* 81:631 (1993).

12. S. Blotnick, G.E. Peoples, M.R. Freeman, T.J. Eberlein, and M. Klagsbrun, T lymphocytes synthesize and export heparin-binding epidermal growth factor-like growth factor and basic fibroblast growth factor, mitogens for vascular cells and fibroblasts: Differential production and release by $CD4^+$ and $CD8^+$ T cells, *Proc. Natl. Acad. Sci. USA* 91:2890 (1994).

13. V. Lindner and M.A. Reidy, Proliferation of smooth muscle cells after vascular injury is inhibited by an antibody against basic fibroblast growth factor, *Proc. Natl. Acad. Sci. USA* 88:3739 (1991).

14. W.H. Burgess, A.M. Shaheen, M. Ravera, M. Jaye, P.J. Donohue, and J.A. Winkles, Possible dissociation of the heparin-binding and mitogenic activities of heparin-binding (acidic fibroblast) growth factor-1 from its receptor-binding activities by site-directed mutagenesis of a single lysine residue, *J. Cell Biol.* 111:2129 (1990).

15. W.H. Burgess, C.A. Dionne, J. Kaplow, R. Mudd, R. Friesel, A. Zilberstein, J. Schlessinger, and M. Jaye, Characterization and cDNA cloning of phospholipase C-gamma, a major substrate for heparin-binding growth factor 1 (acidic fibroblast growth factor)-activated tyrosine kinase, *Mol. Cell. Biol.* 10:4770 (1990).

16. J-K. Wang, G. Gao, and M. Goldfarb, Fibroblast growth factor receptors have different signaling and mitogenic potentials, *Mol. Cell. Biol.* 14:181 (1994).

17. D.K. Morrison, D.R. Kaplan, U. Rapp, and T.M. Roberts, Signal transduction from membrane to cytoplasm: Growth factors and membrane-bound oncogene products increase Raf-1 phosphorylation and associated protein kinase activity, *Proc. Natl. Acad. Sci. USA* 85:8855 (1988).

18. X. Zhan, X. Hu, B. Hampton, W.H. Burgess, R. Friesel, and T. Maciag, Murine cortactin is phosphorylated in response to fibroblast growth factor-1 on tyrosine residues late in the G_1 phase of the BALB/c 3T3 cell cycle, *J. Biol. Chem.* 268:24427 (1993).

19. K.G. Peters, J. Marie, E. Wilson, H.E. Ives, J. Escobedo, M. Del Rosario, D. Mirda, and L.T. Williams, Point mutation of an FGF receptor abolishes phosphatidylinositol turnover and Ca^{2+} flux but not mitogenesis, *Nature* 358:678 (1992).

20. M. Mohammadi, C.A. Dionne, W. Li, N. Li, T. Spivak, A.M. Honegger, M. Jaye, and J. Schlessinger, Point mutation in FGF receptor eliminates phosphatidylinositol hydrolysis without affecting mitogenesis, *Nature* 358:681 (1992).

21. G.T. Williams, A.S. Abler, and L.F. Lau, Regulation of gene expression by serum growth factors, *in*: "Molecular and Cellular Approaches to The Control of Proliferation and Differentiation," G.S. Stein and J.B. Lian, eds., Academic Press, Inc., New York (1992).

22. W.H. Burgess, A.M. Shaheen, B. Hampton, P.J. Donohue, and J.A. Winkles, Structure-function studies of heparin-binding (acidic fibroblast) growth factor-1 using site-directed mutagenesis, *J. Cell. Biochem.* 45:131 (1991).

23. Y. Furukawa, H. Piwnica-Worms, T.J. Ernst, Y. Kanakura, and J.D. Griffin, Cdc2 gene expression at the G_1 to S transition in human lymphocytes, *Science* 250:805 (1990).

24. L.D. Kerr, J.T. Holt, and L.M. Matrisian, Growth factors regulate transin gene expression by c-fos-dependent and c-fos-independent pathways, *Science* 242:1424 (1988).

25. P.J. Donohue, G.F. Alberts, B.S. Hampton, and J.A. Winkles, A delayed-early gene activated by fibroblast growth factor-1 encodes a protein related to aldose reductase, *J. Biol. Chem.* 269:8604 (1994).

26. D.K.W. Hsu, P.J. Donohue, G.F. Alberts, and J.A. Winkles, Fibroblast growth factor-1 induces phosphofructokinase, fatty acid synthase and Ca^{2+}-ATPase mRNA expression in NIH 3T3 cells, *Biochem. Biophys. Res. Commun.* 197:1483 (1993).

27. J.T. Holt, T. Venkat-Gopal, A.D. Moulton, and A.W. Nienhuis, Inducible production of c-fos antisense RNA inhibits 3T3 cell proliferation, *Proc. Natl. Acad. Sci. USA* 83:4794 (1986).

28. K. Nishikura and J.M. Murray, Antisense RNA of proto-oncogene c-fos blocks renewed growth of quiescent 3T3 cells, *Mol. Cell. Biol.* 7:639 (1987).

29. K. Kovary and R. Bravo, The jun and fos protein families are both required for cell cycle progression in fibroblasts, *Mol. Cell. Biol.* 11:4466 (1991).

30. M.A. Brach, H-J. Gruss, C. Sott, and F. Herrmann, The mitogenic response to tumor necrosis factor alpha requires c-jun/AP-1, *Mol. Cell. Biol.* 13:4284 (1993).

31. Y. Shi, H.G. Hutchinson, D.J. Hall, and A. Zalewski, Downregulation of c-myc expression by antisense oligonucleotides inhibits proliferation of human smooth muscle cells, *Circulation* 88:1190 (1993).

32. S. Biro, Y-M. Fu, Z-X. Yu, and S.E. Epstein, Inhibitory effects of antisense oligodeoxynucleotides targeting c-myc mRNA on smooth muscle cell proliferation and migration, *Proc. Natl. Acad. Sci. USA* 90:654 (1993).

33. M.R. Bennett, S. Anglin, J.R. McEwan, R. Jagoe, A.C. Newby, and G.I. Evan, Inhibition of vascular smooth muscle cell proliferation in vitro and in vivo by c-myc antisense oligodeoxynucleotides, *J. Clin. Invest.* 93:820 (1994).

34. M. Simons and R.D. Rosenberg, Antisense nonmuscle myosin heavy chain and c-myb oligonucleotides suppress smooth muscle cell proliferation in vitro, *Circ. Res.* 70:835 (1992).

35. L.S. Mulcahy, M.R. Smith, and D.W. Stacey, Requirement for ras proto-oncogene function during serum-stimulated growth of NIH 3T3 cells, *Nature* 313:241 (1985).

36. W.E. Mercer, D. Nelson, A.B. DeLeo, L.J. Old, and R. Baserga, Microinjection of monoclonal antibody to protein p53 inhibits serum-induced DNA synthesis in 3T3 cells, *Proc. Natl. Acad. Sci. USA* 79:6309 (1982).

37. D. Jaskulski, J.K. DeRiel, W.E. Mercer, B. Calabretta, and R. Baserga, Inhibition of cellular proliferation by antisense oligodeoxynucleotides to PCNA cyclin, *Science* 240:1544 (1988).

38. E. Speir and S.E. Epstein, Inhibition of smooth muscle cell proliferation by an antisense oligodeoxynucleotide targeting the messenger RNA encoding proliferating cell nuclear antigen, *Circulation* 86:538 (1992).

39. M. Simons, E.R. Edelman, J-L. DeKeyser, R. Langer, and R.D. Rosenberg, Antisense c-myb oligonucleotides inhibit intimal arterial smooth muscle cell accumulation in vivo, *Nature* 359:67 (1992).

40. J. Abe, W. Zhou, J. Taguchi, N. Takuwa, K. Miki, H. Okazaki, K. Kurokawa, M. Kumada, and Y. Takuwa, Suppression of neointimal smooth muscle cell accumulation in vivo by antisense cdc2 and cdk2 oligonucleotides in rat carotid artery, *Biochem. Biophys. Res. Commun.* 198:16 (1994).

41. L.F. Lau and D. Nathans, Identification of a set of genes expressed during the G_0/G_1 transition of cultured mouse cells, *EMBO J.* 4:3145 (1985).

42. J.M. Almendral, D. Sommer, H. MacDonald-Bravo, J. Burckhardt, J. Perera, and R. Bravo, Complexity of the early genetic response to growth factors in mouse fibroblasts, *Mol. Cell. Biol.* 8:2140 (1988).

43. E. Boeggeman, A.S. Masibay, P.K. Qasba, and T. Sreevalsan, Identification and partial characterization of genes that are transactivated by different pathways in quiescent mouse cells stimulated with serum, *J. Cell. Physiol.* 145:286 (1990).

44. T. Nikaido, D.W. Bradley, and A.B. Pardee, Molecular cloning of transcripts that accumulate during the late G_1 phase in cultured mouse cells, *Exp. Cell Res.* 192:102 (1991).

45. A. Lanahan, J.B. Williams, L.K. Sanders, and D. Nathans, Growth factor-induced delayed early response genes, *Mol. Cell. Biol.* 12:3919 (1992).

46. S. Vincent, L. Marty, L. LeGallic, P. Jeanteur, and P. Fort, Characterization of late response genes sequentially expressed during renewed growth of fibroblastic cells, *Oncogene* 8:1603 (1993).

47. S.V. Tavtigian, S.D. Zabludoff, and B.J. Wold, Cloning of mid-G_1 serum response genes and identification of a subset regulated by conditional myc expression, *Mol. Biol. Cell* 5:375 (1994).

48. B.H. Cochran, A.C. Reffel, and C.D. Stiles, Molecular cloning of gene sequences regulated by platelet-derived growth factor, *Cell* 33:939 (1983).

49. M. Gomperts, J.C. Pascall, and K.D. Brown, The nucleotide sequence of a cDNA encoding an EGF-inducible gene indicates the existence of a new family of mitogen-induced genes, *Oncogene* 5:1081 (1990).

50. P. Zumstein and C.D. Stiles, Molecular cloning of gene sequences that are regulated by insulin-like growth factor I, *J. Biol. Chem.* 262:11252 (1987).

51. J.A. Fernandez-Pol, D.J. Klos, and P.D. Hamilton, A growth factor-inducible gene encodes a novel nuclear protein with zinc finger structure, *J. Biol. Chem.* 268:21198 (1993).

52. T.H. Lee, G.W. Lee, E.B. Ziff, and J. Vilcek, Isolation and characterization of eight tumor necrosis factor-induced gene sequences from human fibroblasts, *Mol. Cell. Biol.* 10:1982 (1990).

53. V.V. Rangnekar, S. Waheed, T.J. Davies, F.G. Toback, and V.M. Rangnekar, Antimitogenic and mitogenic actions of interleukin-1 in diverse cell types are associated with induction of gro gene expression, *J. Biol. Chem.* 266:2415 (1991).

54. D.E. Sabath, P.L. Podolin, P.G. Comber, and M.B. Prystowsky, cDNA cloning and characterization of interleukin 2-induced genes in a cloned T helper lymphocyte, *J. Biol. Chem.* 265:12671 (1990).

55. C. Beadling, K.W. Johnson, and K.A. Smith, Isolation of interleukin-2-induced immediate-early genes, *Proc. Natl. Acad. Sci. USA* 90:2719 (1993).

56. H. Matsushime, M.F. Roussel, R.A. Ashmun, and C.J. Sherr, Colony-stimulating factor 1 regulates novel cyclins during the G_1 phase of the cell cycle, *Cell* 65:701 (1991).

57. P. Liang and A.B. Pardee, Differential display of eukaryotic messenger RNA by means of the polymerase chain reaction, *Science* 257:967 (1992).

58. J. Welsh, K. Chada, S.S. Dalal, R. Cheng, D. Ralph, and M. McClelland, Arbitrarily primed PCR fingerprinting of RNA, *Nucleic Acids Res.* 20:4965 (1992).

59. D. Bauer, H. Muller, J. Reich, H. Riedel, V. Ahrenkiel, P. Warthoe, and M. Strauss, Identification of differentially expressed mRNA species by an improved display technique (DDRT-PCR), *Nucleic Acids Res.* 21:4272 (1993).

60. Y. Yoshitake and K. Nishikawa, Distribution of fibroblast growth factors in cultured tumor cells and their transplants, *In Vitro Cell. Dev. Biol.* 28A:419 (1992).

61. M. Ii, H. Yoshida, Y. Aramaki, H. Masuya, T. Hada, M. Terada, M. Hatanaka, and Y. Ichimori, Improved enzyme immunoassay for human basic fibroblast growth factor using a new enhanced chemiluminescence system, *Biochem. Biophys. Res. Commun.* 193:540 (1993).

ANTIOXIDANTS AND ENDOTHELIAL EXPRESSION OF VCAM-1: A MOLECULAR PARADIGM FOR ATHEROSCLEROSIS

Russell M. Medford

Division of Cardiology
Department of Medicine
Emory University School of Medicine
Atlanta, Georgia 30322

INTRODUCTION

Atherosclerosis is a complex, multisystem disease whose underlying molecular and cellular mechanisms are poorly understood. Recent insights into the etiology and pathogenesis of this disease suggest that atherosclerosis may be viewed as an inflammatory disease linked to an abnormality in oxidation-mediated signals in the vasculature. The purpose of this brief review is to extend this notion to the molecular level by viewing oxidation-mediated signals as physiological regulators of vascular gene expression that function through specific oxidation-reduction (redox)-sensitive signal transduction pathways and transcriptional regulatory networks. Redox-sensitive regulation of vascular gene expression, especially of genes involved in vascular inflammatory and growth responses, represents an intriguing paradigm for understanding atherogenesis as well as for the development of novel therapeutic treatment regimens, drug design and diagnostic assessment of disease state.

INFLAMMATION AND ATHEROSCLEROSIS

Atherosclerosis is a chronic disease of the arterial intima characterized by the focal accumulation of leukocytes, smooth muscle cells, lipids and extracellular matrix [1]. A central feature, and one of the earliest detectable events in the pathogenesis of the atherosclerotic plaque, is the adherence of mononuclear leukocytes to discrete segments of the arterial endothelium, followed by transformation into lipid-laden macrophages, or "foam cells." In this inflammatory response, leukocyte recruitment to the endothelium is mediated by the interaction of adhesion molecule receptors expressed on the surface of endothelial cells with counter-receptors expressed on immune cells. Endothelial cells play a major role in defining the types of leukocytes recruited (e.g., monocytes, lymphocytes or neutrophils) by selectively expressing specific adhesion molecules (e.g., vascular cell adhesion molecule-1 (VCAM-1), intracellular cell adhesion molecule-1 (ICAM-1) or E-selectin) in response to various inflammatory stimuli [2, 3]. In the earliest atherogenic

Cardiovascular Disease 2
Edited by L.L. Gallo, Plenum Press, New York, 1995

lesions, this involves the localized endothelial expression of VCAM-1 and selective recruitment of mononuclear leukocytes [4].

VCAM-1, a member of the immunoglobulin gene superfamily, mediates leukocyte binding to the endothelial cell through its interaction with its integrin counter-receptor, very late activation antigen-4 (VLA-4) [5]. Due to the selective expression of VLA-4 on monocytes and lymphocytes, but not neutrophils, VCAM-1 plays an important role in mediating mononuclear leukocyte selective adhesion [5-10]. In addition to endothelial-leukocyte interactions, adherence to the endothelium by certain metastatic tumors may be mediated by VCAM-1 [11]. Additional, perhaps novel, roles for VCAM-1 in the immune response as well as in tissue development and differentiation are suggested by its expression in stromal cells of the bone marrow [12] and macrophages [13]. Recently, VCAM-1 and its counter-receptor, VLA-4, were shown to be expressed on both skeletal myoblasts and myotubes in culture and at sites of secondary myogenesis in vivo [14]. Intriguingly, this study suggested a functional role for VCAM-1/VLA-4 in myogenesis by blocking myotube formation in vitro with antibodies to VLA-4 or VCAM-1.

OXIDATIVE MODIFICATION OF LDL HYPOTHESIS

To explain these earliest events in the pathogenesis of atherosclerosis, an oxidative modification hypothesis was proposed [15]. This hypothesis focuses on the modification of low density lipoprotein (LDL) by reactive oxygen species into oxidatively-modified LDL (ox-LDL) as the central initiating and propagating event in atherosclerosis [16]. Monocytes avidly take up ox-LDL through a "scavenger" receptor and are thus transformed into lipid-engorged macrophage-foam cells that are the cellular components of the fatty streak. Ox-LDL may also subsequently participate in other components of atherogenesis such as additional monocyte recruitment, smooth muscle proliferation and vascular injury. Although this "oxidative stress" hypothesis is appealing and serves as a scientific rationale for therapeutic trials of antioxidants in experimental and human atherosclerosis, it should be noted that the regulatory mechanisms by which "oxidation" initiates and propagates the atherogenic lesion are unknown and may involve factors and cellular components in addition to, or other than, LDL [17, 18].

OXIDATIVE SIGNALS AND ENDOTHELIAL EXPRESSION OF VCAM-1

Oxidative stress may play an important role in regulating VCAM-1 gene expression in the vasculature and thus we hypothesize that it may serve as an integrative link between the "oxidative" signals predisposing the vessel wall to atherosclerosis and the early mononuclear leukocyte immune responses characteristic of this earliest stage of the disease. In the early atherogenic lesion, oxidative stress is manifested by the elevated production of reactive oxygen species by endothelial and smooth muscle cells that results in the oxidative modification of LDL [19, 20]. This is likely due in part to both paracrine and autocrine mechanisms by which the cytokines, interleukin-1β (IL-1β) and tumor necrosis factor α (TNFα), derived from both inflammatory and endothelial cells, induce the cellular synthesis of reactive oxygen species [21, 22]. It is in this oxidative milieu that endothelial expression of VCAM-1, but not E-selectin or ICAM-1, and consequent monocyte accumulation are observed in the early atherogenic lesion [4]. These observations suggest that the activation of VCAM-1 gene expression might be distinguished from that of ICAM-1 or E-selectin by its sensitivity to modulation by oxidation-mediated signals.

REDOX-SENSITIVE TRANSCRIPTIONAL REGULATION THROUGH NF-κB

Recent studies of the human VCAM-1 promoter demonstrate that TNFα activation of VCAM-1 transcription in human umbilical vein endothelial (HUVE) cells is dependent on the activation of an NF-κB-like transcriptional regulatory protein [23]. NF-κB is a ubiquitously expressed multisubunit transcription factor whose activation in several cell types by a large and diverse group of agents such as TNFα, IL-1β, bacterial endotoxin lipopolysaccharide (LPS), double-stranded RNA, poly(I:C) (PIC) as well as the oxidant H_2O_2 can be specifically inhibited by antioxidants such as N-acetylcysteine (NAC) and pyrrolidine dithiocarbamate (PDTC) [24-26]. This has led to the hypothesis that oxygen radicals play an important role in the activation of NF-κB through an as-yet undefined reduction-oxidation (redox) mechanism [26]. By extrapolation, oxidative stress in the atherosclerotic lesion may play a role in regulating VCAM-1 gene expression through an NF-κB-like redox-sensitive transcriptional regulatory protein.

VCAM-1 IS A MOLECULAR MARKER OF REDOX-SENSITIVE ENDOTHELIAL CELL GENE EXPRESSION

We have established that the activation of VCAM-1 gene expression in vascular endothelial cells is indeed regulated by a common oxidation-sensitive mechanism [27]. In cultured HUVE cells, the cytokine IL-1β activated VCAM-1 gene expression through a mechanism that is repressed ~90% by the antioxidants PDTC and NAC. Furthermore, PDTC selectively inhibited the induction of VCAM-1, but not ICAM-1, mRNA and protein accumulation by the cytokine TNFα as well as the non-cytokines, LPS and PIC. PDTC also markedly attenuated TNFα induction of VCAM-1-mediated cellular adhesion. PDTC partially inhibited E-selectin gene expression in response to TNFα but not to LPS, IL-1β or PIC. TNFα- and LPS-mediated transcriptional activation of the human VCAM-1 promoter through NF-κB-like DNA enhancer elements and associated NF-κB-like DNA binding proteins was inhibited by PDTC.

In the above studies, we have established that the activation of endothelial cell VCAM-1 gene expression is regulated by a signal transduction pathway that contains a transcriptional NF-κB-like regulatory element sensitive to inhibition by antioxidants. Furthermore, this factor is necessary but not sufficient to activate VCAM-1 gene expression in HUVE cells in response to diverse inducing stimuli. As shown in Figure 1, this redox-sensitive signal transduction pathway does not appear to regulate ICAM-1 and E-selectin gene expression. Thus, we have identified at least two distinct regulatory pathways activated by inflammatory cytokines and a new mechanism by which adhesion molecule genes may be selectively regulated in response to otherwise non-discriminating activating signals. Using the antioxidant PDTC as a molecular probe, we have characterized this new VCAM-1-specific molecular regulatory pathway from the functional expression of cell adhesion to a specific DNA binding protein complex.

These studies strongly support the central tenet of our hypothesis linking oxidative signals to immune responses in the vasculature. Importantly, this places into context our contention that the molecular regulation of VCAM-1 gene expression serves as a key paradigm for an integration of oxidative stress signals and the early immune responses of atherosclerosis. VCAM-1 gene expression in endothelial cells may thus be used as a comparative mechanism to elucidate the physiological role of oxidative signals regulating gene expression in other vascular cell types, such as vascular smooth muscle, exposed to similar pathogenetic signals of atherosclerosis. VCAM-1 expression itself may serve as a molecular marker for oxidation-mediated signal transduction in the vasculature.

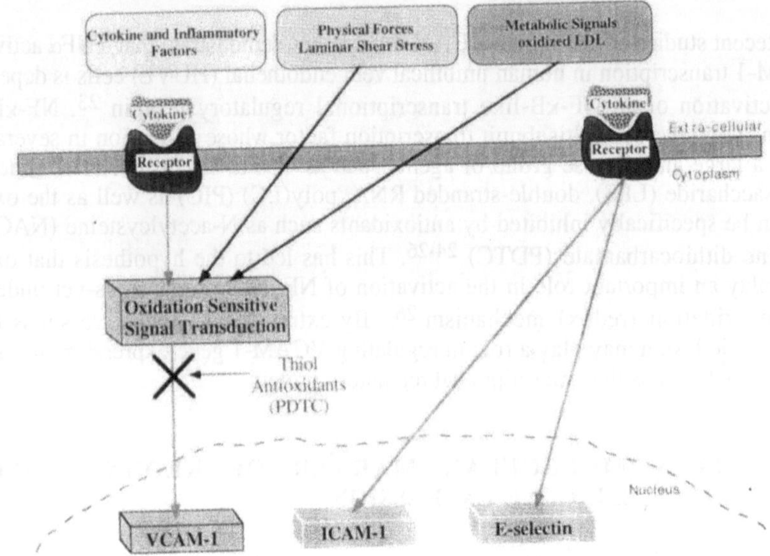

Figure 1. Schema for redox-sensitive differential regulation of VCAM-1 gene expression.

EVIDENCE FOR A COMMON REDOX-SENSITIVE SIGNAL AS THE PHYSIOLOGICAL REGULATOR OF VCAM-1 GENE EXPRESSION

As shown in Figure 1, an important predictor of the redox model of VCAM-1 expression is potentially a common signal pathway utilized by biomechanical forces as well as metabolic factors that influence the pathogenesis of atherosclerosis. The following preliminary studies performed in our laboratory support this prediction.

Regarding biomechanical forces, atherosclerosis exhibits a predilection for focal sites on the vessel wall. Due to vessel geometry, laminar shear stress, an important component of the mechanical force generated by laminar blood flow, is absent at these sites. We established that chronic exposure to laminar shear stress inhibits cytokine-induced, redox-sensitive VCAM-1 gene expression, but not that of ICAM-1 or E-selectin, in a manner similar to that observed for the thiol antioxidant, PDTC [28]. These results were strikingly similar to PDTC and suggest that through a redox-sensitive signal transduction mechanism, biomechanical forces such as laminar shear stress may control the regional predisposition of the vessel wall to inflammatory disease processes such as atherosclerosis.

To establish that pro-oxidant signals could selectively augment VCAM-1 gene expression, we tested whether the naturally occurring pro-oxidant, ox-LDL, augments activation of VCAM-1 expression by inflammatory signals [29]. Dose-response and time-course studies demonstrated that pre-incubation of HUVE cells with ox-LDL enhanced VCAM-1 expression by the cytokine TNFα (100U/ml for 6 hours) by 45% ($p<.05$) compared with either unmodified LDL or no pretreatment. Ox-LDL had no effect on the induction of either ICAM-1 or E-selectin. Essentially similar results were obtained with linoleyl hydroperoxide (13-HPODE, 7.5-30 μM) and lysophosphatidylcholine (lyso-PC, 50 μM), which are significant components of ox-LDL. These results suggest that as long-term regulatory signals, specific phospholipid and oxidized free fatty acid components of ox-LDL augment the ability of vascular endothelial cells to express VCAM-1 in response to a cytokine-mediated inflammatory stimulus. These studies link oxidant signals conferred by ox-LDL to oxidation-sensitive regulatory mechanisms that control the expression of vascular endothelial cell adhesion molecules characteristic of the early inflammatory events in atherosclerosis.

CONCLUSION

VCAM-1 is a member of a class of genes expressed in atherosclerosis whose regulation is characterized by a functional linkage between redox-sensitive modulatory signals and the nuclear regulatory apparatus. A similar mechanism may mediate the molecular regulatory response of endothelial cells to biomechanical forces in the vasculature. PDTC antioxidant inhibitable VCAM-1 gene expression was also used as the experimental paradigm to identify a specific class of free fatty acids that may function as the intracellular oxidant "second messenger" that regulates redox-sensitive gene expression. This would link the metabolic abnormalities of atherosclerosis, such as hypercholesterolemia and ox-LDL, directly with immune-mediated processes through a specific molecular regulatory mechanism controlling redox-sensitive VCAM-1 gene expression. These studies suggest that the therapeutically dominant features of antioxidants in atherosclerosis may be due to alterations in the molecular regulation of gene expression of endothelial, smooth muscle and inflammatory cells. Redox-sensitive regulation of vascular gene expression represents an intriguing paradigm for understanding atherogenesis as well as for the development of novel therapeutic treatment regimens, drug design and diagnostic assessment of disease state.

ACKNOWLEDGMENTS

This research was supported in part by grant HL 48667 from the National Institutes of Health, and a Grant-in-Aid from the American Heart Association (Georgia Affiliate). RMM is an Established Investigator of the American Heart Association.

REFERENCES

1. H. Li, M. Cybulsky, M. Gimbrone, and P. Libby, An atherogenic diet rapidly induces VCAM-1, a cytokine-regulatable mononuclear leukocyte adhesion molecule, in rabbit aortic endothelium, *Arterio. Thromb.* 13:197 (1993).

2. E. Butcher, Leukocyte-endothelial cell recognitioin: Three (or more) steps to specificity and diversity., *Cell.* 67:1033 (1992).

3. J.S. Pober, and R.S. Cotran, What can be learned from the expression of endothelial adhesion molecules in tissues?, *Lab. Invest.* 64:301 (1991).

4. M.I. Cybulsky, and M.A. Gimbrone Jr., Endothelial expression of a mononuclear leukocyte adhesion molecule during atherogenesis, *Science.* 251:788 (1991).

5. M.J. Elices, L. Osborn, Y. Takada, C. Crouse, S. Luhowskyj, M.E. Hemler, and R.R. Lobb, VCAM-1 on activated endothelium interacts with the leukocyte integrin VLA-4 at a site distinct from the VLA-4/fibronectin binding site, *Cell.* 60:577 (1990).

6. B.S. Bochner, F.W. Luscinskas, M.A. Gimbrone Jr., W. Newman, S.A. Sterbinsky, A.C.P. Derse, D. Klunk, and R.P. Schleimer, Adhesion of human basophils, eosinophils, and neutrophils to interleukin 1-activated human vascular endothelial cells: contributions of endothelial cell adhesion molecules, *J. Exp. Med.* 173:1553 (1991).

7. T. Carlos, N. Kovach, B. Schwartz, M. Rosa, B. Newman, E. Wayner, C. Benjamin, L. Osborn, R. Lobb, and J. Harlan, Human monocytes bind to two cytokine-induced adhesive

ligands on cultured human endothelial cells: Endothelial-leukocyte adhesion molecule-1 and vascular cell adhesion molecule-1, *Blood.* 77:2266 (1991).

8. M.N. Oppenheimer, L.S. Davis, D.T. Bogue, J. Ramberg, and P.E. Lipsky, Differential utilization of ICAM-1 and VCAM-1 during the adhesion and transendothelial migration of human T lymphocytes, *J. Immunol.* 147:2913 (1991).

9. A.C. van Dinther Janssen, E. Horst, G. Koopman, W. Newmann, R.J. Scheper, C.J. Meijer, and S.T. Pals, The VLA-4/VCAM-1 pathway is involved in lymphocyte adhesion to endothelium in rheumatoid synovium, *J. Immunol.* 147:4207 (1991).

10. L.C. Burkly, A. Jakubowski, B.M. Newman, M.D. Rosa, R.G. Chi, and R.R. Lobb, Signaling by vascular cell adhesion molecule-1 (VCAM-1) through VLA-4 promotes CD3-dependent T cell proliferation, *Eur. J. Immunol.* 21:2871 (1991).

11. P.I. Martin, R. Mortarini, D. Lauri, S. Bernasconi, M.F. Sanchez, G. Parmiani, A. Mantovani, A. Anichini, and E. Dejana, Heterogeneity in human melanoma cell adhesion to cytokine activated endothelial cells correlates with VLA-4 expression, *Cancer Res.* 51:2239 (1991).

12. K. Miyake, K. Medina, K. Ishihara, M. Kimoto, R. Auerbach, and P.W. Kincade, A VCAM-like adhesion molecule on murine bone marrow stromal cells mediates binding of lymphocyte precursors in culture, *J Cell Biol.* 114:557 (1991).

13. A.E. Koch, J.C. Burrows, G.K. Haines, T.M. Carlos, J.M. Harlan, and S.J. Leibovich, Immunolocalization of endothelial and leukocyte adhesion molecules in human rheumatoid and osteoarthritic synovial tissues, *Lab Invest.* 64:313 (1991).

14. G.D. Rosen, J.R. Sanes, R. LaChance, J.M. Cunningham, J. Roman, and D.C. Dean, Roles for the integrin VLA-4 and its counter receptor VCAM-1 in myogenesis, *Cell.* 69:1107 (1992).

15. D. Steinberg, S. Parthasarathy, T.E. Carew, J.C. Khoo, and J.L. Witztum, Beyond cholesterol: modifications of low-density lipoprotein that increase its atherogenicity., *N. Engl. J. Med.* 320:915 (1989).

16. S. Parthasarathy, S.G. Young, J.L. Witztum, R.C. Pittman, and D. Steinberg, Probucol inhibits oxidative modification of low density lipoprotein, *J. Clin. Invest.* 77:641 (1986).

17. D. Steinberg, Antioxidants in the prevention of human atherosclerosis. Summary of the proceedings of a National Heart, Lung, and Blood Institute Workshop: September 5-6, 1991, Bethesda, Maryland., *Circulation.* 85:2337 (1992).

18. D. Steinberg, Antioxidants and atherosclerosis. A current assessment [editorial], *Circulation.* 84:1420 (1991).

19. C.P. Sparrow, and J. Olszewski, Cellular oxidative modification of low density lipoprotein does not require lipoxygenases, *Proc. Natl. Acad. Sci. U S A.* 89:128 (1992).

20. D.W. Morel, P.E. DiCorleto, and G.M. Chisolm, Endothelial and smooth muscle cells alter low density lipoprotein in vitro by free radical oxidation, *Arteriosclerosis.* 4:357 (1984).

21. B. Meier, H.H. Radeke, S. Selle, M. Younes, H. Sies, K. Resch, and G.G. Habermehl, Human fibroblasts release reactive oxygen species in response to interleukin-1 or tumor necrosis factor-α, *Biochem. J.* 263:539 (1989).

22. B.B. Warner, M.S. Burhans, J.C. Clark, and J.R. Wispe, Tumor necrosis factor-alpha increases Mn-SOD expression: Protection against oxidant injury, *Am. J. Physiol.* 260:L296 (1991).

23. M.F. Iademarco, J.J. McQuillan, G.D. Rosen, and D.C. Dean, Characterization of the promoter for vascular cell adhesion molecule-1 (VCAM-1), *J. Biol. Chem.* 267:16323 (1992).

24. R. Schreck, P. Rieber, and P.A. Baeuerle, Reactive oxygen intermediates as apparently widely used messengers in the activation of the NF-kappa B transcription factor and HIV-1, *EMBO J.* 10:2247 (1991).

25. R. Schreck, B. Meier, D.N. Mannel, W. Droge, and P.A. Baeurle, Dithiocarbamates as potent inhibitors of nuclear factor κB activation in intact cells, *J. Exp. Med.* 175:1181 (1992).

26. R. Schreck, and P. Baeuerle, A role for oxygen radicals as second messengers, *Trends Cell Biol.* 1:39 (1991).

27. N. Marui, M.K. Offermann, R. Swerlick, C. Kunsch, C.A. Rosen, M. Ahmad, R.W. Alexander, and R.M. Medford, VCAM-1 gene transcription and expression is regulated through an antioxidant sensitive mechanism in human vascular endothelial cells, *J. Clin. Invest.* 92:1866 (1993).

28. R. Medford, M. Offermann, and F. Bennett, Inhibition of TNFa induced vascular cell adhesion molecule-1 gene expression in human vascular endothelial and smooth muscle cells using transcriptional factor decoys, *Circulation.* 88:I-77 (1993).

29. B. Khan, S. Parthasarathy, R. Alexander, and R. Medford, Fatty acid hydroperoxides as selective regulators of cytokine activated adhesion molecule expression in human aortic endothelial cells, *Clin. Res.* 42:177 (1994).

MECHANISMS OF POTENTIATION OF PDGF IN ATHEROSCLEROSIS

Xin-Hua Lin[1], Zhao-Yi Wang[1], Hyeong Reh Kim[2], Thomas F. Deuel[1]

[1]Jewish Hospital at Washington University
Department of Medicine
216 So. Kingshighway Blvd.
St. Louis, MO 63110
[2]Wayne State University
Department of Pathology
540 E. Canfield Ave.
Detroit, MI 48201

Among the many processes that contribute to the development of atherosclerosis, the migration and proliferation of smooth muscle cells and fibroblasts and the migration and activation of macrophages are among the most important[1,2]. The responses of normal cells that lead to atherosclerosis are the consequences of directed signals that are initiated by cytokines that are upregulated at sites of developing lesions and by circulating plasma factors, such as oxidized low density lipoproteins (LDLs). Increasingly, evidence suggests that the initial insult may result from the interaction of partially oxidized LDL and scavenger receptors, that, in turn, upregulate a number of cytokine genes whose products function as potent chemoattractants and mitogens for different target cells within the arterial wall[1,3-9]. The PDGF A-chain gene is among the genes that are regularly upregulated in developing atheroma[2]. Its protein product, the homodimeric PDGF A-chain, is a chemoattractant for monocytes, fibroblasts, and smooth muscle cells and a mitogen for fibroblasts and smooth muscle cells[1,10]. In addition, PDGF A upregulates its own transcription (and thus is autoregulatory[11]) and transcription of a number of other potent cytokine genes.

In this manuscript, we describe three lines of investigation relevant to the role of PDGF in the pathogenesis of atherosclerosis. The aim of these investigations is to understand mechanisms by which PDGF activates cells to generate a proliferative response and to understand the mechanisms by which the PDGF gene(s) is upregulated during the pathological process of atherosclerosis. The experiments thus are directed to understanding the bases of the process of atherogenesis, and, with the increase in understanding that may result from this work, it may be possible to provide rational approaches for therapy.

The autocrine hypothesis suggests that upregulation of expression and secretion of growth factors and the subsequent interaction of a secreted growth factor with its receptor establishes the mechanism for the chronic activation of the growth factor receptor, leading to uncontrolled proliferation[12]. However, the reversible upregulation of growth factors that occurs at sites of inflammation and atheroma establishes locally an equally important opportunity to stimulate to cell growth that is directly analogous to transformation. The autocrine mechanism was first shown to be valid in NIH 3T3 cells that overexpress the v-sis oncogene, the viral counterpart of the B-chain of PDGF[13]. It was shown that the secreted growth factor stimulated the autocrine growth of these cells because culture in the presence of anti-PDGF antisera reduced DNA synthesis and thus the growth of these cells, supporting an autocrine growth hypothesis in v-sis (PDGF B). Recently we have shown that the addition of PDGF B to growing NIH 3T3 cells induces transformation of a subpopulation of NIH 3T3 cells. Transformation is reversed when PDGF B is withdrawn. A genetically stable

Cardiovascular Disease 2
Edited by L.L. Gallo, Plenum Press, New York, 1995

program may thus complement PDGF B within the subpopulation of NIH 3T3 cells to induce reversible transformation, whereas other non-transformable populations of NIH 3T3 cells lack this genetic program. Thus, PDGF B treatment alone in the absence of complementation fails to transform[14].

In other experiments, we sought to identify an altered pathway that complemented PDGF B. We demonstrated that chronic stimulation of growing cells with both PDGF A and B complements the anti-apoptotic bcl-2 gene to induce transformation. Control cells that lack overexpression of bcl-2 fail to be transformed by PDGF B or by PDGF A. These results suggest an important model for why PDGF A or PDGF B, when overexpressed within the blood vessel wall, may be important in the proliferative response of atherosclerosis. Thus either growth factor has the potential to complement other gene products that are also upregulated in response to the inflammatory stimuli, thereby amplifying the proliferative response locally to that of more aggressive cell growth. Upregulation of PDGF A and PDGF B therefore may result in the reversible but aggressive growth of the secreting or neighboring cells. Importantly also, by down-regulating the expression of PDGF A, the aggressive growth characteristics of the cells may also be reversed. This result suggests an important and as yet unappreciated mechanism whereby PDGF A or B may contribute directly to abnormal proliferative response of atherosclerotic lesions and contribute to its progression.

A second mechanism that governs the contribution of PDGF to atherosclerosis is its transcriptional activation at sites of lesions. In order to understand mechanisms by which the transcription is regulated, we isolated the PDGF A-chain gene and characterized the regulatory regions within its promoter[15]. Two sites have been identified that appear to be important in the upregulation of PDGF A[11]. The first site is found -468 relative to the site of initiation of transcription of the PDGF A-chain gene. This site contains a typical serum response element (SRE) with the DNA sequence CC(A/T) GG[16]. This sequence in the PDGF A-chain gene is sensitive to S1 in supercoiled plasmids and thus has single stranded character[17]. The SRE binds the serum response factor in a sequence specific manner and, in conjunction with its phosphorylation that is stimulated by serum and by PDGF, thereby mediates the upregulation of PDGF A. The SRE of the PDGF A-chain gene functions to upregulate PDGF A-chain promoter activity in response to serum and to PDGF A; deletion of the PDGF A-chain SRE sequences results in loss of inducibility by serum and PDGF A. When the SRE is reintroduced into its site within the PDGF A-chain promoter or into the TK (HSV thymidine kinase) gene, the PDGF A-chain SRE functions equally as the c-fos SRE in response to PDGF. Mutations within the PDGF A-chain SRE abolish the upregulation of the PDGF A-chain gene in response to serum or to PDGF itself, thus establishing specificity to the response of the SRE and establishing the potential of the SRE to mediate the upregulation of PDGF A through autoregulation by PDGF A itself and perhaps other cytokines as well[18]. This result is important because it establishes one mechanism to increase the levels of expression of PDGF A within the blood vessel wall. Based on our knowledge of cytokines expressed in the blood vessel wall, this mechanism is likely to be the important site of regulation of PDGF A.

The second regulatory element that we have analyzed within the A-chain promoter is a GC rich region within the proximal PDGF A-chain gene that contains three contiguous binding sites for the transcription factor SP1. We observed that this region within the PDGF A-chain gene also binds to the Wilms' tumor gene product WT1[19]. Plasmids that express high levels of WT1 were then analyzed in transient co-transfection assays with the PDGF A-chain promoter-driven chloramphenicol acyl transferase (CAT) reporter plasmid repress the PDGF A-chain promoter activity by 10-fold, suggesting that WT1 functions as a repressor of transcription of the PDGF A-chain gene[19]. However, when the SP1 recognition sites were deleted from the A-chain promoter, a sharp activation of promoter activity was observed when it was co-expressed with WT1[20]. This result led to the identification of a novel WT1 binding sequence ("TCC" motif) 3' to the site of initiation of transcription of the PDGF A-chain gene[21] and that WT1 functions either to activate or to repress PDGF A-chain gene promoter activity, depending upon the number and the position of binding sites. When two sites are present within the promoter region of the PDGF A-chain gene, one 3' and the other 5' to the site of initiation of transcription, WT1 represses promoter function. When either site is present alone, either in the 5' or in the 3' position, WT1 functions as a transcriptional activator. In order to seek reagents that ultimately might be of use in regulating expression of the PDGF A-chain gene, we identified and fine mapped the activation and repression domains within the WT1 protein itself. These domains function independently as long as the DNA binding domain is retained within the WT1 protein itself. Thus, WT1 has the potential to function as a therapeutic approach to control the upregulation of the PDGF A-chain gene at the transcriptional level[21] if issues of its regulatory DNA sequences.

In a second series of experiments directed to understanding how the PDGF A-chain gene is regulated, we have shown that methylation of the PDGF A-chain gene promoter region in vitro blocks transcriptional activity of the PDGF A-chain gene promoter tested in vivo[22]. The repression of promoter function is methylation sequence specific and cell type specific, requires only a single site for suppression of promoter activity, and functions more effectively when in proximity to the TATA box. We also demonstrated methyl DNA binding protein-2 (MeCP-2)[23] can suppress the methylated A-chain gene and activate the unmethylated gene (Lin, et al., unpublished). Finally, by use the methylation inhibitor 5-azacytidine, we established that the endogenous PDGF A-chain gene is methylated in HeLa cells but not into S-44 cells, cells that express low and high levels of the PDGF A-chain gene respectively (Lin, et al., unpublished). These results have established that methylation may be another mechanism to maintain the low level of expression of the PDGF A-chain gene in developing atheromata (Lin, et al., unpublished).

PDGF A not only regulates its own level of transcription but it also induces a number of other cytokine genes, including a recently described growth promoting polypeptide pleiotrophin (PTN)[24]. Pleiotrophin is a chemoattractant for fibroblasts and a weak mitogen for NIH 3T3 cells. It induces process outgrowth from glial progenitor cells and from neurons in primary cultures, suggesting a role in differentiation. Because PTN is upregulated by PDGF, and thus is likely to be upregulated at sites of atherosclerosis, PTN also may be important in the development of atherosclerosis. Its in vitro properties suggest that it may function similarly in vivo. The studies that identified and characterized PTN thus also suggest a site at which the PDGF A-chain gene may be regulated at the level of transcription, and thus sites for new therapeutics.

The studies discussed in this manuscript were performed in order to further understand the basic factors that contribute to atherosclerosis. PDGF is an important cytokine that has been shown regularly to be upregulated at sites of developing atheroma. It is thus implicated because of "guilt by association". It also is implicated because the properties of PDGF described by in vitro experiments are those anticipated for a signaling molecule that is involved in directing cell migration and cell proliferation. Our studies have attempted to further understand how PDGF amplifies the proliferative response, how it is upregulated to the high levels of expression that are regularly seen in developing lesions, and what genes it regulates in order to broaden the range of responses in PDGF stimulated cells. Importantly, these results have broadened the range and added to the specificity of therapeutic approaches to modulate atherogenesis.

REFERENCES

1. Deuel, T.F., Polypeptide Growth Factors: Roles in Normal and Abnormal Cell Growth, *Annu. Rev. Cell Biol.* 3:443-492 (1987).
2. Ross, R., The pathogenesis of atherosclerosis: a perspective for the 1990s. [Review]. *Nature.* 362(6423):801-9 (1993) .
3. Kawahara, R.S., Deng, Z.-W., and Deuel, T.F., "PDGF and the small inducible gene (SIG) family: roles in the inflammatory response", J. Westwick et al, eds., Plenum Press, New York (1991).
4. Deuel, T.F., Senior, R.M., Huang, J.S. and Griffin, G.L., Chemotaxis of Monocytes and Neutrophils to Platelet-Derived Growth Factor, *J. Clin. Invest.* 69:1046-1049 (1992).
5. Krieger, M., Acton, S., Ashkenas, J., Pearson, A., Penman, M. and Resnick, D., Molecular flypaper, host defense, and atherosclerosis. Structure, binding properties, and functions of macrophage scavenger receptors. [Review], *Journal of Biological Chemistry.* 268(7):4569-72 (1993).
6. Brown, M.S. and Goldstein, J.L., Lipoprotein metabolism in the macrophage: implications for cholesterol deposition in atherosclerosis. [Review], *Annual Review of Biochemistry.* 52(223):223-61 (1983).
7. Goldstein, J.L. and Brown, M.S., *in* "The Metabolic Basis of Inherited Disease," Scriver, C.R., et al., Editors. McGraw Hill Book Co., New York (1989).

8. Havel, R.J. and Kane, J.P., *in* "The Metabolic Basis of Inherited Disease," Scriver, C.R., et al., Editors. McGraw Hill Book Co., New York (1989).
9. Steinberg, D., Parthasarathy, S., Carew, T.E., Khoo, J.C. and Witztum, J.L., Beyond cholesterol. Modifications of low-density lipoprotein that increase its atherogenicity [see comments]. [Review], *New England Journal of Medicine.* 320(14):915-24 (1989).
10. Shure, D., Senior, R.M., Griffin, G.L. and Deuel, T.F., PDGF AA homodimers are potent chemoattractants for fibroblasts, neutrophils, and monocytes activated by lymphocytes or cytokines, *Biochem. Biophys. Res. Commun.* 186(3):1510-1514 (1992).
11. Lin, X., Wang, Z., Gu, L. and Deuel, T.F., Functional analysis of the human platelet-derived growth factor A-chain promoter region, *J. Biol. Chem.* 267(35):25614-25619 (1992).
12. Sporn, M.B. and Todaro, G.J., Autocrine secretion and malignant transformation of cells, *N. Engl. J. Med.* 878:30-33 (1980).
13. Huang, J.S., Huang, S.S. and Deuel, T.F., Transforming Protein of Simian Sarcomma Virus Stimulates Autocrine Cell Growth of SSV-Transformed Cells Through Platelet-Derived Growth Factor Cell Surface Receptors,*Cell.* 39:79-87 (1984).
14. Kim, H.-R., Upadhyay, S., Korsmeyer, S. and Deuel, T.F. PDGF B and PDGF A homodimers transform murine fibroblasts depending on the genetic background of the cell, Submitted.
15. Takimoto, Y., Wang, Z., Kobler, K. and Deuel, T.F., Promoter region of the human platelet-derived growth factor A-chain gene, *Proc. Natl. Acad. Sci. USA.* 88:1686-1690 (1991).
16. Treisman, R., Identification of a protein-binding site that mediates transcriptional response of the c-fos gene to serum factors, *Cell.* 46(4):567-74 (1986).
17. Wang, Z., Lin, X.-H., Nobuyoshi, M., Qui, Q. and Deuel, T.F., Binding of single-stranded oligonucleotides to a non-b-form DNA structure results in loss of promoter activity of the platelet-derived growth factor A-chain gene, *J. Biol. Chem.* 267(19):13669-13674 (1992).
18. Paulsson, Y., Hammacher, A., Heldin, C.H. and Westermark, B., Possible positive autocrine feedback in the prereplicative phase of human fibroblasts, *Nature.* 328(6132):715-7 (1987).
19. Wang, Z.-Y., Madden, S.L., Deuel, T.F. and Rauscher, III, F.J., The Wilms' tumor gene product, WT1, represses transcription of the platelet-derived growth factor A-chain gene, *J. Biol. Chem.* 267:21999-22002 (1992).
20. Wang, Z.-Y., Qiu, Q.-Q. and Deuel, T.F., The Wilms' tumor gene product WT1 activates or suppresses transcription through separate functional domains, *J. Biol. Chem.* 268:9172-9175 (1993).
21. Wang, Z., Qiu, Q., Enger, K.T. and Deuel, T.F., A second transcriptionally active DNA-binding site for the Wilms tumor gene product, WT1, *Proc. Natl. Acad. Sci. USA.* 90:8896-8900 (1993).
22. Lin, X., Guo, C., Gu, L. and Deuel, T.F., Site-specific methylation inhibits transcriptional activity of platelet-derived growth factor A-chain promoter, *J. Biol. Chem.* 268(23):17334-17340 (1993).
23. Lewis, J.D., Meehan, R.R., Henzel, W.J., Maurer, F.I., Jeppesen, P., Klein, F. and Bird, A., Purification, sequence, and cellular localization of a novel chromosomal protein that binds to methylated DNA. *Cell.* 69(6):905-14 (1992).
24. Li, Y., Milner, P.G., Chauhan, A.K., Watson, M.A., Hoffman, R.M., Kodner, C.M., Milbrandt, J. and Deuel, T.F., Cloning and expression of a developmentally regulated protein that induces mitogenic and neurite outgrowth activity, *Science.* 250:1690-1694 (1990).

THE REGULATION OF NORMAL AND PATHOLOGICAL ANGIOGENESIS BY VASCULAR ENDOTHELIAL GROWTH FACTOR

Napoleone Ferrara, John E. Park, Claire E. Walder,
Stuart Bunting, and G. Roger Thomas

Department of Cardiovascular Research, Genentech, Inc.,
460 Point San Bruno Boulevard, South San Francisco, CA 94080

A fundamental property of vascular endothelial cells is the ability to proliferate and form a network of capillaries (1, 2). This process, known as "angiogenesis", is prominent during embryonic development and somatic growth but in a normal adult it only takes place following injury or, in a cyclical fashion, in the endometrium and in the ovary (1, 2). Angiogenesis plays a significant role in the pathogenesis of a variety of disorders including cancer, proliferative retinopathies, rheumatoid arthritis or psoriasis. Therefore, inhibition of angiogenesis may constitute an attractive strategy for the treatment of such disorders. Conversely, disorders characterized by inadequate tissue perfusion such as obstructive atherosclerosis and diabetes are expected to benefit from agents able to promote endothelial cell growth and neovascularization

A variety of factors have been identified as potential positive regulators of angiogenesis: aFGF, bFGF, EGF, TGF -α, TGF -β, PGE_2, monobutyrin, TNF-α, PD-ECGF, angiogenin and interleukin-8 (1, 2).

In this article we will attempt to summarize our present knowledge of a recently identified family of directly-acting endothelial cell mitogens and angiogenic factors known as vascular endothelial growth factor (VEGF) or vascular permeability factor (VPF) (reviewed in 3, 4). These factors are products of the same gene and, by alternative exon splicing, may exist in four different isoforms. These isoforms have similar biological activities but differ markedly in their secretion pattern, suggesting that this family of proteins may play multiple roles in the regulation of angiogenesis (5-8). Recent studies point to VEGF as a major regulator of physiological and pathological angiogenesis.

Cardiovascular Disease 2
Edited by L.L. Gallo, Plenum Press, New York, 1995

BIOLOGICAL PROPERTIES OF VEGF

A peculiar aspect of the biology of VEGF is its narrow target cell specificity (9). VEGF is a potent mitogen (ED_{50} 2-10 pM) for endothelial cells derived from small or large vessels but it is apparently devoid of mitogenic activity for other cell types (9-11). VEGF is also able to induce a marked angiogenic response in the chick chorioallantoic membrane (10, 12). VEGF also promotes angiogenesis in a tridimensional *in vitro* model, inducing confluent microvascular endothelial cells to invade a collagen gel and form tube-like structures (13). These findings support the hypothesis that the angiogenic effects of VEGF are direct and do not require the involvement of paracrine mediator(s). These studies also provide evidence for a potent synergism between VEGF and bFGF in the induction of such *in vitro* angiogenic effects (13).

VEGF has the same sequence as VPF, a tumor-derived protein so named because of its ability to promote Evans blue extravasation when injected in the guinea pig skin (14, 15) and proposed to be a specific mediator of the hyperpermeability properties of tumor vessels. It has been suggested that extravasation of fibrinogen and other plasma proteins results in the formation of a fibrin gel that serves as a substrate for endothelial ant tumor growth (16).

VEGF induces expression of the serine proteases urokinase-type and tissue -type plasminogen activators (PA) and also PA inhibitor 1 (PAI-1) in cultured bovine microvascular endothelial cells (17). Furthermore, rhVEGF induces expression of the metalloproteinase interstitial collagenase in human umbilical vein endothelial cells but not in dermal fibroblasts (18). The expression of PAI-1 may serve to regulate and balance the process (17).

VEGF has also been shown to have vasodilator properties, which are mediated by endothelial cell-derived NO (19). This translates in a transient hypotensive effect when the protein is injected *in vivo.*

STRUCTURAL AND GENETIC PROPERTIES OF VEGF

VEGF purified from a variety of species and sources is a basic, heparin-binding, homodimeric glycoprotein of 45,000 daltons (3, 4). These properties are similar to those of the 165-amino acid isoform (see below). VEGF is inactivated by reducing agents, but it is heat-stable and acid-stable. By alternative splicing of mRNA, of a single gene, VEGF may exist as one of four different molecular species, having respectively 121, 165, 189 and 206 amino acids ($VEGF_{121}$, $VEGF_{165}$, $VEGF_{189}$, $VEGF_{206}$) (5, 8, 10). $VEGF_{165}$ is the predominant isoform secreted by a variety of normal and transformed cells. Transcripts encoding $VEGF_{121}$ and $VEGF_{189}$ are detected in the majority of cells and tissues expressing the VEGF gene (5, 10). In contrast, $VEGF_{206}$ is a very rare form, so far identified only in a human fetal liver cDNA library (5). Compared to $VEGF_{165}$, $VEGF_{121}$ lacks 44 amino acids; $VEGF_{189}$ has an insertion of 24 amino acids highly enriched in basic residues and VEGF $_{206}$ has an additional insertion of 17 amino acids. The amino acid sequence of VEGF has limited homology (15-18%) to the A and B chains of platelet-derived growth factor (Leung et al., 1989). The organization of the human VEGF gene has been recently elucidated (5, 8). The VEGF gene is organized in eight exons and the size of its coding region has been estimated to be approximately 14 kb (8). $VEGF_{165}$ lacks the residues encoded by exon 6, while $VEGF_{121}$ lacks the residues encoded by exons 6 and 7.

Hypoxia has been recently identified as an important regulatory mechanism leading to enhanced expression of the VEGF gene, both *in vivo* and *in vitro* (20, 21). Recent studies have shown that similarities exist between the oxygen-sensing mechanisms regulating the expression of the VEGF and the erythropoyetin genes (21). In both cases, gene expression is significantly enhanced by cobalt chloride. Furthermore, the hypoxic induction of both genes is inhibited by carbon monoxide, suggesting the involvement of a heme protein in the process of sensing oxygen levels (21).

THE VEGF ISOFORMS

$VEGF_{121}$ is a weakly acidic polypeptide that does not bind to heparin. In contrast, $VEGF_{165}$ is a basic (isoelectric point. ~8.5), heparin-binding protein . $VEGF_{189}$ and $VEGF_{206}$ are more basic and bind to heparin with even greater affinity (5, 6, 9). Such differences in the isoelectric point have been shown to profoundly affect the targeting of the translated proteins following secretion from the cell (5, 6). $VEGF_{121}$ is secreted as a freely diffusible protein in the conditioned medium of transfected cells. $VEGF_{165}$ is also secreted but a significant fraction remains bound to the cell surface or the extracellular matrix (ECM). The longer isoforms, $VEGF_{189}$ and $VEGF_{206}$, are almost completely sequestered in the ECM (5-7). However, they may be released from the bound state by a variety of agents such as suramin, heparin or heparinase (6). The observation that heparin releases these forms of VEGF suggests that their binding site is represented by heparin-containing proteoglycans, similar to that for bFGF. Furthermore, the long forms may be released by plasmin (6). This physiologically relevant protease is able to cleave $VEGF_{189}$ or $VEGF_{206}$ at the COOH terminus and generate a proteolytic fragment having molecular weight of ~34,000 daltons which is active as an endothelial cell mitogen and as a vascular permeability agent (6). Plasminogen activation and generation of plasmin have been shown to play an important role in the angiogenesis cascade (21). It is possible that this property is not confined to plasmin. It may be that cleavage of VEGF may be brought about by a variety of inflammation-associated proteases. Thus, the VEGF proteins may be made available to endothelial cells by at least two different mechanisms: as freely soluble proteins ($VEGF_{121}$, $VEGF_{165}$) or following protease activation and cleavage of the ECM-bound isoforms.

THE VEGF RECEPTORS

Characteristics of VEGF binding to endothelial cells: Two classes of high affinity VEGF binding sites have been identified on the cell surface of cultured endothelial cells, having K_d values of 10^{-12} and 10^{-11} M, respectively (23, 24). The binding of VEGF to endothelial cells is stimulated by low concentrations of heparin and is inhibited by removal of cell-surface heparan sulfate by heparinases (25). This presumably reflects the heparin-binding nature of VEGF and suggests also that a cell-surface proteoglycan may be required for binding. Lower affinity binding sites on mononuclear phagocytes (K_d ~300-500 pM) have recently been described (26). The molecular nature of these sites remains to be elucidated.

Tissue distribution of high affinity binding sites for VEGF. Ligand autoradiography studies on tissue sections of adult rats revealed that high affinity ^{125}I-VEGF binding sites are localized to the vascular endothelium of large or small vessels but not to other cell types (27, 28). Specific binding colocalized with Factor VIII immunoreactivity and was apparent on both proliferating and quiescent endothelial cells. Binding of ^{125}I-VEGF during development of rat embryos is first detectable in the blood islands of the yolk sac, which contain the earliest progenitors of hematopoeitic and endothelial cells. As the vascular system develops, VEGF binding sites continue to colocalize with the endothelium of blood vessels (28).

The Flt-1 and Flk-1/KDR tyrosine kinases Two tyrosine kinases have been recently identified as putative VEGF receptors. The flt-1 (fms-like-tyrosine kinase) and KDR (kinase domain region) proteins have been shown to bind VEGF with high affinity (29, 30). The overall amino acid sequence identity between the two proteins is 44 %. The murine homologue of KDR is known as flk-1 and shares 85% sequence identity with human KDR (31). Flt-1 and KDR/flk-1 have a single hydrophobic leader peptide, a single transmembrane domain, seven immunoglobulin-like domains in its extracellular domain, and a consensus tyrosine kinase sequence which is interrupted by a kinase-insert domain.

Recently, an alternatively spliced form of flt-1 lacking the seventh immunoglobulin-like domain, the cytoplasmic domain, and transmembrane sequence has been identified in human umbilical vein endothelial cells (32). The encoded protein is expected to have 687 amino acids and is secreted as a soluble protein. This soluble flt-1 protein is able to inhibit VEGF-induced mitogenesis and has been proposed to be a physiological negative regulator of VEGF action.

Characteristics of VEGF binding to flt-1 and KDR/flk-1: Flt-1 has the highest affinity for $rhVEGF_{165}$, with a K_d of approximately 10-20 pM (29). KDR has a somewhat lower affinity for VEGF: the K_d has been estimated to be approximately 75 pM (30). The K_d for binding of $rhVEGF_{165}$ to Flk-1 is 500-600 pM (33, 34). Therefore, it is likely the binding of KDR/Flk-1 to VEGF is partially species-specific. VEGF binding to these receptors was not competed by a structurally related peptide such as PDGF (29, 30).

Affinity cross linking of ^{125}I-rhVEGF$_{165}$ to transfected COS cells expressing KDR/Flk-1 revealed bands of 190-230 kDa (30, 33).

VEGF AS A REGULATOR OF PHYSIOLOGICAL ANGIOGENESIS

The proliferation of blood vessels is crucial for a wide variety of physiological processes such as embryonic development, normal growth and differentiation, wound healing and reproductive functions. The VEGF mRNA is expressed within the first few days following implantation in the giant cells of the trophoblast (28, 35), suggesting a role for VEGF in the induction of vascular growth in the decidua, placenta and vascular membranes. At later developmental stages in the mouse or rat embryos, the VEGF mRNA is expressed in several organs, including heart, vertebral column, kidney, and along the surface of the spinal cord and brain (28, 35). In the developing mouse brain, the highest levels of mRNA expression are associated with the choroid plexus and the ventricular epithelium (35), arguing for a spatial relation between VEGF mRNA expression and angiogenesis (36).

Strong evidence supporting the hypothesis that VEGF may be a physiological regulator of angiogenesis was provided by *in situ* hybridization studies on the ovary (37). Invasion of blood vessels is a prominent aspect of the cyclical development of the corpus luteum. (38). Expression of the VEGF mRNA in the rat ovary is temporally and spatially related to the proliferation of microvessels. Minimal hybridization is detectable in the avascular granulosa cells of preovulatory follicles while a strong hybridization signal is present in the corpus luteum where 50-60% of the total cell population is represented by capillary endothelial cells and pericytes. A similar expression pattern has been recently described in the primate ovary (39).

Interestingly, recent studies provide evidence for the expression of VEGF mRNA in keratinocytes in a healing wound, suggesting the involvement of VEGF in a major pathophysiological process such as wound healing (40).

THE ROLE OF VEGF IN PATHOLOGICAL ANGIOGENESIS

A variety of transformed cell lines express the VEGF mRNA and secrete a VEGF-like protein (12, 41, 42), suggesting that VEGF may facilitate tumor growth through its direct angiogenic effects. Also, recent *in situ* hybridization studies that have shown that the VEGF mRNA is expressed at high level by a variety of human tumors, including renal cell carcinoma, colon carcinoma and several intracranial tumors (20, 43-46). A strong correlation exists between degree of vascularization of the malignancy and VEGF mRNA expression (44-46). The hypothesis that VEGF-expressing cells may have a growth advantage *in vivo* due to stimulation of angiogenesis is supported by previous studies that show that expression of $VEGF_{165}$ or $VEGF_{121}$ confers on a nontumorigenic clone of Chinese hamster ovary cells the ability to proliferate *in vivo* and form vascularized tumors in nude mice (47). However, VEGF expression did not result in a growth advantage *in vitro* for such cells, indicating that their ability to grow *in vivo* was due to paracrine rather than autocrine mechanisms. More direct evidence for a role of VEGF in tumor angiogenesis has been recently made possible by the availability of specific monoclonal antibodies capable of inhibiting VEGF-induced angiogenesis *in vivo* and *in vitro*. Such antibodies exert a dramatic inhibitory effect on the growth of several human tumor cell lines in nude mice (48). However, the antibodies (or VEGF) have no effect on the *in vitro* growth of tumor cells, confirming that VEGF does not act as an autocrine factor for the tumor cells. The density of vessels was markedly decreased in the antibody-treated tumors. More recent studies provide evidence for the involvement of VEGF in the *in vivo* growth of colon carcinoma cells in a nude mouse model of liver metastasis (49) . Treatment with anti-VEGF monoclonal antibodies resulted in a dramatic decrease in the size of liver metastases following injection of tumor cells into the spleen, again confirming that angiogenesis is a critical rate limiting step in tumorigenesis and that VEGF is a major mediator of this process.

Recent studies (50) argue for a role played by VEGF in the pathogenesis of another important angiogenic disease, rheumatoid arthritis (RA). Levels of immunoreactive VEGF were very high (100-1600 ng/ml) in the synovial fluid of RA patients while they were very low or undetectable in the synovial fluid of patients affected by other forms of arthritis or by degenerative joint disease. Furthermore, 50-60% of the endothelial cell chemotactic activity of the RA synovial fluid was blocked by an anti-VEGF antibody, indicating that immunoreactive VEGF was bioactive.

Further disorders where VEGF-induced angiogenesis is expected to play a pathogenic role are proliferative retinopathies. Intraocular neovascularization associated with conditions such as diabetes, occlusion of central retinal veins or prematurity and exposure to oxygen can lead to vitreous hemorrhage, retinal detachment and blindness (1, 2). It is believed that an event common to all of these conditions is the release of diffusible factor(s) from the ischemic retina. VEGF, by its diffusible nature and hypoxia-inducibility, is an attractive candidate. Recent studies where VEGF levels were measured in the ocular fluids of over 150 patients demonstrate a strong correlation between levels of immunoreactive VEGF in the aqueous and vitreous humors and active proliferative retinopathy (51). VEGF levels were undetectable or very low (<0.5 ng/ml) in the eyes of individuals affected by non-neovascular disorders, background diabetic retinopathy or even proliferative diabetic retinopathy in a quiescent stage. In contrast, the VEGF levels were in the range of 3-10 ng/ml in the presence of active proliferative retinopathy associated with diabetes, occlusion of the central retinal vein or prematurity. These findings suggest that ischemia in the retina, regardless of its etiology, may lead to release of VEGF followed by neovascularization.

VEGF AND THERAPEUTIC ANGIOGENESIS

An attractive possibility is that the VEGF protein or gene therapy with a VEGF cDNA may be used to promote revascularization in conditions of insufficient tissue perfusion. For example, chronic limb ischemia, most frequently caused by obstructive atherosclerosis affecting the superficial femoral artery, is associated with a high rate of morbidity and mortality and treatment is currently limited to surgical revascularization or endovascular interventional therapy (52-54). Previous studies where bFGF was administered suggest that an angiogenic therapy may be effective in restoring perfusion following vascular injury (55). Very recent studies indicate that intra-arterial or intra-muscular administration of $rhVEGF_{165}$ may significantly augment perfusion and development of collateral vessels in a rabbit model of hindlimb ischemia (56, 57). These studies provided angiographic evidence of neovascularization in the ischemic limbs. More recently, the hypothesis was tested that the angiogenesis initiated by the administration of $rhVEGF_{165}$ improved muscle function in limbs rendered ischemic by surgical removal of the femoral artery (58). Remarkably, a single intra-arterial injection of $rhVEGF_{165}$ augmented muscle function in this model of peripheral limb ischemia. It is also a characteristic of this model that the ischemic limb cannot augment blood flow in response to oxygen demand during exercise. We found that this exercise-induced hyperemia was significantly improved in ischemic limbs treated with $rhVEGF_{165}$. This improvement in perfusion was however, not seen in resting muscle nor was it apparent in other non ischemic tissues including the contralateral limb. We conclude that the neovascularization and the angiogenesis seen in response to $rhVEGF_{165}$ results in improvements in physiological parameters that indicate that this type of therapy may represent a significant advancement in the treatment of peripheral vascular disease.

Furthermore, it has been suggested that VEGF administration may result in increase in coronary blood flow in a dog model of coronary insufficiency (59). Following occlusion of the left circumflex coronary artery, daily injections of rhVEGF distal to the occlusion resulted in a significant enhancement in collateral blood flow over a period of four weeks.

A further potential therapeutic application of VEGF is the prevention of restenosis following PTA. Between 15% and 75% of patients undergoing PTA for occlusive arterial disease develop restenosis within six months. The frequency of clinical stenosis depends on the size and location of the artery and the definition of stenosis (54). It has been proposed that damage to the endothelium is the crucial event triggering fibrocellular intimal proliferation (54, 60). Therefore, it is tempting to speculate that rapid re-endothelialization promoted by VEGF may prevent the cascade of events leading to neointima formation and restenosis. The specificity of VEGF for endothelial cells may be especially useful for this application.

REFERENCES

1. Folkman, J, Shing, Y. Angiogenesis. J. Biol. Chem. 267:10931 (1992).
2. Klagsbrun, M, D'Amore, PA. Regulators of angiogenesis. Annu. Rev. Physiol. 53:217 (1991).
3. Ferrara, N, Houck, K, Jakeman, L, Leung, DW. Molecular and biological properties of the vascular endothelial growth factor family of polypeptides. Endocr. Rev. 13:18 (1992).
4. Ferrara, N. Vascular endothelial growth factor. Trends Cardiovasc. Med. 3
5. Houck, KA, Ferrara, N, Winer, J, Cachianes, G, Li, B, Leung, DW. The vascular endothelial growth factor family: Identification of a fourth molecular species and characterization of alternative splicing of RNA. Mol. Endocrinol. 5:1806 (1991).
6. Houck, KA, Leung, DW, Rowland, AM, Winer, J, Ferrara, N. Dual regulation of vascular endothelial growth factor bioavailability by genetic and proteolytic mechanisms. J. Biol. Chem. 267:26031 (1992).
7. Park, JE, Keller, G-A, Ferrara, N. The vascular endothelial growth factor (VEGF) isoforms: Differential deposition into the subepithelial extracellular matrix and bioactivity of ECM-Bound VEGF. Molec. Biol. Cell. 4:1317 (1993).
8. Tisher, E, Mitchell, R, Hartmann, T, Silva, M, Gospodarowicz, D, Fiddes, J, Abraham, J. The human gene for vascular endothelial growth factor. J. Biol. Chem. 266:11947 (1991).
9. Ferrara, N, Henzel, WJ. Pituitary follicular cells secrete a novel heparin-binding growth factor specific for vascular endothelial cells. Biochem. Biophys. Res. Commun. 161:851 (1989).
10. Leung, DW, Cachianes, G, Kuang, W-J, Goeddel, DV, Ferrara, N. Vascular endothelial growth factor is a secreted angiogenic mitogen. Science, 246:1306 (1989).
11. Plouet, J, Schilling, J, Gospodarowicz, D. Isolation and characterization of a newly identified endothelial cell mitogen produced by AtT20 cells. EMBO J. 8:3801 (1989).
12. Conn, G, Bayne, M, Soderman, L, Kwok, PW, Sullivan, KA, Palisi, TM, Hope, DA, Thomas, KA. Amino acid and cDNA sequence of a vascular endothelial cell mitogen homologous to platelet-derived growth factor. Proc. Natl. Acad. Sci. USA 87:2628 (1990).

13. Pepper, MS, Ferrara, N, Orci, L, Montesano, R. Potent synergism between vascular endothelial growth factor and basic fibroblast growth factor in the induction of angiogenesis in vitro. Bochem. Biophys. Res. Commun. 189:824 (1992).
14. Connolly, DT, Heuvelman, DM, Nelson, R, Olander, JV, Eppley, BL, Delfino, JJ, Siegel, NR, Leimgruber, RM, Feder, J. Tumor vascular permeability factor stimulates endothelial cell growth and angiogenesis. J. Clin. Invest. 84:1470 (1989).
15. Keck, PJ, Hauser, SD, Krivi, G, Sanzo, K, Warren, T, Feder, J, Connolly, DT. Vascular permeability factor, an endothelial cell mitogen related to platelet derived growth factor. Science 246:1309 (1989).
16. Dvorak, HF. Tumors: wound that do not heal. Similarity between tumor stroma generation and wound healing. New Engl. J. Med. 315:1650 (1986).
17. Pepper MS, Ferrara, N, Orci, L, Montesano, R. Vascular endothelial growth factor (VEGF) induces plasminogen activators and plasminogen activator inhibitor type 1 in microvascular endothelial cells. Biochem. Biophys. Res. Commun. 181:902 (1991).
18. Unemori, E, Ferrara, N, Bauer, EA, Amento, EP: 1992. Vascular endothelial growth factor induces interstitial collagenase expression in human endothelial cells. J. Cell. Physiol. 153:557-562.
19. Ku, DD, Zaleski, JK, Liu, S, Brock, T. Vascular endothelial growth factor induces EDRF-dependent relaxation of coronary arteries. Am. J. Physiol. 265:H586 (1993).
20. Shweiki D, Itin, Soffer, D, Keshet, E: 1992. Vascular endothelial growth factor induced by hypoxia may mediate hypoxia-initiatiated angiogenesis. Nature 359:843-845.
21. Goldberg, MA and Schneider, TJ. Similarities between the oxygen-sensing mechanisms regulating the expression of vascular endothelial growth factor and erythropoietin. J. Biol. Chem. 269:4355 (1994).
22. Mignatti, P, Tsuboi, R, Robbins, E, Rifkin, DB. In vitro angiogenesis on the human amniotic membrane: requirement for basic fibroblast growth factor-induced proteinases. J. Cell Biol. 108:671 (1989).
23. Vaisman, N, Gospodarowicz, D, Neufeld, G: Characterization of the receptors for vascular endothelial growth factor. J. Biol. Chem. 265:19461 (1990).
24. Plouet, J, Moukadiri, HJ. Characterization of the receptors to vasculotropin on bovine adrenal cortex-derived capillary endothelial cell. J. Biol. Chem. 265:22071 (1990).
25. Gitay-Goren, H, Soker, S, Vlodavsky, I, Neufeld, G. The binding of vascular endothelial growth factor to its receptors is dependent on cell surface-associated heparin-like molecules. J. Biol. Chem. 267:6093 (1992).
26. Shen, H, Clauss, M, Ryan, J, Schmidt, AM, Tijburg, P, Borden, L, Connolly, DT, Stern, D, Kao, J. Characterization of vascular permeability factor/vascular endothelial growth factor receptors on mononuclear phagocytes. Blood. 81:2767 (1993).

27. Jakeman, LB, Winer, J, Bennett, GL, Altar, CA, Ferrara, N. Binding sites for vascular endothelial growth factor are localized on endothelial cells in adult rat tissues. J. Clin. Invest. 89:244 (1992).
28. Jakeman, LB, Armanini, M, Phillips, HS, Ferrara, N. Developmental expression of binding sites and mRNA for vascular endothelial growth factor suggests a role for this protein in vasculogenesis and angiogenesis. Endocrinology. 2:913 (1993).
29. deVries, C, Escobedo, JA, Ueno H, Houck, KA, Ferrara, N, Williams, LT. The fms-like tyrosine kinase, a receptor for vascular endothelial growth factor. Science, 255:989 (1992).
30. Terman, BI, Vermazen, MD, Carrion, ME, Dimitrov, D, Armellino, DC, Gospodarowicz, D, Bohlen, P., Identification of the KDR tyrosine kinase as a receptor for vascular endothelial growth factor. Biochem. Biophys. Res. Commun. 34:1578 (1992).
31. Matthews, W, Jordan, CT, Gavin, M, Jenkins, NA, Copeland, NG, Lemischka, IR. A receptor tyrosine kinase cDNA isolated from a population of enriched primitive hematopoietic cells and exhibiting close genetic linkage to c-kit. Proc. Natl. Acad. Sci. USA 88:9026 (1991).
32. Kendell, RL, Thomas, KA. Inhibition of vascular endothelial growth factor bioactivity by an endogenously encoded soluble receptor. Proc. Natl. Acad. Sci. USA. 90:10705 (1993).
33. Millauer, B, Wizigmann-Voos, S, Schnurch, H, Martinez, R, Moller, NP, Risau, W, Ullrich A. High affinity binding and developmental expression suggest Flk-1 as a major regulator of vasculogenesis and angiogenesis. Cell 72:835 (1993).
34. Quinn, T, Peters, KG, deVries C, Ferrara, N, Williams LT. Fetal liver kinase 1 is a receptor for vascular endothelial growth factor and is selectively expressed in vascular endothelium. Proc. Natl. Acad. Sci. USA. 90:7533 (1993.
35. Breier, G, Albrecht, U, Sterrer, S, Risau, W. Expression of vascular endothelial growth factor during embryonic angiogenesis and endothelial cell differentiation. Development, 114:521 (1992).
36. Evans, HM. On the development of aortae, cardinal, umbilical veins, and other blood vessels of vertebrate embryos from capillaries. Anat. Rec. 3:498 (1909).
37. Phillips, HS, Hains, J, Leung, DW, Ferrara, N. Vascular endothelial growth factor is expressed in rat corpus luteum. Endocrinology. 127:965 (1990).
38. Bassett, DL. The changes in vascular pattern of the ovary of the albino rat during the estrous cycle. Am. J. Anat. 73:251 (1943).
39. Ravindranath, N, Little-Ihrig, L, Phillips, HS, Ferrara, N, Zeleznick, AJ. Vascular endothelial growth factor mRNA expression in the primate ovary. Endocrinology. 131:254 (1992).
40. Brown, LF, Yeo, KT, Berse, B, Yeo TK, Senger, DR, Dvorak, HF, vad de Water, L. Expression of vascular permeability factor (vascular endothelial growth factor) by epidermal keratinocytes during wound healing. J. Exp. Med. 176:1375 (1992).

41. Rosenthal, R. Megyesi, JF, Henzel, WJ, Ferrara, N, Folkman, J. Conditioned medium from mouse sarcoma 180 cells contains vascular endothelial growth factor. Growth Factors 4:53 (1990).
42. Senger, D, Perruzzi, CA, Feder, J, Dvorak, HF. A highly conserved vascular permeability factor secreted by a variety of human and rodent tumor cell lines. Cancer Res. 46:5269 (1986).
43. Berse, B, Brown, LF, Van de Vater, L, Dvorak, HF, Senger, DR. The vascular permeability factor (vascular endothelial growth factor) gene is differentially expressed in normal tissues, macrophages and tumors. Molec. Biol. Cell. 3:211 (1992).
44. Plate, KH, Breier, G, Weich, HA, Risau, W. Vascular endothelial growth is a potential tumour angiogenesis factor in vivo. Nature 359: 845 (1992).
45. Berkman, RA, Merrill, MJ, Reinhold, WC, Monacci, WT, Saxena, A, Clark, WC, Robertson, JT, Ali, IU, Oldfield, EH. Expression of the vascular permeability/vascular endothelial growth factor gene in central nervous system neoplasms. J. Clin. Invest. 91:153 (1993).
46. Phillips, HS, Armanini, M, Stavrou, D, Ferrara, N., Westphal, M. Intense focal expression of vascular endothelial growth factor mRNA in human intracranial neoplasms: Association with regions of necrosis. Int. J. Oncology. 2:913 (1993).
47. Ferrara, N, Winer, J, Burton, T, Rowland, A, Siegel, M, Phillips, HS, Terrell, T, Keller, G-A, Levinson, AD. Expression of vascular endothelial growth factor does not promote transformation but confers a growth advantage in vivo to chinese hamster ovary cells. J. Clin. Invest. 91:160 (1993).
48. Kim, KJ, Li, B, Winer, J, Armanini, M, Gillett, N, Phillips, HS, Ferrara, N. Inhibition of vascular endothelial growth factor-induced angiogenesis suppresses tumour growth in vivo. Nature 362:841 (1993).
49. Warren, Rs, Yuan, H, Matli, MR, Gillett, N, Ferrara, N. Regulation by vascular endothelial growth factor of human colon cancer tumorigenesis in a model of experimental liver metastasis. Submitted.
50. Kock, AE, Harlow, L, Haines, GK, Amento, EP, Unemori, EN, Wong, WL, Pope, RM, Ferrara, N. Vascular endothelial growth factor: A cytokine modulating endothelial function in rheumatoid arthritis. J. Immunol. 152:4149 (1994).
51. Aiello, LP, Avery, RL, Arrigg, PG, Keyt, B, Jampel, HD, Shah, ST, Pasquale, LR. Thiems, H, Iwamoto, A, Park, JE, Nguyen, H, Aiello, LP, Ferrara, N, King, GL. Vascular endothelial growth factor (VEGF) in active proliferative diabetic retinopathy and other ischemic neovascular disosrders The N. Engl. J. Med. In press.
52. Topol, EJ. Textbook of Interventional Cardiology: 1990. W.B. Saunders Co. Philadelphia, PA
53. Thompson, RW, D'Amore, PA. Recruitment of growth and collateral circulation. In: Zelenock, GB, D'Alecy, LG, Fantone, JC III, Shlafer, M, Stanley, JC Eds. Clinical

ischemic syndromes: mechanisms and consequences of tissue injury. St. Louis: CV Mosby (1990)

54. Graor, RA, Gray, BH.. Interventional treatment of peripheral vascular disease. In: Young, JR, Graor, RA, Olin, JW, Bartholomew, JR. Eds. Peripheral Vascular Diseases. Mosby, St. Louis, MO (1991)
55. Baffour, R, Berman, J, Garb, JL, Rhee, SW, Kaufman, J, Friedmann, P. Enhanced angiogenesis and growth of collaterals by in vivo administration of recombinant basic fibroblast growth factor in a rabbit model of acute lower limb ischemia: Dose-response effect of basic fibroblast growth factor. J. Vasc. Surg. 16:181 (1992).
56. Takeshita, S, Zhung, L, Brogi, E, Kearney, M, Pu, L-Q, Bunting, S, Ferrara, N, Symes, JF, Isner, JM. Therapeutic angiogenesis: A single intra-arterial bolus of vascular endothelial growth factor augments collateral vessel formation in a rabbit ischemic hindlimb model. J. Clin. Invest. 93:662 (1994).
57. Pu, L-Q, Ferrara, N, Stein, LA, Sniderman, AD, Isner, JM, Symes, JF. Vascular endothelial growth factor induces dose-dependent revascularization in a rabbit model of persistent limb ischemia. Abstract 66th Meeting American Heart Assoc. Nov. 8-11 1993, Atlanta, GA.
58. Walder, CE, Errett, CJ, Bunting, S, Lindquist, P, Ferrara, N, Thomas GR: Vascular endothelial growth factor (VEGF) improves blood flow and muscle function in a chronic ischemic limb model. FASEB J. 8:764 (1994).
59. Banai, S, Shou, M, Jaktlish, MT, Ferrara, N, Epstein, S, Unger E: 1992. Enhancement of coronary collateral flow by intracoronary injection of vascular endothelial growth factor. 41st Sci. Sess. Am. Coll. Cardiol. Dallas, TX, April 12-16, 1992.
60. Essed, CD, Brand, MVD and Becker, AE. Transluminal coronary angioplasty and early restenosis. Br. Heart J. 49:393 (1983).

DEVELOPMENTALLY ASSOCIATED GENE EXPRESSION IN RABBIT VASCULAR SMOOTH MUSCLE CELLS

David K.M. Han and Gene Liau

Department of Molecular Biology
Holland Laboratory
American Red Cross
15601 Crabbs Branch Way
Rockville, MD 20855

INTRODUCTION

The basic process of blood vessel development is not well understood. It is known that vessel formation is initiated by the aggregation of endothelial precursors (angioblasts) as well as by the invasion of endothelial cells into existing tissue.[1] Subsequently, there is recruitment of cells from the surrounding mesenchyme that give rise to smooth muscle cells (SMCs).[2,3] Additionally, in medium and large vessels, there is first continued proliferation of SMCs, organization of these cells into distinct layers separated by elastic laminae, and eventual cessation of growth.[4,3,5,6] This last phase of blood vessel development, which we have termed "developmental maturation," defines the transition of SMCs from a proliferative to a contractile phenotype. Research interest in this phase of vascular development arise from the morphological and biochemical similarities that have been noted between fetal SMCs and SMCs in atherosclerotic plaques.[7,8,9,10,11,12] These similarities have led to the suggestion that SMCs in atherosclerosis and hypertension may recapitulate certain aspects of earlier developmental events.[9,13] To permit a critical analysis of this possibility, it is crucial that we obtain a better understanding of the developmental events associated with the transition of SMCs from a proliferative to a quiescent phenotype. As a first step toward this goal, we have compared these two populations of SMCs and have identified genes that are specifically expressed in the fetal SMC population and not in the adult SMCs.

Cardiovascular Disease 2
Edited by L.L. Gallo, Plenum Press, New York, 1995

MORPHOLOGICAL CHANGES ASSOCIATED WITH AORTIC SMOOTH MUSCLE MATURATION

A complete quantitative histological analysis of late stage prenatal blood vessel development in the rat has been performed by Nakamura.[3] He performed electron microscopic analysis of prenatal development of the thoracic aorta from gestational days 12 to 21. At phase I (gestational day 12), the dorsal aorta consisted of a single layer of endothelial cells loosely surrounded by mesenchymal cells. At phase II (gestational days 13-16), multiple layers of compact SMCs are observed surrounding the endothelium and some of these cells begin to acquire small clusters of myofilaments with dense bodies, rough endoplasmic reticulum, and a discontinuous basal lamina. At phase III (gestational days 17-19), there are 5-8 layers of SMCs and the three major divisions of the vessel wall; intima, media, and adventitia are clearly distinguishable. These SMCs begin to exhibit mature features including, well-developed rough endoplasmic reticulum, massive bundles of myofilaments with dense bodies, and discontinuous basal lamina. At phase IV (gestational days 20 until parturition), elastic lamellae appeared in the lumenal side of the media, and SMCs appear as contractile cells rich in myofilaments. A recent assessment of rat vascular SMC replication rates by BrdU immunocytochemistry is fully consistent with these morphological studies with peak replication occurring between gestational days 13-17.[6] We have been studying blood vessel developmental maturation in the rabbit which has a gestational period of 30 days and the period from days 19-30 is generally equivalent to gestational days 13-21 in the rat. *In vivo* proliferating cell nuclear antigen staining of rabbit aortic SMCs from gestational days 20-29 and at birth indicated that a dramatic reduction in cellular proliferation was observed during this period (Han and Liau, manuscript in preparation). This reduction in growth paralleled an increase in vessel wall elastic fiber content (Fig. 1). At gestational day 20, only weakly-stained, diffuse elastic fibers were present in the aorta. However, by gestational day 29, the elastic fiber content of the vessel wall was greatly increased and well organized. These results clearly illustrate the dramatic tissue remodeling that occurs during this period of blood vessel development.

ISOLATION AND CHARACTERIZATION OF DEVELOPMENTALLY REGULATED GENES IN VASCULAR SMOOTH MUSCLE

The identification of genes specifically down-regulated during the developmental maturation of SMCs may provide additional insights into this important process. We used differential cDNA screening to attempt to identify such genes. Differential screening involves the construction of cDNA libraries from two distinct sources and the subsequent identification of cDNAs encoding mRNAs that are in higher abundance in one source. This is accomplished by hybridization of the cDNA library with ^{32}P-labeled cDNA probes generated from the source of interest. However, one potential problem with such an approach, is that since a major difference between these two SMC populations is their proliferative state, it is likely that many of the known SMC proliferation inducible genes such as c-*fos*, c-*myc*, and thrombospondin will be identified by this technique.[14] We, therefore, chose a strategy that involved initial comparison of fetal aortic smooth muscle with proliferating adult cultured SMCs and subsequent confirmation that the identified clones are also not expressed in adult smooth muscle. A schematic representation of this strategy is shown in Fig. 2. We initially identified a 405 bp cDNA that preferentially hybridized to a 2.3 Kb mRNA in fetal smooth muscle

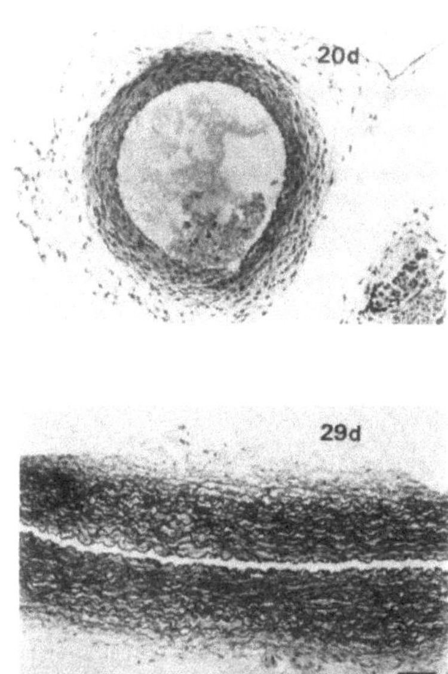

Fig. 1. **Verhoeff's Elastic Staining of the Developing Aortae** - Paraffin-embedded aortic rings from gestational day 20 and 29 animals were stained by Verhoeff's technique which stains elastin dark-blue to black. Bar = 15 μm.

but not in adult SMCs.[15] In addition, we also identified insulin-like growth factor-II (IGF-II) as a gene preferentially expressed in fetal smooth muscle.[15] The expression of F-31, IGF-II, and actin in fetal, newborn and adult aortic smooth muscle is shown in Fig. 3. Expression of F-31 is high in fetal rabbit, somewhat diminished in newborn animals and not detectable in 4-week-old animals. By contrast, the expression of IGF-II is only detected in fetal animals. Actin mRNA level was unchanged during this period. Comparison of the expression of these genes in cultured cells isolated from fetal, newborn, and adult smooth muscle revealed that similar to the *in vivo* situation, F-31 is highly expressed in both fetal and newborn SMCs but not in adult cells whereas IGF-II was expressed only in cultured fetal cells.[15] We used the initial 405 bp cDNA fragment to isolate two larger, overlapping F-31 cDNA fragments of 1.2 and 1.8 Kb. Complete DNA sequencing of 1.9 Kb of this gene and comparison with the GENBANK database revealed that the sequence exhibits a 74% identity with the human H19 gene and is likely its rabbit homologue. Consistent with this, is the finding that the tissue distribution of F-31 and H19 are generally similar. However, there are some discrepancies in the tissue distribution of F-31 mRNA expression in the rabbit and that reported for H19 in the mouse. In particular, expression of F-31 was more prominent in the rabbit adult lung than the skeletal muscle and was undetectable by Northern analysis in the heart.[15] This is in contrast to that reported for H19 in the mouse where expression was detected in the heart and skeletal muscle H19 mRNA level was clearly higher than in the lung.[16] We conclude that the IGF-II and F-31/H19 genes are both developmentally regulated in aortic smooth muscle and that this differential expression is maintained in cultured SMCs isolated from fetal and adult tissue.

POSSIBLE FUNCTION OF F-31/H19 IN VASCULAR SMOOTH MUSCLE DEVELOPMENT

The H19 gene was first isolated as a gene expressed in fetal but not adult mouse liver.[17] Interestingly, H19 expression is also induced during embryonic stem cell

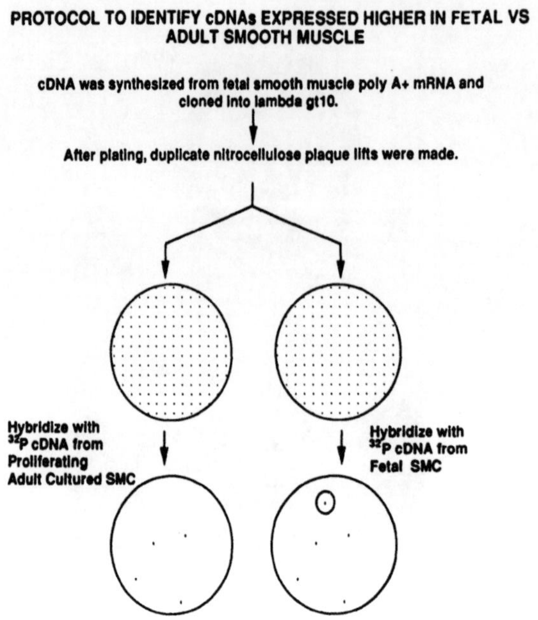

PROTOCOL TO IDENTIFY cDNAs EXPRESSED HIGHER IN FETAL VS ADULT SMOOTH MUSCLE

cDNA was synthesized from fetal smooth muscle poly A+ mRNA and cloned into lambda gt10.

After plating, duplicate nitrocellulose plaque lifts were made.

Hybridize with ^{32}P cDNA from Proliferating Adult Cultured SMC

Hybridize with ^{32}P cDNA from Fetal SMC

Fig. 2. **Schematic Representation of Differential Library Screening.**

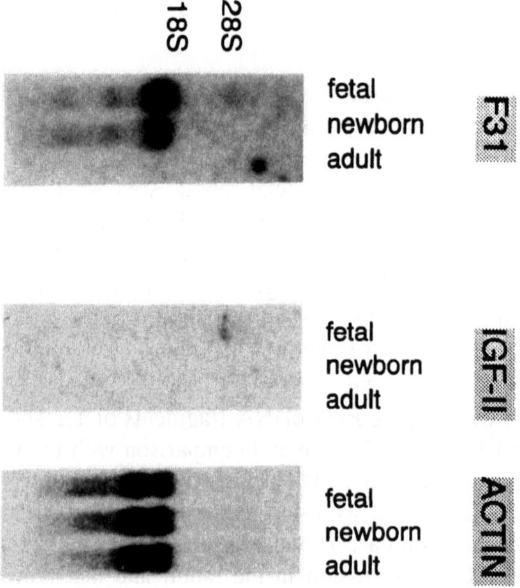

Fig. 3. **Analysis of mRNA Levels of F-31, IGF-II, and Actin in Rabbit Vascular Smooth Muscle *in vivo*** - Rabbit aortic RNA from fetal (gestational day 25), newborn, and adult (4-weeks-old) were isolated and analyzed by Northern blotting. Actin probe recognizes all actin isoforms.

differentiation and C3H10T1/2 conversion to myoblasts.[17,18,16] The H19 gene is transcribed by polymerase II, and exhibit classical properties of a translated mRNA such as RNA splicing and polyadenylation.[19] However, H19 does not contain an open reading frame that is conserved between species, and the H19 RNA is not associated with polyribosomes, suggesting that the active H19 molecule is not a translated protein but rather the RNA itself.[19,20,16,15] Although the function of H19 is unknown, two studies provide direct evidence that H19 may have an important, direct role in cell growth and differentiation. First, it was found that overexpression of the H19 gene in transgenic mice caused prenatal lethality at day 15.[21] Second, H19 overexpression in two embryonal tumour cell lines abrogated the ability of these cells to exhibit anchorage-independent growth and form tumours *in vivo*.[22] Based on its developmentally regulated expression, induction during *in vitro* differentiation, and tumour-suppressor activity, we postulate that F-31/H19 may play an important role in the developmental maturation of SMCs. Interestingly, we have examined the expression of F-31/H19 RNA by *in situ* hybridization and found that F-31/H19 expression was spatially restricted in the prenatal aortic smooth muscle (Han and Liau, manuscript in preparation). In gestational age 25 day aortae, F-31/H19 expression was clearly higher in SMC layers that are more distal to the lumen. Indeed, in newborn animals, expression of F-31/H19 was mostly restricted to the two or three outermost layers of the smooth muscle. This spatial restriction during development is interesting since it is known that SMCs adjacent to the endothelial cell layer are the first cells to acquire a morphologically differentiated phenotype.[3] Our data suggests that loss of F-31/H19 expression correlates with the development of the contractile phenotype. Consistent with this idea, is the recent demonstration that H19, is reexpressed by rat intimal SMCs after vascular injury.[23]

CONCLUSION

We believe these studies, along with parallel studies by others, have defined a new paradigm and present new opportunities in vascular biology research.[11,15,24,23] Specifically, these studies demonstrate that developmentally immature SMCs share a set of molecular markers with intimal SMCs associated with vascular diseases and that a better understanding of events associated with the developmental maturation of vascular smooth muscle may provide important insights for diseases such as atherosclerosis and restenosis. Further characterization of the regulation and function of the F-31/H19 gene will enhance our understanding of the developmental maturation of aortic smooth muscle.

KEY WORDS: Developmental maturation, smooth muscle, gene expression, H19, insulin-like growth factor-II

REFERENCES:

1. Maragoudakis ME, Gullino P, Lelkes PI.(1993). Angiogenesis in health and disease. NATO series. Series A: Life sciences vol. 227.

2. F.J. Manasek, The ultrastructure of embryonic myocardial blood vessels, *Dev. Biol.* 26:42 (1971).

3. H. Nakamura, Electron microscopic study of the prenatal development of thoracic aorta in the rat, *Am. J. Anat.* 181:406 (1988).

4. H.E Karrer, Electron microscopic study of developing chick aorta, *J. Ultrastrucr. Res.* 4:420 (1960).

5. G.K. Owens, M.M. Thompson, Developmental changes in isoactin expression in rat aortic smooth muscle cells *in vivo*, *J. Biol. Chem.* 201:13373 (1986).

6. C.K. Cook, M.C.M. Weiser, P.E. Schwartz, C.L. Jones, R.A. Majack, Developmentally timed expression of an embryonic growth phenotype in vascular smooth muscle cells, *Circ. Res.* 74:189 (1994).

7. G. Gabbiani, O. Kocher, W.S. Bloom, J. Vanderkerckhove, K. Weber, Actin expression in smooth muscle cells of rat aortic intimal thickening, human atheromatous plaque, and cultured rat aortic media, *J. Clin. Invest.* 73:148 (1984).

8. R.P.L. Mosse, G.R. Campbell, Z.L. Wang, J.H. Campbell, Smooth muscle phenotypic expression in human carotid arteries I. Comparison of cells from diffuse intimal thickening adjacent to atheromatous plaques with those of the media, *Lab Invest.* 53:556 (1985).

9. O. Kocher, O. Skalli, D. Cerutti, F. Gabbiani, G. Gabbiani, Cytoskeletal features of rat aortic cells during development. An electron microscopic, immunocytological, and biochemical study. *Circ. Res.* 56:829 (1985).

10. M.A. Glukhova, A.E. Kabakov, M.G. Frid, O.I. Ornatsky, A.M. Belkin, D.N. Mukhin, A.N. Orekhov, V.E. Koteliansky, V.N. Smirnov, Modulation of human aorta smooth muscle cell phenotype: A study of muscle specific variants of vinculin, caldesmon, and actin expression, *Proc. Natl. Acad. Sci. USA* 85:9542 (1988).

11. M.A. Glukhova, M.G. Frid, V.E. Koteliansky, Developmental changes in expression of contractile and cytoskeletal proteins in human aortic smooth muscle, *J. Biol. Chem.* 265:13042 (1990).

12. M. Aikawa, P.N. Sivam, O.-M. Kuro, K. Kimura, K. Nakahara, S. Takewaki, M. Ueda, H. Yamaguchi, Y. Yazaki, Periasamy, R. Nagai, Human smooth muscle myosin heavy chain isoforms as molecular markers for vascular development and atherosclerosis *Circ. Res.* 72:1000 (1993).

13. S.M. Schwartz, G.R. Campbell, J.H. Campbell, Replication of smooth muscle cells in vascular disease, *Circ. Res.* 58:427 (1986).

14. M.F. Janat, G. Liau, Transforming growth factor ß1 is powerful modulator of platelet-derived growth factor action in vascular smooth muscle cells, *J. Cell. Physiol.* 150:232 (1992).

15. D.K.M. Han, G. Liau, Identification and characterization of developmentally regulated genes in vascular smooth muscle cells, *Circ. Res.* 71:711 (1992).

16. F. Poirier, C.-T.J. Chan, P.M. Timmons, E.J. Robertson, M.J. Evans, The murine H19 gene is activated during embryonic stem cell differentiation *in vitro* and at the time of implantation in the developing embryo, *Development* 113:1105 (1991).

17. V. Pachnis, A. Belayew, S.M. Tilghman, Locus unlinked to α-feto-protein under the control of *raf* and *rif* genes, *Proc. Natl. Acad. Sci. USA* 81:5523 (1984).

18. R.L. Davis, H. Weintraub, A.B. Lassar, Expression of a single transfected cDNA converts fibroblasts to myoblasts, *Cell* 51:987 (1987).

19. V. Pachnis, C.I. Brannan, S.M. Tilghman, The structure and expression of a novel gene activated in early mouse embryogenesis, *EMBO J.* 7:673 (1988).

20. C.I. Brannan, E.C. Dees, R.S. Ingram, S.M. Tilghman, The product of the H19 gene may function as an RNA, *Mol. Cell. Biol.* 10:28 (1990).

21. M.E. Brunkow, S.M. Tilghman, Ectopic expression of H19 gene in mice cause prenatal lethality, *Genes Dev.* 5:1092 (1991).

22. Y. Hao, T. Crenshaw, T. Moulton, E. Newcomb, B. Tycko, Tumour-suppressor activity of H19 RNA, *Nature* 365:764 (1993).

23. D.-K. Kim, L. Zhang, V.J. Dzau, R.E. Pratt, H19, a developmentally regulated gene, is reexpressed in rat vascular smooth muscle cells after injury, *J. Clin. Invest.* 93:355 (1994).

24. M.W. Majesky, C.M. Giachelli, M.A. Reidy, S.M. Schwartz, Rat carotid neointimal smooth muscle cells reexpress a developmentally regulated mRNA phenotype during repair of arterial injury, *Circ. Res.* 71:759 (1992).

MOLECULAR BASIS AND PATHOLOGIC CONSEQUENCES OF NEUTROPHIL ADHERENCE TO ENDOTHELIUM

John M. Harlan[1], Robert K. Winn[2], Sam R. Sharar[3] and Amos Etzioni[4].

[1]Department of Medicine
[2]Department of Surgery
[3]Department of Anesthesiology
University of Washington, Seattle, WA 98115
[4]Department of Pediatrics
Rambam Medical Center
Haifa, Israel

MOLECULAR BASIS OF PHAGOCYTE EMIGRATION

The recruitment of leukocytes from the blood stream to extravascular tissue is a critical event in host defense against microbial invasion and in the repair of tissue damage. Studies by intravital microscopy have established a sequence of events involved in phagocyte emigration at sites of inflammation (1). In response to extravascular stimuli such as bacterial-derived chemoattractants or endogenous lipid and peptide mediators, signals that activate both the leukocyte and the endothelial cell are generated. As a consequence of activation, one or both cell types become adhesive leading to transient adhesion of the leukocyte to the vessel wall. The combination of these initial adhesive interactions and the shear forces caused by blood flow results in leukocyte "rolling" along the vessel wall. With further stimulation some of the rolling leukocytes adhere firmly or "stick", and then diapedese between endothelial cells to emigrate to tissue in response to chemoattractants. These adhesive interactions - rolling, sticking, and diapedesis - are mediated by cell surface molecules expressed on the leukocytes and endothelial cells (reviewed in 2 and 3). The majority of surface molecules involved in leukocyte-endothelial adherence can be placed in two categories: leukocyte integrin receptors that interact with ligands on the endothelial cell that are members of the immunoglobulin supergene family, and selectin receptors expressed on both the leukocyte and the endothelial cell that recognize specific carbohydrate counter-structures. Known integrin/immunoglobulin family interactions include: the β_1 integrin VLA-4 (CD49d/CD29) with vascular cell adhesion molecule-1 (VCAM-1, CD106); the β_2 integrins, CD11a/CD18 (LFA-1) with intercellular adhesion molecule-1 (ICAM-1, CD54) or ICAM-2 (CD102); CD11b/CD18 (Mac-1, Mo1, CR3) with ICAM-1, and $\alpha_4\beta_7$ with MAdCAM-1. Endothelial E-selectin (CD62E, ELAM-1), endothelial P-selectin (CD62P,

GMP-140, PADGEM) and leukocyte L-selectin (CD62L, LAM-1, LECAM-1) all recognize sialyl Lewis^x (SLe^x) and other sialylated, fucosylated counter-structures (reviewed in 4 and 5), although the full spectrum of carbohydrate moieties recognized by the various selectins has not been defined. The proteins that bear the carbohydrate ligands for the selectins include the vascular mucins (6) such as PSGL-1 which presents carbohydrates to P- and E-selectin (7) and CD34 (8) and MAdCAM-1 (9) on high endothelial venules which bear an as yet incompletely characterized carbohydrate moiety recognized by murine lymphocyte L-selectin.

Studies *in vitro* and *in vivo* indicate that rolling is mediated by the interaction of selectins with their carbohydrate counter-structures, whereas the interaction of the β_2 integrin receptor complex, CD11/CD18, with ICAM-1 or other endothelial ligands is responsible for neutrophil sticking (reviewed in 1-3). However, there are several important caveats to consider in regards to this current model of neutrophil interaction with endothelium. First, selectin-mediated rolling is not a prerequisite for emigration under conditions of reduced shear forces, i.e., when there is stasis or reduced flow (10). Second, this model was developed for the systemic microcirculation, and its validity in the systemic arterial circulation or the pulmonary microcirculation is untested. In the lung, for example, emigration occurs in the capillaries where neutrophils are in contact with the vessel wall (11), and selectins may therefore not be necessary for initial tethering in this vascular bed.

The final phase of neutrophil-endothelial adhesive interactions during emigration is migration of the neutrophil *between* endothelial cells. This process of diapedesis involves the β_2 integrins and ICAM-1 (or other ligands), and, particularly, the immunoglobulin superfamily member, PECAM-1 (CD31), which is expressed on phagocytes as well as on the endothelial cell (12, 13).

The surface expression of adhesion proteins on endothelium is induced by diverse inflammatory stimuli (reviewed in 3). In cultured human endothelial cells P-selectin is rapidly mobilized from subcellular stores in Weibel-Palade bodies to the luminal surface in response to thrombin, histamine, LTC4/LTD4, and H_2O_2. E-selectin, VCAM-1, and ICAM-1 are induced over a period of hours following stimulation of endothelium by interleukin-1 (IL-1), tumor necrosis factor-alpha (TNF), and lipopolysaccharide. In the mouse P-selectin is also induced over several hours by LPS and cytokines. Interleukin-4 induces VCAM-1 but not E-selectin or ICAM-1, and interferon-gamma induces ICAM-1 and potentiates the induction of E-selectin by other agonists. Stimulation of endothelial cells by a variety of agonists induces surface expression of platelet-activating factor (PAF), a lipid mediator that activates CD11/CD18.

Adhesivity of leukocytes is determined by both quantitative and qualitative changes (reviewed in 3). Activation of neutrophils by chemoattractants (e.g., C5a, bacterial peptide, leukotriene B4, PAF) or by some cytokines (e.g., TNF, granulocyte-macrophage colony stimulating factor, and IL-8) causes shedding of L-selectin, translocation of CD11b/CD18 from secondary or tertiary granules to the plasma membrane, and an increase in the avidity of CD11a/CD18 and CD11b/CD18 for their ligands. Endothelial-derived surface molecules such as PAF and E-selectin may also activate neutrophil CD11b/CD18. The increase in avidity of CD11a/CD18 and CD11b/CD18 heterodimers on neutrophils may be mediated in part by an intracellular lipid mediator (14). Factors regulating the avidity of VLA-4 on mononuclear phagocytes have not been defined.

Selective recruitment of leukocytes subpopulations, e.g., eosinophils to the lung in asthma, results from a combinatorial process involving primary adhesion receptors (selectins), activation-dependent adhesion receptors (leukocyte integrins), and activating agents (e.g., chemoattractants, cytokines, and chemokines) (15, 16).

LEUKOCYTE ADHESION DEFICIENCY SYNDROMES

Studies of genetic deficiency syndromes have provided important insights into the molecular basis and biology of leukocyte emigration. Deficiency of β_2 integrins--leukocyte adhesion deficiency (LAD) type 1--results from heterogeneous mutations in the common β_2-subunit (CD18) (for review see 17). In LAD type 1 neutrophils are unable to stick to endothelium at sites of inflammation, resulting in a profound defect in neutrophil emigration. As a consequence, the severely affected patients suffer from life-threatening bacterial and fungal infections. Mononuclear leukocyte traffic and cell-mediated immune responses are minimally affected in LAD type 1 as circulating leukocytes other than neutrophils express the β_1 integrin receptor VLA-4 that allows firm adhesion to its endothelial ligand, VCAM-1. The phenotype of the LAD type 1 thus results from the unique absence of VLA-4, the alternative integrin receptor, on circulating neutrophils.

Recently, a second LAD syndrome has been identified (18). LAD type 2 results from a defect in fucose metabolism, and the resultant failure to synthesize the fucosylated ligands for the selectin receptors. Neutrophils from the LAD type 2 patients express normal levels of the β_2 integrins and L-selectin, but lack SLe^x or other fucose-containing surface antigens. Neutrophils from these patients do not bind to E-selectin expressed on cytokine-activated cultured endothelial cells (10) or to purified E- or P-selectin, and exhibit a marked decrease in migration to skin chambers or skin windows *in vivo* (19). Consistent with the current paradigm of leukocyte-endothelial adhesive interactions, studies by intravital microscopy demonstrated that under conditions of flow, fluorescein-labeled LAD type 2 neutrophils showed a marked reduction in neutrophil rolling along and subsequent sticking to the venular wall of inflamed rabbit mesentery, but were able to stick and emigrate when flow, i.e., shear force, was reduced (10). Interestingly, the LAD type 2 syndrome appears to have a milder phenotype than LAD type 1, perhaps reflecting selectin-independent mechanisms of adherence under conditions of diminished flow or stasis at sites of inflammation.

PATHOLOGIC CONSEQUENCES OF NEUTROPHIL ADHESION TO ENDOTHELIUM

As clearly illustrated by the LAD syndromes, neutrophil adherence to endothelium is an essential component of host response against bacterial infection and in the repair of tissue damage. Under some circumstances, however, neutrophil-endothelial interactions may contribute to vascular and tissue injury. During the process of adherence and transendothelial migration, neutrophils may release toxic products, e.g., proteases or oxidants, that damage adjacent endothelium, provoking local edema and/or thrombosis. Once emigrated, neutrophils may damage tissue and organs by a similar mechanism. By preventing neutrophil adherence to endothelium and subsequent emigration to tissue, it may be possible to prevent vascular and tissue injury in inflammatory conditions. This approach--"anti-adhesion" therapy--has now been tested in a wide variety of animal models, and has often demonstrated striking protective effects (reviewed in 1, 3). Of particular interest are studies of anti-adhesion therapy in models of ischemia/reperfusion (I/R) injury. While prolonged ischemia clearly induces tissue damage and ultimately death, paradoxically, reperfusion of ischemic tissue may exacerbate damage by triggering an inflammatory response. A number of mediators have been implicated as initiators of reperfusion-associated inflammation, including PAF, IL-8, activated complement, TNF, and leukotrienes. Regardless of the initial stimulus, neutrophils are likely the final mediators of the effector phase of reperfusion injury. This proposal is based upon observation that inhibition of neutrophil adherence to endothelium significantly attenuates vascular tissue injury in models of I/R ranging from myocardial infarction and stroke to digit or limb replantation and frostbite (reviewed in 1). Importantly, blockade of selectins by monoclonal antibodies (20, 21) or oligosaccharides (22) in models of I/R appears to be equally effective as blockade of integrins, consistent with a prerequisite for selectin-mediated rolling for subsequent neutrophil adhesion and emigration in this setting. Interestingly, although a monoclonal antibody to P-selectin prevented ischemia/reperfusion injury in the rabbit ear (21), it failed to prevent neutrophil emigration into the peritoneum in response to instillation of bacterial organisms (23), a setting in which monoclonal antibody to β_2 integrin is fully blocking. It is possible that I/R represents a mild to moderate inflammatory stimulus in which selectin-dependent rolling is necessary, whereas peritoneal inflammation is a more severe inflammatory challenge in which reduced flow or stasis permits selectin-independent emigration. The ability of an anti-P selectin monoclonal antibody to inhibit I/R without impairing neutrophil emigration in response to bacterial infection is consistent with the milder clinical phenotype of LAD type 2 vs LAD type 1. Alternatively, neutrophil emigration into the inflamed peritoneum may involve E-selectin as well as P-selectin. Nevertheless, these results suggest that selectin blockade may be preferable to integrin blockade in settings where infectious complications are a concern, e.g., traumatic shock.

Anti-adhesion therapy can be achieved with agents that block receptor-ligand interaction such as monoclonal antibodies, peptides, oligosaccharides, or, potentially, small molecules. Adhesion can also be inhibited by drugs that target the signaling pathways involved in the regulation of endothelial cell adhesion molecule surface expression or the modulation of leukocyte integrin avidity. Clinical trials testing the safety and efficacy of several of these approaches to anti-adhesion therapy are planned or in progress. Anti-adhesion therapy is clearly a powerful experimental tool to determine the contribution of neutrophils to vascular and tissue injury. Whether it represents a novel approach to the therapy of inflammatory and immune disorders remains to be determined.

REFERENCES

1. JM Harlan, RK Winn, NB Vedder, CM Doerschuk, and CL Rice, In vivo models of leukocyte adherence to endothelium, in: "Adhesion: Its Role in Inflammatory Diseases," W.H. Freeman & Co., ed., New York (1992)
2. TA Springer, Traffic signals for lymphocyte recirculation and leukocyte emigration: the multistep paradigm, *Cell* 76:301 (1994).
3. T Carlos and JM Harlan, Leukocyte and endothelial adhesion molecules, *Blood* (in press, 1994).
4. MP Bevilacqua and RM Nelson, Selectins, *J. Clin. Invest.* 91:379 (1993).
5. LA Lasky, Selectins: interpreters of cell-specific carbohydrate information during inflammation, *Science* 258: 964 (1992).
6. Y Shimizu and S Shaw, Mucins in the mainstream, *Nature* 366:630 (1993).
7. D Sako, X-J Chang, KM Barone, G Vachino, HM White, G Shaw, GM Veldman, KM Bean, TJ Ahern, B Furie, DA Cumming and GR Larsen, Expression cloning of a functional glycoprotein ligand for P-selectin, *Cell* 75: 1179 (1993).
8. S Baumhueter, MS Singer, W Henzel, S Hemmerich, M Renz, SD Rosen and LA Lasky, Binding of L-selectin to the vascular sialomucin, CD34, *Science* 262:436 (1993).
9. EL Berg, LM McEvoy, C Berlin, RF Bargatze, and EC Butcher, L-selectin-mediated lymphocyte rolling on MAdCAM-1, *Nature* 336:695 (1993).
10. UH von Andrian, EM Berger, L Ramezani, JD Chambers, HD Ochs, JM Harlan, JC Paulson, A Etziono and K-E Arfors, In vivo behavior of neutrophils from two patients with distinct inherited leukocyte adhesion deficiency syndromes, *J. Clin. Invest.* 91:2893 (1993).
11. DC Lien, PM Henson, RL Capen, JE Henson, WL Hanson. WW Wagner, Jr., and GS Worthen, Neutrophil kinetics in the pulmonary microcirculation during acute inflammation, *Lab. Invest.* 65:145 (1991).
12. WA Muller, SA Weigl, X Deng and DM Phillips, PECAM-1 is required for transendothelial migration of leukocytes, *J. Exp. Med.* 178:449 (1993).

13. AA Vaporciyan, HM DeLisser, H-C Yan, Mendiguren II, SR Thom, ML Jones, PA Ward and SM Albelda, Involvement of platelet-endothelial cell adhesion molecule-1 in neutrophil recruitment in vivo, *Science* 262:1580 (1993).
14. A Hermanowski-Vosatka, JA van Strijp, WJ Swiggard and SD Wright, Integrin modulating factor-1: A lipid that alters the function of leukocyte integrins, *Cell* 68:341 (1992).
15. EC Butcher, Leukocyte-endothelial cell recognition: three (or more) steps to specificity and diversity, *Cell* 67:1033, (1991).
16. T Schweighoffer, S Shaw, Adhesion cascades: diversity through combinatorial strategies, *Curr. Opin. Cell Biol.* 4:824 (1992).
17. BR Schwartz and JM Harlan, Consequences of deficient granulocyte-endothelium interactions, in: "Vascular Endothelium: Interactions with Circulating Cells," Gordon JL ed., Elsevier Science Publishers B.V., Amsterdam , The Netherlands (1991), pp. 231-252.
18. A Etzioni, M Frydman, S Pollack, I Avidor, ML Phillips, JC Paulson, R Gershoni-Baruch, Brief report: recurrent severe infections caused by a novel leukocyte adhesion deficiency, *N. Engl. J. Med.* 327:1789 (1992).
19. TH Price, HD Ochs, R Gershoni-Baruch, JM Harlan, A Etzioni, In vivo neutrophil and lymphocyte function studies in a patient with leukocyte adhesion deficiency type II, *Blood*, (in press, 1994).
20. X-L Ma, AS Weyrich, DJ Lefer, M Buerke, KH Albertine, TK Kishimoto, and AM Lefer, Monoclonal antibody to L-selectin attenuates neutrophil accumulation and protects ischemic reperfused cat myocardium, *Circulation* 88:649 (1993).
21. RK Winn, D Liggitt, NB Vedder, JC Paulson, and JM Harlan, Anti-P-selectin monoclonal antibody attenuates reperfusion injury to the rabbit ear, *J. Clin. Invest.* 92:2042 (1993).
22. M Buerke, AS Weyrich, Z Zheng, FC Gaeta, MJ Forrest, and AM Lefer, Sialyl Lewis^x-containing oligosaccharide attenuates myocardial reperfusion injury in cats, *J. Clin. Invest.* 93:1140 (1994).
23. SR Sharar, SS Sasaki, LC Flaherty, JC Paulson, JM Harlan, and RK Winn, P-Selectin blockade does not impair leukocyte host defense against bacterial peritonitis and soft tissue infection in rabbits, *J. Immunol.* 151:4982 (1993).

THE DISTRIBUTION OF ADHESION MOLECULES IN NORMAL AND ATHEROSCLEROTIC ARTERIES AND AORTAS

Dinah V. Parums

Royal Postgraduate Medical School
Hammersmith Hospital's Trust
Du Cane Road
LONDON W12 ONN, U.K.

INTRODUCTION

Atherosclerosis is accompanied by lipid accumulation, cellular proliferation and intimal inflammation. T lymphocytes and macrophages, many of which are activated, are present in the fatty streak, the fibrofatty plaque and the complicated or advanced plaque, and have therefore been implicated in atherogenesis.[1-5]

In advanced atherosclerosis, however, in association with medial disruption and neovascularization, chronic inflammation is also present in the adventitia and media of the vessel wall.[6-9] Chronic periaortitis is the term used to describe this triad of advanced atherosclerosis, medial disruption and adventitial chronic inflammation which is particularly common in atherosclerotic abdominal aortic aneurysm walls. It involves a spectrum of chronic inflammation, which in its most severe clinical form includes 'inflammatory aneurysm' or if the aorta is not dilated, 'idiopathic retroperitoneal fibrosis'.[8]

Chronic periaortitis consists of a predominance of B lymphocytes, including plasma cells, which are surrounded by T lymphocytes (mostly CD4-positive) and scattered macrophages. Many of the lymphocytes are present in aggregates or lymphoid follicles with germinal centres. HLA DR expression is abundant, and inflammatory cells are activated and proliferating.[9,10] These findings indicate that this is a local, on-going inflammatory response. Recently, other workers have shown similar inflammatory cell types in adventitial lesions in dilated and undilated aortas.[11]

Adhesion molecules are fundamental in regulating inflammation. They are required for initial leukocyte adhesion to endothelial cells, for specific cellular recruitment into inflammatory tissue sites, for cell-cell interaction and signalling, and for lymphoid organization.[12-14]

E-Selectin (formerly, endothelial leukocyte adhesion molecule-1 (ELAM-1)), intercellular adhesion molecule-1 (ICAM-1), and vascular cell adhesion molecule-1 (VCAM-1) can be upregulated by cytokines during activation, and synthesized *de novo* during inflammation.[15-17] E-Selectin was originally identified as a neutrophil adhesion molecule and was thought to be primarily a mediator of acute inflammatory adhesion.[15-18] It also plays a role in adhesion of memory T cells *in vitro*, and homing *in vivo* in skin inflammation.[19,20] The cell type which binds to E-Selectin may be dependent on the local

Cardiovascular Disease 2
Edited by L.L. Gallo, Plenum Press, New York, 1995

microenvironment under the influence of various cytokines.[21,22] ICAM-1 has been shown to be upregulated in inflammatory responses although it is also present in non-inflamed tissue.[12,23] ICAM-1's co-receptor includes leukocyte function antigen-1 (LFA-1, alß2 integrin), which is constitutively expressed on all leukocytes, although it can also be upregulated.[15,24] The ability of LFA-1 to bind ICAM-1, however, also depends upon qualitative changes resulting from its molecular conformation within the cell membrane.[25] VCAM-1 (also known as INCAM-110) primarily regulates mononuclear cell adhesion, and its co-receptor, very late activation antigen-4 (VLA-4, a4ß1 integrin), is expressed by these cells.[17,26,27]

A greater understanding of adhesion molecules is revealing their immense complexity and importance in inflammatory processes.[14,28,29] Much of our current knowledge of the structure and functions of these molecules derives from *in vitro* work using binding assays, antibody-blocking studies and expression cloning, while the use of monoclonal antibodies in immunohistochemistry has helped to reveal the *in situ* spatial and temporal adhesion molecule expression patterns and has helped in understanding their potential roles *in vivo*.

We have recently shown that E-Selectin and ICAM-1 are expressed by intimal endothelial cells in normal coronary arteries and overlying aortic fatty streaks; that as aortic atherosclerosis develops, ICAM-1 expression is associated with intimal lymphocyte and macrophage populations; that E-Selectin and VCAM-1 expression in the aortic adventitia increases with the severity of the atheromatous plaque and with the development of adventitial inflammatory cell infiltrates.[31] At present, the role of adhesion molecules in the aortic adventitial chronic inflammation associated with advanced atherosclerotic plaques (chronic periaortitis) has not been described.

The objectives of this study were to determine the distribution of cell adhesion molecules ICAM-1, E-Selectin and VCAM-1 in a spectrum of aortic and coronary artery atherosclerotic lesions and normal vessels, in a range of age groups, from fetuses to elderly patients, in order to determine the relationship between adhesion molecule expression and inflammatory cell recruitment in the progression of atherosclerosis.

MATERIALS AND METHODS

Surgical and fresh autopsy (within 24 hours of death) specimens were examined including 105 aortas (17 normal; 26 fatty streaks; 19 diffuse intimal thickening; 8 fibro-fatty plaques and 35 advanced plaques which included 11 cases of chronic periaortitis) and 65 coronary arteries (3 normal; 16 fatty streaks; 12 diffuse intimal thickening; 14 fibro-fatty plaques; 20 advanced plaques) from 47 patients including 5 fetuses (14-19 weeks gestation) and 5 infants (aged 1 day to 7 months) with the remaining cases ranging from 18 to 95 years.

Representative areas of each sample of aortic wall (including the atherosclerotic plaque/intima, media and adventitia) were taken for paraffin embedding and for freezing. Routine histology was carried out on paraffin sections with Haematoxylin and Eosin (H&E) staining to determine the degree of inflammation in the aortic wall and Elastic van Gieson (EVG) staining to assess medial disruption.Immunohistochemistry was carried out on frozen sections from tissue samples which were snap frozen in liquid nitrogen. 8μm cryostat sections were cut, air dried overnight, and fixed in acetone for 10 minutes before immunostaining.

The monoclonal antibodies used in this study are described in **Table 1**. The adhesion molecule specific monoclonal antibodies 13D5 (E-Selectin)[35], 14C11 (ICAM-1)[23] and 4B2 (VCAM-1)[26] were obtained from British Bio-technology Ltd, Oxford, U.K. These antibodies were used at 1 μg/ml as assessed by titration assays on control tonsil sections. The specificity of these antibodies was previousely determined by differential binding of activated endothelial cells, by binding to COS cells transfected with the appropriate adhesion molecule cDNA without cross reactions, and by immunoprecipitation.[36] These antibodies have also been used in immunohistochemistry in reactive human lymph nodes.[48] B-5G10 (CD49d, VLA-4-a-chain), and TMD3-1 (CD11a, LFA-1-a-chain) were obtained from the Fourth Workshop on Human Leukocyte Differentiation Antigens.[37] The remaining antibodies which included: T cell markers 3D4 (CD3)[38], T3-10 (CD4)[39], Tü102 (CD8)[39], and UCHL1 (CD45R0)[40]; B cell markers HD37 (CD19)[41], and 4KB128 (CD22)[41]; macrophage marker EBMII (CD68)[42]; follicular dendritic cell marker R4/23[43]; a marker for MHC Class II molecule (HLA DR a) (TAL.1B5)[44]; endothelial markers to von Willebrand Factor (vWF)(F8/86)[45] and JC70 (CD31)[46], and the basement membrane marker CIV 22 (type IV collagen)[47] were obtained as and used as neat cell culture supernatent from the Nuffield Department of Pathology, Oxford, U.K.

Table 1

Panel of Monoclonal Antibodies

Antibody	Specificity	Association	Reference
13D5 BB19-E6	E-Selectin	activated endothelial cells	Wellicome et al (1990)
14C11 BB19-I1	ICAM-1, CD54	activated endothelial cells, mononuclear cells	Dustin et al (1986)
4B2 BB19-V1	VCAM-1	activated endothelium, follicular dendritic cells	Elices et al (1990)
TMD3-1	LFA-1-a, CD11a	leukocytes Human Leukocyte	IVth Workshop on Diffn. Antigens (1989)
B-5G10	VLA-4-a, CD49d	mononuclear cells	"
F8/86	Factor VIII	most endothelial cells	Naiem et al (1982)
JC70	PECAM, CD31	endothelial cells, platelets, some lymphocytes	Parums et al (1990)
CIV 22	collagen type IV	basement membrane	Odermatt et al (1984)
EBM II	CD68	macrophages	Kelly et al (1988)
3D4	CD3	T cells	Kung et al (1989)
T3-10	CD4	T helper cells	Erber et al (1984)
Tü102	CD8	T cytotoxic/suppressor cells	Erber et al (1984)
UCHL1	CD45RO	T memory cells	Smith et al (1986)
HD37	CD19	B cells	Stein et al (1982)
4KB128	CD22	B cells	Stein et al (1982)
R4/23		follicular dendritic cells, B cells	Naiem et al (1983)
Y2/51	CD61	platelet glycoprotein IIIa	Gatter et al (1988)
1A4	a-smooth muscle actin	smooth muscle cells	Skalli et al (1986)

Single immunostaining was carried out on frozen sections using the alkaline phosphatase anti-alkaline phosphatase (APAAP) method.[49] Double immunostaining was performed in order to observe the cell types associated with specific cell adhesion molecule expression. This was carried out using the three-stage indirect immunoperoxidase technique with nickel enhancement, for the first monoclonal antibody followed by APAAP for the second monoclonal antibody,[50] and by the double immunofluorescent technique using APAAP with Fast-Red followed by indirect immunofluorescence with fluorescein isothiocyanate.[51] (Fast-Red and FITC fluoresce under the same excitation wavelength permitting simultaneous detection of two different antigens.) All sections were counterstained with haematoxylin. Negative controls were carried out by replacing the primary and/or the secondary monoclonal antibody with tris-buffered saline, tissue culture supernatant, or mouse anti-rabbit immunoglobulin (DAKOpatts U.K.) as an irrelevant antibody. Tonsil was used as positive control tissue.

RESULTS

The results are summarised in **Tables 2,3 and 4.**

ICAM-1 was found on both intimal and adventitial vascular endothelium of the normal aorta in all age groups **(Figures 1 and 5)**. E-Selectin and ICAM-1 were expressed by intimal endothelial cells in normal coronary arteries and overlying aortic fatty streaks **(Figures 1 and 2)**; as aortic atherosclerosis developed, ICAM-1 expression was associated with intimal lymphocyte and macrophage populations **(Figures 3 and 6)**;
E-Selectin and VCAM-1 expression in the aortic adventitia increased with the severity of the atheromatous plaque and with the development of adventitial inflammatory cell infiltrates (chronic periaortitis) **(Figure 4)**.

A consistent finding was the presence of E-Selectin on endothelial cells of up to half the vessels throughout the aortic wall and at the base of the atheroma independent of the severity of inflammation. ICAM-1 expression was abundant on many cell types and increased with the severity of chronic inflammation, being the strongest in the germinal centres. VCAM-1 expression was predominant on follicular dendritic cells and also increased with severity of adventitial inflammation (Figure 4). VCAM-1 expression was also detected on endothelial cells within lymphoid follicles.

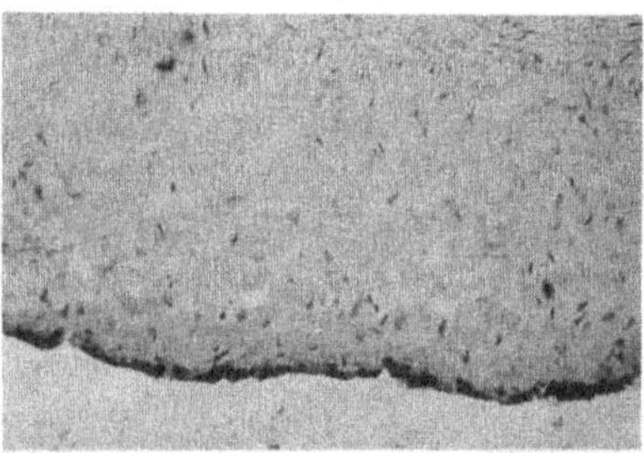

Figure 1
Intimal endothelium of a coronary artery with diffuse intimal thickening staining positively for E-Selectin

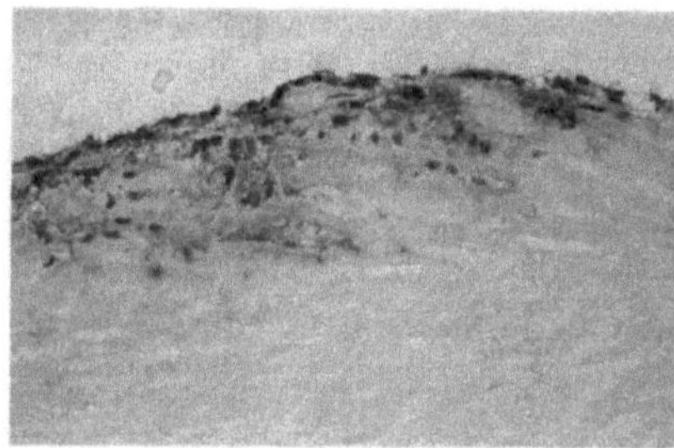

Figure 2
Coronary artery fatty streak showing ICAM-1 staining of cells within the lesion and of the overlying endothelium

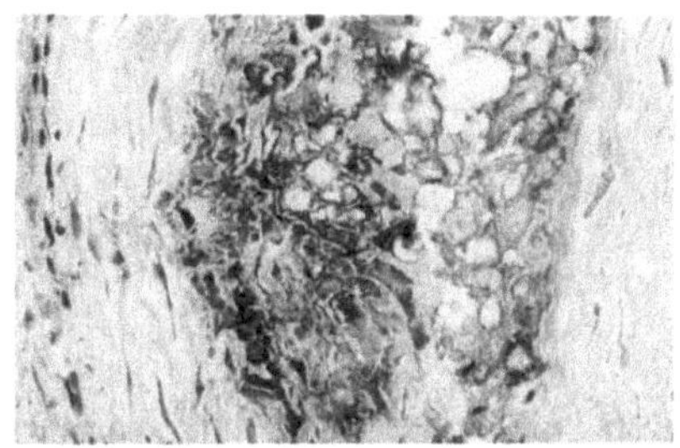

Figure 3
Macrophages and foam cells adjacent to necrotic core of fibro-fatty plaque in a coronary artery stain positively for ICAM-1

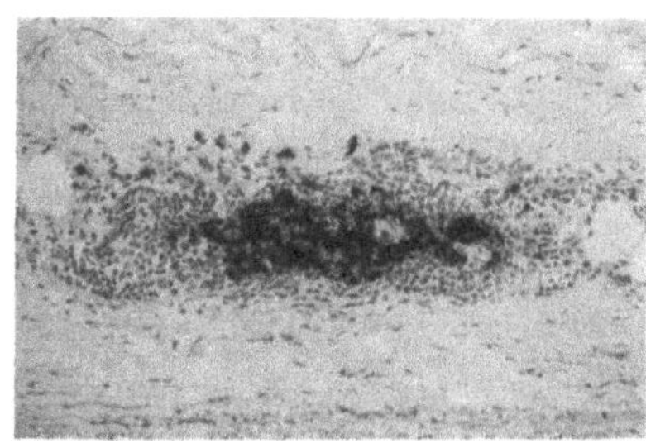

Figure 4
Intense VCAM-1 staining of an aortic adventitial lymphoid aggregate associated with an advanced intimal leasion (not shown)

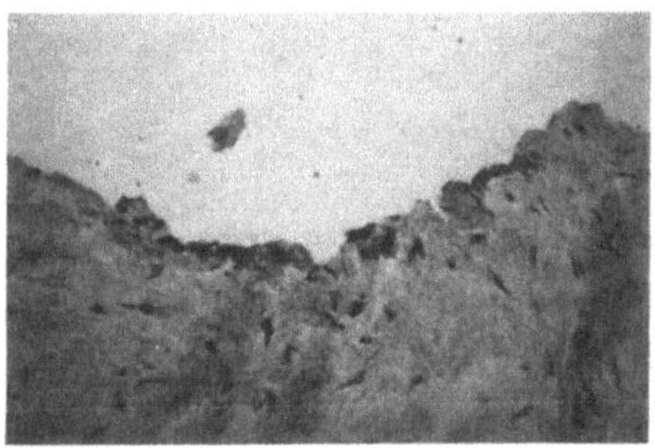

Figure 5
Intimal endothelium of a normal coronary artery shows focal staining for ICAM-1. The staining can be seen to be localised to intact endothelial cells.

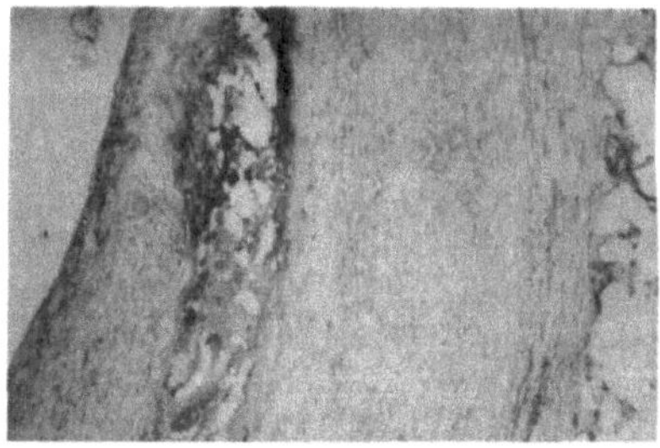

Figure 6
Low power view of a coronary artery fibro-fatty plaque showing ICAM-1 staining in the intima (bottom of the photograph) and in the macrophage-rich areas beneth the fibrous cap (top of the photograph).

Table 2

The Distribution of Adhesion Molecules in Normal Fetal, Infant and Adult Aortas

Patients;
5 aborted fetuses (14 to 19 weeks gestation);
5 infants (age range 1 day to 7 months) and
5 young adults (age range 18 to 32 yrs).

30 aortic specimens were obtained from the abdominal aorta 1) above the bifurcation and 2) from the thoracic aorta adjacent to the origin of the right subclavian artery. 22 aortic specimens containing intact endothelial cells (as determined by CD 31 staining) were studied which included;
13 normal specimens of aortic bifurcation; 9 normal specimens of aortic root and 5 'early' atherosclerotic lesions:

Lesion	Number of Specimens	Distribution of adhesion molecules E-Selectin	ICAM-1	VCAM-1
FETUSES	**9**			
Normal aorta	9	3 - focal intimal & diffuse vasa vasorum staining.	9 - diffuse staining of intima & vasa vasorum.	No staining.
INFANTS	**7**			
Normal aorta	6	4 - focal intimal & diffuse vasa vasorum staining.	6 - diffuse staining of intima & vasa vasorum.	No staining.
Fatty streak	1	1 - strong staining of intimal cells & vasa vasorum.	1 - diffuse staining of intima & vasa vasorum.	No staining.
YOUNG ADULTS	**6**			
Normal aorta	2	2 - focal intimal & diffuse vasa vasorum staining.	2 - diffuse staining of intima & vasa vasorum.	No staining.
Fatty streak	3	3 - strong staining of intimal cells & vasa vasorum.	3 - diffuse staining of intima & vasa vasorum.	1- focal, weak staining of vasa vasorum.
Diffuse intimal thickening	1	1 - strong staining of intimal cells & vasa vasorum.	1- diffuse staining of intima & vasa vasorum.	1 - focal, weak staining of vasa vasorum.

Table 3

The Distribution of Adhesion Molecules in Aortic and Coronary Artery Atherosclerosis

21 patients; 16 M and 5 F; age range 20 - 95 yrs.
65 coronary artery and 72 aortic specimens.

Lesion	Number of Specimens	Distribution of adhesion molecules E-SELECTIN	ICAM-1	VCAM-1
Normal aorta	3	No staining.	Focal, weak intimal endothelial cell staining.	No staining.
Diffuse intimal thickening	30	Staining of vasa vasorum. Coronary artery: staining of intimal endothelium.	Staining of vasa vasorum. Coronary artery: staining of intimal endothelium.	No staining.
Fatty streak	38	Staining of vasa vasorum. Aorta: focal staining of intimal endothelium. Coronary artery: diffuse staining of intimal endothelium.	Staining of intimal endothelial and mononuclear cells. Staining of vasa vasorum.	Focal,weak staining of vasa vasorum.
Fibro-fatty plaque	22	Staining of vasa vasorum. Coronary artery: staining of intimal endothelium.	Staining of intimal endothelial and mononuclear cells. Staining of vasa vasorum.	Focal,weak staining of vasa vasorum and of occasional intimal mononuclear cells.
Advanced plaque	44	Staining of vessels in adventitia and of vessels in the media and intima of inflamed sections.	Strong staining in macrophage-rich areas. Staining of adventitial, medial and intimal vessels. Staining of adventitial lymphoid aggregates.	Weak staining of adventitial and medial vessels, stronger in inflamed sections. Strong staining of adventitial lymphoid aggregates.

Table 4
Adhesion Molecule Expression in the Aortic Wall
in Chronic Periaortitis

11 cases. All were male patients aged between 63 and 83 years (mean 70.2 ± 12.8 yrs).

	E-SELECTIN	ICAM-1	VCAM-1
'Normal' Aorta	a few weakly positive vessels in vasa vasorum and focally on intimal endothelial cells	moderate staining on cells of vasa vasorum; superficial staining in the intima	very few weakly positive endothelial cells in vasa vasorum
Atherosclerotic Aorta			
Non-Inflamed (Case 1)	~one third positive vessels, confined to the outer wall	positive endothelial cells, stronger in outermost adventitia	few moderate to weakly positive focal endothelial cells on larger vessels in adventitia and at base of atheroma
Inflamed (Cases 2-10)	~half vessels strongly positive; no increase with severity of inflammation	increase with severity of inflammation	strong staining reaction from smallest to largest lymphoid aggregates; increase with severity of inflammation, and with size of lymphoid aggregate
cell types	restricted to endothelial cells of capillaries and venules	broad but variable degree of expression: on endothelial cells, mononuclear cells, and other cell types	moderate on some vessels surrounded by mononuclear cells, and weak staining on a few other vessels without surrounding mononuclear cells; strong staining on follicular dendritic cells in the centre of B cell aggregates
associated cell types	associated with macrophages and T cells, not B cells	associated with most cell types in the aortic wall	close proximity to B cells
in lymphoid follicles	no preferential association with lymphoid aggregates	strong germinal centre expression	preferential association with lymphoid aggregates, strong germinal centre expression

CONCLUSION

Adhesion molecule expression by coronary artery and aortic endothelium may represent a physiological property of the endothelium which is up-regulated in atherogenesis. The pattern of expression of the adhesion molecules suggests a role in the initiation and progression of chronic inflammation associated with human atherosclerosis.

THE CYTOKINE NETWORK IN HUMAN ATHEROSCLEROSIS

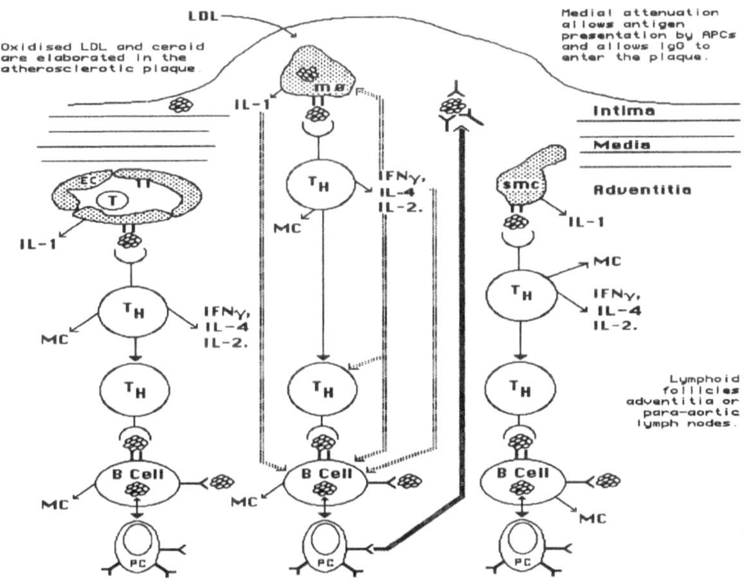

LEGEND:

Antigen presenting cells (APCs): **mø** -macrophage; **EC** -endothelial cell; **smc** -smooth muscle cell. [antigen symbol] -antigen (oxidised LDL/ceroid); **II** -Class II MHC molecule; PC -plasma cell; —< -immunoglobulin molecule (IgG); —(- T lymphocyte receptor; $\mathbf{T_H}$ -T helper lymphocyte; **IFNγ** -interferon; **MC** -memory cell; **IL-1(2)** -interleukin 1(2).

Figure 7

ACKNOWLEDGEMENTS

The author thanks Professor P. J. Morris, Mr P. M. Lamont, Mr J. C. Collin, and Mr R. J. Baigrie of the Nuffield Department of Surgery for providing surgical material; Dr D.E. Roskell, Ms. A.L. Ramshaw and Dr K. Wood for immunohistochemistry techniques; Blackwell Scientific Publications Limited, the editors of *Histopathology*, for allowing black and white reproduction of colour prints 1 to 6 from Ref. 31; Dr A. Gearing and Dr J. Gordon and British Bio-technology Ltd for their donation of the adhesion marker monoclonal antibodies, and Dr K. C. Gatter and Dr D. Y. Mason of the Nuffield Department of Pathology, University of Oxford for their donation of the remaining monoclonal antibodies. This work was supported by the British Heart Foundation.

REFERENCES

1. A.M. Gown, T. Tsukada and R. Ross, Human atherosclerosis II. Immunohistochemical analysis of the cellular composition of human atherosclerotic lesions, *Am J Pathol*.125:191 (1986).

2. L. Jonasson, J. Holm, O. Skalli , G. Bondjers and G.K. Hansson, Regional accumulations of T cells, macrophages, and smooth muscle cells in the human atherosclerotic plaque, *Arteriosclerosis* . 6:131 (1986).

3. G.K. Hansson, J. Holm and L. Jonasson, Detection of activated T lymphocytes in the human atherosclerotic plaque, *Am J Pathol.* 135:169 (1989).

4. A.C. van der Wal, P.K. Das, D.B. van de Berg, C.M. van der Loos and A.E. Becker, Atherosclerotic lesions in humans. *In situ* immunophenotypic analysis suggesting an immune mediated response, *Lab Invest.* 61:166 (1989).

5. P. Libby and G.K. Hansson, Biology of disease. Involvement of the immune system in human atherogenesis: Current knowledge and unanswered questions, *Lab Invest.* 64:5 (1991).

6. C.J. Schwartz and J.R.A. Mitchell, Cellular infiltration of the human arterial adventitia associated with atheromatous plaques, *Circulation* .26:73 (1962).

7. M.J. Mitchinson, Chronic periaortitis and periarteritis, *Histopathology* . 8:589 (1984).

8. D.V. Parums, The spectrum of chronic periaortitis. *Histopathology* . 16:423 (1990).

9. A.L. Ramshaw and D.V. Parums, Immunohistochemical characterization of inflammatory cells associated with advanced atherosclerosis, *Histopathology*.17:543 (1990).

10. A.L. Ramshaw and D.V. Parums, Inflammatory cells in chronic periaortitis are activated and are proliferating, *J Pathol* .163:12A (1991).

11. A.E. Koch, G.K. Haines, R.J. Rizzo, J.A. Radosevich, R.M. Pope, P.G. Robinson and W.H. Pearce, Human abdominal aortic aneurysms. Immunophenotypic analysis suggesting an immune mediated response, *Am J Pathol.*137:1199 (1990).

12. T.A. Springer, Adhesion receptors of the immune system, *Nature*. 346:425 (1990).

13. J.S. Pober and R.S. Cotran, What can be learned from the expression of endothelial -mediated response, *Am J Pathol* .137:1199 (1990).

14. R.O. Hynes and A.D. Lander, Contact and adhesive specificities in the associations, migrations, and targeting of cells and axons, *Cell.* 58:303 (1992).

15. D. Simmons, M.W. Makgoba and B. Seed, ICAM, an adhesion ligand of LFA-1, is homologous to the neural cell adhesion molecule NCAM, *Nature* . 331:624 (1988).

16. M.P. Bevilacqua, S. Stengelin, M.A. Jr. Gimbrone and B Seed, Endothelial leukocyte activation adhesion molecule 1: an inducible receptor for neutrophils related to complement regulatory proteins and lectins, *Science* . 243:1160 (1989).

17. L. Osborn, C. Hesslon, R. Tizard, C. Vassallo, S. Luhowskyj, G. Chi-Rosso and R. Lobb, Direct expression cloning of vascular cell adhesion molecule 1, a cytokine-induced endothelial protein that binds to lymphocytes, *Cell* .59:1203 (1989).

18. J.M. Munro, J.S. Pober and R.S. Cotran, Recruitment of neutrophils in the local endotoxin response: association with *de novo* endothelial expression of endothelial leukocyte adhesion molecule-1, *Lab Invest.* 64:295 (1991).

19. L.J. Picker, T.K. Kishimoto, C.W. Smith, R.A. Warnock and E.C. Butcher, ELAM-1 is an adhesion molecule for skin-homing T cells. *Nature* ,349:796 (1991).

20. Y. Shimizu, S. Shaw, N. Graber, T.V. Gopal, Horgan, G.A. Van Seventer and W. Newman, Activation-independent binding of human memory T cells to adhesion molecule ELAM-1, *Nature* . 349:799 (1991).

21. M.A. Gimbrone, M.S. Obin, A.F. Brock, E.A. Luis, P.E. Hass, C.A. Hébert, Y.K. Yip, D.W. Leung, D.G. Lowe, W.J. Kohr, W.C. Darbonne, K.B. Bechtol and J.B. Baker, Endothelial interleukin-8: a novel inhibitor of leukocyte-endothelial interactions, *Science* . 246:1601 (1989).

22. M.H. Thornhill, S.M. Wellicome, D.L. Mahiouz, J.S.S. Lanchbury, U. Kyan-Aung and D.O. Haskard, Tumor necrosis factor combines with IL-4 or IFN-g to selectively enhance endothelial cell adhesiveness for T cells. The contribution of vascular cell adhesion molecule-1-dependent and independent binding mechanisms, *J Immunol* .146:592 (1991).

23. M.L. Dustin, R. Rothlein, A.K. Bhan, C.A. Dinarello and T.A. Springer, Induction by IL 1 and interferon-g: tissue distribution, biochemistry, and function of a natural adherence molecule (ICAM-1), *J Immunol* .137:245 (1990).

24. M.E.F. Smith and J.A. Thomas, Cellular expression of lymphocyte function associated antigens and intercellular adhesion molecule-1 in normal tissue, *J Clin Pathol* .43:893 (1990).

25. R. Rothlein and T.A. Springer, The requirement for lymphocyte function-associated antigen 1 in homotypic leukocyte adhesion stimulated by phorbol ester, *J Exp Med*.163:1132 (1986).

26. M.J. Elices, L. Osborn, Y. Takada, C. Crouse, S. Luhowsky, M.E. Hemler and R.R. Lobb, VCAM-1 on activated endothelium interacts with the leukocyte integrin VLA-4 at a site distinct from the VLA-4/fibronectin binding site, *Cell* . 60:577 (1990).

27. A.S. Freedman, J.M. Munro, G.E. Rice, M.P. Bevilacqua, C. Morimoto, B.W. McIntyre, K. Rhynhart, J.S. Pober and L.M. Nadler, Adhesion of human B cells to germinal centres *in vitro* involves VLA-4 and INCAM-110, *Science* .249:1030 (1991).

28. A.M. Katz, D. Rosenthal and D.N. Sauder, Cell adhesion molecules. Structure, function, and implication in a variety of cutaneous and other pathologic conditions, *Int J Dermatol* . 30:153 (1991).

29. S. Montefort and S.T. Holgate, Adhesion molecules and their role in inflammation, *Resp Med* .85:91 (1991).

30. R.N. Poston, D.O. Haskard, J.R. Coucher, N.P. Gall and R.R. Johnson-Tidey, Expression of intercellular adhesion molecule-1 in atherosclerotic plaques, *Am J Pathol.* 140:665 (1992).

31. K.M. Wood, M.D. Cadogan, A.L. Ramshaw and D.V. Parums, The distribution of adhesion molecules in human atherosclerosis, *Histopathology* . 22:437 (1993).

32. O.Y. Printseva, M.M. Peclo and A.M. Gown, Various cell types in human atherosclerotic lesions express ICAM-1. Further immunocytochemical and immunochemical studies employing monoclonal antibody 1OF3, *Am J Pathol* . 140:889 (1992).

33. A.C. van der Wahl, P.K. Das, A.J. Tigges and A.E. Becker, Adhesion molecules on the endothelium and mononuclear cells in human atherosclerotic lesions, *Am J Pathol.*141:1427 (1992).

34. C. Page, M. Rose, M. Yacoub and R. Pigott, Antigenic heterogeneity of vascular endothelium, *Am J Pathol.* 141:673 (1992).

35. S.M. Wellicome, M.H.Thornhill, C. Pitzalis, D.S. Thomas, J.J.S. Lanchbury , G.S. Panayi and D.O. Haskard, A monoclonal antibody that detects a novel antigen on endothelial cells that is induced by tumour necrosis factor IL-1 or lipopolysaccharide, *J Immunol.* 144:2558 (1990).

36. D.L. Simmons and L Needham, Cloning cell surface molecules using monoclonal antibodies. In J.L. Gordon ed., *Vascular Endothelium: Interactions with Circulating Cells.* Elsevier Science Publishers 3-29

37. W. Knapp, B. Dörken, W.R. Gilks, E.P. Rieber, R.E. Schmidt, H. Stein and von dem Borne, A.E.G.K.r (eds) Leukocyte Typing IV. White cell differentiation antigens. Oxford University Press, Oxford. (1989)

38. P.C. Kung, G. Goldstein, E.L. Reinherz and S.F. Schlossman. Monoclonal antibodies defining distinctive human T cell surface antigens, *Science* .206:347 (1979).

39. W.N. Erber, A.J. Pinching and D.Y. Mason, . Immunocytochemical detection of T and B cell populations in routine blood smears, *Lancet.* 1:1042 (1984).

40. S.H. Smith, M.H. Brown, D. Rowe, R.E. Callard and P.C.L. Beverley, Functional subsets of human helper-inducer cells defined by a new monoclonal antibody, UCHL1, *Immunology* .58:63 (1986).

41. H. Stein, J. Gerdes and D.Y. Mason, The normal and malignant germinal centre, *Clin Haematol* . 11:531 (1982).

42. P.M.A. Kelly, E. Bliss, J.A. Morton, J. Burns and J.O'D. McGee, Monoclonal antibody EBM/11: high cellular specificity for human macrophages, *J Clin Pathol.* 41:510 (1988).

43. M. Naiem, J. Gerdes, Z. Adulaziz, H. Stein and D.Y. Mason . Production of a monoclonal antibody reactive with human dendritic reticulum cells and its use in the immunohistologic analysis of lymphoid tissue, *J Clin Pathol* .36:167 (1983).

44. A.A. Epenetos, L.G. Borrow, T.E. Adams, C.M. Collins, Isaacson PG and Bodmer WF: A monoclonal antibody that detects HLA-DR region antigen in routinely fixed, wax embedded sections of normal and neoplastic lymphoid tissue, *J Clin Pathol.* 38:12-17 (1985).

45. M. Naiem, J. Gerdes, Z. Adulaziz, C.A. Sunderland, M.J. Allington, H. Stein and D.Y. Mason, The value of immunohistological screening in the production of monoclonal antibodies, *J Immunol Meth.* 50:145 (1982).

46. D.V. Parums, J.L. Cordell, K. Micklem, A.R. Heryet, K.C. Gatter and D.Y. Mason, JC70 : a new monoclonal antibody that detects vascular endothelium associated antigen on routinely processed tissue sections. *J Clin Pathol*.43:752 (1990).

47. B.F. Odermatt, A.B. Lang, J.R. Ruttner, K.H. Wintherhalten and B. Treb, Monoclonal antibodies to human type IV collagen: useful reagents to demonstrate the heterodimeric nature of the molecule, *Proc Natl Acad Sci*. USA. 81:7343 (1984).

48. L.P. Ruco, D. Pomponi, R. Pigott, A.J.H. Gearing, A. Baiochini and C.D. Baroni, Expression and cell distribution of the adhesion molecules ICAM-1, VCAM-1, ELAM-1 and ENDOCAM (CD31) in reactive human lymph nodes and Hodgkin's disease, *Am J Pathol*. 140:1337 (1992).

49. J.L. Cordell, B. Falini, W.N. Erber, A.K. Ghosh, Z. Abdulaziz, S. MacDonald, K.A.F. Pulford, H. Stein, D.Y. Mason, Immunoenzymatic labeling of monoclonal antibodies using immune complexes of alkaline phosphatase and monoclonal anti-alkaline phosphatase (APAAP complexes), *J Histochem Cytochem*.32:219 (1984).

50. D.Y. Mason, A. Abdulaziz, B. Falini, H. Stein, Single and double immunoenzymatic techniques for labeling tissue sections with monoclonal antibodies, *Ann NY Acad Sci*.238:1073 (1983).

51. A.L. Ramshaw and D.V. Parums, Combined immunohistochemical and immunofluorescence method to determine the phenotype of proliferating cell populations, *J Clin Pathol*. 45:1015 (1992).

52. G. Koopman, H.K. Parmentier, H.J. Schuurman, W. Newman and C.J.L.M. Meijer, Pals ST. Adhesion of human B cells to follicular dendritic cells involves both the lymphocyte function-associated antigen 1/intercellular adhesion molecule 1 and very late antigen 4/vascular cell adhesion molecule 1 pathways, *Exp Med*.173:1297 (1991).

53. A.E. Koch, J.C. Burrows, G.K. Haines, T.M. Carlos, J.M. Harlan and S.J. Leibovich, Immunolocalization of endothelial and leukocyte adhesion molecules in human rheumatoid and osteoarthritic synovial tissues. *Lab Invest* 1991;**64;**313-20.

54. G.E. Rice, J.M. Munro, C. Corless and M.P. Bevilacqua, Vascular and nonvascular expression of INCAM-110. A target for mononuclear leukocyte adhesion in normal and inflamed human tissues, *Am J Pathol*.138:385.

55. R.S. Cotran, M.A. Jr Gimbrone, M.P. Bevilacqua, J.L. Madrick and J.S. Pober, Induction and detection of a human endothelial activation antigen *in vivo*. *J Exp Med* 164:661 (1986).

56. J.M. Munro, J.S. Pober and R.S. Cotran, Tumor necrosis factor and interferon-g induce distinct patterns of endothelial activation and associated leukocyte accumulation in skin of *Papio Anubis*. *Am J Pathol*.135:121 (1989).

57. J.S. Pober, M.A. Gimbrone, L.A. LaPierre, D.L. Mendrick, W. Fiers, R. Rothlein and T.A. Springer, Overlapping patterns of activation of human endothelial cells by interleukin 1, tumor necrosis factor, and immune interferon, *J Immunol*.137:1893 (1986).

58. J. Doukas and J.S. Pober, IFN-g enhances endothelial activation induced by tumor necrosis factor but not IL-1, *J Immunol* .145:1727 (1990).

59. A.L. Ramshaw, J.E. Stickland, R,J,Baigrie and D.V. Parums, Cytokine and cytokine receptor expression associated with advanced atherosclerosis, *J Pathol* 164:348A (1991).

60. M.J. Mitchinson, Insoluble lipids in human atherosclerotic plaques, *Atherosclerosis* 45:11 (1982).

61. D.V. Parums, D.L. Brown and M.J. Mitchinson, Serum antibodies to oxidized low-density lipoprotein and ceroid in chronic periaortitis, *Arch Pathol Lab Med* .114:383 (1990).

62. A.N. Orekhov, V.V. Tertov, A.E. Kabakov, I.Y. Adamova, S.N. Pokrovsky and V.N. Smirnov, Autoantibodies against modified low density lipoprotein. Non lipid factor of blood plasma that stimulates foam cell formation, *Arteriosclerosis Thrombosis* . 11:316 (1991).

63. J.T. Salonen, S. Yla-Herttuala, R. Yamamoto, H. Korpela,R. Salonen, K. Nyssinen, W. Palinski and J.L.Witztum, Autoantibody against oxidized LDL and progression of carotid atherosclerosis, *Lancet* .339:883 (1992).

64. D.V. Parums and M.J. Mitchinson, Demonstration of immunoglobulin in the neighbourhood of advanced atherosclerotic plaques, *Atherosclerosis.* 38:211 (1981).

65. D.V. Parums , D.R. Chadwick and M.J. Mitchinson, The localisation of immunoglobulin in chronic periaortitis, *Atherosclerosis* .61:117 (1986).

66. J. Frostegard, A. Haegerstrand, M. Gidlund and J. Nilsson, Biologically modified LDL increases the adhesive properties of endothelial cells, *Atherosclerosis* . 90:119 (1991).

67. M.A. Gimbrone, M.P. Bevilacqua and M.I. Cybulsky, Endothelial-dependent mechanisms of leukocyte adhesion in inflammation and atherosclerosis, *Ann N Y Acad Sci* ..598:77 (1990).

68. M.I. Cybulsky and M.A. Gimbrone, Endothelial expression of a mononuclear leukocyte adhesion molecule during atherogenesis, *Science*. 251:788(1991).

69. J.W. Berman and T.M. Calderon, The role of endothelial cell adhesion molecules in the development of atherosclerosis, *Cardiovasc Pathol* .1:17(1992).

L-SELECTIN REGULATION OF LYMPHOCYTE HOMING AND LEUKOCYTE ROLLING AND MIGRATION

Thomas F. Tedder, Anjun Chen and Pablo Engel

Department of Immunology
Duke University Medical Center
Durham, NC 27710

INTRODUCTION

The recruitment of leukocytes from the blood and lymphatic vascular systems and their extravasation into tissues is critical for host defense against pathogens and response to tissue injury and also contributes to the pathophysiology of many inflammatory diseases. This process is regulated in part by specific leukocyte-endothelial cell interactions with several families of cell adhesion molecules participating in recognition, adhesion, and extravasation[1]. The selectin family of adhesion molecules mediates the initial interactions of leukocytes with endothelium that allows leukocytes to roll along the venular wall[2,3]. The selectins bind to sialomucins which serve as scaffolds for the proper presentation of carbohydrate ligands to the selectins. Subsequently, integrins and immunoglobulin superfamily members interact to arrest leukocyte rolling and mediate "firm" attachment to the vascular endothelium. Multiple leukocyte integrins are known to be involved including LFA-1 (CD11a/CD18), Mo-1/Mac-1 (CD11b/CD18), VLA-4 (CD49d/CD29) and perhaps p150/95 (CD11c/CD18). Several members of the Ig superfamily such as Vascular Cell Adhesion Molecule-1 (VCAM-1, CD106), Intercellular Adhesion Molecule-1 (ICAM-1, CD54), ICAM-2 (CD102) and ICAM-3 (CD50) serve as integrin ligands. The utilization of different selectin, integrin, and immunoglobulin superfamily members by various leukocyte subclasses as they pass through distinct beds of vascular endothelium allows for considerable diversity and specificity in leukocyte migration[4]. L-selectin is a member of the selectin family of adhesion receptors and plays a central role in governing the migration patterns of different leukocyte classes[5,6].

L-SELECTIN STRUCTURE

The selectin family consists of three closely related cell-surface molecules, L- (CD62L), E- (CD62E) and P-selectin (CD62P). Each selectin has a characteristic extracellular region that includes a calcium-dependent lectin domain, an epidermal growth

Cardiovascular Disease 2
Edited by L.L. Gallo, Plenum Press, New York, 1995

factor (EGF)-like domain, and two to nine short consensus repeat (SCR) units homologous to domains found in complement binding proteins[7-11]. L-selectin has two SCR domains, a transmembrane region, and a short cytoplasmic tail[7,8]. The gene for L-selectin is assembled from individual exons encoding distinct structural units[12]. Each of the selectin genes are organized identically, and are arranged in tandem along chromosome 1 in both mice and humans[7,12-15]. The amino-terminal lectin domain is essential for recognition of specific carbohydrate determinants present on the appropriate sialomucin ligands[16-22]. The human L-selectin protein isolated from lymphocytes has a M_r of ~74,000 after reduction and 68,000 under non-reducing conditions[23], whereas that from neutrophils is 90-100,000 after reduction[24]. However, only a single L-selectin cDNA species exists, indicating that cell surface L-selectin undergoes considerable posttranslational processing to become heavily glycosylated[12,21].

L-SELECTIN EXPRESSION

Although structurally similar, the selectins have distinct patterns of expression. L-selectin is constitutively expressed by all classes of leukocytes[24-27]. In contrast, P-selectin is rapidly mobilized to the surface of activated venular endothelium or activated platelets[28-30]. E-selectin is expressed primarily on endothelium following activation with inflammatory cytokines[31,32].

L-selectin was first identified in mice by the MEL-14 mAb[25]. In humans, L-selectin was first identified by cDNA cloning[7]. Existing mAb with unknown target structures, TQ1 and Leu-8, were later shown to reacted with L-selectin cDNA transfected cells[27]. Subsequently, panels of mAb reactive with L-selectin have been generated[21]. All of these mAb demonstrate that L-selectin expression is limited to hematopoietic cells. L-selectin is expressed by a significant proportion of mature thymocytes and a subset of the least mature thymic lymphocytes[27]. In peripheral blood, L-selectin expression correlates with the activation or differentiation status of lymphocytes. The majority of virgin T cells and a subpopulation of memory T cells express L-selectin, but distinct subpopulations of both $CD4^+$ and $CD8^+$ cells lack L-selectin[27]. L-selectin is also expressed by a subpopulation of NK cells although its function on this cell lineage remains unknown. L-selectin is expressed late in B cell ontogeny, but most circulating B cells express L-selectin[33]. T and B lymphocytes exhibit a reversible loss of L-selectin expression after mitogenic stimulation[27]. Concomitant with the decrease in L-selectin expression is an increase in expression of CD2, LFA-1, VLA-4 and LFA-3[27]. L-selectin is expressed at higher frequencies and levels by blood lymphocytes when compared with tissue lymphocytes suggesting that entry into lymphoid tissues may result in the partial or total loss of L-selectin from the cell surface[27]. The regulated expression of L-selectin by distinct subpopulations of lymphocytes may explain why different lymphocyte subsets show marked differences in their migration patterns.

L-selectin is expressed by nearly all blood neutrophils, monocytes, and eosinophils[24,27,34]. Monocytes and neutrophils express similar levels of cell-surface L-selectin as lymphocytes, but activation of these cell types during their isolation usually results in some loss of the molecule[35]. L-selectin is expressed more or less continuously throughout myeloid differentiation in the bone marrow and myeloid progenitor cells at all levels of maturation express L-selectin[24,36]. Early erythroid progenitor cells (BFU-E) also express L-selectin, although mature erythrocytes do not express L-selectin. Thus, L-selectin may play a broad role in the trafficking of various leukocyte lineages.

L-SELECTIN FUNCTION

Stamper and Woodruff first demonstrated that the binding of mouse lymphocytes to specialized high endothelial cells present in postcapillary venules (HEV) of lymph nodes was a specific event[37]. In this *in vitro* assay, frozen sections of lymph nodes were placed on glass slides and overlayered with lymphocytes while the slides were rotating. Lymphocytes attached to the HEV, but not to other portions of the tissue. L-selectin was then structurally identified by the MEL-14 mAb which blocks the binding of mouse lymphocytes to HEV of peripheral lymph nodes *in vitro* and inhibits lymphocyte homing to peripheral lymph nodes *in vivo*[25]. These studies also revealed that L-selectin primarily mediated lymphocyte attachment to peripheral and mesenteric lymph nodes, but not to Peyer's patch HEV. Lymphocyte binding to Peyer's patch HEV was subsequently shown to be controlled by the $\alpha4\beta7$ integrin[38]. This has lead to the hypothesis that lymphocyte entry into secondary lymphoid organs is initiated by the binding of specific lymphocyte "homing" receptors with their appropriate "vascular addressin" ligands located on the luminal surface of lymph nodes and Peyer's patch HEV[39]. More recent studies have confirmed that human L-selectin also mediates lymphocyte binding to peripheral lymph node HEV[40-42]. Further, a recombinant L-selectin/IgG heavy chain chimera protein binds to peripheral lymph node HEV[43,44].

L-selectin also mediates the binding of leukocytes at sites of tissue injury and inflammation. Neutrophil binding to HEV from inflamed peripheral lymph nodes is inhibited by the MEL-14 mAb, and activated neutrophils which have lost L-selectin expression do not bind to HEV and do not home to inflammatory sites *in vivo*[26,45]. In addition, intravenous administration of the MEL-14 mAb inhibits accumulation of neutrophils in inflammatory lesions[45]. When neutrophil, monocyte, lymphocyte and eosinophil binding to cytokine-treated human umbilical cord endothelial cells is assessed under non-static or rotating conditions to recapitulate blood flow, anti-L-selectin mAb significantly inhibit cell attachment[34,35,46,47]. Similarly, transfection of L-selectin negative cell lines with L-selectin cDNA confers the ability to bind to activated endothelial cells[47]. These data directly demonstrate the presence of an L-selectin ligand on the surface of activated endothelium. Since L-selectin binding in this assay is completely dependent on the prior activation of the endothelial cells with proinflammatory mediators, the L-selectin ligand must be inducibly expressed on endothelial cells. In contrast to L-selectin, P-selectin expressed by activated endothelial cells and platelets binds myeloid cells and a subset of T cells[17,29,30]. E-selectin also mediates adhesion of myeloid cells and a subset of memory T cells to activated endothelium[31,32,48,49].

LYMPHOCYTE HOMING AND L-SELECTIN DEFICIENT MICE

Mice lacking detectable expression of cell surface L-selectin have been generated by homologous recombination using embryonic stem cells[50]. Two striking features of the L-selectin deficient mice are the significant reduction in the number of resident lymphocytes in peripheral lymph nodes and that lymphocytes from these mice are completely inhibited in their ability to attach to peripheral lymph node-HEV in *in vitro* assays. Both short term and long term *in vivo* trafficking experiments also revealed that lymphocytes from L-selectin deficient mice are unable to home into peripheral lymph nodes of normal mice. There is also a significant reduction in the number of histologically distinct HEV observed in peripheral lymph nodes of L-selectin-deficient mice. This phenomenon mimics the rapid decrease in lymphocyte adherence to HEV in lymph nodes deprived of afferent lymphatic vessels[51-53]. These studies clearly demonstrate the essential role that L-selectin plays in lymphocyte binding to HEV and subsequent entry into peripheral lymph nodes.

The recirculation and homing of L-selectin deficient lymphocytes into mesenteric lymph nodes of normal mice is also significantly inhibited, but not as completely as with peripheral lymph nodes. A dual homing specificity for mesenteric lymph nodes involving L-selectin and additional receptors confirms previous *in vitro* studies[54-56]. HEV of mesenteric lymph nodes have also been found to express high levels of both the peripheral lymph node addressin recognized by the MECA-79 antibody[57] and the mucosal addressin MadCAM-1 (mucosal addressin cell adhesion molecule 1)that is preferentially expressed by Peyer's patch HEV[56]. MadCAM-1 contains immunoglobulin-like domains and a mucin-like domain[58] and is a ligand for the $\alpha4\beta7$ integrin[59]. MadCAM-1 isolated from mesenteric lymph nodes can also serve as a ligand for L-selectin *in vitro*[60]. The presence of MadCAM-1 in mesenteric lymph nodes may explain why lymphocyte migration to mesenteric lymph nodes in L-selectin deficient mice was less severely inhibited than migration to peripheral lymph nodes which lack this ligand.

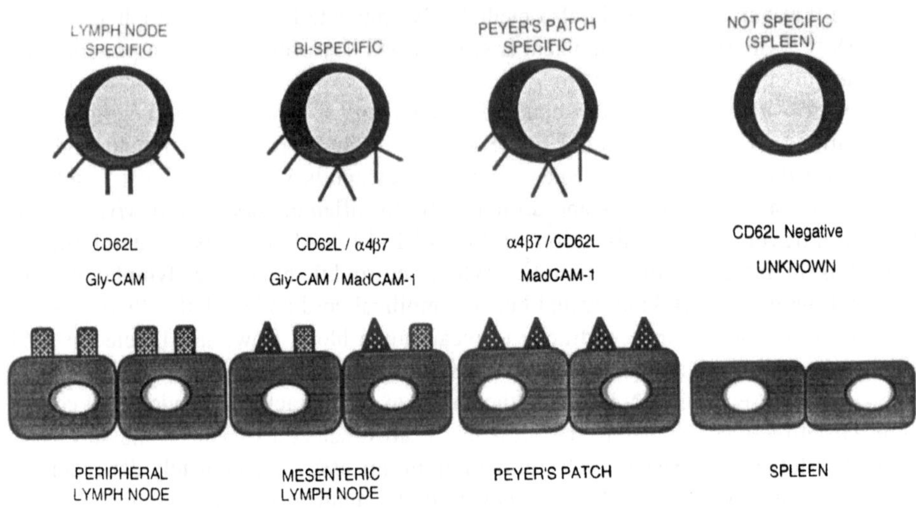

Figure 1. Role of L-selectin in the homing of lymphocytes to peripheral and mesenteric lymph nodes, Peyer's patch and spleen. L-selectin (CD62L) mediates all lymphocyte attachment to peripheral lymph node HEV by binding to a cell surface ligand termed Gly-CAM. In combination with other adhesion receptors, L-selectin contributes to lymphocyte homing to mesenterid nodes. L-selectin also contributes to Peyer's patch homing, perhaps by binding to the MadCAM-1 vascular addressin. In contrast, spleen does not have HEV and a loss of L-selectin expression appears to result in a preferential recruitment of lymphocytes to that tissue.

L-selectin has not been regarded as important for lymphocyte homing to gut-associated lymphoid tissues since the MEL-14 antibody does not block lymphocyte binding to Peyer's patch HEV or inhibit lymphocyte migration into Peyer's patches[25,38,61]. Similarly, treatment of HEV with sialidase inactivates peripheral lymph node HEV adhesive ligands, but has no effect on lymphocyte attachment to Peyer's patch HEV[55,62]. Further, an L-selectin-IgG chimera fails to stain Peyer's patch HEV in most instances or block lymphocyte binding to Peyer's patch endothelium[44]. In addition, lymphocyte attachment to Peyer's patch HEV is specifically and completely blocked by antibodies against α_4 or β_7 integrin chains[63] demonstrating a major role for this integrin in directing leukocyte trafficking to the gut. However, mouse Peyer's patch HEVs express low but functional levels of the addressin recognized by the MECA-79 antibody[57] and mouse lymphocyte homing into Peyer's patches can be partially inhibited by Fab fragments of the MEL-14 antibody[64]. Short term (1 h) *in vivo* homing experiments in L-selectin deficient mice demonstrated a reduction in the ability of lymphocytes to migrate into Peyer's patches, clearly revealing that L-selectin is involved in lymphocyte homing to the gut (Fig. 1). Appropriate carbohydrates decorating the mucin-like domain of MadCAM-1 may serve as binding sites for L-selectin on Peyer's patch HEV as previously suggested[58].

A deficiency in L-selectin expression also results in an increased number of resident lymphocytes in the spleen and a greater tendency of cells to migrate to the spleen. Although the spleen is one of the most important organs for lymphocyte recirculation and more lymphocytes pass through the spleen than recirculate through all lymph nodes, little is known about the molecular mechanisms involved in lymphocyte homing to and entry into the spleen[65]. Specific "homing" receptors or vascular "addressins" for spleen have not been defined, but they clearly differ from those by regulate lymphocyte migration to lymph nodes, Peyer's patch or tonsil. Thus, the spleen may serve as a reservoir for cells that are not able to specifically localize in other secondary lymphoid tissues (Fig. 1). This suggests that lymphocyte migration can also be regulated by the specific loss of adhesion receptors such as L-selectin, rather than just by the expression or acquisition of adhesion receptors specific for particular lymphoid tissues. The rapid down regulation of L-selectin may thus be an effective mechanism for increasing the number of lymphocytes that enter the spleen and perhaps Peyer's patches.

LEUKOCYTE ROLLING IS MEDIATED BY L-SELECTIN

Minutes after a tissue is injured circulating leukocytes begin to interact with the vascular endothelium by rolling along the vessel wall[66]. In mesenteric venules, leukocyte rolling reaches a peak 20-40 min after exteriorization of the tissue and remains fairly constant over at least 2 hours[67]. The finding that the adhesion of leukocytes to cytokine-activated endothelial cells is reduced by L-selectin blocking antibodies under non-static conditions[35,46,47,68], suggests that L-selectin is involved in the earliest adhesive interactions that permit leukocyte rolling. In agreement with this, leukocyte rolling *in vivo* is reduced by intravascular infusion of an antibody against L-selectin or L-selectin-IgG[69,70]. In addition, transfection of the human L-selectin cDNA into a cell line that does not express P- or E-selectin ligands demonstrated that L-selectin alone can mediate leukocyte rolling independent of P- or E-selectin[71]. These *in vivo* experiments also demonstrated that rolling at the earliest time points is likely to be mediated by P-selectin and at later time points by the induction of E-selectin expression. However, L -selectin is the major component of rolling at intermediate time points in this system. The frequency of rolling leukocytes in L-selectin deficient mice is reduced by ~70%[50], clearly demonstrating that L-selectin is involved in this process. However, the defect is incomplete suggesting that other adhesion molecules participate in this process. Leukocyte rolling is also almost completely absent in the venules of the exposed mesentery of P-selectin deficient mice[72]. Both L- and P-selectin function are therefore necessary to promote normal levels of rolling, with each receptor interacting with its own ligand on the opposing cell to help guarantee sufficient adhesive interactions to resist shear stress. Since P-selectin can be mobilized from intracellular storage granules to the cell surface within minutes following endothelial activation it is likely to be involved in the early phases of rolling. Subsequent rolling (20 to 120 min) is likely to be dominated by L-selectin. The requirement for intact L- and P-selectin function for optimal neutrophil rolling and the fact that most lymphocytes do not express functional ligands for P-selectin may at least partially explain why a much higher fraction of normal neutrophils roll in comparison with lymphocytes, and why lymphocytes but not neutrophils leave the circulation via HEV of lymph nodes.

Consistent with the decrease in leukocyte rolling in L-selectin deficient mice, there is a severe reduction in the ability of leukocytes to migrate into the peritoneal cavity subsequent to administration of an inflammatory agent[50]. L-selectin neutralizing antibodies or a soluble L-selectin-IgG chimera also inhibit neutrophil entry into the peritoneal cavity with a time course of neutrophil extravasation very similar to that obtained in the L-selectin deficient mice[43]. A similar defect with almost identical kinetics is observed in P-selectin deficient mice[72]. Additionally, recruitment of neutrophils into an inflamed peritoneum at 4 h is partially blocked by an E-selectin neutralizing antibody[73], suggesting that neutrophil influx at later time points is influenced by an E-selectin dependent pathway. These kinetic differences which result from the differential use of selectin family members may at least partially explain why neutrophils emigrate into inflammatory sites prior to the entry of monocytes and lymphocytes. The finding that leukocyte rolling and neutrophil extravasation at sites of inflammation are impaired in L-selectin deficient mice supports the multi-step model of leukocyte transmigration in which firm leukocyte-endothelial cell interactions are preceded by weak rolling interactions mediated by the selectins.

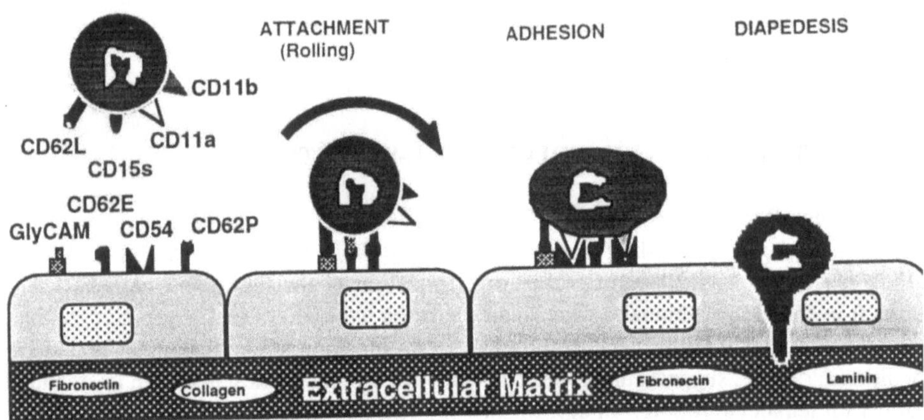

Figure 2. Neutrophil interactions with endothelial cells at sites of inflammation. Neutrophil entry into tissues is preceded first by rolling along the endothelial surface, a process mediated by the selectins. L-selectin (CD62L) binds to a ligand designated as GlyCAM while E-selectin (CD62E) and P-selectin (CD62P) bind to CD15s-related ligands on the surface of the neutrophil. This is followed by firm adhesion mediated through the integrins binding to immunoglobulin superfamily members. This is followed by diapedesis between endothelial cells.

The above data suggest the following model of leukocyte recruitment during inflammation (Fig. 2). Proinflammatory cytokines released by cells within inflamed tissues induce changes in the surrounding endothelium, including the *de novo* expression of ligands for leukocyte adhesion molecules: P-selectin, E-selectin, VCAM-1, and the L-selectin ligand. Adhesion of circulating leukocytes to activated endothelium is initiated by L-selectin binding to its ligand, and this initial interaction is stabilized by interaction of VLA-4 with VCAM-1 (for lymphocytes) or P- and E-selectin with sialylated CD15-related ligands (for neutrophils). Once firm attachment of the leukocyte to the luminal surface of the endothelial cell is stabilized, the cells migrate to intercellular junctions between adjacent endothelial cells. This movement probably relies on combinations of L-selectin, the CD11/CD18 proteins and VLA-4. Monocyte rolling, arrest and spreading on activated vascular endothelium under flow is mediated via sequential action of L-selectin, β_1-

integrins, and β_2-integrins[74]. A role for L-selectin in leukocyte migration across the endothelial cell surface towards junctional borders is suggested by the finding of a specific L-selectin epitope associated with cell motility is concentrated along the leading tip of extended processes on pseudopods and ruffles[75]. Additionally, L-selectin reorganizes into extending processes at sites of lymphocyte contact with endothelium[76].

REGULATION OF L-SELECTIN FUNCTION

Leukocyte rolling and adhesion to endothelium are dynamic processes that involve multiple adhesion processes and the active participation of the cells involved. L-selectin does not merely mediate the passive "adsorption" of leukocytes to receptors on the endothelial surface since L-selectin function involves a change in receptor affinity following cellular activation[41] and requires an intact cytoplasmic domain[77]. The increase in ligand binding activity for lymphocytes and neutrophils results directly from an increase in the affinity of the receptor for ligand which peaks at ~5 min and rapidly declines to baseline levels thereafter. The lectin domain alone of L-selectin primarily appears to mediate ligand binding[19]. However, some studies suggest that the EGF-like domain may be involved in regulating the affinity of the lectin domain[21,41]. Cooperativity between the lectin and EGF-like domains would lead to a higher avidity interaction, an important consideration given the assumption that L-selectin in its native conformation on unactivated leukocytes may have a low affinity for ligand. It is likely that the cytoplasmic domain of L-selectin mediates cytoskeletal associations which regulate the affinity change in the receptor that occurs with cellular activation. Affinity regulation of L-selectin by lineage-specific signals may account, at least in part, for why a much higher fraction of normal neutrophils than lymphocytes utilize L-selectin for rolling, and conversely, why lymphocytes but not neutrophils leave the circulation via HEV of lymph nodes.

L-selectin can also be rapidly shed from the leukocyte surface following cellular activation[42,78,79]. This may serve as a mechanism for the rapid modulation of leukocyte recirculation patterns and may regulate the migration of specific lymphocyte subpopulations[23]. Although the precise mechanism of shedding is unknown, it appears to result from activation-induced changes in the conformation of the L-selectin protein which exposes nascent sites on L-selectin that are then susceptible to cleavage by proteases[42]. Shed L-selectin (sL-selectin) from both lymphocytes and neutrophils is present at high levels (1.6 ± 0.8 μg/ml) in human plasma[80,81]. sL-selectin from plasma inhibits L-selectin-specific attachment of lymphocytes to endothelium in a dose dependent manner with leukocyte attachment completely inhibited at sL-selectin concentrations of 8 to 15 μg/ml. Elevated levels of serum sL-selectin have been observed in some clinical situations such as AIDS and leukemia[80]. Immunohistochemical staining of endothelium in lymph nodes and at sites of inflammation with L-selectin-directed mAb suggests that sL-selectin may bind to the luminal surface of vascular endothelium *in vivo*. The presence of serum sL-selectin with functional activity indicates a potential role for sL-selectin in the regulation of leukocyte attachment to endothelium. In this case, circulating sL-selectin may serve as a buffer system preventing leukocyte rolling at sites of sub-acute inflammation. However, once a threshold of inflammation is achieved leukocytes entering the site of inflammation may be able to effectively displace the sL-selectin present on the endothelial surface and bind specifically. Consistent with the observation that sL-selectin may bind activated endothelium, significantly decreased levels of sL-selectin have been observed in serum of patients with adult respiratory distress syndrome[82]. In fact, reduced levels of sL-selectin is the best available prognostic indicator for the development of adult respiratory distress syndrome in at-risk patient groups. It is important to note that the cellular activation signals which lead to the increase in affinity of L-selectin, in most cases, are the same signals which lead to the shedding of the molecule. The two processes, however, exhibit different kinetics. The upregulation of L-selectin affinity occurs rapidly (seconds to minutes), whereas the shedding requires minutes to hours[41]. The combination of regulated receptor function along with regulated levels of soluble receptor in the circulation are likely to play major roles in directing leukocyte migration.

CONCLUSIONS AND L-SELECTIN-DIRECTED THERAPIES

L-selectin serves an important role in leukocyte-endothelial attachment at sites of inflammation as clearly illustrated in L-selectin deficient mice. These studies provide a platform for further characterization of the molecular mechanisms of inflammation and provide a rational for use of antagonists of L-selectin function as therapeutic agents. Inhibition of L-selectin function is likely to have a dramatic effect on the progression of inflammatory responses. As an example of this, blocking L-selectin function completely inhibits ischemia-reperfusion injury in an animal model[83]. An understanding of the events that regulate inflammation will take considerable further effort, but should provide ample opportunities for therapeutic intervention.

ACKNOWLEDGMENTS

This work was supported by grants from the National Institutes of Health, HL-50985, AI-26872 and CA-54464. T. F. T. is a Scholar of the Leukemia Society of America.

REFERENCES

1. T.A. Springer, Traffic signals for lymphocyte recirculation and leukocyte emigration: the multistep paradigm, *Cell* 76:301 (1994).
2. L.A. Lasky, Selectins: Interpreters of cell-specific carbohydrate information during inflammation, *Science* 258:964 (1992).
3. M.P. Bevilacqua, and R.M. Nelson, Selectins, *J. Clin. Invest.* 91:379 (1993).
4. E.C. Butcher, Leukocyte-endothelial cell recognition: three (or more) steps to specificity and diversity, *Cell* 67:1033 (1991).
5. G.S. Kansas, O. Spertini, and T.F. Tedder, Leukocyte adhesion molecule-1 (LAM-1): structure, function, genetics and evolution., *in*: "Cellular and Molecular Mechanisms of Inflammation: Vascuclar Adhesion Molecules," Cochrane, C. and M. Gimbrone Jr., eds., Academic Press, Orlando, (31).
6. T.F. Tedder, W. Luscinskas, and G.S. Kansas, Regulation of leukocyte migration by L-selectin: mechanisms, domains and ligands, *Behring Inst. Mitt.* 92:165 (1993).
7. T.F. Tedder, T.J. Ernst, G.D. Demetri, C.M. Isaacs, D.A. Adler, and C.M. Disteche, Isolation and chromosomal localization of cDNAs encoding a novel human lymphocyte cell-surface molecule, LAM1: Homology with the mouse lymphocyte homing receptor and other human adhesion proteins, *J. Exp. Med.* 170:123 (1989).
8. L.A. Lasky, M.S. Singer, T.A. Yednock, D. Dowbenko, C. Fennie, H. Rodriguez, T. Nguyen, S. Stachel, and S.D. Rosen, Cloning of a lymphocyte homing receptor reveals a lectin domain, *Cell.* 56:1045 (1989).
9. M.H. Siegelman, M. van de Rijn, and I.L. Weissman, Mouse lymph node homing receptor cDNA clone encodes a glycoprotein revealing tandem interaction domains, *Science.* 243:1165 (1989).
10. G.I. Johnston, R.G. Cook, and R.P. McEver, Cloning of GMP-140, a granule membrane protein of platelets and endothelium: sequence similarity to proteins involved in cell adhesion and inflammation, *Cell* 56:1033 (1989).

11. M.P. Bevilacqua, S. Stengelin, M.A. Gimbrone Jr., and B. Seed, Endothelial leukocyte adhesion molecule 1: an inducible receptor for neutrophils related to complement regulatory proteins and lectins, *Science* 243:1160 (1989).

12. D.C. Ord, T.J. Ernst, L.J. Zhou, A. Rambaldi, O. Spertini, J.D. Griffin, and T.F. Tedder, Structure of the gene encoding the human leukocyte adhesion molecule-1 (TQ1, Leu-8) of lymphocytes and neutrophils, *J. Biol. Chem.* 265:7760 (1990).

13. M.L. Watson, S.F. Kingsmore, G.I. Johnston, M.H. Siegelman, M.M. Le Beau, R.S. Lemons, N.S. Bora, T.A. Howard, I.L. Weissman, R.P. McEver, and M.F. Seldin, Genomic organization of the selectin family of leukocyte adhesion molecules on human and mouse chromosome 1, *J. Exp. Med.* 172:263 (1990).

14. G.I. Johnston, G.A. Bliss, P.J. Newman, and R.P. McEver, Structure of the human gene encoding granule membrane protein-140, a member of the selectin family of adhesion receptors for leukocytes, *J. Biol. Chem.* 265:21381 (1990).

15. T. Collins, A. Williams, G.I. Johnston, J. Kim, R. Eddy, T. Shows, M.A. Gimbrone Jr., and M.P. Bevilacqua, Structure and chromosomal location of the gene for endothelial-leukocyte adhesion molecule 1, *J. Biol. Chem.* 266:2466 (1991).

16. Y. Imai, D.D. True, M.S. Singer, and S.D. Rosen, Direct demonstration of the lectin activity of $gp90^{mel}$, a lymphocyte homing receptor, *J. Cell Biol.* 111:1225 (1990).

17. E. Larsen, T. Palabrica, S. Sajer, G.E. Gilbert, D.D. Wagner, B.C. Furie, and B. Furie, PADGEM-dependent adhesion of platelets to monocytes and neutrophils is mediated by a lineage-specific carbohydrate, LNF III (CD15), *Cell* 63:467 (1990).

18. M.L. Phillips, E. Nudelman, F.C.A. Gaeta, M. Perez, A.K. Singhal, S.-I. Hakomori, and J.C. Paulson, ELAM-1 mediates cell adhesion by recognition of a carbohydrate ligand, Sialyl-Le^x, *Science.* 250:1130 (1990).

19. G.S. Kansas, K.B. Saunders, K. Ley, A. Zakrzewicz, R.M. Gibson, B.C. Furie, B. Furie, and T.F. Tedder, A role for the epidermal growth factor-like domain of P-selectin in ligand recognition and cell adhesion, *J. Cell Biol.* 124:609 (1994).

20. B.R. Bowen, C. Fennie, and L.A. Lasky, The MEL-14 antibody binds to the lectin domain of the murine peripheral lymph node homing receptor, *J. Cell Biol.* 110:147 (1990).

21. O. Spertini, G.S. Kansas, K.A. Reimann, C.R. Mackay, and T.F. Tedder, Functional and evolutionary conservation of distinct epitopes on the leukocyte adhesion molecule-1 (LAM-1) that regulate leukocyte migration., *J. Immunol.* 147:942 (1991).

22. G.S. Kansas, O. Spertini, L.M. Stoolman, and T.F. Tedder, Molecular mapping of functional domains of the leukocyte receptor for endothelium, LAM-1, *J. Cell Biol.* 114:351 (1991).

23. T.F. Tedder, T. Matsuyama, D.M. Rothstein, S.F. Schlossman, and C. Morimoto, Human antigen-specific memory T cells express the homing receptor necessary for lymphocyte recirculation, *Eur. J. Immunol.* 20:1351 (1990).

24. J.D. Griffin, O. Spertini, T.J. Ernst, M.P. Belvin, H.B. Levine, Y. Kanakura, and T.F. Tedder, GM-CSF and other cytokines regulate surface expression of the leukocyte adhesion molecule-1 on human neutrophils, monocytes, and their precursors, *J. Immunol.* 145:576 (1990).

25. W.M. Gallatin, I.L. Weissman, and E.C. Butcher, A cell-surface molecule involved in organ-specific homing of lymphocytes, *Nature* 304:30 (1983).

26. M.A. Jutila, L. Rott, E.L. Berg, and E.C. Butcher, Function and regulation of the neutrophil MEL-14 antigen in vivo: comparison with LFA-1 and MAC-1, *J. Immunol.* 143:3318 (1989).

27. T.F. Tedder, A.C. Penta, H.B. Levine, and A.S. Freedman, Expression of the human leukocyte adhesion molecule, LAM1. Identity with the TQ1 and Leu-8 differentiation antigens, *J. Immunol.* 144:532 (1990).

28. S.-C. Hsu-Lin, C. Berman, B. Furie, D. August, and B. Furie, A platelet membrane protein expressed during platelet activation and secretion, *J. Biol. Chem.* 259:9121 (1984).

29. R.P. McEver, J.H. Beckstead, K.L. Moore, L. Marshal-Carlson, and D.F. Bainton, GMP-140, a platelet alpha granule membrane protein, is also synthesized by vascular endothelial cells and is localized in Weibel-Palade bodies., *J. Clin. Invest.* 84:92 (1989).

30. J.G. Geng, M.P. Bevilacqua, K.L. Moore, T.M. McIntyre, S.M. Prescott, J.M. Kim, G.A. Bliss, G.A. Zimmerman, and R.P. McEver, Rapid neutrophil adhesion to activated endothelium mediated by GMP-140, *Nature* 343:757 (1990).

31. M.P. Bevilacqua, J.S. Pober, D.L. Mendrick, R.S. Cotran, and M.A. Gimbrone Jr., Identification of an inducible endothelial-leukocyte adhesion molecule, *Proc. Natl. Acad. Sci. USA* 84:9238 (1987).

32. F.W. Luscinskas, M.I. Cybulsky, J.M. Kiely, C.S. Peckins, V.M. Davis, and M.A. Gimbrone Jr., Cytokine-activated human endothelial monolayers support enhanced neutrophil transmigration via a mechanism involving both endothelial-leukocyte adhesion molecule-1 and intercellular adhesion molecule-1, *J. Immunol.* 146:1617 (1991).

33. G.S. Kansas, and M.O. Dailey, Expression of adhesion structures during B cell development in man, *J. Immunol.* 142:3058 (1989).

34. E.F. Knol, F. Tackey, T.F. Tedder, D.A. Klunk, C.A. Bickel, and B.S. Bochner, Comparison of human eosinophil and neutrophil adhesion to endothelial cells under non-static conditions: the role of L-selectin, *J. Immunol.* (in press):(1994).

35. O. Spertini, F.W. Luscinskas, M.A. Gimbrone Jr., and T.F. Tedder, Monocyte attachment to activated human vascular endothelium *in vitro* is mediated by Leukocyte Adhesion Molecule-1 (L-selectin) under non-static conditions., *J. Exp. Med.* 175:1789 (1992).

36. G.S. Kansas, M.J. Muirhead, and M.O. Dailey, Expression of the CD11/CD18, LAM-1, and CD44 adhesion molecules during myeloid and erythroid differentiation in man, *Blood.* 76:2483 (1990).

37. H.B. Stamper Jr., and J.J. Woodruff, Lymphocyte homing into lymph nodes: in vitro demonstration of the selective affinity of recirculating lymphocytes for high-endothelial venules, *J. Exp. Med.* 144:828 (1976).

38. B. Holzmann, B.W. McIntyre, and I.L. Weissman, Identification of a murine Peyer's patch-specific lymphocyte homing receptor as an integrin molecule with an α chain homologous to human VLA-4α, *Cell* 56:37 (1989).

39. L.J. Picker, and E.C. Butcher, Physiological and molecular mechanisms of lymphocyte homing, *Annu. Rev. Immunol.* 10:561 (1992).

40. T.K. Kishimoto, M.A. Jutila, and E.C. Butcher, Identification of a human peripheral lymph node homing receptor: a rapidly down-regulated adhesion molecule, *Proc. Natl. Acad. Sci. USA* 87:2244 (1990).

41. O. Spertini, G.S. Kansas, J.M. Munro, J.D. Griffin, and T.F. Tedder, Regulation of leukocyte migration by activation of the leukocyte adhesion molecule-1 (LAM-1) selectin, *Nature* 349:691 (1991).

42. O. Spertini, A.S. Freedman, M.P. Belvin, A.C. Penta, J.D. Griffin, and T.F. Tedder, Regulation of leukocyte adhesion molecule-1 (TQ1, Leu-8) expression and shedding by normal and malignant cells, *Leukemia* 5:300 (1991).

43. S.R. Watson, C. Fennie, and L.A. Lasky, Neutrophil influx into an inflammatory site inhibited by a soluble homing receptor-IgG chimera, *Nature* 349:164 (1991).

44. S.R. Watson, Y. Imai, C. Fennie, J.S. Geoffrey, S.D. Rosen, and L.A. Lasky, A homing receptor-IgG chimera as a probe for adhesive ligands of lymph node high endothelial venules, *J. Cell Biol.* 110:2221 (1990).

45. D.M. Lewinsohn, R.F. Bargatze, and E.C. Butcher, Leukocyte-endothelial cell recognition: evidence of a common molecular mechanism shared by neutrophils, lymphocytes, and other leukocytes, *J. Immunol.* 138:4313 (1987).

46. O. Spertini, F.W. Luscinskas, G.S. Kansas, J.M. Munro, J.D. Griffin, M.A. Gimbrone Jr., and T.F. Tedder, Leukocyte adhesion molecule-1 (LAM-1, L-selectin) interacts with an inducible endothelial cell ligand to support leukocyte adhesion and transmigration., *J. Immunol.* 147:2565 (1991).

47. H.R. Brady, O. Spertini, W. Jimenez, B.M. Brenner, P.A. Marsden, and T.F. Tedder, Neutrophils, monocytes and lymphocytes bind to cytokine-activated kidney glomerular endothelial cells through L-selectin (LAM-1) *in vitro*, *J. Immunol.* 149:2437 (1992).

48. L.J. Picker, T.K. Kishimoto, C.W. Smith, R.A. Warnock, and E.C. Butcher, ELAM-1 is an adhesion molecule for skin-homing T cells, *Nature* 349:796 (1991).

49. Y. Shimizu, S. Shaw, N. Graber, T.V. Gopal, K.J. Horgan, G.A. Van Seventer, and W. Newman, Activation-independent binding of human memory T cells to adhesion molecule ELAM-1, *Nature* 349:799 (1991).
50. M.L. Arbones, D.C. Ord, K. Ley, H. Radich, C. Maynard-Curry, D.J. Capon, and T.F. Tedder, Lymphocyte homing and leukocyte rolling and migration are impaired in L-selectin (CD62L) deficient mice, *Immunity* 1:247 (1994).
51. H.R. Hendriks, I.L. Eestermans, and E.C. Hoefsmit, Depletion of macrophages and disappearance of postcapillary high endothelial venules in lymph nodes deprived of afferent lymphatic vessels, *Cell Tissue Res.* 211:375 (1980).
52. H.R. Hendriks, A.M. Duijvestijn, and G. Kraal, Rapid decrease in lymphocyte adherence to high endothelial venules in lymph nodes deprived of afferent lymphatic vessels, *Eur. J. Immunol.* 17:1691 (1987).
53. R.E. Mebius, P.R. Streeter, J. Breve, A.M. Duijvestijn, and G. Kraal, The influence of afferent lymphatic vessel interruption on vascular addressin expression, *J. Cell Biol.* 115:85 (1991).
54. E.C. Butcher, R.G. Scollay, and I.L. Weissman, Organ specificity of lymphocyte migration: mediation by highly selective lymphocyte interaction with organ-specific determinants on high endothelial venules, *Eur. J. Immunol.* 10:556 (1980).
55. S.D. Rosen, M.S. Singer, Y.A. Yednock, and L.M. Stoolman, Involvement of sialic acid on endothelial cells in organ-specific lymphocyte recirculation, *Science.* 228:1005 (1985).
56. P.R. Streeter, E.L. Berg, B.N. Rouse, R.F. Bargatze, and E.C. Butcher, A tissue-specific endothelial cell molecule involved in lymphocyte homing, *Nature (Lond.)* 331:41 (1988).
57. R.F. Bargatze, P.R. Streeter, and E.C. Butcher, Expression of low levels of peripheral lymph node-associated vascular addressin in mucosal lymphoid tissues: possible relevance to the dissemination of passaged AKR lymphomas, *J. Cell Biochem.* 42:219 (1990).
58. M.J. Briskin, L.M. McEvoy, and E.C. Butcher, MadCAM-1 has homology to immunoglobulin and mucin-like adhesion receptors and to IgA1., *Nature* 363:461 (1993).
59. C. Berlin, E.L. Berg, M.J. Briskin, D.P. Andrew, P.J. Kilshaw, B. Holzmann, I.L. Weissman, A. Hamann, and E.C. Butcher, $\alpha 4\beta 7$ integrin mediates lymphocyte binding to the mucosal vascular addreessin MadCAM-1, *Cell* 74:185 (1993).
60. E.L. Berg, L.M. McEvoy, C. Berlin, R.F. Bergatze, and E.C. Butcher, L-selectin-mediated lymphocyte rolling on MadCAM-1, *Nature* 366:695 (1993).
61. J.S. Geoffroy, and S. Rosen, Demonstration that a lectin-like receptor (gp90mel) directly mediates adhesion of lymphocytes to high endothelial venules of lymph nodes, *J. Cell Biol.* 109:2463 (1989).
62. S.D. Rosen, S.I. Chi, D.D. True, M.S. Singer, and T.A. Yednock, Intravenously injected sialidase inactivates attachment sites for lymphocytes on high endothelial venules, *J. Immunol.* 142:1895 (1989).
63. M.C. Hu, D.T. Crowe, I.L. Weissman, and B. Holzmann, Cloning and expression of mouse integrin $\beta_p(\beta_7)$: A functional role in Peyer's patch-specific lymphocyte homing, *Proc. Natl. Acad. Sci. USA* 89:8254 (1992).
64. A. Hamann, D. Jablonski-Westrich, P. Jonas, and H. Thiele, Homing receptors reexamined: mouse LECAM-1 (MEL-14 antigen) is involved in lymphocyte migration into gut-associated lymphoid tissue, *Eur. J. Immunol.* 21:2925 (1991).
65. R. Pabst, The spleen in lymphocyte migration, *Immunol. Today* 9:43 (1988).
66. E. Fiebig, K. Ley, and K.-E. Arfors, Rapid leukocyte accumulation by "spontaneous" rolling and adhesion in the exteriorized rabbit mesentery, *Int. J. Microcirc. Clin. Exp.* 10:127 (1991).
67. A. Atherton, and G.V.R. Born, Quantitative investigation of the adhesiveness of circulating polymorphonuclear leukocytes to blood vessel walls, *J. Physiol.* 222:447 (1972).
68. C.W. Smith, T.K. Kishimoto, O. Abbass, B. Hughes, R. Rothlein, L.V. McIntire, E. Butcher, and D.C. Anderson, Chemotactic factors regulate lectin adhesion molecule 1 (LECAM-1)-dependent neutrophil adhesion to cytokine-stimulated endothelial cells *in vitro*, *J. Clin. Invest.* 87:609 (1991).
69. K. Ley, P. Gaehtgens, C. Fennie, M.S. Singer, L.A. Lasky, and S.D. Rosen, Lectin-like cell adhesion molecule 1 mediates leukocyte rolling in mesenteric venules *in vivo.*, *Blood* 77:2553 (1991).

70. U.H. von Andrian, J.D. Chambers, L.M. McEvoy, R.F. Bargatze, K.-E. Arfors, and E.C. Butcher, Two-step model of leukocyte-endothelial cell interaction in inflammation: distince roles for LECAM-1 and the leukocyte ß2 integrins in vivo., *Proc. Natl. Acad. Sci. USA* 88:7538 (1991).
71. K. Ley, T.F. Tedder, and G.S. Kansas, L-selectin can mediate leukocyte rolling in untreated mesenteric venules in vivo independent of E- or P-selectin, *Blood* 82:1632 (1993).
72. T.N. Mayadas, R.C. Johnson, H. Rayburn, R.O. Hynes, and D.D. Wagner, Leukocyte rolling and extravasation are severely compromised in P selectin-deficient mice, *Cell* 74:541 (1993).
73. M.S. Mulligan, J. Varani, M.K. Dame, C.L. Lane, C.W. Smith, D.C. Anderson, and P.A. Ward, Role of endothelial-leukocyte adhesion molecule (ELAM-1) in neutrophil-mediated lung injury in rats, *J. Clin. Invest.* 88:1396 (1991).
74. F.W. Luscinskas, G.S. Kansas, H. Ding, P. Pizcueta, B. Schleiffenbaum, T.F. Tedder, and M.A. Gimbrone Jr., Monocyte rolling, arrest and spreading on IL-4-actvivated vascular endothelium under flow is mediated via sequential action of L-selectin, β_1-integrins, and β_2-integrins, *J. Cell Biol.* 125:1417 (1994).
75. L.M. Pilarski, E.A. Turley, A.R.E. Shaw, W.M. Gallatin, M.P. Laderoute, R. Gillitzer, I.G.R. Beckman, and H. Zola, FMC46, a cell protrusion-associated leukocyte adhesion molecule-1 epitope on human lymphocytes and thymocytes., *J. Immunol.* 147:136 (1991).
76. S.J. Rosenman, A.A. Ganji, T.F. Tedder, and W.M. Gallatin, *Syn*-capping of human T lymphocyte adhesion/activation molecules and their redistribution during interaction with endothelial cells, *J. Leuk. Biol.* 53:1 (1993).
77. G.S. Kansas, K. Ley, J.M. Munro, and T.F. Tedder, Regulation of leukocyte rolling and adhesion to HEV through the cytoplasm domain of L-selectin, *J. Exp. Med.* 177:833 (1993).
78. T.K. Kishimoto, M.A. Julita, E.L. Berg, and E.C. Butcher, Neutrophil Mac-1 and MEL-14 adhesion proteins inversely regulated by chemotactic factors, *Science* 245:1238 (1989).
79. T.M. Jung, W.M. Gallatin, I.L. Weissman, and M.O. Dailey, Down-regulation of homing receptors after T cell activation, *J. Immunol.* 141:4110 (1988).
80. O. Spertini, B. Schleiffenbaum, C. White-Owen, P. Ruiz Jr., and T.F. Tedder, ELISA for quantitation of L-selectin shed from leukocytes in vivo, *J. Immunol. Methods* 156:115 (1992).
81. B.E. Schleiffenbaum, O. Spertini, and T.F. Tedder, Soluble L-selectin is present in human plasma at high levels and retains functional activity, *J. Cell Biol.* 119:229 (1992).
82. S.C. Donnelly, C. Haslett, I. Dransfield, C.E. Robertson, D.C. Carter, I.S. Grant, and T.F. Tedder, Altered levels of soluble L-selectin adhesion receptor in plasma is correlated with development of the adult respiratory distress syndrome (ARDS) in at-risk patient groups, *Lancet* (in press, 1994).
83. D. Mihelcic, B. Schleiffenbaum, T.F. Tedder, J.M. Harlan, and R.K. Winn, Inhibition of leukocyte L-selectin function with monoclonal antibody attenuates reperfusion injury to the rabbit ear, *Blood* (in press, 1994).

L-SELECTIN, A LECTIN-LIKE RECEPTOR INVOLVED IN NORMAL LYMPHOCYTE RECIRCULATION AND INFLAMMATORY LEUKOCYTE TRAFFICKING

Steven D. Rosen

Department of Anatomy and Program in Immunology
University of California
San Francisco, CA 94143-0452

L-selectin is a calcium-dependent lectin-like receptor[1,2] that is widely distributed on all leukocytes in the blood and belongs to the newly emerging selectin family of cell-cell adhesion proteins, which includes E-selectin and P-selectin[3]. Each selectin contains an amino-terminal C-type lectin domain, followed by an EGF-like motif, short consensus repeats (SCR) similar to those found in complement-regulatory proteins, a transmembrane domain, and a short cytoplasmic tail. These three proteins functions as lectins by virtue of a calcium-type lectin domain[4] at the amino terminus of each. The selectins are involved in a wide array of leukocyte-endothelial and leukocyte-platelet interactions in the blood vascular compartment (Table 1).

L-selectin was discovered as a lymphocyte homing receptor[5], which is involved in the organ-specific adherence of blood-borne lymphocytes to high endothelial venules (HEV) in lymph nodes. This adhesive interaction initiates the movement of lymphocytes into lymph nodes during the constitutive process of lymphocyte recirculation. However, the widespread distribution of L-selectin on all classes of leukocytes underlies a broader role for L-selectin in inflammatory leukocyte trafficking in a variety of acute and chronic sites[2]. With respect to cardiovascular disease, it has recently been shown that either an L-selectin antibody or a specific carbohydrate that binds to L-selectin provides substantial protection to myocardium in reperfusion-injury models[6,7]. ThIS protection is attributable to the inhibition of neutrophil binding to the endothelium in ischemic blood vessels. A role for L-selectin in chronic indications has been most clearly demonstrated in the nonobese diabetic mouse (NOD) model of diabetes[8]. HEV-like vessels are induced in the inflamed islets of these mice and ligands for L-selectin are detectable on these vessels. Moreover, an antibody to

L-selectin substantially prevents both leukocyte infiltration and the development of diabetes[9].

Table 1.Distribution and Regulation of Selectins[10]

Designations	Distribution	Regulation
L-selectin, CD62L, LECAM-1, LAM-1, $gp90^{MEL}$	on all circulating leukocytes, including subpopulations of lymphocytes	subject to complex regulation upon activation of leukocytes
P-selectin, CD62P, GMP-140, PADGEM	a-granules of platelets, megakaryocytes, endothelial cells (Weibel-Palade granules)	rapidly elicited to surface of platelets and endothelial cells by thrombin, histamine etc;
E-selectin, CD62E, ELAM-1	on activated endothelial cells	trancriptionally induced by IL-1, TNF-α, LPS

Considerable attention has been devoted to the identification of the endothelial ligands for L-selectin. In the mouse, the HEV-associated ligands have been identified as sulfated, sialylated, and fucosylated glycoproteins of ≈50 kDa and ≈90 kDa (originally called Sgp50 and Sgp90)[11]. These glycoproteins were identified by precipitation of a lysate of metabolically-labeled mouse lymph nodes with a recombinant form of L-selectin. The 50 kDa ligand, now designated as GlyCAM-1, has been cloned and revealed to be a mucin-like glycoprotein containing ≈70% of its mass as O-linked carbohydrate chains[12]. GlyCAM-1 is present in serum in high concentrations[13] but also may have a peripheral membrane association with the apical plasma membrane of endothelial cells. This high serum level raises the possibility that GlyCAM-1 can serve as a modulator of L-selectin dependent leukocyte adhesion. It is suggested that the polypeptide core of this ligand functions in the presentation of clustered carbohydrate chains to the lectin domain of L-selectin. More recently, Sgp90 has also been shown to have a mucin-like organization with the demonstration that it corresponds to the previously identified sialomucin known as CD34[14]. This glycoprotein, unlike GlyCAM-1, has a classical transmembrane domain and is believed to function strictly as a pro-adhesive ligand for L-selectin.

Sialylation and probably fucosylation of HEV-ligands are required for their avid interaction with L-selectin[15]. Furthermore, the sulfation of GlyCAM-1 and CD34 is extensive. Recent work, based on the use of chlorate as a metabolic inhibitor of sulfation, has established that sulfation is required for ligand activity[16]. A monoclonal antibody, known as MECA 79, has been described[17], which stains HEV in lymph nodes and HEV-like vessels that are induced in a large number of inflammatory sites in both mouse and human[8,18]. The epitope for this function-blocking antibody is present on both GlyCAM-1 and CD34[11]. Sialylation is not necessary for the epitope. However, the inhibition of sulfation results in a complete loss of the epitope (S. Hemmerich and S. Rosen, unpublished observations). Sulfation, therefore represents a key modification of HEV-ligands and that is necessary for recognition by both L-selectin and MECA 79. Given the staining pattern of MECA 79 at many sites of chronic inflammation, the sulfation requirement may also extend to extralymphoid endothelial ligands for L-selectin.

Early studies[1] of the carbohydrate specificity of E- and P-selectin uncovered a recognition motif common to both receptors, the sialylated and fucosylated tetrasaccharide known as sialyl Lewis x or sLe^x:

$$\text{Sia}\alpha 2 \rightarrow 3\text{Gal}\beta 1 \rightarrow 4(\text{Fuc}\alpha 1 \rightarrow 3)\text{GlcNAc}$$

Later, L-selectin was also shown to bind sLe^x-containing oligosaccharides[15,19], and it is now suspected that the biological ligands for the selectins present variants of sLe^x. In support of this possibility, structural analysis of the actual carbohydrate chains of GlyCAM-1[20,21] has revealed the presence of a 6'-sulfated modification of sLe^x as a major capping group:

$$\text{Sia}\alpha 2 \rightarrow 3(\text{SO}_4\text{-6})\text{Gal}\beta 1 \rightarrow 4(\text{Fuc}\alpha 1 \rightarrow 3)\text{GlcNAc}$$

This structure would accommodate the requirements for sialic acid, sulfate, and fucose. It is predicted that this compound will be superior to sLe^x as a carbohydrate ligand for L-selectin. If so, the design of anti-inflammatory therapeutics may be facilitated.

ACKNOWLEDGMENT

The research from the author's laboratory was supported by grants from the NIH (GM23547) and Genentech Inc.

REFERENCES

1. L. A. Lasky, Selectins: Interpreters of cell-specific carbohydrate information during inflammation, *Science* 258:964 (1992)

2. S. D. Rosen, L-selectin and its biological ligands, *Histochemistry* 100:185 (1993)

3. M. P. Bevilacqua, and R. M. Nelson, Selectins, *J. Clin. Invest.* 91:379 (1993)

4. K. Drickamer, and M. E. Taylor, Biology of animal lectins, *Annu. Rev. Cell Biol.* 9:237 (1993)

5. W. Gallatin, I. Weissman, and E. Butcher, A cell-surface molecule involved in organ-specific homing of lymphocytes., *Nature* 304:30 (1983)

6. X. Ma, A. S. Weyrich, D. J. Lefer, M. Buerke, K. H. Albertine, T. K. Kishimoto, and A. M. Lefer, Monoclonal antibody to L-selectin attenuates neutrophil accumulation and protects ischemic reperfused cat myocardium, *Circulation* 88:649 (1993)

7. M. Buerke, A. S. Weyrich, Z. Zheng, F. Gaeta, M. J. Forrest, and A. M. Lefer, Sialyl Lewis x-containing oligosaccharide attenuates myocardial reperfusion injury in cats, *J. Clin. Invest.* 93:1140 (1994)

8. A. Hänninen, C. Taylor, P. R. Streeter, L. S. Stark, J. M. Sarte, J. A. Shizuru, O. Simell, and S. A. Michie, Vascular addressins are induced on islet vessels during insulitis in nonobese diabetic mice and are involved in lymphoid binding to islet endothelium, *J. Clin. Invest.* 92:2509 (1993)

9. X.-D. Yang, N. Karin, R. Tisch, L. Steinman, and H. O. McDevitt, Inhibition of insulitis and prevention of diabetes in nonobese diabetic mice by blocking L-selectin and very late antigen 4 adhesion receptors, *Proc. Natl. Acad. Sci.(USA)* 90:10494 (1993)

10. T. A. Springer, Traffic signals for lymphoycte recirculation and leukocyte emigration: the multistep paradigm, *Cell* 76:301 (1994)

11. Y. Imai, M. S. Singer, C. Fennie, L. A. Lasky, and S. D. Rosen, Identification of a carbohydrate-based endothelial ligand for a lymphocyte homing receptor, *J. Cell Biol.* 113:1213 (1991)

12. L. A. Lasky, M. S. Singer, D. Dowbenko, Y. Imai, E. J. Henzel, C. Fennie, N. Gillett, S. R. Watson, and S. D. Rosen, An endothelial ligand for L-selectin is a novel mucin-like molecule, *Cell* 69:927 (1992)

13. M. Brustein, G. Kraal, R. E. Mebius, and S. R. Watson, Identification of a soluble form of a ligand for the lymphocyte homing receptor, *J.Exp. Med.* 176:1415 (1992)

14. S. Baumhueter, M. S. Singer, W. Henzel, S. Hemmerich, M. Renz, S. D. Rosen, and L. A. Lasky, Binding of L-Selectin to the Vascular Sialomucin, CD34, *Science* 262:436 (1993)

15. Y. Imai, L. A. Lasky, and S. D. Rosen, Further characterization of the interaction between L-selectin and its endothelial ligands, *Glycobiology* 2:373 (1992)

16. Y. Imai, L. A. Lasky, and S. D. Rosen, Sulphation requirement for GlyCAM-1, an endothelial ligand for L-selectin, *Nature* 361:555 (1993)

17. P. R. Streeter, B. T. N. Rouse, and E. C. Butcher, Immunohistologic and functional characterization of a vascular addressin involved in lymphocyte homing into peripheral lymph nodes, *J.Cell Biol.* 107:1853 (1988)

18. S. A. Michie, P. R. Streeter, P. A. Bolt, E. C. Butcher, and L. J. Picker, The human peripheral lymph node vascular addressin, *Amer. J. Path.* 143:1688 (1993)

19. C. Foxall, S. R. Watson, D. Dowbenko, C. Fennie, L. A. Lasky, M. Kiso, A. Hasegawa, D. Asa, and B. K. Brandley, The three members of the selectin receptor family recognize a common carbohydrate epitope, the sialyl Lewisx oligosaccharide, *J. Cell Biol.* 117:895 (1992)

20. S. Hemmerich, C. R. Bertozzi, H. Leffler, and S. D. Rosen, Identification of the sulfated monosaccharides of GlyCAM-1, an endothelial derived ligand for L-selectin, *Biochemistry* 33:4820 (1994)

21. S. Hemmerich, and S. D. Rosen, 6'-sulfated, sialyl Lewis X is a major capping group of GlyCAM-1, *Biochemistry* 33:4830 (1994)

NEW PERSPECTIVES IN P-SELECTIN BIOLOGY

Barbara C. Furie and Bruce Furie

Center for Hemostasis and Thrombosis Research
Division of Hematology-Oncology
New England Medical Center
Departments of Medicine and Biochemistry
Tufts University School of Medicine
Boston Massachusetts 02111

INTRODUCTION

P-selectin is a cell adhesion molecule that is an integral membrane protein of the alpha granules of resting platelets [1,2] and the Weibel-Palade bodies of endothelial cells[3,4]. Stimulation of these cells results in translocation of P-selectin to the plasma membrane where it serves as a receptor for leukocytes[5,6]. P-selectin is a member of the selectin family of adhesion molecules. It shares a common domain structure with other family members, E-selectin and L-selectin. These proteins contain a lectin domain, an EGF domain, a variable number of concensus repeats, a transmembrane domain, and a cytoplasmic tail[7,8,9,10].

The selectins play an important role as receptors in cell-cell interaction binding neutrophils and monocytes to the activated endothelium and activated platelets. Following an inflammatory stimulus, leukocytes are known to roll along the vessel wall in a selectin mediated process[11,12,13,14]. This phenomenon indicates that the selectins play a role in inflammation by mediating rolling as a first step in firm attachment of leukocytes to the vessel wall and their emigration out of the vasculature[2,15]. Our studies have focused more on the role that, P-selectin, in particular, may play in thrombogenesis[16]. This brief summary will discuss our current results regarding this function of P-selectin as well as several other areas of P-selectin biology.

ROLE OF P-SELECTIN IN THROMBOSIS

We have previously shown, using an arteriovenous shunt model in baboons, that leukocytes accumulate within an experimental thrombus in a P-selectin mediated process[16].

When anti-P-selectin antibodies are used to block leukocyte accumulation in the graft, deposition of fibrin within the thrombus is similarly inhibited. Platelets have long been known to induce procoagulant activity on mononuclear cells[17], a finding which we confirmed. We investigated whether the induction of tissue factor in the mononuclear cell population upon contact with activated platelets was a P-selectin mediated event. The induction of tissue factor activity was inhibited with antibodies directed against tissue factor or blocking antibodies against P-selectin[18]. Purified P-selectin stimulated tissue factor expression in a dose-dependent manner. Although monocytes are the only members of the mononuclear cell population known to synthesize tissue factor, we confirmed that they were the cells responsible for the activity by using isolated unstimulated monocytes[18] to demonstrate that CHO cells expressing P-selectin but not untransfected CHO cells or E-selectin expressing CHO cells stimulated tissue factor expression in the monocytes. Using reverse transcriptase/PCR, we demonstrated that unstimulated monocytes had no detectable tissue factor mRNA while monocytes stimulated with CHO-P-selectin contained tissue factor mRNA. Thus the binding of monocytes to P-selectin bearing cells at sites of inflammation or vascular injury may be important for the initiation of thrombogenesis. We have continued to explore the nature of the P-selectin ligand on mononuclear cells which fosters this interaction.

THE P-SELECTIN LIGAND

The list of cells bearing a P-selectin ligand includes neutrophils and monocytes[6], a small subpopulation of T-lymphocytes [19], and a number of malignant cells[20]. Lex (Lewis x)[21], and sialic acid[22] have been described as components of the P-selectin ligand. However, SLex, though required for P-selectin interaction on cell surfaces, is not sufficient for high affinity binding. A protein component also appears to be required[23,24]. A putative P-selectin ligand of 120,000 molecular weight has been identified from myeloid cells by blotting techniques[25]. Sulfatides have been implicated as P-selectin ligands[26] as has L-selectin[27]. More recently, a cDNA has been isolated that encodes the protein component of the P-selectin ligand. This protein, identified as PSGL-1 (P-Selectin Glycoprotein Ligand-1), has a large number of sites for O-linked glycosylation and is indeed heavily glycosylated[28]. PSGL-1, a mucin-like, integral membrane protein, has a monomeric molecular weight of 110,00 and is homodimeric. The protein contains an extracellular domain made up of 15 repeating units that are rich in threonine, a transmembrane domain and a cytoplasmic domain. Cotranfection of COS cells with the cDNAs for PSGL-1 and 1/3,4 fucosyltransferases yields a biologically active ligand. Polyclonal antibodies raised against PSGL-1 inhibit P-selectin mediated binding to HL60 cells.

To ascertain the importance of PSGL-1 as an *in vivo* ligand for P-selectin we have intiated efforts to develop a murine model of PSGL-1 defiency. A murine homolog of human PSGL-1 has been cloned. The deduced amino acid sequence of the murine PSGL-1 has an open reading frame encoding a protein of 397 amino acids, 5 residues shorter than human PSGL-1 [29]. The murine and human PSGL-1 show an overall homology of 67% and identity of 50% and contain similar domain structures including a signal peptide, propeptide, homologous repeat units, transmembrane region and cytoplasmic tail. There are 15 threonine-rich decameric repeats in the human PSGL-1, but only ten such repeats in murine PSGL-1. Instead of threonines as potential O-linked glycosylation sites in the repeat region as in human PSGL-1, the murine PSGL-1 repeat sequence contains threonine and serine residues. Compared to the three potential N-glycosylation sites in human PSGL-1, the murine PSGL-1 carries only one conserved site and an additional unique site. The murine PSGL-1 functions as a ligand for human P-selectin in a cell binding assay. Furthermore, we have established by Northern analysis that WEHI-3 cells, a murine myeloid leukemia cell line, synthesize the PSGL-1 mRNA and that these cells express the P-selectin ligand in a functional form that interacts with human P-selectin. PU-5 cells, also contains the PSGL-1 mRNA but do not bind P-selectin; the defect in this PSGL-1 function may be the absence of a co-receptor or incomplete posttranslational processing.

Other mucins have been described on the surface of leukocytes, and their relationship to selectin ligands remains unknown. CD68 is a 110,000 molecular weight transmembrane protein that is expressed on monocytes and macrophages[30]. The cDNA for human CD68 predicts a heavily glycosylated extracellular domain with numerous potential N-linked and O-linked glycosylation sites; it is distinct from PSGL-1. The function of this protein is unknown. To test the hypothesis that CD68 may be an E-selectin or P-selectin ligand, the CD68 cDNA was cloned and cotransfected into COS cells with cDNA for a 1/3,4 fucosyltransferase. COS cells expressing CD68 alone or co-expressing CD68 and SLex did not acquire an enhanced ability to bind to CHO cells expressing P-selectin or E-selectin as compared to COS cells expressing SLex only. Cotransfection of COS cells with PSGL-1 and fucosyltransferase, in contrast, enhanced COS cell binding to CHO-P-selectin. These experiments suggest that CD68 is not a ligand for either E-selectin or P-selectin. In addition to exploring the nature of the P-selectin ligand we have continued to explore the ligand binding site of P-selectin itself.

STRUCTURE-FUNCTION RELATIONSHIPS IN P-SELECTIN

The P-selectin literature is fraught with contradictions regarding the nature of potential ligands. We believe that the basis for these differences has to do, at least in part, with the density of P-selectin and P-selectin ligand used in the various assay systems. To determine the relationship of P-selectin density on the membrane surface to the cell adhesion properties of a cell, we have performed quantitative experiments. Purified P-selectin was incorporated, at known concentrations, into lipospheres, uniform micro-glass beads encapsulated with phospholipid containing P-selectin[31]. The binding of lipospheres containing varying concentrations of P-selectin to HL60 cells was analyzed. A critical P-selectin density of about 50 molecules per μm^2 is required to sustain cell adhesion; maximal cell adhesion is observed at 100-150 molecules per μm^2. A similar analysis of the surface density of P-selectin on CHO cell clones indicates optimal HL60 cell adhesion at about 150-200 molecules per μm^2.

While the role of the lectin domain of P-selectin in ligand binding appears clear the role of the EGF domain remains ambiguous. We have tried in a series of experiments to clarify the role of the EFG domain in ligand binding. In collaboration with Dr. Tom Tedder and Geoffrey Kansas, we have prepared chimeras of P-selectin and L-selectin to determine the specific functions of the domains of these selectins. P2L contains the P-selectin lectin domain on an L-selectin background; P3L contains the lectin and EGF domain of P-selectin on an L-selectin background; P4L contains the lectin, EGF and consensus repeats of P-selectin on an L-selectin background. Stably transfected CHO cell lines expressing these chimeras as well as wild type P-selectin and L-selectin were established and methatrexate amplified clones selected to provide cell lines expressing similar selectin surface density. The ability of the chimeras and wild type proteins to bind to HL60 cells was assessed at a surface density of selectin or selectin chimera of 150 molecules per μm^2. P-selectin but not L-selectin expressing CHO cells bound to HL60 cells. The CHO cells expressing the selectin chimeras P3L and P4L bound to HL60 cells well while the P2L chimera expressing CHO cells bound HL60 cells much less well. We conclude that, although the lectin domain is sufficient to support cell adhesion, the lectin-EGF domain represents the cell recognition unit on P-selectin.

The 40 amino acid P-selectin EGF domain was prepared by solid phase synthesis, folded and its structure verified by sequence analyses. We have solved the three dimensional structure of the P-selectin EGF domain by 2D NMR spectroscopy (unpubl. data, D. Sanford, B.C. Furie, B. Furie). The EGF structure is homologous to the murine EGF, the single most unique feature of the P-selectin EGF being the close spatial relationship of the N- and C-termini of the polypeptide, in contrast to other, larger EGF domains.

We have expressed the lectin domain of P-selectin in *E. coli*, isolated the peptide and folded it in the presence of $CaCl_2$. The oxidized lectin domain, purified by HPLC yields a single band on SDS gel. The role of P-selectin domains in cell adhesion was evaluated by using the isolated EGF domain and lectin domain as competitors in a cell adhesion assay based upon the binding of P-selectin expressing CHO cells to HL60 cells. The P-selectin lectin domain was a potent inhibitor of cell adhesion; half-maximal inhibition was observed at 200 nM[32]. In contrast, the P-selectin EGF domain showed no inhibition even at concentrations up to 2 mM. These results taken with our results that the chimera of L-selectin containing the P-selectin EGF domain is capable of binding HL60 cells[33], unlike L-selectin, and that the recognition unit on P-selectin is the lectin-EGF domain, suggest that the P-selectin EGF domain does not make direct contact with the P-selectin ligand but rather influences the structure of the adjacent lectin domain, whether from P-selectin or L-selectin, to enable it to bind to the P-selectin ligand.

PHOSPHORYLATION OF P-SELECTIN

P-selectin is phosphorylated upon platelet activation with αthrombin, the thrombin receptor peptide (SFLLR), epinephrine, ADP or collagen[34]. As a lower limit, the molar ratio of phosphate to P-selectin in activated platelets is 6:10 at 15 seconds, indicating the very significant rapid phosphorylation of P-selectin. P-selectin phosphorylation following thrombin stimulation is detectable at 5 sec, with maximum incorporation observed at 15-30 seconds. Phosphoamino acid analysis of the phosphorylated P-selectin revealed the presence of phosphoserine, phosphothreonine and phosphotyrosine. Phosphotyrosine and phosphothreonine rapidly disappeared 60 sec following platelet activation. The C-terminal peptide consisting of the cytoplasmic domain contained the phosphoamino acids. We have recently extended these studies to show that P-selectin in unstirred thrombin-activated platelets undergoes phosphorylation but not dephosphorylation. The rapid phosphorylation and selective dephosphorylation of tyrosine and threonine on the cytoplasmic domain of P-selectin following platelet activation may be important for P-selectin function; e.g. inside-out signalling and signal transduction within platelets.

Stimulation of primary cultures of human umbilical vein endothelial cells with thrombin similarly leads to rapid phosphorylation of P-selectin. P-selectin undergoes phosphorylation and dephosphorylation, with peak phosphorylation at about 1-2 minutes. A second phosphorylation event occurs at about 8-10 minutes. This is also followed by dephosphorylation.

ACKNOWLEDGMENTS

We gratefully acknowledge our collaborators and students who have contributed very significantly to this work. They include A. Celi, R. Gibson, G. Kansas, R. Lorenzet, D. Sanford, G. Larsen, C. Crovello, T. Tedder, J. Yang, S.J. Freedman, J. Galipeau, E. Gottlieb and T. Palabrica as well as earlier students who helped to define the biology of P-selectin.

LITERATURE CITED

1. C.L. Berman, E. Yeo, J.D. Wencel-Drake, B.C. Furie, M.H. Ginsberg, and B. Furie, A platelet alpha granule membrane protein that is incorporated into the plasma membrane during activation. Characterization and subcellular localization of PADGEM protein, *J. Clin. Invest.* 78:130(1986).

2. P.E. Stenberg, R.P. McEver, M.A. Shuman, Y.V. Jacques, D.F. Bainton, A platelet alpha-granule membrane protein (GMP140) is expressed on the plasma membrane after activation, *J. Cell Biol.* 101:880-886(1985).

3. R. Bonfanti, B.C. Furie, B. Furie, and D.D. Wagner, PADGEM is a component of Weibel-Palade bodies in endothelial cells, *Blood.*73:1109(1989).

4. R.P. McEver, J.H. Beckstead, K.L. Moore, L. Marshall-Carlson, and D.F. Bainton, GMP-140, a platelet alpha-granule membrane protein, is also synthesized by vascular endothelial cells and is localized in Weibel-Palade bodies, *J. Clin. Invest.*84:92(1989).

5. E. Larsen, A. Celi, G. Gilbert, B.C. Furie, J. Erban, R. Bonfanti, D.D. Wagner, and B. Furie, PADGEM protein: A receptor that mediates the interaction of activated platelets with neutrophils and monocytes,*Cell.*59:305(1989).

6. J.G. Geng, M.P. Bevilacqua, K.L. Moore, T.M. McIntyre, S.M. Prescott, J.M. Kim, G.A. Bliss, G.A. Zimmerman, and R.P. McEver, Rapid neutrophil adhesion to activated endothelium mediated by GMP-140,*Nature.*343:757(1990).

7. T.F. Tedder, C.M. Isaacs, T.J. Ernst, G.D. Demetri, D.A. Adler, and C.M. Disteche, Isolation and chromosomal localization of cDNAs encoding a novel human lymphocyte cell surface molecule, LAM-1. Homology with the mouse lymphocyte homing receptor and other human adhesion proteins,*J. Exp. Med.* 170:123(1989).

8. L.A. Lasky, M.S. Singer, T.A. Yednock, D. Dowbenko, C. Fennie, H. Rodriguez, T. Nguyen, S. Stachel, and S.D. Rosen, Cloning of a lymphocyte homing receptor reveals a lectin domain, *Cell*.56:1045(1989).

9. M.P. Bevilacqua, S. Stengalin, M.A.Gimbrone Jr, and B. Seed, Endothelial leukocyte adhesion molecule 1: an inducible receptor for neutrophils related to complement regulatory proteins and lectins, *Science*.243:1160(1989).

10. G.I. Johnston, R.G. Cook, and R.P. McEver, Cloning of GMP-140, a granule membrane protein of platelets and endothelium: Sequence similarity to proteins involved in cell adhesion and inflammation, *Cell*.56:1033(1989).

11. M.B. Lawrence, and T.A. Springer, Leukocytes roll on a selectin at physiologic flow rates: distinction from and prerequisite for adhesion through integrins, *Cell*.65:859(1991).

12. K. Ley, P. Gaehtgens, C. Fennie, M.S. Singer, L.A. Lasky, and S.D. Rosen, Lectin-like cell adhesion molecule 1 mediates leukocyte rolling in mesenteric venules *in vivo*, *Blood*.77:2553(1991).

13. U.H., von Andrian, J.D. Chambers, E.L. Berg, S.A. Michie, D.A. Brown, D. Karolak, L. Ramezani, E.M. Berger, K.-E. Arfors, and E.C. Butcher, L-selectin mediates neutrophil rolling in inflamed venules through sialyl Lewis-dependent and independent recognition pathways, *Blood*.82:182(1993)

14. T.N. Mayadas, R.C. Johnson, H. Rayburn, R.O. Hynes, and D.D. Wagner, Leukocyte rolling and extravasation are severely compromised in P-selectin-deficient mice, *Cell*74:541(1993).

15. T.A. Springer, Traffic signals for lymphocyte recirculation and leukocyte emigration: the multistep paradigm, *Cell*.76:301(1994).

16. T. Palabrica, R. Lobb, B.C. Furie, M. Aronowitz, C. Benjamin, Y.-M. Hsu, S.A. Sajer, and B. Furie, Leukocyte accumulation which promotes fibrin deposition is mediated in vivo by P-selectin (CD62) on adherent platelets, *Nature*.359:848(1992).

17. J. Niemetz, and A.J. Marcus, The stimulatory effect of platelets and platelet membranes on the procoagulant effect of leukocytes, *J. Clin. Invest*.54:1437(1974).

18. A. Celi, B.C. Furie, R. Lorenzet, G. Pellegrini, and B. Furie, P-selectin induces the expression of tissue factor on monocytes, *Proc. Natl. Acad. Sci. USA*. In press (August 1994).

19. K.L. Moore, L.F. Thompson, P-selectin (CD62) binds to subpopulations of human memory T lymphocytes and natural killer cells, *Biochem Biophys Res Commun*.186:173(1990).

20. J.P. Stone, and D.D.Wagner, P-selectin mediates adhesion of platelets to neuroblastoma and small cell lung cancer, *J. Clin. Invest*.92:804(1993).

21. E. Larsen, T. Palabrica, S.A. Sajer, G.E. Gilbert, D.D. Wagner, B.C. Furie, and B.Furie, PADGEM-dependent adhesion of platlets to monocytes and neutrophils is mediated by a lineage-specific carbohydrate, LNF III (CD15), *Cell*63:467(1990).

22. L. Corral, M.S. Singer, B.A. Macher, and S.D. Rosen, Requirement for sialic acid on neutrophils in a GMP-140 (PADGEM) mediated adhesive interaction with activated platelets, *Biochem Biophys Res Commun*.172:1349(1990).

23. G. Larsen, D. Sako, T. Ahern, M. Shaffer, J. Erban, S.A. Sajer, R.M. Gibson, D.D. Wagner, B.C. Furie, and B.Furie, P-selectin and E-selectin: Distinct but overlapping leukocyte ligand specificities, *J. Biol. Chem.*266:11104(1992).

24. K.L. Moore, A. Varki, and R.P. McEver, GMP-140 binds to a glycoprotein receptor on human neutrophils: evidence for a lectin-like interaction, *J Cell Biol.*112:491(1991).

25. K.L. Moore, N.L. Stults, S. Diaz, D.F. Smith, R.D. Cummings, A. Varki, and R.P.McEver, Identification of a specific glycoprotein ligand for P-selectin (CD62) on myeloid cells, *J Cell Biol.*118:445(1992).

26. A. Aruffo, W. Kolanus, G. Walz, P. Fredman, and B. Seed, CD62/P-selection recognition of myeloid and tumor cell sulfatides, *Cell.*67:35(1991).

27. L.J. Picker, R.A. Warnock, A.R. Burns, C.M. Doerschuk, E.L Berg, and E.C. Butcher, The neutrophil selectin LECAM-1 presents carbohydrate ligands to the vascular selectins ELAM-1 and GMP-140, *Cell.*66:921(1991).

28. D. Sako, X.-J. Chang, K.M. Barone, G. Vachino, G. Shaw, T. Veldman, K.M. Bean , T.J. Ahern, B. Furie, D.A. Cumming, and G.R. Larsen, Expression cloning of a functional glycoprotein ligand for P-selectin, *Cell.*75:1179(1993).

29. J. Yang, B.C. Furie, and B. Furie, Cloning of a murine homolog of the human P-selectin glycoprotein ligand-1, *Circulation*, in press (Abstract).

30. C.L. Holness, and D.L. Simmons, Molecular cloning of CD68, a human macrophage marker related to lysosomal glycoproteins, *Blood.*81:1607(1993).

31. G.E. Gilbert, D. Drinkwater, S. Barter, and S.B. Clouse, Specificity of phosphatidylserine-containing membrane binding sites for factor VIII. Studies with model membranes supported by glass microspheres (lipospheres), *J. Biol. Chem.*267:15861(1992).

32. S.J. Freedman, B. Furie, and B.C. Furie, The lectin domain but not the EGF domain is a potent inhibitor of P-selectin-mediated cellular adhesion, *Blood.*82 (Suppl. 1):341A (1993).

33. G.S. Kansas, K. Ley, A. Zakrzewicz, R.M. Gibson, B.C. Furie, B. Furie, and T.F. Tedder, A direct role in cell adhesion for the EFG-like domain of P-selectin, *J. Biol. Chem. 1245:609(1994).*

34. C.S. Crovello, B.C. Furie, and Furie B, Rapid phosphorylation and selective dephosphorylation of P-selectin accompanies platelet activation, *J. Biol. Chem.*268:14590(1993).

RAT P-SELECTIN MEDIATES NEUTROPHIL-PLATELET INTERACTIONS VIA TWO SITES (23-30, 76-90) LOCATED ON ITS LECTIN-LIKE DOMAIN.

E. Chignier[1], Marie-Hélène Sparagano[1], Lilian McGregor[1], Annie Thillier[2], Dorothée Pellecchia[2], Marie-Pierre Reck[2] and John McGregor[2]

[1]INSERM U 331, Faculté de Médecine Alexis Carrel and [2]Institut Pasteur de Lyon, France

INTRODUCTION

P-selectin, (GMP-140/PADGEM, or CD62-P),[1,2,3,4] has been shown to mediate adhesion of human leukocytes to activated platelets or endothelial cells.[5,6,7] This adhesive receptor is a member of the selectin family which includes E-selectin (ELAM-1) and L-selectin (LAM-1, LECAM).[8] The three members of the selectin family share extensive homologies in their lectin-, EGF- and complement-like domains. The lectin-like domain in human P-selectin appears to play a major role in mediating P-selectin interaction with leukocytes. Peptides derived from the lectin-like domain have been shown to block neutrophil interaction to thrombin-activated platelets[9], endothelial cells or isolated P-selectin. We have previously shown [10] that LYP-20, an anti P-Selectin monoclonal antibody (MAb), inhibits human platelet aggregation and platelet-monocyte interactions. Moreover, LYP-20 recognises P-Selectin expressed by thrombin-activated rat platelets.[11] Rats are extensively used as *in vivo* models of vascular injury. However, at this stage, little is known on the distribution and function of rat P-Selectin.

Cardiovascular Disease 2
Edited by L.L. Gallo, Plenum Press, New York, 1995

The purpose of this study was to identify functional sites on P-Selectin involved in rat leukocyte-platelet interactions. To perform this work we have cloned rat P-selectin.[12] Lectin and EGF-like regions of human and rat P-Selectin were observed.to share strong homologies.[12] *In vitro* work, as a prelude to *in vivo* work, was performed on rat platelets-leukocyte interactions to identify the rat P-selectin functional sites interacting with neutrophils using synthetic peptides derived from functional domains.

RESULTS

Rosetting inhibition by anti-P-selectin MoAb

The anti-P-selectin MoAb (LYP20) inhibits thrombin-activated rat platelets binding to rat neutrophils in very significant manner (58% of inhibition, $p<0.001$) (Fig. 1).

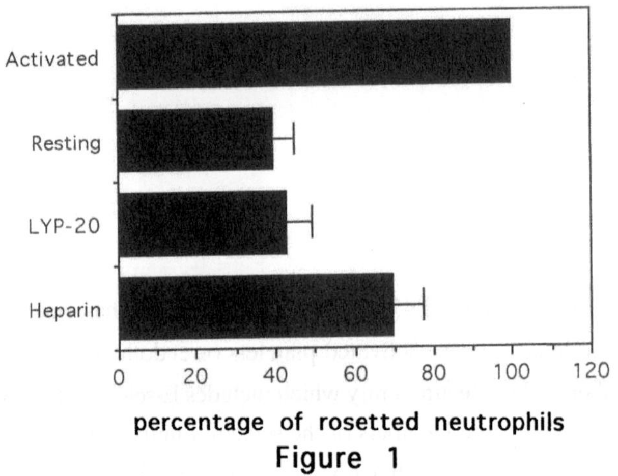

Figure 1

Effect of an anti-P-selectin MoAb (LYP-20) and heparin on platelets-PMN interactions. Resting or thrombin-activated (0.3U/ml final concentration) washed platelets ($2x10^8$ platelets/ml) were incubated with an anti-P-selectin monoclonal antibody (LYP-20, 60µg/ml, final concentration) or heparin (40U final concentration) before incubation with neutrophils (2 to $3x10^7$ cells/ml). The results, that are representative of 6 experiments, are expressed as the percentage of neutrophils bearing platelet rosettes. Bars correspond to the percentage (mean±SD) of neutrophils binding two or more platelets, each count was done in triplicate by different observers. LyP-20 inhibits the interaction between activated platelets with the neutrophils in a significant manner ($p<0.001$). Heparin also inhibits significatively this interaction ($p<0.05$).

Human P-selectin has previously been shown to bear a heparin-binding site, so we investigated the interaction of thrombin-activated rat platelets with rat neutrophils in presence or absence of heparin. The results show that heparin at a final concentration of 80U/ml inhibits rat thrombin-stimulated platelet/rat neutrophils rosetting (36% of inhibition, $p<0.01$), indicating that also in rats the interaction of P-selectin expressed by activated platelets with neutrophils is mediated by a sulfated glycan, or that the binding site of the glycan is proximal or identical to that of CD15 receptor in neutrophils.

Rosetting inhibition by peptides from the lectin-like P-selectin domains.

To investigate whether the lectin-like region of P-selectin molecule is implicated in rat's cell interactions as in humans, a series of peptides corresponding to the lectin-like domain of the P-selectin molecule[13] were tested for their ability to inhibit platelet rosetting to neutrophils. They were synthesized in the laboratory on an ABI 431 (Applied Biosystem Synthesis, Paris, France) using Fmoc (N-(9-fluorenyl)methoxycarbonyl) chemistry. Synthesized peptides were purified by high performance liquid chromatography (HPLC) and their composition were confirmed by amino-acid analysis followed by mass spectroscopy (Hewlet PAckard 5989 spectroscope, equipped with an electrospray, Palo Alto, USA). These peptides correspond to residues 23-30 (YTDLVAIQ, Mw 1046), 51-61 (IGIRKNNKTWT, Mw 1508) and 76-90 (ADNEPNNKRNNEDC, Mw 1870). Control peptides were prepared for the 51-61 and 76-90 residues which were identical to the P-selectin original peptides, but the sequences were scrambled (RIKNJINTKG for the 51-61 and ANPENCNDKDENRN for the 76-90 modified peptides), so that no aminoacids were located in the same position or adjacent to the same residues as in the authentic sequence. At a concentration of 1.3 mM peptides 23-30 (A) and 76-90 (C), but not 51-61 (B), from the lectin-like domain had an inhibitory effect (A=33% and C=46%, $p<0.05$; B=17%, p=0,2) (Fig. 2).

The scambled forms of peptide A had no inhibitory effect in activated platelet/neutrophil interactions. Using a combination of peptides A+C at 1.3 mM ($p<0.001$) we increased the inhibition of platelets binding to neutrophils. The Ic_{50} of peptides A + C was 0.11mM (results not shown).

DISCUSSION

Our findings show: I) A MAb (LYP-20) directed against human P-selectin is capable of inhibiting rat PMN-activated platelets interaction. Moreover, heparin is a potent inhibitor of such cell-cell interaction. II) Peptides 23-30 and 76-90 obtained from the lectin-like domain of

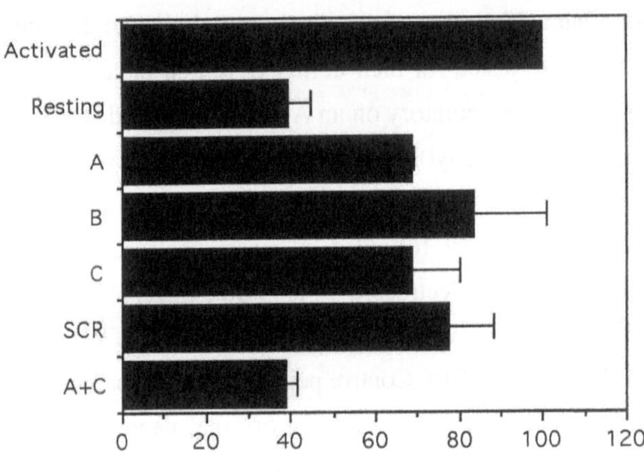

Figure 2

Effect of peptides derived from the P-selectin lectin-like domain on platelets-PMN interactions. Neutrophils were incubated with peptides prior to incubation with platelets. Results shown are representative of 8 experiments (mean±SD). Lectin-like derived peptides, but not scrambled form from peptide A (23-30), partially inhibited the platelets/PMN interaction ($p<0.05$ for isolated peptides A and C). The main effect was obtained when neutrophils were incubated with two functionnal (A+C) peptides ($p<0.001$). Peptide A corresponds to 23-30 AA sequence, peptide A corresponds to B: 51-60 AA sequence, and peptide C corresponds to 76-90 AA sequence of the human P-selectin molecule.

P-selectin inhibit rat activated platelet-neutrophils interactions. The combination of the two different peptides (23-30, 76-90) increased the inhibitory effects with low concentrations (0.11mM), suggesting that the whole lectin region could be necessary for cell-cell interaction.

In our Laboratory Sparagano et al (manuscript in preparation) isolated and sequenced the rat cDNA P-selectin. It was observed that the lectin-like domain of rat P-selectin has 15 out of 120 amino acids that are different, when compared to its human counterpart. These results show an important homology (74%) between rat and human P-selectin lectin-like domains.

The lectin-like domain is known to interact with sialylated CD15.[14,15] Moore et al [16] demonstrated that human neutrophils constitutively express a glycoprotein receptor for P-selectin bearing sialic acid residues. Our results, suggest that rat neutrophil presents a receptor binding to P-selectin, which is either identical or very similar to the human one.[17] The increased inhibitory effect we obtained by associating the residue 23-30 to the residue 76-90 is possibly related with the fact that 23-30 may corresponds to one of the Ca^{2+} binding sites in the P-selectin molecule.[6] Platelet P-selectin plays a role not only in the recruitment of neutrophils and monocytes to hemorrhagic and inflammatory sites but, potentially, in the activation of these leukocytes.[18]

The inhibition of rat platelet-PMN interactions by synthetic peptides may prove to be effective in *in vivo* models leading to new therapies in human inflammatory syndromes. The effect of these P-selectin peptides are currently tested *in vivo* in rats .

Acknowledgements

Part of this work was supported by grants from ARC (subvention 6585) Ligue Nationale Française Contre le Cancer and Mutuelle Générale Education Nationale. Marie-Hélène Sparagano is the recipient of a Fellowship from GEHT (Groupe d'Etudes sur l'Hemostase et la Thrombose, France) founded by KABI-PHARMACIA.

KEYWORDS

P-Selectin, monoclonal antibodies, neutrophils, platelets, peptides, rats

REFERENCES

1- S.C. Hsu-Lin, C. L.Berman, B. C.Furie, D.August and B.Furie, A platelet membrane protein expressed during platelet activation and secretion. Studies using a monoclonal antibody specific for thrombin-activated platelets, *J. Biol. Chem.* 259:9121 (1984).

2- P.E. Stenberg, R.P. McEver, M.A. Shuman, Y.V. Jacques and D.F.Bainton, A platelet alpha-granule membrane protein (GMP-140) is expressed on the plasma membrane after activation, *J. Cell Biol.* 101:880 (1985)

3- W. Knapp, B.Dorken, P. Rieber, R.E. Schimidt, H. Stein and A.E.G.Kr. von dem Borne, CD antigens, *Blood* 74:1448 (1989)

4- R.P. McEver, J.H. Beckstead, K.L. Moore, L. Marshall-Carlson and D.F. Bainton, GMP-140, a platelet alpha-granule membrane protein, is also synthesized by vascular endothelial cells and is localized in Weibel-Palade bodies, *J. Clin. Invest.* 84: 92 (1989)

5- E. Larsen, Celi A., G.E. Gilbert, B.C. Furie, J.K. Erban, R. Bonfanti, D.D. Wagner and B.Furie, PADGEM protein: a receptor that mediates the interaction of activated platelets with neutrophils and monocytes, *Cell* 59:305 (1989).

6- S.A. Hamburger and R.P. McEver, GMP-140 mediates adhesion of stimulated platelets to neutrophils, *Blood* 75: 550 (1990).

7- M.H. Rinder, J.L. Bonan, C.S. Rinder, K.A. Ault and B.R. Smith, Activated and unactivated platelet adhesion to monocytes and neutrophils, *Blood* 78: 1760 (1991).

8- T.A. Springer, Adhesion rceptors of the immune system, *Nature* 346:425 (1990).

9- J.G. Geng, G.A. Heavner and R.P. McEver, Lectin domain peptides from selectins interact with both cell surface ligands and Ca^{2+} ions, *J Biol. Chem.* 267:19846 (1992).

10- S. Parmentier, L. McGregor, B. Catimel, L.L. Leung and J.L. McGregor, Inhibition of platelets functions by a monoclonal antibody (LyP-20) directed against a granule membrane glycoprotein (GMP-140/PADGEM), *Blood* 77:1734 (1991).

11- Winocour P. D., Chignier E., Parmentier S. and McGregor J. L. (1992) A member of the selectin family (GMP-140/PADGEM) is expressed on thrombin-stimulated rat platelets in vitro *Comp. Biochem. Physiol.* **102A**, 265-271.

12- M.H. Sparagano, E. Chignier, A. Thillier, O. Gayet and J.L. McGregor, Isolation and sequencing of Rat P-selectin cDNA, *Thromb. Haemostas.*, 69:568 (1993).

13- G.I. Johnston, R.G Cook.and R.P. McEver, Cloning of GMP-140, a granule membrane protein of platelets and endothelium: sequence similarity to proteins involved in cell adhesion and inflammation,*Cell* 56: 1033 (1989).

14- J.B. Lowe, L.M. Stoolman, R.P. Nair, R.D Larsen, T. L. Berhend and R.M. Marks, ELAM-1-dependent cell adhesion to vascular endothelium determined by a transfected human fucosyltransferase cDNA. *Cell* 63: 475 (1990).

15- E.L. Berg, J.Magnani, R.A.Warnock, M.K. Robinson and E.C. Butcher, Comparison.of L-selectin and E-selectin ligand specifities: the L-selectin can bind the E-selectin ligands sialyl-Le^x and sialyl-Le^a Biochem, *Biophys. Res. Commun.* 184: 1048 (1992).

16- K.L. Moore, A. Varki and R.P. McEver, GMP-140 binds to a glycoprotein receptor on human neutrophils: evidence for a lectin-like interaction, *J. Cell Biol.* 112: 491 (1991).

17- M.P Skinner, C.M. Lucas, G.F.Burns, C.N. Chesterman and M.C.Berndt, GMP-140 binding to neutrophils is inhibited by sulfated glycans, *J. Biol. Chem.* 266: 5371 (1991).

18- K. Nagata, T. Tsuji, N. Todoroki, Y. Katagiri, K. Tanoue, H. Yamazaki, N. Hanai and T. Irimura, Activated platelets induce superoxide anion release by monocytes and neutrophils through P-selectin (CD62), *J. Immunol.*, 151:3267 (1994)

REGULATION OF VASCULAR FIBRINOLYSIS BY TYPE 1 PLASMINOGEN ACTIVATOR INHIBITOR

Dietmar Seiffert, Benien E. van Aken,
and David J. Loskutoff

Department of Vascular Biology
The Scripps Research Institute
La Jolla, CA 92037
U.S.A.

INTRODUCTION

Vascular hemostasis results from the regulated interaction of the coagulation and fibrinolytic systems. Any imbalance in these systems leads to a tendency to develop a bleeding diathesis or to an increased risk of thrombosis. Thrombotic events have been implicated in the pathogenesis of cardiovascular disease, including myocardial infarction and stroke. The recent success of thrombolytic therapy in acute myocardial infarction has drawn considerable attention to the role of the fibrinolytic system in vascular disease.

FIBRINOLYTIC SYSTEM

Degradation of fibrin, the key step of intravascular fibrinolysis, normally results from the action of plasmin, a serine protease of broad specificity which circulates in plasma as the inactive proenzyme plasminogen (1,2). Plasmin can also hydrolyze proteins of the extracellular matrix, either directly, or indirectly by activation of latent collagenases (3,4). These later properties of the enzyme are believed to be important in cellular migration and invasion, key events in the development of vascular lesions (5,6).

Two types of plasminogen activators (PAs) can be distinguished, tissue-type PA (tPA) and urinary-type PA (uPA) (7). The catalytic efficiency of plasminogen activation by tPA is drastically increased in the presence of fibrin (8). Thus, tPA appears to be the primary PA of the intravascular fibrinolytic system (7) and inactivation of the tPA gene

Cardiovascular Disease 2
Edited by L.L. Gallo, Plenum Press, New York, 1995

in mice impairs clot lysis (9). Urinary-PA is believed to be the primary PA of the extravascular PA system (7). Surprisingly, inactivation of the uPA gene also appears to promote fibrin deposition (9). This later finding points to an additional role of uPA in intravascular fibrinolysis. However, fibrin deposition in these uPA-deficient mice could also result from deficient tissue remodelling rather than from the reduced thrombolytic potential (9).

In contrast to plasminogen which is present at relatively high and unchanging concentrations in most body fluids (1,2), PAs are trace proteins and their biosynthesis is highly regulated (10). Thus, changes in the availability and activity of PAs must contribute significantly to the dynamic regulation of the fibrinolytic system. The activity of PAs is controlled by PA inhibitors (PAIs) (11,12). Although four molecules with PAI activity have been identified, considerations of rate constants, in vivo inhibitor concentrations, and the presence of naturally occurring PA/PAI-1 complexes, suggest that PAI-1 is the primary physiological inhibitor of plasminogen activation (12). Indeed, disruption of the PAI-1 gene in mice appears to induce a mild hyperfibrinolytic state and a greater resistance to venous thrombosis (13,14), while reduced PAI-1 activity in man results in a hemorrhagic tendency with delayed rebleeding after surgery or trauma (15,16).

PAI-1 appears to be synthesized in an active form, but is unstable in solution and rapidly decays into an inactive form upon secretion (12). The majority of PAI-1 in blood is active and circulates in complex with the adhesive glycoprotein vitronectin (Vn) (17,18). The binding of PAI-1 to Vn stabilizes the inhibitor in the active conformation, thus increasing its biological half-life (17,19). Vn is also the primary PAI-1 binding protein in the endothelial cell extracellular matrix (20,21). These observations suggest that Vn serves to stabilize and concentrate PAI-1 in tissues, thereby functioning as a cofactor of the fibrinolytic system.

EXPRESSION OF PAI-1 AND VN IN NORMAL AND DISEASED VASCULAR TISSUES

The plasma levels of PAI-1 were increased in some patient groups with cardiovascular disease, including young survivors of myocardial infarction (22). These studies suggested that PAI-1 is a "risk factor" for atherosclerotic vascular disease. However, a number of studies failed to demonstrate any correlation between plasma PAI-1 levels and vascular disease (22). While such measurements of PAI-1 in plasma are relatively straight forward and important, they remain incomplete and may not reflect the actual situation within the vessel wall. We therefore began to analyze PAI-1 gene expression in both normal and inflamed murine vascular tissue. Nuclease protection analysis revealed a relatively high concentration of PAI-1 mRNA in the murine aorta (23). In fact, of eleven different murine organs studied, the aorta contained the highest concentration of PAI-1 specific mRNA (23). In situ hybridization studies localized PAI-1 mRNA primarily to the medial layer of the muscular wall of murine vessels (24). Although cultured endothelial cells produce large amounts of PAI-1 (12), endothelial cells in vivo were entirely negative (24). Interestingly, intraperitoneal injection of endotoxin seemed to reduce the level of PAI-1 mRNA in the medial layer of the vessel wall, and induce it in the endothelial cells (24). In contrast, transforming growth factor beta primarily induced PAI-1 gene expression in the medial layer of the vessel wall (24).

These results indicate that vascular PAI-1 gene expression is highly regulated, and raise the possibility that PAI-1 biosynthesis also may be regulated in human vascular disease.

To begin to investigate PAI-1 in vascular disease, we evaluated PAI-1 gene expression within segments of atherosclerotic human arteries (25). Total RNA was isolated from eleven severely diseased and five relatively normal human arteries obtained from patients undergoing reconstructive surgery for aortic occlusive or aneurysmal disease, and was analyzed for PAI-1 gene expression by Northern blotting. We found that in severely atherosclerotic vessels, PAI-1 mRNA levels were significantly increased compared with normal or mildly affected arteries, and that in most instances, the level of PAI-1 mRNA was correlated with the degree of atherosclerosis (25). In situ hybridization analysis revealed an abundance of PAI-1 mRNA-positive cells within the thickened intima of atherosclerotic arteries, mainly around the base of the plaques, and in cells scattered within the necrotic material (25). We recently collected specimens from 23 additional patients undergoing reconstructive vascular surgery, and again localized PAI-1 mRNA to smooth muscle cells in the proximity of the plaque, and to macrophages scattered around the necrotic core. Moreover, immunohistochemical studies localized PAI-1 antigen to smooth muscle cells in the media close to the plaque, and it was detected diffusely distributed throughout the necrotic core (van Aken et al., unpublished observation). These observations have been confirmed recently (26). The actual PAI-1 binding structures/proteins in the vessel wall have not yet been identified, although Vn would seem to be a likely candidate.

The expression of Vn in the vessel wall has not yet been studied in great detail. Vn staining was noted in intimal thickenings and fibrous plaques of atherosclerotic arteries in association with collagen bundles, elastic fibers, and cell debris in the vicinity of elastin (27,28). Moreover, a monoclonal antibody raised against extracts of atherosclerotic lesions, was shown to detect Vn (29). In a recent study, Vn was found to co-localize with PAI-1 in extracellular areas of early atherosclerotic lesions (26). In general, we were unable to detect Vn mRNA in the atherosclerotic vessel wall, although in control experiments, a strong signal was demonstrable in normal human liver sections. In contrast, Vn antigen was present in necrotic plaque material, and in the extracellular matrix of the media and elastic fibers of the adventitia. Taken together, these results raise the possibility that Vn present in the plaque is derived from the plasma rather than from local synthesis by cells of the vessel wall. PAI-1 and Vn antigen appear to co-localize in some areas of the extracellular matrix and acellular plaque (van Aken et al., unpublished observation). However, in other areas (e.g., the luminal part of the plaque), PAI-1 and Vn were detected in distinctly different patterns. These observations raise the possibility that PAI-1 may also bind to structures distinct from Vn in the diseased vascular tissue.

SUMMARY

In summary, studies of the expression of PAI-1 in vascular tissues suggest that the level of PAI-1 gene expression is correlated with the severity of the atherosclerotic lesions. This finding raises the possibility that local increased expression of PAI-1 leads to an imbalance of the fibrinolytic system. The resulting fibrinolytic shut-down would be expected to promote local fibrin deposition and may be causally related to the pathogenesis of atherosclerotic vascular disease.

REFERENCES

1. J.H. Henkin, P. Marcotte, and H. Yang, The Plasminogen-Plasmin System. *Progress in Cardiovascular Diseases.* 34:135 (1991).
2. K.C. Robbins, The plasminogen-plasmin enzyme system, in: "Haemostasis and Thrombosis," R.W. Colman et al., eds., J.B. Lippincott Company, Philadelphia, Toronto (1987).
3. L.A. Liotta, R.H. Goldfarb, R. Brundage, G.P. Siegal, V. Terranova, and S. Garbisa, Effect of plasminogen activator (urokinase), plasmin, and thrombin on glycoprotein and collagneous components of basement membranes. *Cancer Res.* 41:4629 (1982).
4. J.L. Gross, D. Moscatelli, E.A. Jaffer, and D.B. Rifkin, Plasminogen activator and collagenase production by cultured capillary endothelial cells. *Cell Biol.* 95:974 (1982).
5. K. Dano, P.A. Andreasen, J. Grondahl-Hansen, P. Kristensen, L.S. Nielsen, and L. Skriver, Plasminogen activators, tissue degradation, and cancer. *Adv Cancer Res.* 44:139 (1985).
6. O. Saksela and D.B. Rifkin, Cell-associated plasminogen activation: Regulation and physiological functions. *Ann Rev Cell Biol.* 4:93 (1988).
7. F. Bachmann, Plasminogen Activators, in: "Haemostasis and Thrombosis," R.W. Colman et al., eds., Lippincott Company, Philadelphia-Toronto (1987).
8. M. Hoylaerts, D.C. Rijken, H.R. Linjen, and D. Collen, Kinetics of the activation of plasminogen by human tissue plasminogen activator: Role of fibrin. *J Biol Chem.* 257:2912 (1982).
9. P. Carmeliet, L. Schoonjans, L. Kieckens, B. Ream, J. Degen, R. Bronson, R. de Vos, J.J. van den Oord, D. Collen, and R.C. Mulligan, Physiological consequences of loss of plasminogen activator gene function in mice. *Nature.* 368:419 (1994).
10. R.D. Gerard and R.S. Meidell, Regulation of tissue plasminogen activator expression. *Annu Rev Physiol.* 51:245 (1989).
11. D.J. Loskutoff, Regulation of PAI-1 gene expression. *Fibrinolysis.* 5:197 (1991).
12. D.J. Loskutoff, M. Sawdey, and J. Mimuro, Type 1 plasminogen activator inhibitor. *Prog Hemost Thromb.* 9:87 (1989).
13. P. Carmeliet, L. Kieckens, L. Schoonjans, B. Ream, A. van Nuffelen, G. Prendergast, M. Cole, R. Bronson, D. Collen, and R.C. Mulligan, Plasminogen activator inhibitor-1 gene-deficient mice I. Generation by homologous recombination and characterization. *J Clin Invest.* 92:2746 (1993).
14. P. Carmeliet, J.M. Stassen, L. Schoonjans, B. Ream, J.J. van den Oord, M. de Mol, R.C. Mulligan, and D. Collen, Plasminogen activator inhibitor-1 gene deficient mice II. Effects on Hemostasis, Thrombosis, and Thrombolysis. *J Clin Invest.* 92:2756 (1993).
15. W.P. Fay, A.D. Shapiro, J.L. Shih, R.R. Schleef, and D. Ginsburg, Complete deficiency of plasminogen-activator inhibitor type 1 due to a frame-shift mutation. *N Engl J Med.* 327:1729 (1992).
16. R.R. Schleef, D.L. Higgins, E. Pillemer, and L.J. Levitt, Bleeding diathesis due to decreased functional activity of type 1 plasminogen activator inhibitor. *J Clin Invest.* 83:1747 (1989).
17. P.J. Declerk, M. de Mol, M.-C. Alessi, S. Baudner, E.-P. Paques, K.T. Preissner, G. Mueller-Berghaus, and D. Collen, Purification and characterization of a plasminogen activator inhibitor 1 binding protein from human plasma: Identification as a multimeric form of S protein (vitronectin). *J Biol Chem* 263:15454 (1988).
18. B. Wiman, T. Lindahl, and A. Almqvist, Evidence for a discrete binding protein of plasminogen activator inhibitor in plasma. *Thromb Haemost.* 59:392 (1988).
19. D. Seiffert and D.J. Loskutoff, Kinetic analysis of the interaction between type 1 plasminogen activator inhibitor and vitronectin and evidence that the bovine inhibitor binds to a thrombin-derived amino-terminal fragment of bovine vitronectin. *Biochem Biophys Acta.* 1078:23 (1991).
20. D. Seiffert, N.N. Wagner, and D.J. Loskutoff, Serum-derived vitronectin influences the pericellular distribution of type 1 plasminogen activator inhibitor. *J Cell Biol.* 111:1283 (1990).
21. K.T. Preissner, J. Grulich-Henn, H.J. Ehrlich, P. Declerck, C. Justus, D. Collen, H. Pannekoek, and G. Mueller-Berghaus, Structural requirements for the extracellular interaction of plasminogen activator inhibitor 1 with endothelial cell matrix-associated vitronectin. *J Biol Chem.* 265:18490 (1990).
22. B. Wiman and A. Hamsten, Impaired fibrinolyis and risk of thromboembolism. *Progress in Cardiovascular Diseases.* 34:179 (1991).
23. M. Sawdey and D.J. Loskutoff, Regulation of murine type 1 plasminogen activator inhibitor gene expression in vivo. *J Clin Invest.* 88:1346 (1991).

24. M. Keeton, Y. Eguchi, M. Sawdey, C. Ahn, and D.J. Loskutoff, Cellular localization of type 1 plasminogen activator inhibitor messenger RNA and protein in murine renal tissue. *Am J Pathol.* 142:59 (1993).
25. J. Schneiderman, M. Sawdey, M. Keeton, G.M. Bordin, E.F. Bernstein, R.B. Dilley, and D.J. Loskutoff, Increased type 1 plasminogen activator inhibitor gene expression in atherosclerotic human arteries. *Proc Natl Acad Sci USA.* 89:6998 (1992).
26. F. Lupu, G.E. Bergonzelli, D.A. Heim, E. Cousin, C.Y. Genton, F. Bachmann, and E.K.O. Kruithof, Localization and production of plasminogen activator inhibitor-1 in human healthy and atherosclerotic arteries. *Arteriosclerosis and Thrombosis.* 13:1090 (1993).
27. F. Niculescu, H.G. Rus, D. Porutiu, V. Ghiurca, and R. Vlaicu, Immunoelectron-microscopic localization of S-protein/vitronectin in human atherosclerotic wall. *Atherosclerosis.* 78:197 (1989).
28. C. Guettier, N. Hinglais, P. Bruneval, M. Kazatchkine, J. Bariety, and J.-P. Camilleri, Immunohistochemical localization of S protein/vitronectin in human atherosclerotic versus arteriosclerotic arteries. *Virchows Archiv A Pathol Anat.* 414:309 (1989).
29. R. Sato, Y. Komine, T. Imanaka, T. Takano. Monoclonal antibody EMR1a/212D recognizing site of deposition of extracellular lipid in atherosclerosis. Isolation and Characterization of a cDNA clone for the antigen. *J Biol Chem.* 265:21232 (1990).

LOCAL INCREASE IN PAI-2 ON STIMULATION OF MONOCYTES WITH MODIFIED LDL

Helen M.Ritchie,[1] Alec Jamieson,[2] and Nuala A. Booth[1]

[1]Department of Molecular and Cell Biology
Marischal College
University of Aberdeen
Scotland, AB9 1AS
[2]Zeneca Pharmaceuticals
Alderley Park
Macclesfield
Cheshire,
United Kingdom, SK10 4TG

INTRODUCTION

Fibrin deposistion is a feature of inflammatory diseases including atherosclerosis. Fibrin is a component of the atherosclerotic plaque and it is thought to act as a chemoattractant for infiltrating cells, as a scaffold for migrating cells and as a lipid binding surface, therefore promoting accumulation in the artery wall. (reviewed in[1]) Thrombus formation at the atherosclerotic plaque leading to myocardial infarction is a major cause of death. The balance of the fibrinolytic and coagulation systems is therefore important in the local environment of the artery wall . Tissue factor (TF) is present in the atherosclerotic plaque[2] and decreased fibrinolytic activity may be attributed to increased levels of plasminogen activator inhibitor types 1 (PAI-1) and/or 2 (PAI-2) leading to persistence of fibrin already deposited as a result of TF activity.

The monocyte is a central cell type in atherogenesis and has the ability to affect both coagulation and fibrinolytic systems. Activated monocytes have been shown to produce urokinase (u-PA)[3,4], tissue-type plasminogen activator (t-PA)[5], PAI-1[6], PAI-2 and TF[7,8], depending on the stimulus. Lipopolysaccharide (LPS) has been shown to increase levels of PAI-2 and to induce TF expression in monocytes[7,8]. Cytokines such as tumour necrosis factor-α and interleukin-1ß, both important cytokines implicated in inflammation, also increase synthesis of PAI-2 by monocytes[9,10], whereas PAI-1 synthesis can be stimulated by transforming growth factor ß[11]. PAI-2 is the major PAI produced by monocytes and u-PA the major PA[3].

Modified low density lipoprotein (LDL) has a key role in the development of the atherosclerotic plaque, where uptake of modified LDL, primarily by the monocyte but also by the smooth muscle cell, leads to foam cell formation. Oxidised LDL has also been found have many properties different from native LDL and can act as a chemoattractant for circulating monocytes and yet inhibits migration of macrophages from the atherosclerotic plaque[12]. In addition modified LDL can stimulate production of cytokines[13] and adhesion of monocytes to endothelial cells[14]. Modified LDL has also been found to affect fibrinolysis and coagulation by inducing TF in monocytes[15,16] and endothelial cells[17] in addition to increasing PAI-1 synthesis from endothelial cells[18].

Cardiovascular Disease 2
Edited by L.L. Gallo, Plenum Press, New York, 1995

We demonstrate that stimulation of peripheral blood monocytes with LDL modified by acetylation, copper oxidation and minimal modification leads to an increased secretion of PAI-2 above control levels. This increase in PAI-2 is specific as PAI-1 levels show no change and u-PA is undetectable in both control and stimulated monocytes.

METHODS

Monocyte Isolation and Culture

Peripheral blood from healthy individuals was collected into tri-sodium citrate. Mononuclear cells were isolated by centrifugation on Ficoll-Paque (Pharmacia). Monocytes were further purified by adherence to 24-well tissue culture plates (Nunc) by incubation at 37°C for 1hour in a 5% CO_2 environment before aspiration of non-adherent cells. Monocytes were found to constitute > 90% of the population using the non specific esterase stain[19] (Sigma). Monocytes (5×10^6/ml) were incubated in RPMI (Gibco), supplemented with 4 mM glutamine, 50µg/ml gentamycin and treated as appropriate with 50µg/ml LDL, 10ng/ml LPS or 1µg/ml tunicamycin for specified times. Supernatant was removed and stored at -70°C. Monocyte cell extracts were made by scraping cells from the plate into phosphate buffered saline and stored at -70°C for assay. All medium was sterile and all plasticware used was pyrogen-free.

LDL Isolation and Preparation

LDL (d=1.019 to 1.063g/ml) was isolated from normal human plasma by sequential ultracentrifugation. LDL was then acetylated using acetic anhydride as described[20]. Copper oxidation of LDL was performed by dialysis of LDL against 5µM $CuSO_4$, 150mM NaCl, 10mM Tris, pH7.4 for 24hrs at 4°C. Oxidation was terminated by dialysis into 1mM EDTA, 150mM NaCl, 10mM Tris, pH7.4 for 24hrs at 4°C . Minimal modification of LDL using soybean lipoxygenase and phospholipase A_2 was carried out as described[21]. LDL was sterilised using a 0.2µm filter and checked for endotoxin contamination by the limulus amoebocyte lysate assay (Sigma). Modification of LDL was verified using a conjugated diene assay[22] and by agarose gel electrophoresis[23].

Measurement of PAI-1 and PAI-2 Antigen

PAI-1[24] as in the case of PAI-2, was measured quantiatively using a specific ELISA. Microtitre plates (96 well, Costar) were coated overnight at 4°C with a monoclonal antibody to PAI-2 at 2µg/ml (MAI-21, Biopool). Plates were blocked and samples or standard added to wells for 2 hours at 37°C. Recombinant PAI-2, kindly provided by Delta Biotechnology, was diluted to give a standard curve of 12.4ng/ml-0.78ng/ml. The IgG fraction of a rabbit polyclonal antibody to PAI-2 (10µg/ml) was incubated for 2 hours at 37°C, followed by a goat anti rabbit IgG linked to alkaline phosphatase (Boehringer Mannheim, diluted 1/750). Colour was developed using p-nitrophenyl phosphate (2mg/ml, Sigma) for 30minutes at 37°C. Absorbance of 405nm was read using a Titertek spectrophotometer.

Measurement of u-PA Antigen

A specific ELISA was used and 96 well plates (Costar) were coated for 3hours at 37°C with a monoclonal antibody to u-PA (2.5µg/ml Muk-1, Biopool). The plate was blocked and samples and standard incubated overnight at room temperature. u-PA was diluted to give a standard curve of 21ng/ml-0.65ng/ml. A rabbit polyclonal antibody to u-PA conjugated to horseradish peroxidase (1.3µg/ml) was incubated at room temperature for 3 hours. Colour was developed by incubation at room temperature for 15 minutes with 100mM sodium acetate citrate buffer, pH6.0, urea hydrogen peroxide and TMB. The reaction was stopped with 2.5M sulphuric acid and absorbance read at 450nm using a Titertek spectrophotometer.

RESULTS AND DISCUSSION

Modication of LDL resulted in all forms of modified LDLs migrating further than native LDL on an agarose gel. Measurement of conjugated dienes showed that AcLDL had the same level as native, minimally modifed showed a 1-2 fold increase and CuLDL had a 5-6 fold increase over native LDL.

All forms of modified LDL were seen to stimulate production of PAI-2 into the culture supernatant of monocytes (Figure 1). Acetylated LDL (AcLDL) and copper oxidised LDL (CuLDL) stimulated similar levels of PAI-2 which were greatly elevated over control levels, whereas minimally modified LDL (MMLDL) stimulated lower levels of PAI-2 although still significantly increased above control.

The increase in PAI-2 in culture supernatant was confirmed by zymography following non-denaturing gel electrophoresis, a semi quantitative technique[25](data not shown). The inhibitory band was neutralized by an antibody to PAI-2.

PAI-2 is found in an 47kDa non-glycosylated intracellular form and as a 60kDa glycosylated secreted protein[26]. LDL has been reported to be cytotoxic to monocytes[27] depending on the oxidation conditions, so increased PAI-2 in the supernatant could result from release of intracellular PAI-2 rather than de novo synthesis. This explanation was ruled out by measurement of intracellular PAI-2. It did not fall as supernatant PAI-2 rose. LDL stimulation of monocytes in the presence of tunicamycin, which blocks protein glycosylation, did not lead to the increased PAI-2 in the supernatant suggesting that PAI-2 must enter the secretory pathway on LDL stimulation. Northern blotting of total RNA isolated from control and LDL stimulated monocytes, probed for PAI-2 mRNA, shows an increase in PAI-2 mRNA over control indicating that upregulation is at the level of transcription.

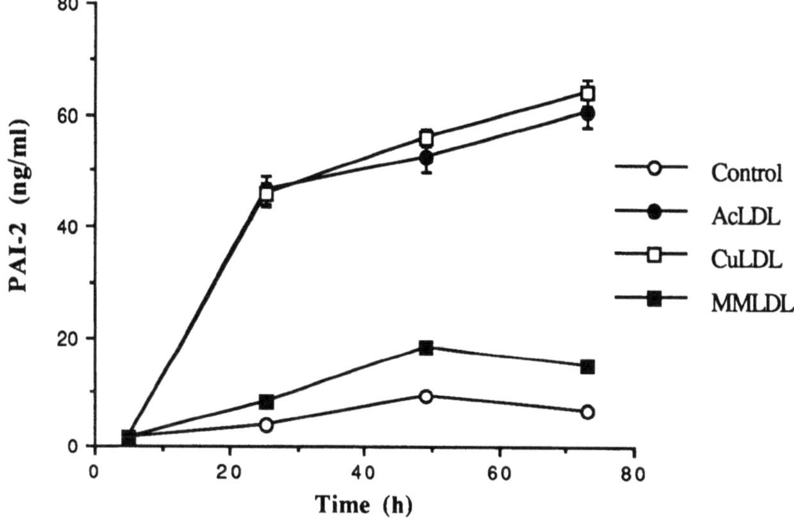

Figure 1. Time course of PAI-2 synthesis following stimulation by modified LDL. Values represent means ±S.E.M. from duplicate wells. Statistical analysis was carried out using student's t-test and $p>0.05$.

Levels of PAI-2 in culture supernatant of LDL stimulated monocytes varied depending on the donor used. Variability was noted in both control and LDL stimulated samples. Thus the differences were not due entirely to different preparations of LDL but also reflected variability in the monocytes themselves. This was also verified by measuring levels of PAI-2 in culture supernatant of monocytes stimulated with 10ng/ml lipopolysaccharide (LPS) and again PAI-2 production was variable in both control and stimulated samples. However, despite varibility, modified LDLs and LPS did stimulate PAI-2 production from peripheral blood monocytes significantly above control. Donor variability in PAI-2 production is a phenomenon that has been shown by others[7] and has been suggested to play a role in differential fibrin persistence in patients[26]. In addition, the ratio of intracellular to extracellular PAI-2 has been reported to differ, but we find that, on LDL stimulation, intracellular levels of PAI-2 consistently remained similar to control.

As PAI-1 is produced from monocytes, levels of PAI-1 were also measured quantitatively by ELISA and found to be very low (0.313ng/ml - 2.5ng/ml). PAI-1 levels in supernatant do not increase above control levels. In fact, MMLDL was seen to decrease PAI-1 production compared to control. The response of the monocyte to MMLDL was distinct from that to AcLDL and CuLDL in that AcLDL and CuLDL stimulated PAI-2 synthesis with no effect on PAI-1, but MMLDL modulated both PAI-2 and PAI-1.

Cell supernatants and lysates were also assayed for u-PA by ELISA and no u-PA was detectable. Monocytes usually produce low levels of u-PA, so supernatants and lysates were studied by SDS-PAGE followed by zymography[28], which can detect less than a nanogram of active u-PA; no u-PA was detectable. This indicates that the increase seen in PAI-2 is not accompanied by a detectable increase in u-PA.

We have shown that peripheral blood monocytes respond to modified LDL by increased synthesis of PAI-2, but the monocyte also exists in the plaque as a macrophage and foam cell. Lipid loading of macrophages to foam cells is associated with increased expression of TF[29], therefore study of these cell types may ascertain which stage in monocyte development is important in determining the balance of the fibrinolytic system during atherogenesis.

We have demonstrated that acetylated, copper oxidised and minimally modified LDL can stimulate increased production of PAI-2 from peripheral blood monocytes, without affecting PAI-1 or u-PA, and thus causing an imbalance of the fibrinolytic system. PAI-2 was detected by a specific ELISA developed for this purpose. Analysis of cell lysate PAI-2 indicated that the increase in supernatant PAI-2 was not due to release of cell associated PAI-2 caused by cytotoxicity of the modified LDL, verified by incorporation of tunicamycin into the system and northern blotting for PAI-2 mRNA.

The effect of oxidised LDL on the synthesis of PAI-1 from endothelial cells has been reported[18] and atherosclerotic arteries have been shown to have increased level of PAI-1[30,31]. PAI-1 is synthesised by both endothelial cells and smooth muscle cells, but the role of PAI-2 in the atherosclerotic plaque has not been well studied. The monocyte is the major infiltrating cell type in the plaque and activation results in increased PAI-2. Therefore PAI-2 production may be important in the local environment of the artery wall. Intracellular PAI-2 may act as a reservoir of PAI that can be released when appropriate or even contribute to the necrotic core of the advanced atherosclerotic plaque in later stages of the disease. Recent studies have demonstrated the presence of PAI-2 in atherosclerotic plaques (L.A. Robbie, N.A. Booth and B. Bennett, unpublished data).

In summary, our study demonstrates enhanced production of PAI-2 from monocytes stimulated with different forms of modified LDL including acetylated, copper oxidised and minimally modified. In the local environment of the artery wall, the effect of PAI-2 produced from activated monocytes may result in persistence in fibrin deposits and thus contribute to thrombus stability.

REFERENCES

1. D. Collen and I. Juhan-Vague, Fibrinolysis and atherosclerosis, *Seminars in Thrombosis and Haemostasis*. 14:180 (1988)

2. J.N. Wilcox, K.M. Smith, S.M. Schwartzand D. Gordon, Localization of tissue factor in the normal vessel wall and in the atherosclerotic plaque, *Proc. Natl. Acad. Sci. USA* 86:2839 (1989)

3. J.-D. Vassalli, J.-M. Dayer, A. Wohlwend and D. Belin, Concomitant secretion of prourokinase and of a plasminogen activator-specific inhibitor by cultured human monocytes-macrophages, *J. Exp. Med*. 159:1653 (1984)

4. J.A. Hamilton, G.A. Whitty, H. Stanton and A. Meager, Effects of macrophage-colony stimulating factor on human monoytes: induction of expression of urokinase-type plasminogen activator, but not of secreted prostaglandin E_2, interleukin-6, interleukin-1, or tumour necrosis factor-α, *J. Leukoc. Biol.* 53:707 (1993)

5. P.H. Hart, D.R. Burgess, G.F. Vitti and J.A. Hamilton, Interleukin-4 stimulates human monocytes to produce tissue-tpye plasminogen activator, *Blood* 74:1222 (1989)
6. J.C. Castellote, E. Grau, M.A. Linde, N. Pujol-Moix and M. Ll. Rutllant, Detection of both type 1 and type 2 plasminogen activator inhibitors in human monocytes, *Thrombosis and Haemostasis* 63:67 (1990)
7. B.S. Schwartz, M.C. Monroe and E.G. Levin, Increased release of plasminogen activator inhibitor type 2 accompanies the human mononuclear cell tissue factor response to lipopolysaccharide, *Blood* 71:734 (1988)
8. B.S. Schwartz and J.D. Bradshaw, Differential regulation of tissue factor and plasminogen activator inhibitor by human mononuclear cells, *Blood* 74:1644 (1989)
9. M.R. Gyetko, C.C. Wilkinson and R.G.Sitrin, Monocyte urokinase expression: modulation by interleukins, *J. Leuk. Biol.* 53:598 (1993)
10. M.R. Gyetko, S.B. Shollenberger and R.G. Sitrin, Urokinase expression in mononuclear phagocytes: cytokine-specific modulation by interferon-γ and tumor necrosis factor-α, *J. Leuk. Biol.* 51:256 (1992)
11. J.A. Hamilton, G.A. Whitty, J. Wojta, M. Gallichio, K. McGrath and G. Ianches, Regulation of plasminogen activator inhibitor-1 levels in human monocytes, *Cellular Immunology* 152:7 (1993)
12. M.T. Quinn, S. Parthasarathy, L.G. Fong and D. Steinberg, Oxidatively modified low density lipoproteins: A potential role in recruitment and retention of monocyte/macrophaged during atherogenesis, *Proc. Natl. Acad. Sci. USA* 84:2995 (1987)
13. T.B. Rajavashisth, A. Andalibi, M.C. Territo, J.A. Berliner, M. Navab, A.M. Fogelman and A.J. Lusis, Induction of endothelial cell expression of granulocyte and macrophage colony-stimulating factors by modified low density lipoproteins, *Nature* 344:254 (1990)
14. J.A. Berliner, M.C. Territo, L. Almada, A. Carter, E. Shafonsky and A.M. Fogelman, Minimally modified low density lipoprotein stimulates monocyte endothelial interactions, *J. Clin. Invest.* 85:1260 (1990)
15. P. Schuff-Werner, G. Claus, V.W. Armstrong, H. Kostering and D. Seidel, Enhanced procoagulant activity (PCA) of human monocytes/macrophages after in vitro stimulation with chemically modified LDL, *Atherosclerosis* 78:109 (1989)
16. K. Brand, C.L. Banka, N. Mackman, R.A. Terkeltaub, S.-T. Fan and L.K. Curtiss, Oxidised LDL enhances lipopolysaccharide- induced tissue factor expression in human adherent monocytes, *Arterioscler. Thromb.* 14:790 (1994)
17. T.A. Drake, K. Hannani, H. Fei, S. Lavi and J.A. Berliner, Minimally oxidised low-density lipoprotein induec tissue factor expression in cultured human endothelial cells, *American Journal of Pathology* 138:601 (1991)
18. E. Tremoli, M. Camera, P. Maderna, L. Sironi, L. Prati, S. Colli, F. Piovella, F. Bernini, A. Corsini and L. Mussoni, Increased synthesis of plasminogen inhibitor-1 by cultured human endothelail cells exposed to native and modified LDLs, *Arteroiscler. Thromb.* 13:338 (1993)
19. S.B. Tucker, R.V. Pierre and R.E. Jordan, Rapid identification of monocytes in a mixed mononuclear cell preparation, *J. Immun. Methods* 14:267 (1977)
20. S.K. Basu, J.L. Goldstein, R.G.W. Anderson and M.S. Brown, Degradation of cationized low density lipoprotein and regulation of cholesterol metabolism in homozygous familial hypercholesterolemia fibroblasts, *Proc. Natl. Acad. Sci. USA* 73:3178 (1976)
21. C.P. Sparrow, S. Parthasarathy and D. Steinberg, Enzymatic modification of low density lipoprotein by purified lipoxygenase plus phospholipaseA_2 mimics cell-mediated oxidative modification, *J. Lipid Research* 29:745 (1988)
22. F.P. Corongiu and A. Milia, An improved and simple method for determining diene conjugation in autoxidised polyunsaturated fatty acids, Chem-Biol Interact 44:289 (1983)
23. R.P. Noble, Electrophoretic separation of plasma lipoproteins in agarose gel, *J Lipid Research* 9:693 (1968)
24. I.R. MacGregor and N.A. Booth, An enzyme-linked immunosorbent assay (ELISA) used to study the cellular secretion of endothelial plasminogen activator inhibitor (PAI-1), *Thrombosis and Haemostasis* 59:68 (1988)
25. A. Reith, B. Bennett and N.A. Booth, Plasminogen activator inhibitor (PAI-2) in pregnancy plasma and ovarian cysts, studied by zymography after non-denaturing gel electrophoresis, *Fibrinolysis* 3:159 (1989)

26. D. Belin, A. Wohlwend, W-D. Schleuning, E.K.O. Kruithof and J-D. Vassalli, Facultative polypeptide translocation allows a single mRNA to encode the secreted and cytosolic forms of PAI-2, *EMBO J.* 8:3287 (1989)
27. V.C. Reid, M.J. Mitchison and J.N. Skepper, Cytotoxicity of oxidised low-density lipoprotein to mouse peritoneal macrophages: an ultrastructural study, *J. Pathology* 171:321 (1993)
28. N.A. Booth, J.A. Anderson and B. Bennett, Plasminogen activators in alcoholic cirhossis:demonstration of increased tissue type and urokinase type activator, *J. Clin Pathology* 37:772 (1984)
29. P. Lesnik, M. Rouis, S. Skarlatos, H.S. Kruth, M.J. Chapman, Uptake of exogenous free cholesterol induces upregulation of TF expression in human monocyte derived macrophages, *Proc. Natl. Acad. Sci. USA* 89:10370 (1992)
30. J. Schneiderman, M.S. Sawdey, M.R. Keeton, G.M. Bordin, E.F. Bernstein, R.B. Dilley and D.J. Loskutoff, Increased type 1 plasminogen activator inhibitor gene expression in atherosclerotic human arteries, *Proc. Natl. Acad. Scs. USA* 89:6998 (1992)
31. F. Lupu, G.E. Bergonzelli, D.A. Heim, E. Cousin, C.Y. Genton, F. Bachmann and E.K.O. Kruithof, Localization and production of plasminogen activator inhibitor-1 in human healthy and atheroscleroric arteries, Arterioscler Throm 13:1090 (1993)

STRUCTURE, FUNCTION AND REGULATION OF THE UROKINASE RECEPTOR

Francesco Blasi

Dipartimento di Genetica e Biologia dei Microrganismi
University of Milano and Unit of Molecular Genetics
DIBIT, H.S. Raffaele Scientific Institute,
via Olgettina 60, 20132 Milan, Italy

INTRODUCTION

The receptor for urokinase plasminogen activator (uPAR) acts as an anchorage site for uPA on the cell surface where it stimulates activation of pro-uPA and plasminogen. Thereby, uPAR regulates cell surface proteolytic activity, an activity which has been abundantly shown to be important in fibrinolysis, neo-angiogenesis, cell migration and in conditioning the invasive behaviour of cancer cells[1]. In line with its migration-regulating activity, activation of plasminogen can activate latent TGF-β[2]; alternatively, also prohepatocyte growth factor, a protein belonging to the plasminogen family, endowed with growth factor and migration-stimulating activity, can be directly activated proteolytically by uPA[3]. Finally, direct binding of uPA to its receptor without proteolytic activation of other proteins, can stimulate, through as yet undefined messengers, several cellular functions like mitogenesis, cell adhesion, chemotaxis, cell migration[1].

The uPA receptor, uPAR, is a three-domains protein that belongs to the Ly6 family of peptides, characterized by a highly conserved cysteine-spacing and disulfide bond pattern[4,5]. At the genomic level, each of the three domains is coded for by two exons, in a pattern that is also conserved throughout the family. The two 5'located exons of the uPAR gene encode the amino-terminal, ligand-binding domain (87 residues) of the protein (Soravia, E., Helin, K., Suh, T.T., Degen, J.L. and Blasi, F., unpublished). However, the other two domains also participate into the binding, since mutant receptors containing in addition to the binding domain only one of the two other domains, display an at least 100 fold reduced affinity for the ligand despite being correctly exposed on the cell surface (L. Riittinen, F. Blasi and F. Fazioli, unpublished).

uPAR is a highly glycosylated protein. The carbohydrate sidechain present in the ligand-binding domain contributes to the affinity for the ligand. In fact, substitution of Asn52 with a Gln residues results in an about 8-fold decrease in ligand affinity. Conversely, uPAR hyper-glycosylation caused by phorbol esters also results in a decrease in affinity[6,7].

The membrane location of uPAR is dependent on the presence at the carboxy terminus of a glycosyl-phosphatidylinositol (GPI) anchor[8], again a conserved feature in the Ly6 family of proteins. The information for the GPI anchor resides in the last 32 amino acid residues of the nascent uPAR, which are in fact removed during biosynthesis and substituted by the glycolipid anchor. The last 8 carboxy-terminal residues encoded on the mRNA contain the anchor signal, while the anchor attachment site is located at residues ser282/gly283[9]. Mutations in these two areas of the uPAR cDNA result in a ligand-binding, intracellular, soluble receptor[10]. Pulse-chase immunoprecipitation experiments show that the protein is processed early in the biosynthetic phase to reach the cell surface where it is attached through a glycolipid anchor and hence is releasable by phosphatidylinositol-specific phospholipase C (P. Limongi, M. Resnati, F. Blasi and F. Fazioli, unpublished data).

Despite its GPI-anchorage, which should route it at the apical side of the cells, uPAR is present at the focal and cell-to-cell contacts, i.e. at baso-lateral areas. This location appears to depend on the presence of its ligand, uPA[11]. Ligand-binding causes uPAR to move more freely within the membrane, and appears to be the basis of its signal-transducing ability. Signal transduction in fact correlates with clustering of uPAR at the cell surface, which depends on the presence of the ligand (F. Fazioli, M. Guttinger and F. Blasi, unpublished). Ligand-binding results in activation of both serine-threonine[12] and tyrosine-protein kinases and in the stimulation of cell migration of monocytic cell lines in vitro (F. Fazioli, S. Valcamonica, M. Resnati and F. Blasi, unpublished). Intermediate in this process appears to be the phosphorylation of cytoskeletal[12] and other proteins. Little information is available on the mitogenic activity, another signal-transducing function assigned to uPA and uPAR.

As long as it remains active, receptor-bound uPA remains at the cell surface and is not internalized nor degraded. However, once uPA is complexed with one of its specific inhibitors (plasminogen activator inhibitors type-1 and type 2, PAI-1 and PAI-2, or protease nexin, PN-1), a profound change in the properties of uPAR occurs, and the complexes are internalized and degraded[13]. Similarly, complexes of uPA with the plant toxin saporin are also internalized and degraded through uPAR[14]. However, the amino terminal fragment of uPA, ATF, is not internalized, whereas uPA inhibited by low molecular weight compounds like DFP and also pro-uPA are internalized at a very slow rate. The process of internalization of the uPA:inhibitors and uPA:saporin complexes depends on the previous binding to uPAR, as it can be inhibited by the uPA antagonists ATF and DFP-uPA, by anti-uPAR antibodies and does not occur when uPAR is released by PI-PLC[15]. However, ligands or antibodies recognizing a different receptor can inhibit degradation of uPA:inhibitors complexes, although not preventing their binding to the cell surface. The second receptor involved is the trans-membrane α 2 macroglobulin receptor/LDL receptor-related protein (LRP)[16]. The process requires formation of a tetrameric complex (uPAR:uPA:inhibitor:LRP) on the cell surface, which is then internalized (M. Conese and F. Blasi, unpublished data). The internalization route is not yet known, but may occur through coated pits, since LRP internalizes other ligands through this pathway. The formation of tetrameric complexes is probably dependent on direct binding of the ligands to both uPAR and LRP. Binding of uncomplexed uPA and PAI-1 to LRP has a low affinity, while that of the uPA:PAI-1 complex is higher, although not as high as for uPAR (A. Nykjaer and J. Gliemann, unpublished). The physiological role of the uPAR/LRP mediated internalization of uPA:inhibitors complexes is still not known. In the absence of uPAR, LRP could still mediate the clearance of uPA:inhibitors complexes and of pro-uPA, but only at very high, non-physiological, concentration. LRP alone, therefore, is likely to be relevant in pharmacological conditions, when the concentration of, i.e. pro-uPA, is very high.

The uPA/uPAR system is very complex at the molecular level: the various molecules of this system, uPA, uPAR, PAIs, must be capable of direct or indirect interaction with a variety of other molecules: i.e. inhibitors, other receptors (like LRP), serine- and tyrosine-kinases and others. A major question therefore arises: how are these interactions regulated to ensure proper function? So far, the information accumulated suggest that this system undergoes a strict, and intricated, cross-regulation both at the level of gene expression and of activity. One case is the PMA-dependent differentiation of the monocyte-like U937 cells to macrophages. Here cells initially adhere to the plastic dish and subsequently display a variety of macrophage markers and functions. Binding of uPA to uPAR is required for the PMA induction of adhesion to the plastic dish. The effect is negatively controlled by the inhibitors PAI-1 and PAI-2[17]. During differentiation, expression of the genes for uPA, uPAR, PAI-1 and PAI-2 are induced and the corresponding proteins over-expressed. One would therefore expect fully differentiated U937 cells to internalize and degrade uPA:PAI-1 even more efficiently than undifferentiated cells. However, upon prolonged PMA treatment, the efficiency of internalization and degradation of the uPA:PAI-1 complexes is not increased, rather it is almost completely blocked. This appears to be due to a differential effect of PMA on the surface expression of the uPAR and LRP proteins: while uPAR is increased the latter essentially disappears. The data suggest that the expression of the two receptors is coordinated, being increased at the time when adhesion is beginning to occur and ensuring adequate internalization of the uPA:inhibitors complexes; once differentiation has occurred, surface expression of LRP is drastically reduced and therefore internalization/degradation of complexes no longer occurs (M. Conese, U. Cavallaro, D. Olson, M. R. Soria and F. Blasi, unpublished).

The scope of this regulation however, is still not clear, and may be dependent on the specific environment and on the function regulated by the binding of uPA and derivatives to uPAR; i.e. proteolysis of specific cell-cell bonds, activation of signal transducing factors or direct signal transduction. To approach this question in a more physiological and relevant environment, the expression of the various genes involved in internalization/degradation has been studied during the first stage of embryo implantation in mouse by in situ hybridization to detect the site of synthesis, immunohistochemistry to detect the proteins, and in situ zymography to detect the presence of functional enzyme. Each component of the system displays a regional expression in the apical pole of the implanting zone in an area that can be divided in three adjacent layers: the innermost one, the embryo and the extraembryonic trophoblast; the external layer, decidua, with active neoangiogenic activity, and the intermediate layer directly surrounding the implanting pole of the embryo. Synthesis of uPA is confined to the ectodermal portion of the embryo, in particular the extraembryonic trophoblast while uPAR mRNA is found at very low levels in this same layer, but is very abundant in the most external, decidual, layer. Cells from the same layer also actively synthesize LRP. Finally, PAI-1 synthesis is confined to the intermediate layer, in contact with the extraembryonic trophoblast. By employing immunodetection, uPA is found not only at the site of synthesis, but also throughout the intermediate and decidual layers, i.e. in contact with PAI-1, uPAR and LRP. The data suggest that uPA may have different functions and fate in the different layers. Indeed, in situ zymography shows that uPA activity is restricted to its site of synthesis. These data suggest a tight regulation of the expression of uPA and PAI-1 gene at the embryo implantation area, the scope of which would be to restrict uPA activity mostly to the area of synthesis. The data also assign to the neighbouring areas the task to inhibit uPA outside of the area of synthesis and subsequently (in space) to degrade and clear it. The data also suggest that the expression of uPA, PAI-1, uPAR and possibly even α2MR genes may be inter-connected (T. Teesalu, F. Blasi and D. Talarico, unpublished). The search for the connecting mechanism is now under way. Also, the exact role of this whole system during embryo implantation is not clear. The fact that uPA$^{-/-}$ mice are fertile, suggests that whatever its function, uPA is not essential for embryo implantation and can be substituted. At the same time, the presence of such a complex regulation suggests an important function for which different molecules can substitute each other.

uPA and uPAR appear to be important to regulate the invasiveness of human cancer cells both in vitro and in vivo. This comes from a very large number of data accumulated in several years[1] which show that: 1. Tumors express more uPA and PAI-1 than normal tissues, and the level of uPA and PAI-1 is of important prognostic relevance. 2. In cell culture systems, uPA activity can be directly correlated to the degradation of the extracellular matrix. 3. In in vitro model systems, increase of uPA expression causes an increase in the invasive and metastatic ability of tumor cells. 4. Conversely, the block of uPA activity by expression of PAI-1 or by addition of anti-uPA antibodies, results in a drastic decrease of the invasive phenotype. 5. Finally, block of the uPA-uPAR interaction also drastically decreases the invasive phenotype of human tumor cells in nude mice.

The large number of positive indications obtained in experimental models and the enthusiasm surrounding these new ideas have prompted a search for anti-uPAR reagents to be used as diagnostic or therapeutic tools. The practical use of the inhibition of the uPA-uPAR interaction for anti-metastatic therapy of human tumors will therefore be subjected to test in a, hopefully, near future.

REFERENCES

1. Fazioli, F . and Blasi, F. Urokinase-type plasminogen activator and its receptor: new target for anti-metastatic therapy? TiPS, 15, 25-29, (1994).
2. Odekon, L. E . , Blasi , F. and Rifkin, D. B. A requirement for receptor-bound urokinase in plasmin-dependent cellular conversion of latent TGF-β to TGF-β. J. Cell. Physiol. 158, 396-407 (1994).
3. Naldini, L., Tamagnone, L., Vigna, E., Sachs, M., Hartmann, G., Birchmeier, W., Daikuhara, Y., Tsubouchi, H., Blasi, F. and Comoglio, P.M. . Extracellular proteolytic cleavage by urokinase is required for activation of hepatocyte growth factor/scatter factor. EMBO J., 11, 4825-4833, (1992) .
4. Behrendt, N., Ploug, M., Patthy, L., Houen, G., Blasi, F. & Dano, K. The ligand-binding domain of the cell surface receptor for urokinase-type plasminogen activator. J. Biol. Chem., 266, 7842-7847, (1991).
5. Ploug, M., Kjalke, M., Ronne, E., Weidle, U., Hoyer-Hansen, G. and Dano, K. Localization of the disulfide bonds in the NH2-terminal domain of the cellular receptor for human urokinase-type plasminogen activator. A domain structure belonging to a novel superfamily of glycolipid-anchored membrane proteins. J. Biol. Chem., 268, 17539-17546 (1993).
6. Moller, L.B., Pöllänen, J., Ronne, E., Pedersen, N. and Blasi, F. N-linked glycosylation of the ligand binding domain of the human urokinase receptor contributes to the affinity to its ligand. J. Biol. Chem . 268, 11152-11159, (1993).
7. Picone, R, Kajtaniak, EL, Nielsen, LS, Behrendt, N, Mastronicola, MR, Cubellis, MV, Stoppelli, MP, Dano, K & Blasi, F. Phorbol ester regulates synthesis and affinity of urokinase receptors in monocyte-like U937 cells. J. Cell Biol., 108, 693-702, (1989).
8. Ploug, M., Ronne, E., Behrendt, N., Jensen, A.L., Blasi, F. & Dano, K. Cellular Receptor for urokinase plasminogen activator: carboxyterminal processing and membrane anchoring by glycosylphosphatidylinositol. J. Biol. Chem.. 266, 1926-1933, (1991).
9. Moller, L.B., Ploug, M. and Blasi, F. Structural requirements for glycosyl-phosphatidylinositol anchor attachment in the cellular receptor for urokinase plasminogen activator. Eur. J. Biochem. 208, 493-500, (1992).
10. Masucci, M.T., Pedersen, N. and Blasi, F. A ligand-binding mutant form of the human urokinase plasminogen activator receptor. J. Biol. Chem., 266, 8655-8658 (1991).

11. Myöhanen, H.T., Stephens, R.W., Hedman, K., Tapiovaara, H., Ronne, E., Hoyer-Hansen, G., Dano, K. and Vaheri, A. Distribution and lateral mobility of the urokinase-receptor complex at the cell surface. J Histochem. Cytochem. 41, 1291 -1301 (1993).
12. Busso, N., Masur, S.K., Lazega, D., Waxman, S. and Ossowski, L. Induction of cell migration by pro-urokinase binding to its receptor: possible mechanisms for signal transduction in human epithelial cells. J. Biol. Chem.. (1994) in press.
13. Cubellis, M.V., Wun, T.-C and Blasi, F. Receptor-mediated internalization and degradation of urokinase is caused by its specific inhibitor PAI-1. EMBO J., 9, 1079-1085, (1990).
14. Cavallaro, U., del Vecchio, A., Lappi, D.A. and Soria, M.R. A conjugate between human urokinase and saporin, a type-1 ribosome-inactivating protein, is selectively cytotoxic to urokinase receptor-expressing cells. J. Biol. Chem., 268, 23186-23190, (1993).
15. Olson, D., Pöllänen, J., Hoyer-Hansen, G., Ronne, E., Sakaguchi, K., Wun, T.-C., Appella, E., Dano, K. and Blasi, F. Internalization of the urokinase:plasminogen acftivator inhibitor type-1 complex is mediated by the urokinase receptor. J. Biol. Chem., 267, 9129-9133, (1992).
16. Nykjaer, A., Petersen, C.M., Moller, B., Jensen, P.H., Moestrup, S.K., Holtet, T.L., Etzerodt, M., Thoghersen, H.C., Munch, M., Andreasen, P.A. and Gliemann, J. Purified $\alpha 2$ macroglobulin receptor/LDL receptor related protein binds urokinase-plasminogen activator inhibitor type-1 complex. Evidence that $\alpha 2$ macroglobulin receptor mediates cellular degradation of urokinase receptor-bound complexes. J. Biol. Chem., 267, 14543-14546, (1992).
17. Waltz, D.A., Sailor, L.Z. and Chapman, H.A. Cytokines induce urokinase-dependent adhesion of human myeloid cells. A regulatory role for plasminogen activator inhibitors. (1993) J. Clin. Invest. 91, 1541 - 1552.

ROLE OF THE LDL RECEPTOR-RELATED PROTEIN IN PROTEINASE AND LIPOPROTEIN CATABOLISM

Dudley K. Strickland[1], Suzanne E. Williams[1], Maria Z. Kounnas[1], W. Scott Argraves[1], Ituro Inoue[2], Jean-Marc Lalouel[2], and David A. Chappell[3]

[1]Biochemistry Department, Holland Laboratory, American Red Cross, Rockville, MD 20855, [2]Howard Hughes Medical Institute, University of Utah, Salt Lake City, Utah 84132, and the [3]Department of Medicine, University of Iowa, Iowa City, Iowa 52242

INTRODUCTION

Proteinases play an important role in biological processes and consequently their activity is carefully regulated. This often occurs by reaction of the proteolytic enzyme with specific inhibitors. Removal of the inhibited proteinase is then accomplished by its interaction with cell surface receptors which mediate its internalization and subsequent degradation. The LDL receptor-related protein/α_2M receptor (LRP) is large cell surface receptor that mediates the removal of proteinases and proteinase-inhibitor complexes[1-5]. In addition, this receptor plays an important role in the hepatic clearance of certain apolipoprotein E- and lipoprotein lipase-enriched lipoproteins[6-8]. Thus, LRP serves a unique role in biology by virtue of its capacity to mediate the cellular uptake of both proteinases and lipoproteins.

THE LDL RECEPTOR FAMILY

LRP is a member of the LDL receptor family. The domain organization of known members of this family of receptors is depicted in Figure 1. The LDL receptor[9] contains a cytoplasmic domain, a transmembrane domain, an O-linked sugar domain, a cluster of cysteine-rich epidermal growth factor-like repeats, five copies of a tetrapeptide sequence of tyrosine-tryptophan-threonine-aspartic acid, and a second cysteine-rich cluster containing seven complement-like repeats. The cytoplasmic tail contains an asparagine-proline-X-tyrosine sequence which targets this receptor to clathrin-coated pits[10]. The growth factor and tetrapeptide repeats appear necessary for the receptor to undergo an acid-dependent conformational change which releases ligands within the endosomes allowing the unoccupied receptor to recycle back to the cell surface[11]. The complement-like repeats are responsible for binding apolipoproteins (apo) B-100 and E[12].

The most recently discovered member of the LDL receptor family is the very low

Cardiovascular Disease 2
Edited by L.L. Gallo, Plenum Press, New York, 1995

density lipoprotein (VLDL) receptor[13]. The primary sequence of this receptor is similar to the LDL receptor[9,13]. Further, the structure and organization of the human VLDL receptor gene[14] is also similar to that of the human LDL receptor gene[15]. Both genes have almost complete conservation of exon and intron positions, except that the VLDL receptor gene contains an extra exon that encodes an additional cysteine-rich repeat sequence within the ligand binding domain. A variant form of the VLDL receptor which lacks the O-linked sugar domain has been identified[14].

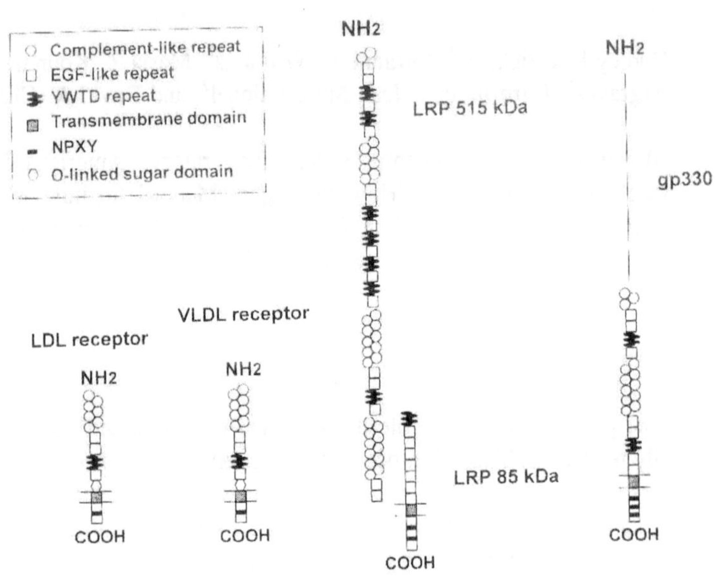

Figure 1. Domain organization of LDL receptor family members

LRP[16] is considerably larger than either the LDL receptor or the VLDL receptor. The O-linked sugar domain of the LDL receptor is replaced by six growth factor like repeats in LRP. LRP contains a total of 22 growth factor repeats and 31 complement-type repeats arranged into four clusters, and is synthesized as a single chain precursor that is cleaved into a 515 kDa heavy chain and an 85 kDa light chain[17]. The larger subunit, which contains the ligand binding regions, associates non-covalently with the smaller subunit.

A fourth member of the LDL receptor family is a molecule termed glycoprotein 330 (gp330)[18]. The complete structure of this molecule has not been reported at this time. Based on the partial primary sequences of rat[19] and human[20] gp330 that are available, it is evident that gp330 shares many structural features with other members of this family. These include the cysteine-rich growth factor repeats, complement-type repeats, and repeats containing the tetrapeptide sequence tyrosine-tryptophan-threonine-aspartic acid. Finally, a 95 kDa vitellogenin receptor from chicken oocytes has been identified[21]. Partial amino-acid sequencing has confirmed that this molecule is a member of the LDL receptor family.

Together, members of this receptor family play an important role in the catabolism of lipoproteins, proteinases, and proteinase-inhibitor complexes. The LDL receptor mediates the catabolism of apolipoprotein B and/or apolipoprotein E containing lipoproteins and plays a key role in cholesterol homeostasis[22]. The VLDL receptor binds apoE-containing lipoproteins, such as VLDL, β-VLDL, and IDL[13]. This receptor is abundant in several tissues, including muscle and adipose tissues, but is not expressed in the liver, and is thought to play a key role in the catabolism of triglyceride-rich lipoproteins[13]. LRP, which is widely distributed and found in many cell types[51], mediates the removal of proteinases, proteinase-inhibitor complexes, and certain apoE- and lipoprotein lipase-enriched lipoproteins[1-8]. While the function of gp330 is not known at this time, this receptor binds many, but not all, of the ligands that interact with LRP[19,23]. The distinct distribution of gp330 on a restricted group of epithelial cells[24] suggests a unique role for this receptor.

REGULATION OF PLASMA AND CELL SURFACE PROTEINASE LEVELS BY LRP

Plasminogen is a proenzyme that is activated to form the serine proteinase, plasmin. Plasmin activation occurs locally without significant systemic plasmin formation. This is due to the local release of plasminogen activators and inhibitors, and the effective removal of these molecules from the circulation. Studies have confirmed that LRP is capable of binding and mediating the internalization of both tissue-type plasminogen activator[3] (tPA) and urokinase-type plasminogen activator[2] (uPA), and thus LRP may play an important role in their hepatic removal.

In addition to its role in mediating the hepatic clearance of uPA and tPA from the circulation, LRP also appears to be involved in the regulation of cell surface uPA activity. uPA is synthesized as a proenzyme (pro-uPA), that is activated to the two chain molecule. Both forms of uPA bind to the urokinase receptor (uPAR) which is a single chain protein with three homologous domains that is anchored to the cell surface via a glycosyl-phosphatidylinositol-anchor[25,26]. The association of uPA with its receptor, along with the simultaneous binding of plasminogen to the cell surface, provides a potent cell surface plasmin generating mechanism[27]. Thus uPA and its receptor have been implicated in several processes requiring proteinase activity, such as pericellular proteolysis, cell migration, and tissue remodeling[28].

Cell surface uPA activity is regulated by PAI-1 which reacts rapidly with uPA to form a complex that is rapidly internalized and degraded[29]. The mechanism by which this occurs was unknown until Nykjaer *et al.*[4] determined that LRP is responsible for this process. This was demonstrated by using a 39 kDa protein that was identified when it co-purified with LRP during affinity chromatography [1,30]. This molecule, termed the receptor-associated protein (RAP), binds to LRP[31], gp330[32], and the VLDL[33] receptor with high affinity and to the LDL receptor with weaker affinity.[34] When bound to these receptors, RAP completely blocks their ability to bind ligands. Utilizing RAP as an LRP antagonist, Nykjær et al.[4] demonstrated that the internalization of ^{125}I-uPA:PAI-1 complexes in monocytes is inhibited by RAP. Further, they demonstrated that an amino-terminal fragment representing residues 1-135 of uPA (ATF) also blocks the internalization of ^{125}I-labeled uPA:PAI-1 complexes. Since ATF prevents the association of uPA with uPAR, this suggests that binding of the complex to uPAR is also required for this process. The presentation of uPA:PAI-1 complexes by uPAR to LRP for lysosomal degradation represents a mechanism that is somewhat analogous to the participation of cell surface proteoglycans in the LRP-mediated catabolism of lipoprotein lipase[8] (see below), and the participation of cell surface proteoglycans in the presentation of basic fibroblast growth factor to its cell surface receptor[35].

In addition to internalizing uPA:PAI-1 complexes, LRP is also capable of binding and internalizing pro-uPA[2]. Pro-uPA can bind directly to LRP, with an affinity that is 15 to 20-fold weaker than that measured for the uPA:PAI-1 complex. Its binding is prevented by RAP and like most LRP ligands, its binding requires calcium. ^{125}I-Pro-uPA is rapidly internalized and subsequently degraded in Hep G2 cells, a human hepatoma cell line. The internalization and degradation of ^{125}I-pro-uPA are blocked by RAP and anti-LRP[2] antibodies, confirming that LRP is responsible for this process. Studies[2] indicate that at low uPA concentrations initial binding of uPA to uPAR is the favored pathway for uPA metabolism. It appears that uPA, initially bound to uPAR, is readily transferred to LRP for subsequent internalization and degradation. This process appears to require dissociation of uPA from uPAR and seems to be more efficient than would be predicted from consideration of the binding affinity of uPA for LRP and uPAR measured in *in vitro* binding studies.

In summary, it has become apparent that LRP is an important receptor for mediating the lysosomal degradation of uPA and inhibited uPA bound to the cell surface. Thus LRP may act in concert with uPAR in regulating cell surface plasmin generation.

ROLE OF LRP IN THE CATABOLISM OF LIPOPROTEIN LIPASE

Lipoprotein lipase is a heparin-binding enzyme that hydrolyzes triglycerides in lipoproteins to form smaller, cholesterol-rich remnant particles[36,37]. These remnant particles are then removed from the circulation by receptor-mediated uptake in the liver. The LDL receptor is responsible for the normal hepatic removal of remnant particles. Homozygous familial hypercholesterolemia patients[38], who have severe deficiencies of LDL receptors, do not have elevated levels of remnants. Therefore, another receptor must be able to mediate the clearance of these particles, and LRP is a candidate for this receptor[54]. Based on the observation that chylomicrons are only taken up following lipolysis, Felts et al.[39] proposed that lipoprotein lipase attached to chylomicron remnants might be a signal that allows the liver to specifically recognize these particles. Beisiegel et al.[40] demonstrated that lipoprotein lipase enhances the binding of chylomicron remnants to fibroblasts, and demonstrated that lipoprotein lipase can be crosslinked to LRP, suggesting that LRP may be responsible for the lipoprotein lipase promoted degradation of lipoproteins. Consequently, we investigated the role of LRP in this process.

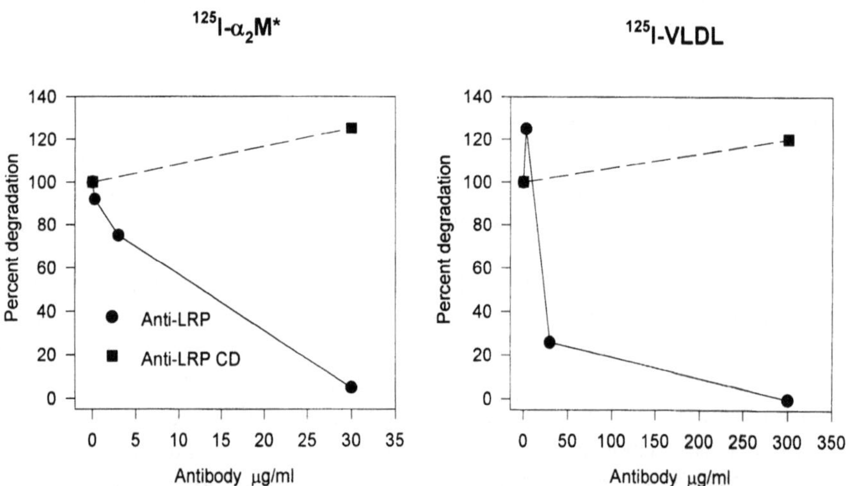

Figure 2. Anti-LRP IgG blocks the lipoprotein lipase promoted degradation of ^{125}I-VLDL in cultured fibroblasts. FH fibroblasts were preincubated with 5 nM lipoprotein lipase at 4 °C. Following washing, 2.8 nM ^{125}I-α_2M* (left panel) or 10 µg/ml of ^{125}I-S_f 100-400 lipoproteins (right panel) and various concentrations of anti-LRP or anti-LRP cytoplasmic domain antibodies (anti-LRP CD) were added, and incubated with the cells for 5 h at 37 °C. Following incubation, degradation was measured as described[8].

Lipoprotein lipase binds to purified LRP, and, interestingly, activated α_2M (α_2M^*) partially competes for its binding[41]. To study the role of LRP in the catabolism of lipoprotein lipase, fibroblasts deficient in the LDL receptor were utilized. Lipoprotein lipase is rapidly internalized and degraded in these cells. The degradation is blocked by affinity purified antibodies against LRP, and by RAP, indicating that LRP is responsible for this process[41]. We also observed that lipoprotein lipase induced the binding, uptake, and degradation of ^{125}I-labeled normal human VLDL[8]. This effect was blocked by antibodies against lipoprotein lipase, but not by antibodies against apolipoprotein E or apolipoprotein B, indicating that lipoprotein lipase is mediating the binding of these lipoproteins to cells. To examine the role of LRP in mediating the degradation of lipoprotein lipase-enriched ^{125}I-VLDL, affinity purified antibodies against LRP were utilized. The results of these experiments, shown in Figure 2, demonstrate that degradation was blocked by antibodies against LRP, indicating that LRP is responsible for mediating this process. A control antibody, prepared against the cytoplasmic domain of LRP, had little effect on the degradation of these lipoproteins. Interestingly, anti-LRP antibodies had little impact on the cell-surface binding of these lipoprotein lipase-enriched VLDL particles, indicating that other cell surface molecules participate in the initial binding of these complexes to the cell. Studies have implicated cell-surface proteoglycans as participants in lipoprotein lipase induced lipoprotein catabolism[42,43], and thus the effect of heparinase-treatment on this process was examined[8]. Incubation of cells with heparinase reduced the surface binding, uptake, and degradation of lipoprotein lipase-mediated catabolism of lipoproteins. The catabolism of α_2M^*, which is known to be mediated by LRP, is unaffected by this treatment. Thus, proteoglycans enhance the lipoprotein lipase mediated cellular uptake and degradation of lipoproteins, but are not required for the binding of these lipoproteins to purified LRP. It seems likely that proteoglycans serve to concentrate lipoprotein lipase on the cell-surface, thereby promoting lipoprotein lipase's interaction with LRP (Figure 3). This role of surface proteoglycans is somewhat analogous to the presentation of basic fibroblast growth factor with its receptor, and the role of uPAR in presenting uPA and uPA:PAI-1 complexes for internalization by LRP.

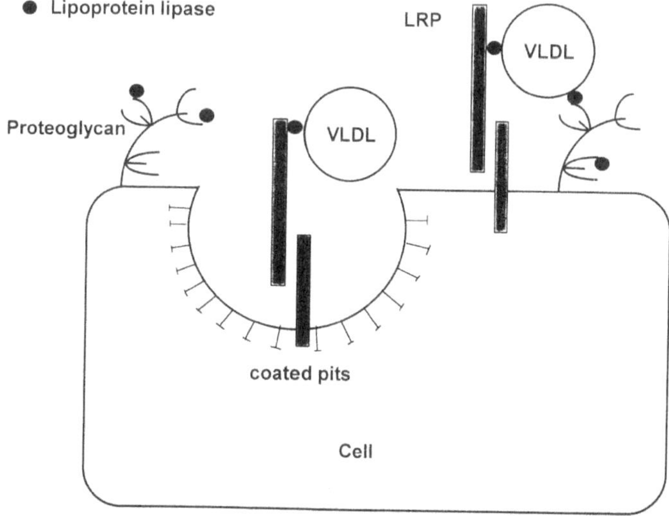

Figure 3. Model for the role of proteoglycans in facilitating the uptake of lipoprotein lipase-enriched VLDL by LRP.

STRUCTURAL REQUIREMENTS FOR LIPOPROTEIN LIPASE BINDING TO LRP

Lipoprotein lipase exists both as a 48,300 dalton monomer and a 96,600 dalton noncovalently associated homodimer[44,45], and is a member of a family of structurally and functionally related enzymes that include pancreatic lipase and hepatic lipase. The three dimensional crystallographic structure of human pancreatic lipase has been solved[46] and shows that this molecule consists of two domains, an N-terminal domain that is dominated by a central parallel β-sheet structure, and a smaller C-terminal domain that forms a β sandwich structure. Recently, a three-dimensional model of lipoprotein lipase was constructed[47] based upon two recently determined x-ray crystal structures of pancreatic lipase.

Our strategy to localize an LRP-binding region was guided by analysis of primary sequence similarities among lipoprotein lipase, RAP, and apolipoprotein E. A weak similarity was noted between the carboxyl-terminal domain of RAP, the carboxyl-terminal domain of lipoprotein lipase, and the LDL receptor-binding domain of apolipoprotein E. To localize the portion of lipoprotein lipase that is responsible for interacting with LRP, fragments of lipoprotein lipase were expressed in bacteria. A fragment of human lipoprotein lipase containing the C-terminal domain (residues 313-448, designated LPLC) which lacks the catalytic site, was able to bind to LRP [48]. Purified LRP bound specifically to microtiter wells coated with lipoprotein lipase or LPLC with K_D values of 0.7 and 1 nM, respectively.

Examination of the three-dimensional structure for the carboxyl-terminal domain of lipoprotein lipase reveals a cluster of positively charged residues at one end of the domain, and a hydrophobic loop spanning residues 387-394 at the opposite end. This loop, which contains Trp^{393} and Trp^{394} (see Figure 4), is not present in pancreatic lipase. To examine the role of various LPLC residues that may interact with LRP, site directed mutagenesis experiments were performed. The results of these experiments are summarized in Table I. Mutation of Lys^{407} to Ala reduced the affinity of LPLC for LRP by approximately 8-fold. Further, removal of ten residues from the carboxyl-terminus of LPLC reduced the affinity by approximately 5-fold. Mutation of other basic residues had a slight effect on the affinity of this interaction (Table I).

Table I. Apparent affinity of purified LRP for immobilized LPLC and LPLC mutants

Protein	$^1K_{D,app}$ nM
LPLC	1.0
$LPLC^{K413A,K414A}$	1.9
$LPLC^{R405A}$	2.0
$LPLC^{W393A,W394A}$	2.4
$^2LPLC^{\Delta 439\text{-}448}$	5.5
$LPLC^{K407A}$	7.7

[1]Measured using an ELISA in which LPLC or the LPLC mutant is immobilized in microtiter wells as described[48]

[2]LPLC deletion mutant in which residues 439 to 448 have been deleted.

Lys^{407} is located amongst a cluster of basic residues on the front face of the carboxyl-terminal domain (Figure 4). It is possible that clusters of positively charged residues, at least in part, may mediate the high affinity interaction of LRP with its ligands. Several LRP ligands, such as activated α_2M^{49} and PE exotoxin A[50] depend on arginine and lysine residues

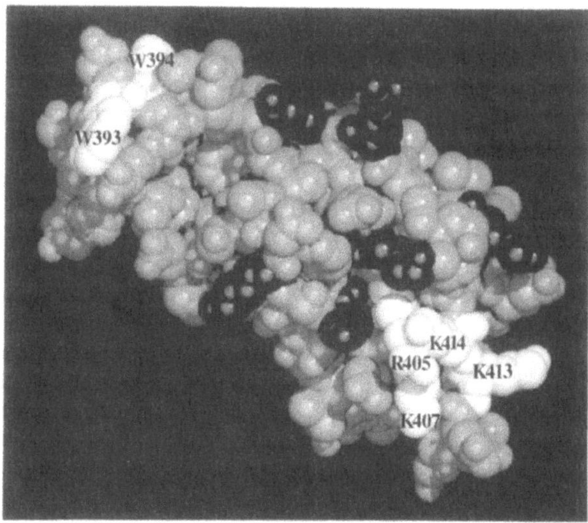

Figure 4. Molecular model for the carboxyl-terminal domain of lipoprotein lipase. Mutated residues are displayed in white, while other basic residues are depicted in black.

for interaction with LRP. Thus LRP, and likely other LDL receptor family members as well., appear to interact with molecules containing clusters of positively charged residues.

We noted[48] that LPLC also binds to ^{125}I-labeled human normal triglyceride-rich lipoproteins (Figure 5A), and promotes their binding to purified LRP (Figure 5B) and to cultured cells[53]. In examining the model of the carboxyl-terminal domain of lipoprotein lipase, it was speculated that the hydrophobic loop could represent a lipoprotein binding site[48]. This was confirmed when we mutated Trp^{393} and Trp^{394} to alanine, and demonstrated that the mutated molecule was unable to bind to lipoproteins (Figure 5A), or promote their interaction with LRP (Figure 5B). Thus it appears that the lipoprotein binding region of LPLC is localized on the opposite end of the domain from the LRP binding site, and adequately explains the ability of LPLC to simultaneously bind LRP and lipoproteins.

SUMMARY

LRP is a multiligand receptor that plays a unique role in biology, mediating the cellular internalization of both lipoproteins and proteinases. The finding[52] that LRP is expressed in smooth muscle cells and macrophages of early and advanced atherosclerotic lesions, suggests that this receptor may contribute to the formation of foam cells. This process may be accelerated by lipoprotein lipase, which can promote the cellular internalization of lipoproteins in a process mediated by LRP.

In addition to the importance of lipoprotein lipase in the catabolism of lipoproteins, studies on this molecule have offered some insight into the structural determinants that are responsible for its interaction with LRP. Since LRP interacts with many unrelated ligands, an understanding of the structural determinants that confer high affinity binding of these ligands to this receptor is an intriguing problem in structural biology. An examination of the molecular model of lipoprotein lipase has suggested that clusters of charged residues exist on the molecules surface, and our mutagenesis studies reveal that several of these residues appear important in LPR interaction.

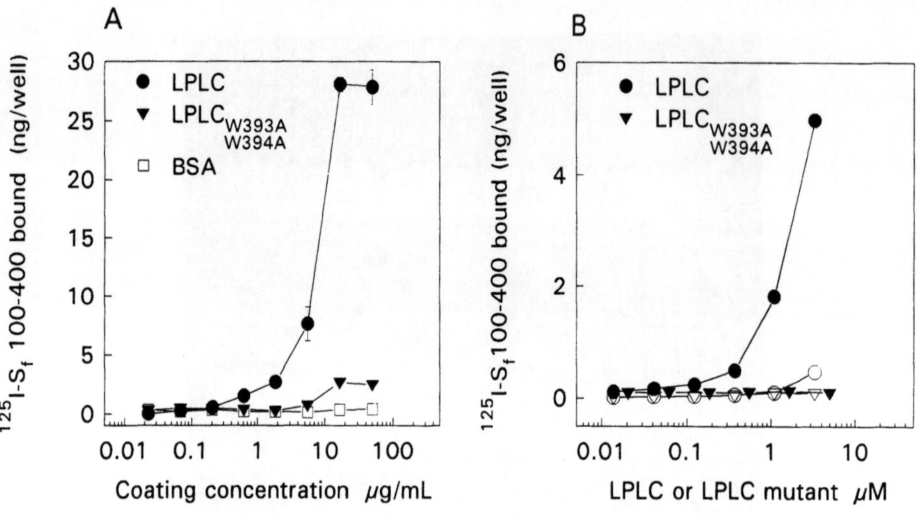

Figure 5. LPLC binds lipoproteins and promotes their binding to LRP. A, Microtiter wells were coated with increasing concentrations of His-LPLC (•), $LPLC^{W393A,394A}$ (▾), or BSA (□). 5 µg/ml of ^{125}I-S_f 100-400 lipoproteins were added, and after incubating overnight at 4 °C, the wells were washed, and the bound radioactivity removed with 0.1 N NaOH and counted. B, microtiter wells, coated with purified LRP (closed symbols) or BSA (open symbols), were incubated with 5 µg/ml of ^{125}I-S_f 100-400 lipoproteins and increasing amounts of His-LPLC (•) and $LPLC^{W393A,394A}$ (▾). After washing, the bound radioactivity was removed with 0.1 N NaOH and counted.

ACKNOWLEDGMENTS

This work was supported by grants GM42581, HL50784, HL49264, HL07413, HL45753, and DK42912. SEW and MZK are recipients of Individual National Research Service Awards. J-ML is an Investigator of the Howard Hughes Medical Institute.

REFERENCES

1. Ashcom J D, Tiller S E, Dickerson K, Cravens J L, Argraves W S, Strickland D K. The human α_2-macroglobulin receptor: identification of a 420-kD cell surface glycoprotein specific for the activated conformation of α_2-macroglobulin. *J Cell Biol* 1990; 110: 1041-1048.
2. Kounnas M Z, Henkin J, Argraves W S, Strickland D K. Low density lipoprotein receptor-related protein/α_2-macroglobulin receptor mediates cellular uptake of pro-urokinase. *J Biol Chem* 1993; 268: 21862-21867.
3. Bu G, Williams S, Strickland D K, Schwartz A L. Low density lipoprotein receptor-related protein/α_2-macroglobulin receptor is an hepatic receptor for tissue-type plasminogen activator. *Proc Natl Acad Sci USA* 1992; 89: 7427-7431.
4. Nykjær A, Petersen C M, Moller B, Jensen P H, Moestrup S K, Holtet T L, Etzerodt M, Thogersen H C, Munch M, Andreasen P A, Gliemann J. Purified α_2-macroglobulin receptor/LDL receptor-related protein binds urokinase•plasminogen activator inhibitor type-1 complex. Evidence that the α_2-macroglobulin receptor mediates cellular degradation of urokinase receptor-bound complexes. *J Biol Chem* 1992; 267: 14543-14546.
5. Orth K, Madison E L, Gething M -J, Sambrook J F, Herz J. Complexes of tissue-type plasminogen activator and its serpin inhibitor plasminogen-activator inhibitor type 1 are internalized by means of the low density lipoprotein receptor-related

protein/α_2-macroglobulin receptor. *Proc Natl Acad Sci USA* 1992; 89: 7422-7426.

6. Kowal R C, Herz J, Goldstein J L, Esser V, Brown M S . Low Density lipoprotein receptor-related protein mediates uptake of cholesteryl esters derived from apoprotein E-enriched lipoproteins. *Proc Natl Acad Sci USA* 1989; 86: 5810-5814.
7. Beisiegel U, Weber W, Ihrke G, Herz J, Stanley K K . The LDL-receptor related protein, LRP, is an apolipoprotein E binding protein. *Nature* 1989; 341: 162-164.
8. Chappell D A, Fry G L, Waknitz M A, Muhonen L E, Pladet M W, Iverius P -H, Strickland D K . Lipoprotein lipase induces catabolism of normal triglyceride-rich lipoproteins via the low density lipoprotein receptor-related protein/ α_2-macroglobulin receptor *in vitro*. A process facilitated by cell-surface proteoglycans. *J Biol Chem* 1993; 268: 14168-14175.
9. Yamamoto T, Davis G C, Brown M S, Schneider W J, Casey M L, Goldstein J L, Russel D W . The human LDL receptor: A cysteine-rich protein with multiple Alu sequences in its mRNA. *Cell* 1984; 39: 27-38.
10. Chen W-J, Goldstein J L, Brown M S . NPXY, a sequence often found in cytoplasmic tails, is required for coated pit-mediated internalization of the low density lipoprotein receptor. *J Biol Chem* 1990; 265: 3116-3123.
11. Davis C G, Goldstein J L, Sudhof T C, Anderson R G W, Russell D W, Brown M S . Acid-dependent ligand dissociation and recylcing of LDL receptor mediated by growth factor homology region. *Nature* 1987; 326: 760-765.
12. Russell D W, Brown M S, Goldstein J L . Different combinations of cyteine-rich repeats mediate binding of LDL receptor to different proteins. *J Biol Chem* 1989; 264: 21682-21688.
13. Takahashi S, Kawarabayasi Y, Nakai T, Sakai J, Yamamoto T . Rabbit very low density lipoprotein receptor: a low density lipoprotein receptor-like protein with distinct ligand specificity. *Proc Natl Acad Sci U S A* 1992; 89: 9252-9256.
14. Sakai J, Hoshino A, Takahashi S, Miura Y, Ishii H, Suzuki H, Kawarabayasi Y, Yamamoto T . Structure, Chromosome Location, and Expression of the human very low density lipoprotein receptor gene. *J Biol Chem* 1994; 269: 2173-2182.
15. Sudhof T C, Goldstein J L, Brown M S, Russell D W . The LDL receptor gene: A mosaic of exons shared with different proteins. *Science* 1985; 228: 815-822.
16. Herz J, Hamann U, Rogne S, Myklebost O, Gausepohl H, Stanley K K . Surface location and high affinity for calcium of a 500kDa liver membrane protein closely related to the LDL-receptor suggest a physiolocical role as lipoprotein receptor. *EMBO J* 1988; 7: 4119-4127.
17. Herz J, Kowal R C, Goldstein J L, Brown M S . Proteolytic processing of the 600 kD low density liprotein receptor related protein LRP occurs in a trans-Golgi compartment. *EMBO J* 1990; 9: 1769-1776.
18. Kerjaschki D, Farquhar M G . The pathogenic antigen of Heymann nephritis is a membrane glycoprotein of the renal proximal tubule brush border. *Proc Natl Acad Sci U S A* 1982; 79: 5557-5561.
19. Kounnas M Z, Strickland D K, Argraves W S . Glycoprotein 330, a member of the LDL receptor family, binds lipoprotein lipase *In Vitro*. *J Biol Chem* 1993; 268: 14176-14181.
20. Korenberg J R, Argraves K M, Chen X-N, Tran H, Strickland D K, Argraves W S . Chromosomal Localization of Human Genes for the LDL Receptor Family Member Glycoprotien 330 and its Associated Protein RAP. *Genomics* 1994; 22: 88-93
21. Barber D L, Sanders E J, Aebersold R, Schneider W J . The receptor for yolk lipoprotein deposition in the chicken oocyte. *J Biol Chem* 1991; 266: 18761-18770.
22. Brown M S, Goldstein J L . A receptor-Mediated Pathway for Cholesterol Homeostasis. *Science* 1986; 232: 34-47.

23. Willnow T E, Goldstein J L, Orth K, Brown M S, Herz J. LDL receptor related protein and gp330 bind similar ligands, including plasminogen activator-inhibitor complexes and lactoferrin, an inhibitor of chylomicron remnant clearance. *J Biol Chem* 1992; 267: 26172-26180.
24. Zheng G, Bachinsky D R, Stamenkovic I, Strickland D K, Brown D, Andres G, McCluskey R T. Organ distribution in rats of two members of the low-density lipoprotein receptor gene family, gp330 and LRP/$\alpha_2$macroglobulin receptor, and the receptor-associated protein (RAP). *J Histochem Cytochem* 1994; 42: 531-542.
25. Roldan A L, Cubellis M V, Masucci M T, Behrendt N, Lund L R, Dan: K, Appella E, Blasi F. Cloning and expression of the receptor for human urokinase plasminogen activator, a central molecule in cell surface, plasmin dependent proteolysis *EMBO J* 1990; 9: 467-474.
26. Ploug M, Ronne E, Behrendt N, Jensen A L, Blasi F, Dano K. Cellular receptor for urokinase plasminogen activator. Carboxyl-terminal processing and membrane anchoring by glycosyl-phosphatidylinositol. *J Biol Chem* 1991; 266: 1926-1933.
27. Ellis V, Dano K. The urokinase receptor and the regulation of cell surface plasminogen activation. *Fibrinolysis* 1992; 6 Suppl. 4: 27-34.
28. Blasi F. Urokinase and Urokinase receptor: A paracrine/autocrine system regulating cell migration and invasiveness. *BioEssays* 1993; 15: 105-111.
29. Cubellis M V, Wun T C, Blasi F. Receptor-mediated internalization and degradation of urokinase is caused by its specific inhibitor PAI-1. *EMBO J* 1990; 9: 1079-1085.
30. Jensen P H, Moestrup S K, Gliemann J. Purification of the human placental α2-macroglobulin receptor. *FEBS Lett* 1989; 255: 275-280.
31. Williams S E, Ashcom J D, Argraves W S, Strickland D K. A novel mechanism for controlling the activity of α_2-macroglobulin receptor/low density lipoprotein receptor-related protein. Multiple regulatory sites for 39-kDa receptor-associated protein. *J Biol Chem* 1992; 267: 9035-9040.
32. Kounnas M Z, Argraves W S, Strickland D K. The 39-kDa receptor-associated protein interacts with two members of the low density lipoprotein receptor family, α_2-macroglobulin receptor and glycoprotein 330. *J Biol Chem* 1992; 267: 21162-21166.
33. Battey F, Gafvels M E, Fitzgerald D J, Argraves W S, Chappell D A, Strauss III J F, Strickland D K. The 39 kDa Receptor Associated Protein Regulates Ligand Binding by the Very Low Density Lipoprotein Receptor. *J Biol Chem* 1994; in press:
34. Medh J D, Fry G L, Bowen S L, Pladet M W, Strickland D K, Chappell D A. The 39 kDa Receptor-Associated Protein Modulates Lipoprotein Catabolism by Binding to LDL receptors. *J Biol Chem* 1994; submitted
35. Yayon A, Klagsbrun M, Esko J D, Leder P, Ornitz D M. Cell surface, heparin-like molecules are required for binding of basic fibroblast growth factor to its high affinity receptor. *Cell* 1991; 64: 841-848.
36. Nilsson-Ehle P,Garfinkel,A.S.,and Schotz,M.C. *Annu Rev Biochem* 1980; 49: 667-693.
37. Auwerx J, Leroy P, Schoonjans K. Lipoprotein lipase: Recent contributions from molecular biology. *Crit Rev Clin Lab Sci* 1992; 29: 243-268.
38. Goldstein, J.L. and Brown, M.S. Familiar hypercholesterolemia. In: *The Metabolic Basis of Inherited Disease*, edited by Scriver, C.R., Beaudet, A.L., Sly, W.S. and Valle, D. New York: McGraw-Hill Publishing Company, 1989, p. 1215-1250.
39. Felts J M, Itakura H, Crane R T. The mechanism of assimilation of constituents of chylomicrons, VLDL and remnants. *Biochem Biophys Res Comm* 1975; 66: 1467-1475.
40. Beisiegel U, Weber W, Bengtsson-Olivecrona G. Lipoprotein lipase enhances the

binding of chylomicrons to low density lipoprotein receptor-related protein. *Proc Natl Acad Sci USA* 1991; 88: 8342-8346.
41. Chappell D A, Fry G L, Waknitz M A, Iverius P -H, Williams S E, Strickland D K . The low density lipoprotein receptor-related protein/α_2-macroglobulin receptor binds and mediates catabolism of bovine milk lipoprotein lipase. *J Biol Chem* 1992; 267: 25764-25767.
42. Eisenberg S, Sehayek E, Olivecrona T, Vlodavsky I . Lipoprotein lipase enhances bindng of lipoproteins to Heparin Sulfate on Cell Surfaces and Extracellular Matrix . *J Clin Invest* 1992; 90: 2013-2021.
43. Williams K J, Fless G M, Petrie K A, Snyder M L, Brocia R W, Swenson T L . Mechanism by which lipoprotein lipase alters cellular metabolism of lipoprotein(a), LDL, and nascent lipoproteins. *J Biol Chem* 1992; 267: 13284-13292.
44. Iverius P-H, Ostlund-Lindqvist A M . LPL from bovine milk. Isolation procedure, chemical characterization and molecular weight analysis. *J Biol Chem* 1976; 251: 7791-7795.
45. Osborne J C, Bengtsson-Olivecrona G, Lee N S, Olivecrona T . Studies on Inactivation of lipoprotein lipase: Role of the dimer to monomer dissociation. *Biochemistry* 1985; 24: 5606-5611.
46. Winkler F K,D'Arcy,A.,and Hunziker,W. Structure of human pancreatic lipase. *Nature* 1990; 343: 771-774.
47. van Tilbeurgh H, Roussel A, Lalouel J-M, Cambillau C . Lipoprotein Lipase. Molecular Model Based on the Pancreatic Lipase X-Ray Structure: Consequences for heparin binding and Catalysis. *J Biol Chem* 1994; 269: 4626-4633.
48. Williams S E, Inoue I, Tran H, Fry G L, Pladet M W, Iverius P-H, Lalouel J-M, Chappell D A, Strickland D K . The Carboxyl-terminal Domain of Lipoprotein Lipase Binds to the Low Density Lipoprotein Receptor-related Protein/α2-Macroglobulin receptor (LRP) and mediates Binding of Normal Very Low Density Lipoproteins to LRP. *J Biol Chem* 1994; 269: 8653-8658.
49. Sottrup-Jensen L, Gliemann J, Van Leuven F . Domain structure of human α2-macroglobulin: characterization of a receptor-binding domain obtained by digestion with papain. *FEBS* 1986; 205: 20-24.
50. Jinno Y, Chaudhary V K, Kondo T, Adhya S, Fitzgerald D J, Pastan I . Mutational analysis of Domain I of *Pseudomonas* Exotoxin A. *J Biol Chem* 1988; 263: 13203-13207.
51. Kounnas M Z, Haudenschild C C, Strickland D K, Argraves W S . Immunological Localization of glycoprotein 330, Low Density Lipoprotein Receptor Related Protein and 39 kDa Receptor Associated Protein in Embryonic Mouse Tissue. *In Vivo* 1994; in press:
52. Luoma J, Hiltunen T, Särkioja T, Moestrup S K, Gliemann J, Kodama T, Nikkari T, Ylä-Herttuala S . Expression of α_2-macroglobulin receptor/low density lipoprotein receptor-related protein and scavenger receptor in human atherosclerotic lesions. *J Clin Invest* 1994; 93: 2014-2021.
53. Chappell, D.A., Inoue,I., Fry,G.L., Pladet,M.W., Bowen,S.L., Iverius,P-H, Lalouel,J-M., and Strickland, D.K. Cellular catabolism of Normal Very Low Density Lipoproteins via the Low Density Lipoprotein Receptor-related Protein/α_2-Macroglobulin receptor is Induced by the C-terminal domain of lipoprotein lipase. J. Biol. Chem. 1994;269: 18001-18006
54 Willnow T E, Sheng Z, Ishibashi S, Herz J . Inhibition of Hepatic Chylomicron Remnant Uptake by Gene Transfer of a Receptor Antagonist. *Science* 1994; in press:

MOLECULAR BIOLOGY OF TF

Thomas S. Edgington and Wolfram Ruf

Vascular Cell Molecular Biology Unit
The Department of Immunology
The Scripps Research Institute, La Jolla, CA 92037

INTRODUCTION

About half of deaths and major morbidity of the populations of the advanced Western nations have as the underlying or penultimate process vascular pathology associated with and attributable to activation of the coagulation-thrombogenic molecular pathways. Whereas the general principles, and majority of molecular players in the thrombogenic pathways are known and biochemically characterized, the major challenge continues to be safe and effective prevention and intervention in coronary, cerebral and other thrombotic diseases.

To intervene at the beginning of the thrombogenic pathways is an attractive rational approach to pathogenetic intervention for therapy. However, this is not readily approachable with the current knowledge base. One of the major current challenges is the unravelling of the molecular and structural biology of the "initiation complex" on vascular cells that drives the coagulation-thrombogenic cascades [1-4]. To interrupt one or more of the complex interactions of TF with other molecules in the initiation complex is conceptually one of the most attractive approaches. To do so requires that both the assembly and function of the TF directed initiation complex be understood and that it be resolved in atomic detail at the three dimensional level. Progress has been realized along the path to these goals. Both the rules of assembly [1,5-7] and the crystallographic solutions to three dimensional structure have been advanced [8-11].

Tissue Factor (TF) is the initiating molecule that assembles the "initiation complex" for these cascades *in vivo* [12-14]. TF is a cell surface transmembrane protein with an extracellular domain organized as a tandem pair of modules, folded in the general architecture of the fibronectin type III repeats. TF belongs structurally to the Cytokine Receptor Family such as growth hormone receptor, interferon gamma receptor, interleukin 4 receptor and others [3,15]. TF is a high affinity receptor and cofactor for the serine protease factor VIIa (VIIa) and its zymogen form factor VII (VII). Current understanding of the structural basis of function of TF has progressed from a combination of immunochemical, binding, functional, chemical cross-linking, mutational, and crystallographic analyses aided by computational modelling. Integration of these data provide a model for the structure and the molecular mechanics of both receptor and cofactor functions of TF and the assembled initiation complex.

In vivo investigation of selected anti-TF monoclonal antibodies and of TF pathway inhibitor have demonstrated effective blocking of initiation of the coagulation protease cascades in primates, supporting the hypothesis that TF is indeed the major initiator *in vivo* in

Cardiovascular Disease 2
Edited by L.L. Gallo, Plenum Press, New York, 1995

the normal physiologic state as well as in response to endotoxin and infection [12-14]. A firm understanding of the structural chemistry and molecular biology of TF as now rapidly evolving provides the opportunity for effective knowledge-guided computational design of therapeutic antagonists to the assembly and function of the TF initiation complex.

Thrombogenesis

The thrombogenic process may generally be considered to arise from a local imbalance of the proteolytic enzyme cascades of the procoagulant coagulation protease cascade, the attenuating effects of the anti-coagulation cascade, and the thrombolytic effects of the fibrinolytic systems. These systems involve a large number of proteins of the plasma and the cells that they interact with in the blood and the vessel wall.

Selected cell types actively regulate the initial triggering of these enzyme cascades by the expression of initiating cell surface receptors which interact with serine proteases and their zymogen precursors, e.g. TF which binds VII and VIIa to drive the thrombogenic kinetics, thrombomodulin which is a receptor for thrombin that diverts the bound thrombin from a prothrombotic molecule to an anti-coagulant molecule as reviewed elsewhere in this volume, and the assembly of the plasminogen activators on cells to mediate thrombolysis. The expression of a small number of initiating cell surface molecules efficiently modifies the local extracellular vascular environment.

Cell surfaces play a pivotal role in many of the reactions necessary to the initiation of transcriptional and cell surface protein assembly events for initiation and driving of the thrombogenic processes. Engagement of specific cognate receptors on the endothelium, vascular smooth muscle cells and leukocytes by cytokines, bacterial lipopolysaccharide (LPS), oxidative events associated with the pathology of atherosclerosis, and presumably other unidentified agonists result in a complex series of transcriptional events culminating in vessel wall injury, thrombosis, stenosis, and inflammation. These involve the release of NF-κB related transcription factors from the cytoplasm of these cells, nuclear translocation and induction of transcription and biosynthesis of TF and other inflammatory genes.

Initiation of the Protease Cascades

Triggering by TF of the extrinsic and intrinsic limbs of the revised coagulation pathways[2] is responsible for the activation of coagulation *in vivo*, both basal activation [12] and that in response to infectious events [13,14]. As illustrated in Figure 1, the cell surface initiation complex TF•VIIa directly activates the zymogen factor X (X) by hydrolysis of a single peptidyl bond and shedding of the fifty-two amino acid activation peptide. The resultant serine protease factor Xa (Xa) not only assembles with its cofactor factor Va on the surface of activated cells to directly drive the extrinsic limb of the unified coagulation cascade but also binds to the cells surface receptor EPR-1 [16,17] to signal cellular responses [18]. Xa also proteolytically converts the zymogen factor IX (IX) to the first intermediate derivative factor IXα [19], which is then efficiently activated proteolytically by the TF•VIIa cell surface complex to the active serine protease factor IXa (IXa). IXa is the first protease in the intrinsic limb of the cascades. The specific inhibitor of the TF pathway (TFPI) [20-22] and anti-thrombin III in the presence of glycosaminoglycans [23] attenuate the kinetics of the TF-dependent driven pathways thereby counteracting intravascular thrombosis. Consequently, any change in this regulated balance due to *de novo* synthesis of TF by vascular cells has to be considered a major influence on thrombogenesis.

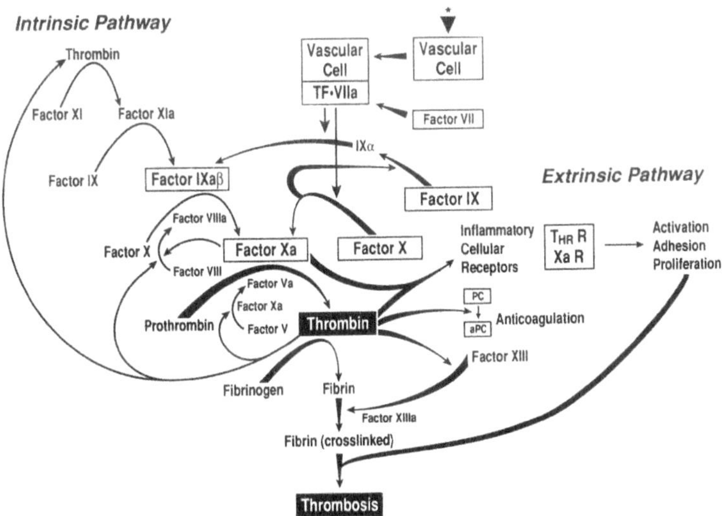

Fig. 1. The unified coagulation-thrombogenic cascade. Both the extrinsic and intrinsic limbs of the cascade are initiated and driven by the TF•VIIa initiation complex. Positive feedback occurs via thrombin activation of factor XI to provide an independent activation of the intrinsic limb. The cellular effects are mediated by thrombin and Xa binding to their respective receptors.

CELLULAR EXPRESSION OF TF

The human TF gene, circa 12.4 kbp, is resident on chromosome 1 at p21-p22. Transcription results in a 2.3 kb mRNA which is translated into a 295 residue precursor which is processed into a mature 219 residue extracellular domain, a 23 residue transmembrane region and a 21 residue intracellular domain (for review [3]). The extracellular domain possesses three N-linked glycosylation moieties which are variable in structure and are responsible for considerable charge heterogeneity from variable sialation. The cytoplasmic domain of all species of TF possesses a single highly reactive cysteine residue which is thioester bonded to stearate or palmitate [24] which may assist in anchoring TF in the plasma membrane. Serines in the cytoplasmic domain can be phosphorylated by a protein kinase C dependent mechanism [25].

TF is rapidly transported to the cell surface following synthesis and post-translational modifications [26]. TF expressed on the cell surface is accessible and competent to bind ligand [26,27]; however, the proteolytic activity of the assembled TF•VIIa initiation complex can be modulated, apparently by events which alter the organization of the cell membrane [27,28].

Constitutive Expression of TF

Constitutive expression of TF is observed in specific cell types which are bronchial mucosa and alveolar epithelial cells in the lung [29]; the astroglia, notably the Bergmann astrocytes in the cerebellum, in the central nervous system [30]; cardiac myocytes, renal glomerular mesangial and epithelial cells, the epithelial layers of the body, both mucosal and cutaneous [29]. TF is further found at boundaries between organs, such as in fibrous organ capsules of liver, spleen and kidney, as well as in the adventitia of arteries and venules. The localization of TF expression is consistent with a function as an initiating molecule to arrest bleeding. This is further supported by the very poor expression of TF surrounding blood vessels of joints, skeletal muscle, and the dermis of the skin, major sites of bleeding in hemophilic patients.

Pathological Expression in Thromboembolic Disease

Endothelial cells and monocytes are the only cells within the vascular compartment that can be induced to express TF. Cells of monocytic type are induced to express TF by antigen stimulation of the cellular immune pathways, mediated by certain cytokines or alternatively by direct contact with activated T helper lymphocytes [31,32]. Endothelial cells also can be stimulated to express TF by selected cytokines, and by gram negative bacterial lipopolysaccharide (LPS), the pathogenic initiator of the gram-negative sepsis syndrome. Exposure of primates to LPS results in TF expression *in vivo* which appears to be entirely responsible for the resulting activation of the coagulation system [13,14].

Disseminated intravascular coagulation (DIC) is an important pathophysiological contributor to the lethality of the sepsis syndrome. This was directly demonstrated by administration of a neutralizing anti-TF monoclonal antibody to baboons in which a lethal dose of *E. coli* were administered intravenously [14]. Lethality was prevented in all animals, and the morbidity of this lethal sepsis syndrome was reduced. Despite the profound stimulation with LPS in this model, generalized expression of TF by the vascular endothelium was not detected immunohistochemically. Rather, TF was observed in the endothelium confined to sinusoids in the marginal zone of the spleen, where TF positive macrophages were found [33]. Enhanced expression of TF was observed by alveolar epithelial cells and glomerular epithelium, both of which are localized in organs with increased fibrin deposition during the generalized Schwartzman reaction. While not excluding a more generalized expression of TF at undetectable levels by the endothelium or circulating monocytes, it appears that a locally confined triggering of coagulation by TF may be operative in this form of disseminated coagulopathy.

The pathogenetic importance of TF dependent initiation of the coagulation protease pathways is further supported by the modulating effect of the specific physiologic inhibitor of the TF pathway, TFPI, in animal models of septic shock [34]. Although questioned by some, the initiation complex does appear to be required for the continued driving of the coagulation cascade since arrest of the TF initiation complex *in vivo* by specific monoclonal antibody arrests further prothrombin activation (in preparation).

A variety of thromboembolic complications are frequently associated with clinical neoplasia. Solid tumors of lung, pancreas, breast and colon are not infrequently associated with chronic DIC and Trousseau's syndrome (for review see [35]). Whereas the expression of TF by solid tumors, such as colon carcinomas, may reflect preservation of a normal phenotypic marker of differentiated non-transformed cells, other tumors appear to express TF in concert with the neoplastic phenotype. Whereas melanocytes do not express TF, increased cellular levels of TF appear to follow the progression to the neoplastic phenotype and progression of melanoma to the metastatic phenotype in animal models [36]. The access of tumor cells, or tumor cell derived vesicles, to the vasculature can be linked to local, and even generalized activation of the coagulation protease cascade.

A variety of pathogenic infectious organisms induce TF expression by vascular cells, and this is interpreted as the causal basis for the associated thromboembolic complication. *Plasmodium falciparum* induces TF expression in monocytes [37] and *rickettsia conorii* [38] or *herpes simplex virus* [39,40] in endothelial cells. Induction of endothelial TF is suggested to change the hemostatic balance at the endothelial-blood interface in occlusive vascular disease.

Monocyte derived foam cells in atherosclerotic plaques express TF [41]; and the specific inhibitor of the TF pathway (TFPI) prevents re-occlusion after experimental thrombolysis [42]. A role of TF is suggested for the acute thrombotic complications of vascular disease. TF has to be considered responsible for activation of the coagulation pathways under various pathologic conditions, most of them associated with significant morbidity and frequent lethality.

STRUCTURE OF TF

Structure based sequence comparison algorithms, combined with analysis of consensus sequences for protein families, suggested that the TF extracellular domain likely shares an immunoglobulin type protein folding architecture common to the cytokine receptor superfamily [15]. Based on conserved consensus sequence and an identical disulfide bonding pattern in the extracellular domains, TF appears to be more closely related to the receptors for the interferons than to other members of the cytokine receptor superfamily. The circular dichroism spectrum of TF indicated a predominance of β-structure [43] consistent with the predicted seven β-strand fold of the extracellular domains of the cytokine receptors. Indeed, the established three dimensional structure of the receptor for growth hormone [44], a member of the cytokine receptor superfamily, supports the structural prediction. The predictions have recently been validated by solution of the three dimensional solution of the structure of TF [9,10] and of the interferon gamma receptor (unpublished). The two modules of TF are folded into seven β-strands aligned in two β-sheets, similar to the architecture of the fibronectin type III repeats.

FUNCTION OF TF

Role of Phospholipid in the Proteolytic Efficiency of the Initiation Complex

TF binds VIIa in a binary complex to form a catalytically active cofactor•enzyme complex. Full proteolytic activity however requires the presence of a phospholipid bilayer containing phosphatidylserine. VIIa, as well as IX and X the substrates for the TF•VIIa complex, contain γ-carboxyglutamic acid (Gla) domains. The charge dependent phospholipid surface interactions of the substrates X and IX influence the efficiency of their activation. In contrast the critical role of the VIIa Gla domain may be in the interaction with TF [45]. Since macromolecular substrates are poorly attacked in the absence of a phospholipid membrane, and further enhanced in the presence of negatively charged phospholipid membrane [5,6,46], the role of substrate-phospholipid interactions was analyzed using a soluble extracellular domain truncation mutant of TF which allowed analysis of the relative role of membrane association of substrate versus cofactor. Phospholipid bound factor X was shown to be preferentially activated by the complex of VIIa with the soluble TF cell surface domain [5]. Detailed kinetic analysis of factor X activation by the membrane anchored TF•VIIa complex further supported the preferred utilization of membrane bound substrate [6] similar to other membrane assembled enzyme-cofactor complexes. Preferential utilization of membrane associated substrate results in a useful mechanism by which cells can modulate TF function. This is observed in studies implicating a subset of TF•VIIa complexes on tumor cells as responsible for efficient activation of protein substrates [27], although all the cell surface TF is capable of binding VIIa and supporting enhanced hydrolysis of small peptidyl substrates by the bound VIIa. We conclude that changes in the cell surface organization including exposure of negatively charged phospholipid or mimics of such a charged patch on the cell surface must be considered as a significant modulator of the generation of proteolytic activity triggered by cell surface TF.

Catalytic Function of the TF•VIIa Complex

Assembly of VIIa with TF as a binary complex results in functional changes in the VIIa that result in increased catalytic efficiency for cleavage of appropriate substrates as well as enhanced susceptibility to anti-thrombin III [47,48] and the Kunitz-type inhibitor TFPI [49]. Whereas VIIa in the presence of Ca^{2+} poorly hydrolyses small peptidyl [7,45,50,51] and is virtually inactive against protein substrates [5,46], the specific enzymatic activity is enhanced by many orders of magnitudes in the presence of TF [5]. The rate increase observed for small substrates is at least 10-fold less than for macromolecular substrate. The cleavage of small substrates is largely unaffected by deletion of the Gla-domain of VIIa, whereas proteolysis of macromolecular substrate by the desGla-VII•TF complex is severely reduced compared to TF•VIIa with a fully Ca^{2+}-saturated Gla domain [45]. Thus, efficient proteolysis is not simply dependent on changes in the active site environment of VIIa, but appears to involve additional interactions between protein substrate and structures on the surface of the enzyme•cofactor complex. Notably in respect to a general paradigm for cofactor function, TF may contribute to recognition of macromolecular substrate. This is based on the identification of a particularly interesting type of monoclonal antibodies to the TF extracellular domain. Whereas most antibodies to TF block function by inhibiting VIIa binding, this class I_N antibodies, represented by TF8-5G9, were much more potent in inhibiting TF induced coagulation of plasma compared to antibodies of comparable affinity directed to other epitopes on TF [52]. Inhibition by anti-TF antibody TF8-5G9 did not require dissociation of the TF•VIIa complex, and kinetic analysis demonstrated a mode of inhibition which was competitive with macromolecular substrate factor X [52]. Activation of the alternative protein substrates, factors IX and VII, is also blocked by this class of monoclonal antibody, whereas the cleavage of small peptidyl substrates is not [53]. These class I_N antibodies rapidly inhibit preformed TF•VIIa complex by binding to the substrate assembly site on the TF•VIIa complex responsible for assembly of substrate to form the ternary complex [52].

The epitope recognized by class I_N monoclonal antibodies which block substrate assembly was localized to the carboxyl terminal structural module (residues 106-219) of TF [54]. Functional residues in this module were further localized by mutational analysis. These demonstrated that certain residues in the sequence span 151-174, in particular Lys-165 and Lys-166, were associated with a functional site [55,56]. The functional defect caused by alanine replacements for certain of these residues was not a result of diminished binding of the ligand protease VIIa [57]. Catalytic complexes formed by these mutant TFs and VIIa possessed normal catalytic activity against small peptidyl substrates indicating that the active site of VIIa becomes catalytically competent upon assembly of VIIa with the mutant TF molecules. However, proteolytic activation of macromolecular substrate factor X was diminished. The selective reduction of proteolytic function of the TF•VIIa complex upon mutational exchange of TF residues is consistent with direct interactions between this predicted surface loop of TF and protein substrates though not excluding more complex contributions through formation of a collision structure including residues from both TF and VIIa, or an induced allosteric effect on VIIa. Macromolecular substrate assembly with enzyme•cofactor complexes is expected to involve multiple amino acid side chain interactions occurring within a contiguous surface area formed by charged residues of both enzyme [58] and cofactor. The molecular mechanism by which class I_N antibodies rapidly inhibit the TF•VIIa complex can therefore be deduced as a result of steric hindrance of a bioactive site on the TF protein surface required for assembly with protein substrates.

Binding of Cognate Ligand by TF

TF binds VIIa with very high affinity, with a K_d as low as 2.7-7 pM [59-61]. Binding of VIIa to TF predominantly involves protein-protein interactions [61]. The zymogen VII binds to TF and is rapidly converted when bound to TF [62-64] leading to the likelihood that conformational changes in the activation region resulting in stabilization of the transition state of the sessile peptide bond. The TF bound VII is activated by selected proteases including factors IXa and Xa, the products formed by the TF•VIIa complex. VIIa bound to TF efficiently activates the zymogen VII; and this cleavage is enhanced by association of VII with positively charged surfaces [51] or by the assembly of VII on TF [65,66]. The latter virtually requires membrane anchoring of TF [66] suggesting that proximity of TF molecules in a lipid bilayer [67] may allow activation of VII bound to TF by an adjacent TF•VIIa complex. It is presently unclear whether the "back-activation" by the product proteases IXa and Xa or rather the "auto-activation" by TF•VIIa represent the predominant activation pathway for VII bound to TF as would occur in the vascular exposure of blood to TF expressing cells.

Both structural modules of TF contribute to the high affinity binding of VIIa. Residues constituting the surface structure of the contact site, analyzed from crystallographic analysis [11], are contributed mostly by the amino terminal (N) module but also a few come from the carboxyl terminal (C) module [9-11]. The N-module of TF was initially implicated in the binding of VIIa based on epitope assignment of inhibitory antibodies to the sequence 40-71 [54] and chemical cross-linking of VIIa to TF residues 44-84 [43]. Interactions of the C-module of TF with VIIa was suggested by diminished VIIa binding and function following structurally disruptive mutations of selected residues of the C-module [68] as well as by chemical cross-linking of VIIa with residues in the sequence 129-169 in TF [43]. Further insight into the regions which mediate ligand assembly was obtained from structurally conservative alanine scanning mutagenesis [61,69]. Aligned on the predicted structural model of TF, several residues which are critical for binding of VIIa were found to be clustered at the boundary of the two structural modules. Key binding residues Lys-20, Ile-22, Asp-58 and Thr-60 form a cluster in the amino-terminal module and the ligand site may extend to more distant residues as Trp-45, Lys-48 and Tyr-78. In the carboxyl-terminal module, residues Arg-135 and Phe-140 provide one contact site [61] and these residues are spatially close to the binding cluster in the amino-terminal module. The side chains of these residues may thus be embedded in a contiguous area for ligand binding.

Notably, the regions which appear to mediate ligand recognition in TF overlap significantly with the ligand interface of the structurally related growth hormone receptor defined by the crystallographic solution of the complex of the growth hormone with its receptor [44]. This suggests that the structurally related receptors in the cytokine receptor family not only adopt a similar architecture, but also share preference for ligand binding at the inter-domainal boundary. Variability and ligand specificity for different members of the cytokine receptor superfamily result from the specific sequence differences of surface loops that collide at this site, analogous to antigen recognition by the hypervariable regions of immunoglobulins. Unlike the four helix bundle cytokine proteins, the VIIa interaction with TF is mediated predominantly by its β-structured epidermal growth factor-like domains. The exact residues in these VIIa domains which constitute the interface for interaction with TF are currently not established. However, point mutations in the protease domain of VIIa have affected VIIa binding to TF as well [70] consistent with the proposed dual interaction between these proteins.

SELECTIVE MOLECULAR INTERVENTION IN VIVO

TF is synthesized *de novo* and inappropriately expressed by intravascular cells in a variety of acute and life threatening diseases. Since its function is to initiate, rather than to amplify the coagulation protease pathways, specific inhibition of TF provides an opportunity to selectively intervene in excessive activation. In such experiments neutralizing monoclonal to TF have demonstrated therapeutic efficacy. A primate model of lethal gram negative septic shock syndrome was utilized. Prevention lethality was achieved without evidence of diminished hemostasis [13,14].

There are two mechanistic approaches to neutralization of the initiation complex. Blocking the binding of VIIa by a specific inhibitor directed to the ligand interface is one mechanism to interfere with TF function. This form of inhibitor will only react with free TF which may result in slow inhibition with residual TF•VIIa complexes at equilibrium due to the competition of VIIa and inhibitor for the same site on TF. The second type of neutralization is through blocking of the substrate interactive site on TF by class I_N monoclonal antibody. This does not require dissociation of the TF•VIIa complex, thus resulting in more rapid inhibition. This rapid inhibition of the TF•VIIa complex function is indeed effective *in vivo* as demonstrated by the potency of TF8-5G9 to completely block LPS triggered coagulation in non-human primates [12,13].

Rapidly advancing knowledge of the specific functional role of regions and specific residues in TF from the variety of experimental approaches coupled with the three dimensional structure of TF has significantly advanced the ability to initiate knowledge-based inhibitor design. In addition, new insight into the structural basis for cofactor functions promise general insight into protein chemistry.

ABBREVIATIONS

Abbreviations: tissue factor, TF; factor VII, VII; factor VIIa, VIIa; bacterial lipopolysaccharide, LPS; disseminated intravascular coagulation, DIC; γ-carboxyglutamic acid, Gla.

ACKNOWLEDGEMENTS

We thank Barbara Parker for assistance with manuscript preparation. This is publication 8941-IMM from The Scripps Research Institute and was supported by grant P01-HL-16411 from the National Heart Lung and Blood Institute.

REFERENCES

1. Ruf, W. and T. S. Edgington, Structural biology of tissue factor, the initiator of thrombogenesis in vivo, *FASEB J.* 8:385 (1994).

2. Davie, E. W., K. Fujikawa, and W. Kisiel, The coagulation cascade: Initiation, maintenance, and regulation, *Biochemistry* 30:10363 (1991).

3. Edgington, T. S., N. Mackman, K. Brand, and W. Ruf, The structural biology of expression and function of tissue factor, *Thromb. Haemost.* 66:67 (1991).

4. Edgington, T. S., N. Mackman, S. -T. Fan, and W. Ruf, Cellular immune and cytokine pathways resulting in tissue factor expression and relevance to septic shock, *Nouv. Rev. Fr. Hematol.* 34(Suppl):S13 (1992).

5. Ruf, W., A. Rehemtulla, J. H. Morrissey, and T. S. Edgington, Phospholipid-independent and -dependent interactions required for tissue factor receptor and cofactor function, *J. Biol. Chem.* 266:2158 (1991).

6. Krishnaswamy, S., K. A. Field, T. S. Edgington, J. H. Morrissey, and K. G. Mann, Role of the membrane surface in the activation of human coagulation factor X, *J. Biol. Chem.* 267:26110 (1992).

7. Krishnaswamy, S., The interaction of human factor VIIa with tissue factor, *J. Biol. Chem.* 267:23696 (1992).

8. Ruf, W., E. A. Stura, R. J. LaPolla, R. Syed, T. S. Edgington, and I. A. Wilson, Purification, sequence and crystallization of an anti-tissue factor Fab and its use for the crystallization of tissue factor, *J. Crystal Growth* 122:253 (1992).

9. Harlos, K., D. M. A. Martin, D. P. O'Brien, E. Y. Jones, D. I. Stuart, I. Polikarpov, A. Miller, E. G. D. Tuddenham, and C. W. G. Boys, Crystal structure of the extracellular region of human tissue factor, *Nature* 370:662 (1994).

10. Muller, Y. A., M. H. Ultsch, R. F. Kelley, and A. M. De Vos, Structure of the extracellular domain of human tissue factor: Location of the factor VIIa binding site, *Biochemistry* 33:10864 (1994).

11. Ruf, W., C. R. Kelly, J. R. Schullek, D. M. A. Martin, I. Polikarpov, C. W. G. Boys, E. G. D. Tuddenham, and T. S. Edgington, Energetic contributions and topographical organization of ligand binding residues of tissue factor, Submitted:(1994).

12. ten Cate, H., K. A. Bauer, M. Levi, T. S. Edgington, R. D. Sublett, S. Barzegar, B. L. Kass, and R. D. Rosenberg, The activation of factor X and prothrombin by recombinant factor VIIa is mediated by tissue factor, *J. Clin. Invest.* 92:1207 (1993).

13. Levi, M., H. ten Cate, K. A. Bauer, T. van der Poll, T. S. Edgington, H. R. Büller, S. J. H. van Deventer, C. E. Hack, J. W. Ten Cate, and R. D. Rosenberg, Inhibition of endotoxin-induced activation of coagulation and fibrinolysis by pentoxifylline or by a monoclonal anti-tissue factor antibody in chimpanzees, *J. Clin. Invest.* 93:114 (1994).

14. Taylor, F. B., Jr., A. Chang, W. Ruf, J. H. Morrissey, L. B. Hinshaw, R. Catlett, K. Blick, and T. S. Edgington, Lethal E.coli septic shock is prevented by blocking tissue factor with monoclonal antibody, *Circ. Shock* 33:127 (1991).

15. Bazan, J. F., Structural design and molecular evolution of a cytokine receptor superfamily, *Proc. Natl. Acad. Sci. USA* 87:6934 (1990).

16. Altieri, D. C. and T. S. Edgington, Identification of effector cell protease receptor-1: A leukocyte-distributed receptor for the serine protease factor Xa, *J. Immunol.* 145:246 (1990).

17. Altieri, D. C., Molecular cloning of effector cell protease receptor-1, a novel cell surface receptor for the protease factor Xa, *J. Biol. Chem.* 269:3139 (1994).

18. Altieri, D. C. and S. J. Stamnes, Protease-dependent T cell activation: Ligation of effector cell protease receptor-1 (EPR-1) stimulates lymphocyte proliferation, *Cell. Immunol.* 155:372 (1994).

19. Lawson, J. H. and K. G. Mann, Cooperative activation of human factor IX by the human extrinsic pathway of blood coagulation, *J. Biol. Chem.* 266:11317 (1991).

20. Rao, L. and S. I. Rapaport, Studies of a mechanism inhibiting the initiation of the extrinsic pathway of coagulation, *Blood* 69:645 (1987).

21. Girard, T. J., R. Eddy, R. L. Wesselschmidt, L. A. MacPhail, K. M. Likert, M. G. Byers, T. B. Shows, and G. J. Broze,Jr., Structure of the human lipoprotein-associated coagulation inhibitor gene, *J. Biol. Chem.* 266:5036 (1991).

22. Broze, G. J., Jr., Tissue factor pathway inhibitor and the revised hypothesis of blood coagulation, *Trends Cardiovasc. Med.* 2:72 (1992).

23. Broze, G. J., Jr., K. Likert, and D. Higuchi, Inhibition of factor VIIa/tissue factor by antithrombin III and tissue factor pathway inhibitor, *Blood* 82:1679 (1993).

24. Bach, R., W. H. Konigsberg, and Y. Nemerson, Human tissue factor contains thioester--linked palmitate and stearate on the cytoplasmic half-cystine, *Biochemistry* 27:4227 (1988).

25. Zioncheck, T. F., S. Roy, and G. A. Vehar, The cytoplasmic domain of tissue factor is phosphorylated by a protein kinase C-dependent mechanism, *J. Biol. Chem.* 267:3561 (1992).

26. Drake, T. A., W. Ruf, J. H. Morrissey, and T. S. Edgington, Functional tissue factor is entirely cell surface expressed on lipopolysaccharide-stimulated human blood monocytes and a constitutively tissue factor-producing neoplastic cell line, *J. Cell Biol.* 109:389 (1989).

27. Le, D. T., S. I. Rapaport, and L. V. M. Rao, Relations between factor VIIa binding and expression of factor VIIa/tissue factor catalytic activity on cell surfaces, *J. Biol. Chem.* 267:15447 (1992).

28. Bach, R. and D. B. Rifkin, Expression of tissue factor procoagulant activity: Regulation by cytosolic calcium, *Proc. Natl. Acad. Sci. USA* 87:6995 (1990).

29. Drake, T. A., J. H. Morrissey, and T. S. Edgington, Selective cellular expression of tissue factor in human tissues. Implications for disorders of hemostasis and thrombosis, *Am. J. Pathol.* 134:1087 (1989).

30. Eddleston, M., J. C. de la Torre, M. B. A. Oldstone, D. J. Loskutoff, T. S. Edgington, and N. Mackman, Astrocytes are the primary source of tissue factor in the murine central nervous system - a role for astrocytes in cerebral hemostasis, *J. Clin. Invest.* 92:349 (1993).

31. Gregory, S. A. and T. S. Edgington, Tissue factor induction in human monocytes: Two distinct mechanisms displayed by different alloantigen responsive T cell clones, *J. Clin. Invest.* 76:2440 (1985).

32. Fan, S. -T. and T. S. Edgington, Clonal analysis of mechanisms of murine T helper cell collaboration with effector cells of macrophage lineage, *J. Immunol.* 141:1819 (1988).

33. Drake, T. A., J. Cheng, A. Chang, and F. B. Taylor,Jr., Expression of tissue factor, thrombomodulin, and E-selectin in baboons with lethal E.coli sepsis, *Am. J. Pathol.* 142:1458 (1993).

34. Creasey, A. A., A. C. K. Chang, L. Fiegen, T. C. Wün, F. B. Taylor,Jr., and L. B.

Hinshaw, Tissue factor pathway inhibitor (TFPI) reduces mortality from *E. coli* septic shock, *J. Clin. Invest.* 91:2850 (1993).

35. Rao, L. V. M., Tissue factor as a tumor procoagulant, *Cancer Metastasis Rev.* 11:249 (1992).

36. Mueller, B. M., R. A. Reisfeld, T. S. Edgington, and W. Ruf, Expression of tissue factor by melanoma cells promotes efficient hematogenous metastasis, *Proc. Natl. Acad. Sci. USA* 89:11832 (1992).

37. Pernod, G., B. Polack, F. Peyron, A. Luisy, L. Kolodie, P. Ambroise-Thomas, and F. Santoro, Monocyte tissue factor expression induced by *Plasmodium falciparum*--infected erythrocytes, *Thromb. Haemost.* 68:111 (1992).

38. Teysseire, N., D. Arnoux, F. George, J. Sampol, and D. Raoult, von Willebrand factor release and thrombomodulin and tissue factor expression in *Rickettsia conorii*-infected endothelial cells, *Infect. Immun.* 60:4388 (1992).

39. Etingin, O. R., R. L. Silverstein, H. M. Friedman, and D. P. Hajjar, Viral activation of the coagulation cascade: Molecular interactions at the surface of infected endothelial cells, *Cell* 61:657 (1990).

40. Key, N. S., G. M. Vercellotti, J. C. Winkelmann, C. F. Moldow, J. L. Goodman, N. L. Esmon, C. T. Esmon, and H. S. Jacob, Infection of vascular endothelial cells with herpes simplex virus enhances tissue factor activity and reduces thrombomodulin expression, *Proc. Natl. Acad. Sci. USA* 87:7095 (1990).

41. Wilcox, J. N., K. M. Smith, S. M. Schwartz, and D. Gordon, Localization of tissue factor in the normal vessel wall and in the atherosclerotic plaque, *Proc. Natl. Acad. Sci. USA* 86:2839 (1989).

42. Haskel, E. J., S. R. Torr, K. C. Day, M. O. Palmier, T. -C. Wun, B. E. Sobel, and D. R. Abendschein, Prevention of arterial reocclusion after thrombolysis with recombinant lipoprotein-associated coagulation inhibitor, *Circulation* 84:821 (1991).

43. Ruf, W. and T. S. Edgington, Two sites in the tissue factor extracellular domain mediate the recognition of the ligand factor VIIa, *Proc. Natl. Acad. Sci. USA* 88:8430 (1991).

44. De Vos, A. M., M. Ultsch, and A. A. Kossiakoff, Human growth hormone and extracellular domain of its receptor: Crystal structure of the complex, *Science* 255:306 (1992).

45. Ruf, W., M. W. Kalnik, T. Lund-Hansen, and T. S. Edgington, Characterization of factor VII association with tissue factor in solution. High and low affinity calcium binding sites in factor VII contribute to functionally distinct interactions, *J. Biol. Chem.* 266:15719 (1991).

46. Bom, V. J. J. and R. M. Bertina, The contribution of Ca^{2+}, phospholipids and tissue-factor apoprotein to the activation of human blood-coagulation factor X by activated factor VII, *Biochem. J.* 265:327 (1990).

47. Rapaport, S. I. and L. V. M. Rao, Initiation and regulation of tissue factor-dependent blood coagulation, *Arterioscler. Thromb.* 12:1111 (1992).

48. Lawson, J. H., S. Butenas, N. Ribarik, and K. G. Mann, Complex-dependent inhibition of factor VIIa by antithrombin III and heparin, *J. Biol. Chem.* 268:767 (1993).

49. Pedersen, A. H., O. Nordfang, F. Norris, F. C. Wiberg, P. M. Christensen, K. B. Moeller, J. Meidahl-Pedersen, T. C. Beck, K. Norris, U. Hedner, and W. Kisiel, Recombinant human extrinsic pathway inhibitor. Production, isolation, and characterization of its inhibitory activity on tissue factor-initiated coagulation reactions, *J. Biol. Chem.* 265:16786 (1990).

50. Lawson, J. H., S. Butenas, and K. G. Mann, The evaluation of complex-dependent alterations in human factor VIIa, *J. Biol. Chem.* 267:4834 (1992).

51. Pedersen, A. H., T. Lund-Hansen, H. Bisgaard-Frantzen, F. Olsen, and L. C. Petersen, Autoactivation of human recombinant coagulation factor VII, *Biochemistry* 28:9331 (1989).

52. Ruf, W. and T. S. Edgington, An anti-tissue factor monoclonal antibody which inhibits TF•VIIa complex is a potent anticoagulant in plasma, *Thromb. Haemost.* 66:529 (1991).

53. Fiore, M. M., P. F. Neuenschwander, and J. H. Morrissey, An unusual antibody that blocks tissue factor/factor VIIa function by inhibiting cleavage only of macromolecular substrates, *Blood* 80:3127 (1992).

54. Ruf, W., A. Rehemtulla, and T. S. Edgington, Antibody mapping of tissue factor implicates two different exon-encoded regions in function, *Biochem. J.* 278:729 (1991).

55. Rehemtulla, A., W. Ruf, D. J. Miles, and T. S. Edgington, The third Trp-Lys-Ser (WKS) tripeptide motif in tissue factor is associated with a function site, *Biochem. J.* 282:737 (1992).

56. Ruf, W., D. J. Miles, A. Rehemtulla, and T. S. Edgington, Tissue factor residues 157-167 are required for efficient proteolytic activation of factor X and factor VII, *J. Biol. Chem.* 267:22206 (1992).

57. Ruf, W., D. J. Miles, A. Rehemtulla, and T. S. Edgington, Cofactor residues lysine 165 and 166 are critical for protein substrate recognition by the tissue factor•factor VIIa protease complex, *J. Biol. Chem.* 267:6375 (1992).

58. Ruf, W., Factor VIIa residue Arg^{290} is required for efficient activation of the macromolecular substrate factor X, *Biochemistry* in press:(1994).

59. Waxman, E., J. B. A. Ross, T. M. Laue, A. Guha, S. V. Thiruvikraman, T. C. Lin, W. H. Konigsberg, and Y. Nemerson, Tissue factor and its extracellular soluble domain: The relationship between intermolecular association with factor VIIa and enzymatic activity of the complex, *Biochemistry* 31:3998 (1992).

60. Ruf, W., D. J. Miles, A. Rehemtulla, and T. S. Edgington, Mutational analysis of receptor and cofactor function of tissue factor, *Methods Enzymol.* 222:209 (1993).

61. Schullek, J. R., W. Ruf, and T. S. Edgington, Key ligand interface residues in tissue factor contribute independently to factor VIIa binding, *J. Biol. Chem.* 269:19399 (1994).

62. Nemerson, Y. and D. Repke, Tissue factor accelerates the activation of coagulation factor VII: The role of a bifunctional coagulation cofactor, *Thromb. Res.* 40:351 (1985).

63. Rao, L. V. M. and S. I. Rapaport, Activation of factor VII bound to tissue factor: A key early step in the tissue factor pathway of blood coagulation, *Proc. Natl. Acad. Sci. USA* 85:6687 (1988).

64. Yamamoto, M., T. Nakagaki, and W. Kisiel, Tissue factor-dependent autoactivation of human blood coagulation factor VII, *J. Biol. Chem.* 267:19089 (1992).

65. Nakagaki, T., D. C. Foster, K. L. Berkner, and W. Kisiel, Initiation of the extrinsic pathway of blood coagulation: evidence for the tissue factor dependent autoactivation of human coagulation factor VII, *Biochemistry* 30:10819 (1991).

66. Neuenschwander, P. F. and J. H. Morrissey, Deletion of the membrane anchoring region of tissue factor abolishes autoactivation of factor VII but not cofactor function. Analysis of a mutant with a selective deficiency in activity, *J. Biol. Chem.* 267:14477 (1992).

67. Roy, S., L. R. Paborsky, and G. A. Vehar, Self-association of tissue factor as revealed by chemical cross-linking, *J. Biol. Chem.* 266:4665 (1991).

68. Rehemtulla, A., W. Ruf, and T. S. Edgington, The integrity of the CYS^{186}-CYS^{209} bond of the second disulfide loop of tissue factor is required for binding of factor VII, *J. Biol. Chem.* 266:10294 (1991).

69. Ruf, W., J. R. Schullek, M. J. Stone, and T. S. Edgington, Mutational mapping of functional residues in tissue factor: Identification of factor VII recognition determinants in both structural modules of the predicted cytokine receptor homology domain, *Biochemistry* 33:1565 (1994).

70. O'Brien, D. P., K. M. Gale, J. S. Anderson, J. H. McVey, G. J. Miller, T. W. Meade, and E. G. D. Tuddenham, Purification and characterization of factor VII 304-Gln: A variant molecule with reduced activity isolated from a clinically unaffected male, *Blood* 78:132 (1991).

THROMBIN RECEPTOR:STRUCTURE AND FUNCTION

Kenji Ishii, Ji Chen, Maki Ishii, Thien-Khai H. Vu, Robert E. Gerszten, Tania Nanevicz, Ling Wang, and Shaun R. Coughlin

Cardiovascular Research Institute, University of California
San Francisco, CA 94143-0524

INTRODUCTION

Thrombin is a multifunctional serine protease generated at sites of vascular injury. While it is best known for its ability to cleave fibrinogen and trigger fibrin formation, thrombin is also a powerful agonist for a variety of cellular responses. First and foremost, thrombin is the most potent activator of platelets in vitro. Thrombin is also chemotactic for monocytes, and is mitogenic for lymphocytes, fibroblasts, and vascular smooth muscle cells. Thrombin acts on the vascular endothelium to stimulate production of prostacyclin, plasminogen activator inhibitor, and the potent smooth muscle cell mitogen platelet-derived growth factor. Thrombin also induces neutrophil adherence to the vessel wall by an endothelial-dependent mechanism, probably by causing surface expression of GMP-140 on the endothelial cell and directly activates neutrophils themselves. Teleologically, these disparate functions of thrombin may be unified by viewing thrombin as an orchestrator of the response to vascular injury or wounding, mediating not only hemostatic but inflammatory and proliferative or reparative responses.[1]

A recently cloned thrombin receptor has provided a framework for understanding how thrombin talks to cells and appears to account for many of thrombin activities.[2] Structure-activity studies with this receptor have revealed a novel proteolytic mechanism of receptor activation. Moreover, this receptor has provided new tools for defining the role of thrombin-induced cell activation in vivo and may represent a new target for antithrombotic and other therapies.

HOW DOES A PROTEASE TALK TO A CELL?

Thrombin Receptor Structure

The deduced structure of thrombin receptor shows that it is a member of the seven transmembrane domain receptor family. Its amino terminal exodomain includes a putative thrombin cleavage site resembling the known thrombin cleavage site in protein C. Carboxyl to this site was a sequence resembling the carboxyl tail of hirudin, a leech proteinknown to interact with thrombin's anion-binding exosite. These analogies suggested that thrombin might interact with the receptor with an unique mechanism (Figure 1).[3] Thrombin binds to and cleaves the receptor's extracellular amino terminal domain to unmask a new amino terminal. This new amino terminal then functions as a tethered peptide ligand,binding to as yet undefined sites within the body of the receptor to effect receptor activation. A synthetic peptide mimicking the new amino terminus is a full agonist for the receptor, bypassing receptor cleavage. The first 5 amino acids of the receptor's agonist peptide domain

Cardiovascular Disease 2
Edited by L.L. Gallo, Plenum Press, New York, 1995

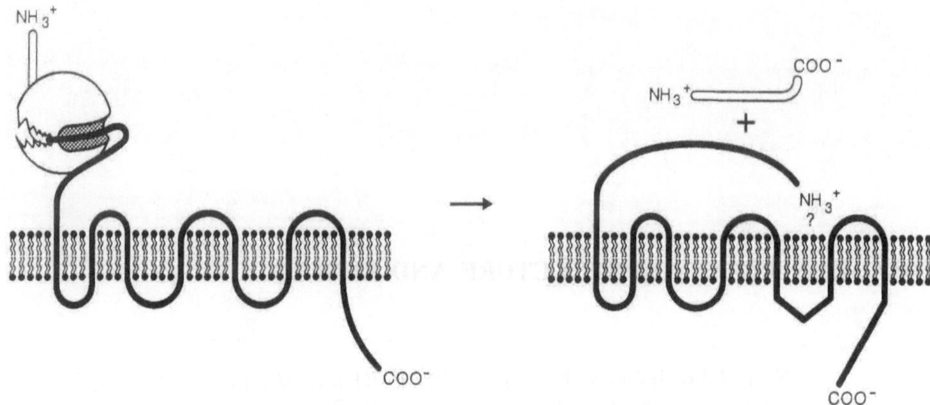

Figure 1. Model of thrombin receptor activaiton. Thrombin, the sphere in this figure, binds to its receptor's extracellular amino terminal extension. After binding to the amino terminal extension, thrombin cleaves the receptor at the LDPR/S cleavage site (junction between open and filled receptor segments), releasing an inactive fragment of the receptor's amino terminus (open fragment) and exposing a new amino terminus. This newly unmasked amino terminus then functions as a tethered peptide ligand, binding to as yet undefined sites within the body of the receptor to effect receptor activation. As shown, this binding event presumably translates into a conformational change in the receptor's cytoplasmic face, effecting G-protein activation. (reprinted with permission from NATURE (3)).

(SFLLR) are sufficient to specify agonist activity, and polypeptide with the six amino acids has the most potent activity.[4,5]

The protonated amino group of the Ser1 and the Phe2 side chain are vital for agonist function; the Leu4 and Arg5 side chains play less important roles. The importance of the Ser1's protonated amino group is particularly appealing, because this group is created by receptor cleavage. This may explain in part how the agonist peptide domain's activity is masked when the receptor is in the uncleaved state .

Tethered Ligand Hypothesis

Recently described constitutively active mutations in seven transmembrane domain receptors suggest that activation of these molecules includes release from a tonically inhibited state. Those findings suggest the possibility that the thrombin receptor's amino terminal exodomain may serve a tonic inhibitory function and a binding interaction involving the SFLLRN sequence is critical for releasing from tonic inhibition. Receptor activation might occur when cleavage of the amino terminus by thrombin or competition by free SFLLRN peptide disrupted these inhibitory binding interactions. To exclude this alternative to the tethered ligand hypothesis (Figure 1) mutant receptor lacking the amino terminal exodomain was tested in Xenopus laevis oocytes. This mutant did not show constitutive activation; basal ^{45}Ca release from Xenopus oocytes expressing this receptor was indistinguishable from that of oocytes expressing wild type receptor.[6] This mutant receptor did not respond to thrombin, but did respond to agonist peptide. These observations refute "release from tonic inhibition" hypothesis and confirm the "tethered ligand" hypothesis.

LIGAND BINDING DETERMINATION SITES

G-protein coupled receptors for catecholamines and some other small ligands are activated when agonists bind to the transmembrane region of the receptor. However, the docking interaction through which peptide agonists activate their receptors were less well known. To identify a thrombin receptor that differed sufficiently from the human to aid our structure-activity studies, we isolated a cDNA encoding the Xenopus laevis homologue of the human thrombin receptor.[7] The tethered ligand domains of the human and Xenopus receptors were strikingly different. SFLLRN for human and TFRIFD for Xenopus. As expected synthetic peptides mimicking these domains were full agonists for their respective receptors. However, each agonist peptide showed surprising selectivity for the receptor of its own species. The human agonist peptide was -1,500 fold more potent at the human receptor than at the Xenopus receptor, and the xenopus peptide was -30 fold more potent at the Xenopus than at the human receptor. These observations indicated that the receptor domains that define agonist specificity might be identified using human-Xenopus chimaeric receptors. Because the human agonist peptide functioned poorly at the Xenopus receptor, we began by building human sequences into the Xenopus receptor and looking for gain of responsiveness to the human agonist peptide.

Replacing xenopus receptor's extracellular face with that of human caused a remarkable switch in receptor specificity with a nearly 3-log gain of potency for the human agonist and 3-log loss of potency for the xenopus agonist. Substitution of individual human receptor domains showed that the amino terminal exodomain residues 76-93 and extracellular loop 2 are important determinants of thrombin receptor agonist specificity. The dependency of agonist binding on particular surface residues in some of the seven transmembrane receptor such as neurokinin receptors and growth hormone releasing factor receptors suggest that this paradigm may operate in many peptide receptors.

HOW DOES A PROTEASE ELICIT CONCENTRATION-DEPENDENT RESPONSES?

Like other important signaling molecules, thrombin effects concentration-dependent and graded responses in its target cells. Classical ligands effect concentration-dependent responses via graded receptor occupancy. How can thrombin, acting as an enzyme rather than a classical ligand, effect concentration-dependent responses? Specifically, even low amounts of thrombin might eventually cleave and activate all cell surface receptors. A recent study used antibodies which distinguished the naive receptor from the cleaved and activated form and followed the rate of thrombin receptor cleavage on intact cells.[8] The rate of receptor cleavage was proportional to thrombin concentration over the physiologic range. Thus cells must be able to distinguish high from low thrombin concentration by reading the rate of receptor cleavage. Cumulative phosphoinositide hydrolysis in response to thrombin correlated best with absolute receptor cleavage during a given time interval, less well with the integral of receptor cleavage as a function of time. When ongoing thrombin receptor cleavage was interrupted with hirudin, phosphoinositide hydrolysis ceased in parallel. These data suggest that each activated thrombin receptor generates a quantum of second messenger, then "shuts off". Thrombin concentration would then determine the rate of receptor activation and therefore the rate of second messenger generation.

Second messenger levels and magnitude of the cellular response are presumably then determined by the balance between the rates of second messenger generation and clearance. This is unlike the case for classical ligands, because in case of thrombin receptor, there is no "equilibrium" binding and the cell cannot use graded receptor occupancy to effect graded responses. This formulation of thrombin receptor signaling has several implications. First, it places great importance on the mechanisms that terminate thrombin receptor signaling. Second, it suggests that thrombin receptor antagonists need only slow thrombin receptor activation enough that clearance of second messengers outstrips their generation to effectively block signaling. This notion may encourage attempts at antagonist development.

THROMBIN RECEPTOR SHUT OFF MECHANISM

The formulation of thrombin receptor signaling outlined above suggests that the thrombin receptor's "shut off" mechanism is a critical determinant of the gain of the system and thereby of thrombin responsiveness. Termination of signaling by the seven transmembrane domain signaling molecules such as rhodopsin and ß-adrenergic receptor is effected by receptor phosphorylation, predominantly by G-protein coupled receptor kinases. Because G-protein coupled receptor kinases appear to phosphorylate only liganded receptors, they are excellent candidates for mediating the shut off of activated thrombin receptors. For these reasons we addressed the possible role of G-protein coupled receptor kinases in terminating thrombin receptor signaling.

We first addressed the role of receptor kinases in terminating thrombin receptor signaling by directly examining receptor phosphorylation. Thrombin receptor antiserum immuno-precipitated a ^{32}P-labeled protein from thrombin-treated but not untreated receptor-expressing cells. Activation of signaling pathways downstream of the thrombin receptor by phorbol ester or calcium ionophore caused much less receptor phosphorylation than that elicited when the receptor itself was activated by thrombin, and the adenylate cyclase activator forskolin did not stimulate receptor phosphorylation.[9] These data suggested that protein kinase C activation could contribute to thrombin receptor phosphorylation. Down regulation of protein kinase C by long term pretreatment with phorbol ester blocked receptor phosphorylation in response to phorbol ester or calcium ionophore but not in response to thrombin. These data are consistent with the hypothesis that a BARK-like kinase mediates most of the phosphorylation of activated thrombin receptors.

If a BARK-like kinase were a physiologic mediator of thrombin receptor shut off, overexpressing it might be expected to inhibit thrombin receptor signaling. Co-expression of BARK2 (beta-adrenergic receptor kinase 2) with the thrombin receptor in Xenopus oocytes markedly inhibited thrombin receptor signaling at EC_{50} concentrations of thrombin. Strikingly, the highly related kinase BARK1 was much less effective than BARK2 in inhibiting thrombin signaling.

Thrombin receptor sensitivity to inhibition by BARK2 required the presence of specific serine and threonine residues in the receptor's cytoplasmic domain. When serines and threonines in the receptor's C-tail were substituted with alanine, the receptor was insensitive to BARK2. By contrast, receptors with alanine substitutions at other cytoplasmic serine and threonine remained sensitive to inhibition by BARK2. These data suggest that BARK2 inhibits thrombin receptor signaling by phosphorylating the receptor's carboxyl tail at serine/threonine residues.

These studies demonstrate that the thrombin receptor becomes rapidly phosphorylated upon activation in a manner consistent with the action of a G-protein coupled receptor kinase. Moreover, our functional studies show that the G-protein coupled receptor kinase BARK2 can inhibit thrombin receptor signaling, likely by phosphorylating serine/threonine residues in the receptor's carboxyl tail. These data suggest that thrombin receptor phosphorylation by a BARK2-like receptor kinase may in part mediate the initial shut off of thrombin receptor signaling.

POSSIBLE ROLES IN CARDIOVASCULAR DISEASE

While the role of thrombin receptor activation in human hemostasis and thrombosis remains to be rigorously demonstrated, the receptor's clear role in mediating thrombin-induced platelet activation, the importance of thrombin in platelet-dependent models of arterial thrombosis, and the efficacy of anti-thrombin therapy for unstable angina all suggest that thrombin receptor activation will play an important role.

Given thrombin's known actions on inflammatory and mesenchymal cells and its generation at sites of vascular injury, it is tempting to postulate a role for thrombin receptor activation in inflammatory and proliferative responses. Recently, robust thrombin receptor expression was noted in human atherosclerotic plaques, apparently by smooth muscle cells, mesenchymal-appearing cells of unknown origin, and macrophages.[10] In the context of thrombin's known actions on these cells, the high and selective expression of thrombin receptor in atherosclerotic lesions suggests a possible role for thrombin receptor activation in restenosis and possibly in atherogenesis itself.

REFERENCES

1. J.W. Fenton II, Regulation of thrombin generation and functions. Semin. Thromb. Homeost. 14: 234 (1988).
2. T.K. Vu, D.T. Hung , V.I. Wheaton, and S.R. Coughlin, Molecular cloning of a functional thrombin receptor reveals a novel proteolytic mechanism of receptor activation. Cell 64: 1057 (1991).
3. T.K. Vu, V.I. Wheaton, D.T. Hung, and S.R. Coughlin, Domains specifying thrombin-

receptor interaction. Nature 353: 674 (1991).

4. R.M. Scarborough, M. Naughton, W. Teng, D.T. Hung, J. Rose, T.K. Vu, V.I. Wheaton, C.W. Turck, and S.R. Coughlin, Tethered ligand agonist peptides: structural requirements for thrombin receptor activation reveal mechanism of proteolytic unmasking of agonist function. J. Biol. Chem. 267: 13146 (1992).
5. R.R.J. Vassallo, T. Kieber-Emmons, K. Cichowski, L.F. Brass,Structure-function relationships in the activation of platelet thrombin receptors by receptor-derived peptides. J. Biol. Chem. 267 (9): 6081 (1992).
6. J. Chen, M. Ishii, L. Wang, K. Ishii, and S.R. Coughlin, Thrombin receptor activation: confirmatin of the intramolecular tethered liganding hypothesis and discovery of an alternative intermolecular liganding mode. J. Biol. Chem. 269: (1994) in press.
7. R.E. Gerszten, J. Chen, M. Ishii, K. Ishii, L. Wang, T. Nanevicz, C. Turck, T.K. Vu,and S.R. Coughlin, The thrombin receptor's specificity for agonist peptide is defined by its extracellular surface--not by its transmembrane domains. Nature 368: 648 (1994).
8. K. Ishii, L. Hein, B. Kobilka, and S.R. Coughlin, Kinetics of thrombin receptor cleavage on intact cells. Relation to signaling. J. Biol. Chem. 268 (13): 9780 (1993).
9. K. Ishii, J. Chen, M. Ishii,W.J. Koch, N.J. Freedman, R.J. Lefkowitz, and S.R. Coughlin, Inhibiton of thrombin receptor signaling by a G-protein coupled receptor kinase. Functinal specificity among G-proein coupled receptor kinases. J. Biol. Chem. 269 (2): 1125 (1994).
10. N.A. Nelken, S.J. Soifer, J. O'Keefe, T.K. Vu, I.F. Charo, and S.R. Coughlin, Thrombin receptor expression in normal and atherosclerotic human arteries. J. Clin. Invest. 90: 1614 (1992).

ANTIPLATELET EFFECTS OF DIRECT ACTING THROMBIN INHIBITORS AND PLATELET GPIIb/IIIa ANTAGONISTS: COMPARATIVE ANALYSIS

Shaker A. Mousa and Thomas M. Reilly

The DuPont Merck Pharmaceutical Company
Cardiovascular Diseases Div., Exp. Station,
Wilmington, DE 19880-0400

ABSTRACT

The present investigation was undertaken to compare the effects of different direct thrombin inhibitors including hirudin, DUP 714 (Ac-D-Phe-Pro-boro Arg) and PPACK (Phe-Pro-Arg-chloromethyl ketone) with different platelet GPIIb/IIIa antagonists including linear Arginine-Glycine-Aspartate (RGD), cyclic RGD, peptidomimetic and non-peptide inhibitors on various platelet functions including: (a) platelet aggregation, (b) fibrinogen binding, (c) platelet plasminogen activator inhibitor type-1 (PAI-1) release, (d) platelet serotonin release, and (e) platelet intracellular Ca^{++} mobilization. Hirudin, DUP 714 and PPACK were effective in inhibiting all of the responses (a-e) mediated by thrombin, with IC_{50}s ranging from 0.002 to 0.9 uM depending upon the concentration of thrombin used, but not those mediated by other platelet agonists including collagen, ADP, and arachidonic acid. In contrast, linear RGD, constrained cyclic RGD (XM648), an RGD peptidomimetic (YY751), a non-peptide GPIIb/IIIa antagonist were effective in blocking fibrinogen binding and platelet aggregation mediated by any of the agonists. The potency of XL111, XM648 and YY751 in inhibiting thrombin-mediated platelet aggregation (a) and fibrinogen binding (b) ranged from IC_{50} of 0.05 to 1.0 uM in (a) and IC_{50} of 0.01 to 0.005 uM in (b). However, all platelet GPIIb/IIIa antagonists tested were ineffective in blocking either (c), (d) or (e). Flow cytometric analysis using dual fluorescence markers for the platelet GPIIb/IIIa membrane receptors (FITC-labeled cyclic RGD analog, XL086) and for the alpha granule (PE-monoclonal antibody for P-selectin) demonstrated a dissociation between the platelet GPIIb/IIIa receptors by inhibitors and platelet granular secretion. These data suggest that GPIIb/IIIa receptor blockade as compared with direct inhibition of thrombin, does not inhibit intracellular Ca^{++} mobilization signal transduction. These data also show that GPIIb/IIIa antagonists, in contrast to thrombin inhibitors, inhibit platelet aggregation mediated by different agonists. The universal anti-aggregatory efficacy of the GPIIb/IIIa antagonists and efficacy of the thrombin inhibitors against

Cardiovascular Disease 2
Edited by L.L. Gallo, Plenum Press, New York, 1995

thrombin-induced platelet secretion suggest potential benefits of their combinations in various thrombotic disorders.

INTRODUCTION

Intravascular thrombosis is one of the most frequent pathological events and major cause of morbidity and mortality in western civilization (ref). Factors which stimulate thrombosis include vascular damage, stimulation of platelets, and activation of the coagulation cascade. Platelet activation and subsequent aggregation leads to the exposure of phospholipid on platelet surface which facilitate the activation of factor X and factor II (ref.). Platelet activation and the resulting aggregation has been shown to be associated with various pathological conditions including cardiovascular and cerebrovascular thromboembolic disorders such as unstable angina, myocardial infarction, transient ischemic attack, stroke and atherosclerosis.[1-5] The contribution of platelets to these disease processes stems from their ability to form aggregates or platelet thrombi as a consequence of arterial wall injury.[6] Injury of blood vessel walls could occur either acutely or chronically by various pathophysiological processes. Platelets are then activated by a number of activators or agonists that are released from within platelets or from the injured arterial walls, with the subsequent adherence, aggregation to the disrupted vessel surface and resultant formation of an occlusive thrombus in the lumen of the vessel.[7-9] A number of agonists, generated at the interface between the vessel wall and circulating blood at the site of vascular injury, have been shown to activate platelets.[1-5] These include ADP, epinephrine, thromboxane A2 and thrombin in the fluid phase, collagen and other components of the extracellular matrix in the subendothelium.[2] It has also been demonstrated that platelets are a fairly rich source of the fast-acting plasminogen activator inhibitor type 1 (PAI-1), serotonin, different coagulation factors and many other important biomolecules stored in the platelet granules.[9,10] Agents which activate blood platelets and induce platelet aggregation are also known to induce the release of platelet alpha or dense granules.[9]

The platelet glycoprotein (GP) GPIIb/IIIa membrane receptor (fibrinogen receptor) represents the final common pathway for platelet aggregation following activation by any agonist compounds capable of inhibiting the GPIIb/IIIa receptor in theory have the advantages over other anti-platelet drugs in blocking aggregation in response to all agonists. A number of different GPIIb/IIIa receptor antagonists are currently undergonig preclinical and clinical evaluation. Current antiplatelet drugs are mainly effective against one of the many platelet activators. These drugs include aspirin, which blocks cyclooxygenases[9]; ticlopidine, which acts against thromboxane A_2[16]; and hirudin, which acts against thrombin.[17] In general, these thrombotic processes are thrombin-mediated, platelet-dependent and irresponsive to aspirin and heparin therapies.[18,19] Thrombin represents the most potent platelet agonist as well as the final enzyme in the coagulation cascade responsible for converting fibrinogen to fibrin (ref). Many thrombolic processes in the coronary and cerebral arteries are thrombin-mediated, platelet-dependent and unresponsive to aspirin and heparin therapies [(18,19)]. The emphasis of thrombin in thrombolic processes has triggered an intense search for direct acting thrombin inhibitors as potential antithrombotic agents.

A number of novel thrombin inhibitors have been preclinically investigated and some are in clinical trials such as hirudin a potent and specific inhibitor of human thrombin, with a Ki of 2×10^{-14} M.[20,21] Additionally, a number of synthetic analogs containing the tripeptide (phe-pro-Arg) that interacts with the active site has been developed.[22-25] In contrast to Argatroban which is a reversible and competitive inhibitor (Ki=19nM), PPACK (D-ph-pro-Arg-CH2CL) binds covalently and inactivates the enzyme irreversibly.[22-25] Peptides containing a boronic acid derivative of arginine exhibit a similar spectrum of activities with high affinity approaching that of hirudin.[26] The in vivo antithrombotic activity of direct thrombin inhibitors has been examined in a wide variety of venous and arterial thrombosis models.[24,25] However, there have been few direct comparisons of the different types of specific thrombin inhibitors in the same model. Similarly, a number of novel GPIIb/IIIa antagonists including: 7E3 (chimeric monoclonal antibody for platelet GPIIb/IIIa), Integrelin (cyclic KGD), DMP 728 (cyclic RGD), MK383 (non-peptide) and many others have been preclinically and clinically investigated.[27]

The present study was undertaken to compare the antiplatelet effects of different platelet GPIIb/IIIa receptor antagonists versus different thrombin inhibitors.

MATERIALS AND METHODS

Reagents. Adenosine 5'-diphosphate (ADP), collagen from calf skin, epinephrine bitartrate, human alpha - thrombin (specific activity = 3000 NIH units/mg) and ristocetin were obtained from Sigma Chemical Company (St. Louis, MO). Arachidonate was purchased from Nu Chek Prep (Elysian, MN). ^{125}I-fibrinogen (26.5 μCi/mg) was obtained from DuPont NEN (Boston, MA). RGD and RGDS were obtained from Peninsula Laboratories (Belmont, CA). Other reagents used in this laboratory but not specifically mentioned were obtained from Sigma Chemical Company (St. Louis, MO). The PE labeled P-selectin monoclonal antibody was ordered from (Becton-Dickinson, Mountain View, DMP 728, Cyclic [D-2-aminobutyryl -N2-methyl -L-arginyl -glycyl- L-aspartyl -3-(aminomethyl-benzoic acid] methanesulfonic acid, was synthesized at The DuPont Merck Pharmaceutical. XL086 a cyclic [D-Lys-N2-methyl -L-arginyl -glycyl-L -aspartyl -3-(aminomethyl-benzoic acid] conjugated with fluorescin isothiocyanate (FITC) at the D-Lys position was synthesized at The DuPont Merck Pharmaceutical Co. (Wilmington, DE). DUP 714, XL111, XM648, and YY751 were also synthesized at The DuPont Merck Pharmaceutical (Wilmington, DE).

Platelet Aggregation (Light Transmittance) Assay. Venous blood was obtained from the arm of healthy human donors who were drug-and aspirin-free for at least two weeks prior to blood collection. Blood was collected into citrated Vacutainer tubes. The blood was centrifuged for 15 minutes at 150 x g at room temperature, and platelet-rich plasma (PRP) was removed. The remaining blood was centrifuged for 15 minutes at 1500 x g at room temperature, and platelet-poor plasma (PPP) was removed. For thrombin-induced aggregation studies washed platelets were used. Samples were assayed on an aggregometer (PAP-4 Platelet Aggregation Profiler), using PPP as the blank (100% transmittance). 200 μl of PRP was added to each micro test tube, and transmittance was set to 0%. 20 μl of various agonists including ADP, collagen, arachidonate, epinephrine, thrombin or their combination were added to each tube to initiate platelet aggregation. The aggregation profiles were plotted (% transmittance versus time). The results are expressed as % inhibition of agonist-induced platelet aggregation. For IC_{50} evaluation, the test compounds were added at various concentrations prior to the activation of the platelets.

Fibrinogen Receptor Binding Assay. Binding of ^{125}I-fibrinogen to platelets was performed as previously described[28] with some modifications as described below. Human PRP was applied to a Sepharose column for the purification of platelet fractions. Aliquots of platelets (5 x 10^8 cells) along with 1 mM calcium chloride were added to removable 96 well plates prior to the activation of the human gel purified platelets (h-GPP). Activation of h- GPP was achieved using ADP, collagen, arachidonate, epinephrine, and/or thrombin in the presence of the ligand, ^{125}I-fibrinogen. The ^{125}I-fibrinogen bound to the activated platelets was separated from the free form by centrifugation and then counted on a gamma counter. For an IC_{50} evaluation, the test compounds were added at various concentrations prior to the activation of the platelets.

Platelet Granular Secretion:

Measurement of Platelet Plasminogen Activator Inhibitor Release. Human PRP was prepared as previously described, divided into aliquots and incubated for 5 minutes at 37°C with thrombin (5 IU/ml). At the end of incubation, the aliquots were centrifuged for 15 minutes at 1500 x g and at room temperature to pellet the platelets. The plasma was removed, divided into duplicate aliquots, and frozen at -40°C until assayed for PAI antigen levels. Control samples of PRP (to measure baseline PAI-1 release) and PRP treated with 1% Triton X-100 (to measure total platelet PAI-1) were simultaneously prepared. PAI-1 antigen levels were measured using a sandwich ELISA with two murine monoclonal antibodies generated against PAI-1.[29]

Platelet Dense Granular Secretion of Serotonin. Human platelet-rich plasma (PRP) was prepared as previously described and washed according to the methods described by Derian and Friedman[30] and by Sage and Rink.[31] PRP was incubated with 200 nM of DMP 728 or saline for 15 min at 37°C, and then incubated with serotonin (0.5 mg/ml) for another 15 min at 37°C (to load the platelets with exogenous serotonin). After incubation, platelets were washed 2x with saline buffer and then resuspended. Triton X-100 (1%), thrombin (10 IU/ml) or saline was added to the platelet suspension for 15 min at 37°C, for the release of serotonin. The release reaction was then stopped by pelleting out the platelets. Aliquots from the supernatants were then injected onto the HPLC, employing the following method: Mobile Phase: 0.1 M Na Acetate, 0.1 M Na Citrate, 0.63 mM Na Octane Sulfonate, 0.15 mM EDTA, and 12% Methanol; Column: Nova-Pak C_{18}; Detection: Electrochemical detector employing +0.8 V vs.Ag/AgCl electrode; Flow: 1.0 ml/min. Human washed platelets not preloaded with serotonin were also stimulated in the absence or presence of the test agent with thrombin, arachidonic acid, or saline. The reaction was then terminated by pelleting out the platelets. The supernatant was injected onto the previously described HPLC system for determination of released serotonin.

Measurement of Platelet Intracellular Ca^{2+} Mobilization. Washed platelets used for the measurement of $[Ca^{2+}]_i$ mobilization were prepared according to procedures described by Derian and Friedman[30] and by Sage and Rink.[31] In brief, platelets from PRP were sedimented for 15 min at 800 x g and 20°C, then washed twice in a modified Tyrode's buffer containing BSA 0.35%, apyrase (20 ug/ml), PGE_1 (5 ng/ml) and EGTA (1 mM) at pH 6.5. Washed platelets were suspended in HEPES-buffered saline (HBS) (145 mM NaCl, 5 mM KCl, 1 mM $MgCl_2$, 10 mM glucose and 10 mM HEPES, pH 7.4) at a concentration of 8×10^8 cells/ml, and incubated with 4 mM Fura-2AM for 40 min at 20°C with gentle shaking. Platelets were then washed twice with HBS containing apyrase and EGTA (pH 6.5) and resuspended in HBS (pH 7.4) at a concentration of 2×10^8 cells/ml. Fluorescence measurements were recorded on a PTI Deltascan Fluorometer (South Brunswick, NJ) using settings of 339 nm for excitation and 500 nm for emission. The $[Ca^{2+}]_i$ mobilization assay was performed at 37°C. Results are expressed as % inhibition of the maximal response to thrombin (0.5 IU/ml).

Platelet Activation and immunofluorescent labeling: 5 μl of PRP (~10^6 cells) was diluted 10 fold with modified Hank's balanced salt solution (138 mM NaCl, 0.3 mM Na_2HPO_4, 4 mM $NaHCO_3$, 5 mM KCl, 0.3 mM KH_2PO_4, 2 mM $CaCl_2$, 10 mM HEPES). Platelets were activated with 10^{-4} M ADP for 10 min and then fixed with 2%

paraformaldehyde for another 10 min. To study the effect of the platelet GPIIb/IIIa antagonist, DMP 728 and hirudin on thrombin-induced degranulation, the platelets were pre-incubated with either DMP 728 or hirudin for 10 min prior to the addition of the thrombin. Alpha granule secretion in fixed platelets was measured by flow cytometry using an antibody for P-selectin, CD62-PE, an established marker of platelet a granules.[32] In addition, platelet GPIIb-IIIa was labeled using FITC-conjugated cyclic RGD peptide, XL086 (DuPont Merck Pharmaceutical Co.). After 20 min of incubation the samples were analyzed in the flow cytometer.

Flow cytometry: Platelets were analyzed in a FACScan flow cytometer (Becton Dickinson, Mountain View, CA). Forward scatter, side scatter and fluorescence signals were obtained by four decade logarithmic amplifiers.[32] Fluorescence signals is gated by appropriate forward and side scatter to exclude contaminating erythrocytes and debris. A platelet gate was set around events that was above the forward scatter threshold and 2,000 gated events were collected. Data acquired were stored on list mode and analyzed using LYSIS II software (Becton Dickinson, Mountain View, CA).

RESULTS

Effects on Thrombin-Induced Platelet Aggregation. Washed human platelets were used to compare the inhibitory effects of different thrombin inhibitors including: DUP 714, PPAK and hirudin versus different platelet GPIIb/IIIa antagonists including XL111, XM648 and YY751, on thrombin induced-platelet aggregation. All thrombin inhibitors specifically inhibited thrombin-induced platelet aggregation, but not the aggregation induced by any other agonists (Table 1). In contrast, the platelet GPIIb/IIIa antagonists tested inhibited platelet aggregation induced by any agonist with comparable IC_{50} ranges for any of the platelet GPIIb/IIIa antagonists tested against all agonists (Table 1). Linear RGDS demonstrated an anti-aggregatory efficacy against all agonists tested with an IC_{50} of 100-120 μM.

Table 1. Anti-aggregatory effects of different thrombin inhibitors versus GPIIb/IIIa antagonists against agonist-induced human platelet aggregation.

	IC_{50}(μM)					
	Thrombin Inhibitors			GPIIb/IIIa Antagonists		
Agonist	Hirudi	PPAK	SUP 714	XL111	XM648	YY751
Thrombin	0.07	0.12	0.15	ND	0.75	0.28
ADP	---	---	---	4.40	0.50	0.16
TEAC	---	---	---	4.00	0.85	0.30

Thrombin was used at 0.5 IU/ml; Adenosine diphosphate (ADP) at 100 uM; Collagen at 20 ug/ml; TEAC a mixture of thrombin (T) 0.001 IU/ml, epinephrine (E) at 10 uM, ADP (A) at 10 uM, and Collagen (C) at 20 ug/ml, -- = Inactive at concentrations up to 10 uM. ND = non-determined.

Effects on Thrombin-Induced Fibrinogen Binding to Gel Purified Platelets: In human gel purified platelets (h-GPP), the inhibitory effect of the different thrombin inhibitors or the different GPIIb/IIIa antagonists against thrombin-induced ^{125}I-fibrinogen binding to h-GPP was determined. In purified platelet systems, DUP 714 (IC_{50} = 3 nM) was somewhat more potent in inhibiting thrombin-induced fibrinogen binding to GPP than hirudin (IC_{50} = 20 nM). A very sharp submaximal range of inhibitory responses for both DUP 714 and hirudin was noted. In contrast, the platelet GPIIb/IIIa antagonists tested inhibited platelet aggregation in response to all agonists with comparable IC_{50} ranges for any antagonists tested against all different agonists (Table 2). Linear RGDS inhibited ^{125}I-fibrinogen binding to h-GPP against all agonists with an IC_{50} of 10-15 μM.

Table 2. Effects of different thrombin inhibitors versus GPIIb/IIIa antagonists against agonist-induced human platelet fibrinogen binding

	IC_{50}(μM)					
	Thrombin Inhibitors			GPIIb/IIIa Antagonists		
Agonist	Hirudi	PPAK	SUP 714	XL111	XM648	YY751
Thrombin	0.02	0.005	0.003	2.30	0.038	0.007
AEAAC	---	---	---	2.00	0.026	0.006

Thrombin was used at 0.5 IU/ml; AEAAC a mixture of ADP (A) at 10 uM, epinephrine (E) at 10 uM, Arachidonic acid (AA) at 10 μM, and Collagen (C) at 20 ug/ml.
-- = Inactive at concentrations up to 10 uM.

Effects on Thrombin-Induced Platelet PAI-1 Secretion

In human PRP, the effect of various thrombin inhibitors are GPIIb/IIIa receptor antagonists on platelet PAI-1 release was determined. A concentration dependent inhibitory effect of DUP 714, PPAK and hirudin against thrombin induced platelet PAI-1 secretion was demonstrated (Table 3). All thrombin inhibitors platelet PAI-1 release were induced by thrombin, but not by any other agonist (Table 3). In contrast, none of the platelet GPIIb/IIIa antagonists showed any significant effects on thrombin mediated platelet PAI-1 release (Table 3).

Table 3. Effects of different thrombin inhibitors versus GPIIb/IIIa antagonists against agonist-induced human platelet PAI-1 release

	IC_{50}(μM)					
	Thrombin Inhibitors			GPIIb/IIIa Antagonists		
Agonist	Hirudi	PPAK	SUP 714	XL111	XM648	YY751
Thrombin	0.2	0.5	0.9	---	---	---

Thrombin was used at 0.5 IU/ml, -- = Inactive at concentrations up to 10 μM.

Effects on Thrombin-Induced Platelet Serotonin Secretion

The various inhibitors were next evaluated for these effects on thrombin-induced platelet secretion in PRP. A concentration dependent inhibitory effect of DUP 714, PPAK and hirudin against thrombin induced platelet serotonin secretion was demonstrated (Table 4). All thrombin inhibitors inhibited platelet serotonin release induced by thrombin but not by any other agonist (Table 4). Platelet GPIIb/IIIa antagonists did not show any significant effects on thrombin mediated platelet serotonin release (Table 4). Furthermore, the effect of different concentrations of hirudin, PPAK, RGDS, and GPIIb/IIIa monoclonal antibody on thrombin (0.5 IU/ml) - induced PAI-1 and serotonin release from human platelets was determined. A concentration dependent inhibitory effects were demonstrated with the direct thrombin inhibitors as compared to a lack of any significant effects of the GPIIb/IIIa antagonists tested (Figure 1).

Table 4. Effects of different thrombin inhibitors versus GPIIb/IIIa antagonists against agonist-induced human platelet serotonin release.

	$IC_{50}(\mu M)$					
	Thrombin Inhibitors			GPIIb/IIIa Antagonists		
Agonist	Hirudi	PPAK	SUP 714	XL111	XM648	YY751
Thrombin	0.2	0.3	0.6	---	---	---

Thrombin was used at 0.5 IU/ml, -- = Inactive at concentrations up to 10 uM.

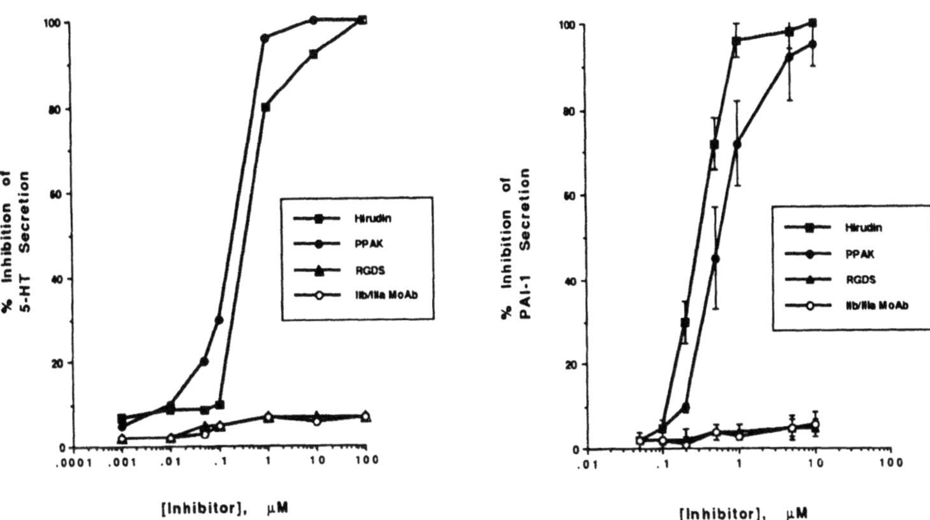

Figure 1. Effects of different concentrations of hirudin, PPAK or RGDS, GPIIb/IIIa monoclonal antibody on thrombin (0.5 IU/ml) - induced 5-HT (serotonin) and PAI-1 release from human platelets. Data are expressed as the mean ($\pm$ SEM) percent inhibition (n = 3).

Effects on Thrombin-Induced Intracellular Calcium ($[Ca^{++}]_i$) Mobilization

Washed platelets loaded with the fluorescent calcium chelator Fura-2AM were used to compare the inhibitory effects of different thrombin inhibitors and platelet GPIIb/IIIa antagonists against thrombin (0.5 IU/ml) induced intracellular calcium mobilization (Table 5). As with the other thrombin-induced responses, thrombin inhibitors inhibited platelet intracellular calcium mobilization induced by thrombin but not by any other agonist (Table 5). Again, platelet GPIIb/IIIa antagonists did not show any significant effects on thrombin mediated platelet $[Ca^{++}]_i$ release (Table 5). Comparative tracings are shown illustrating the inhibitory effects of hirudin on thrombin (0.5 IU/ml) - induced platelet aggregation and platelet $[Ca^{++}]_i$ release as compared to the effects of GPIIb/IIIa antagonist under the same conditions (Figure 2).

Table 5. Effects of different thrombin inhibitors versus GPIIb/IIIa antagonists against agonist-induced human platelet serotonin release.

	IC_{50}(μM)					
	Thrombin Inhibitors			GPIIb/IIIa Antagonists		
Agonist	Hirudi	PPAK	SUP 714	XL111	XM648	YY751
Thrombin	0.2	0.3	0.6	---	---	---
Arachidonic Acid	---	---	---	---	---	---

Thrombin was used at 0.5 IU/ml and Arachidonic acid at 100 μM.
-- = Inactive at concentrations up to 10 uM, ND = non-determined.

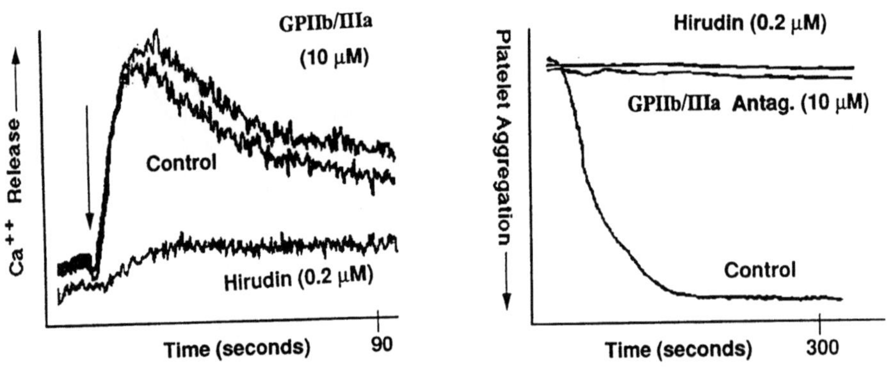

Figure 2. Representative tracings demonstrating the effects of hirudin versus GPIIb/IIIa antagonist on thrombin (0.5 IU/ml) - induced platelet aggregation and $[Ca^{++}]_i$ release from human platelets.

Flow Cytometric Analysis: Data were analyzed from dual parameter histograms of XL086 and P-selectin binding to platelets. The level of "degranulation" was measured by the percentage of cells labeled by both P-selectin and XL086. Quadrants were set with non-activated platelet sample labeled with XL086 and P-selectin. Without ADP activation, only 1% of the platelets sampled showed P-selectin staining above baseline (Fig. 3A) and 98.9% of the total platelets sampled were labeled by XL086, i.e. platelets in the lower right quadrant, with mean fluorescent intensity of 381. ADP activation increased P-selectin positive cells from 1% to 46.3%. Activation also increased XL086 labeling per platelet, as indicated by the increased in mean fluorescent intensity of XL086 by 21% (Fig. 3B). This increase in platelet XL086 binding probably resulted from an increase in GPIIb/IIIa receptors on the platelet surface. Addition of GPIIb/IIIa receptor antagonist, YY751 displaced XL086 binding as indicated by the decreased in fluorescence intensity to 73. However, YY751 did not decrease the percentage of P-selectin positive cells; this actually increased from 46.3% to 60.3% (Fig. 3C). In contrast, hirudin inhibited thrombin-induced expression of platelet GPIIb/IIIa as well as P-selectin (Fig. 3D).

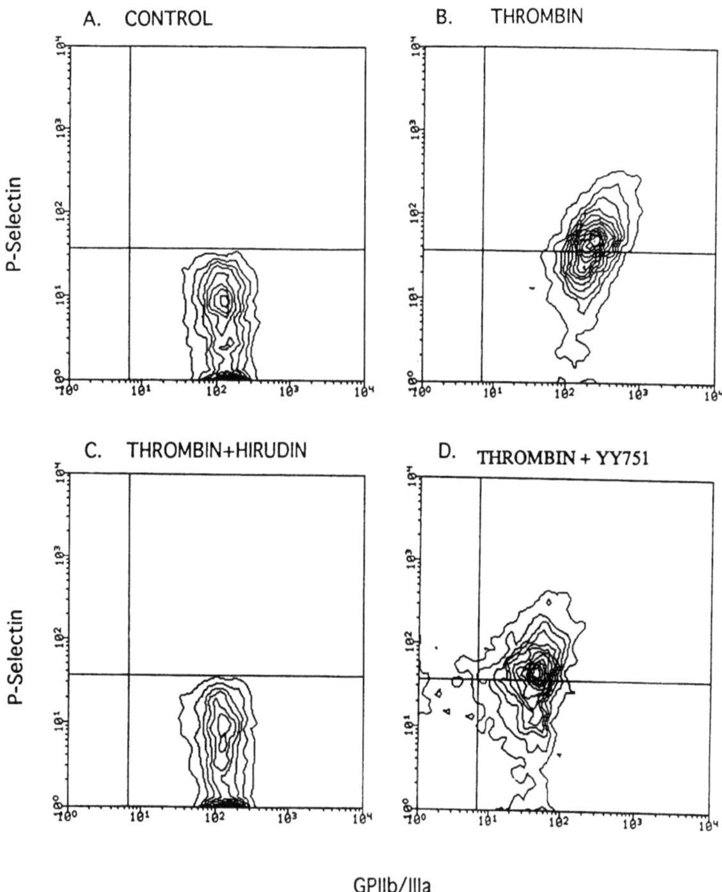

Figure 3. Dual parameter contour plot showing XL086 (FL1, x-axis) and P-selectin (FL2, y-axis) binding to platelets. (A) Control ; (B) thrombin activated. Note the increase in cell population in the upper right quadrant indicating an increase in P-selectin positive cells; (C) YY751 treated and thrombin activated sample. Note the cell population only shifted to the left indicating lower XL086 binding with no decrease in P-selectin positive cells. (D) Hirudin treated and thrombin activated sample. Note cell population shifted to the left indicating lower XL086 binding with a decrease in P-selectin positive cells.

DISCUSSION

Among the key factors leading to thrombosis are vessel wall injury, platelet activation and activation of the coagulation cascades. Initially platelets attach to subendothelial cells, and subsequently spread, expressing functional receptors for adhesive molecules. The process of platelet recruitment is mediated by various agonists including: thrombin, generated on phospholipid platelet surfaces; ADP, secreted from platelet granules; and thromboxane A2, generated from within the platelets. The recent appreciation of thrombin's key role in promoting platelet-dependent, arterial thrombosis suggests the importance of evaluating the antiplatelet effects of direct acting thrombin inhibitors as compared to platelet GPIIb/IIIa antagonists which act at the final common pathway for platelet aggregation.[11,12] In this study, we have compared the antiplatelet activity of various thrombin inhibitors including DUP 714, a member of a new class of potent synthetic thrombin inhibitors[22,23,26,] of hirudin, the 65 amino acid polypeptide which represents the most potent and selective thrombin inhibitor yet identified[17] and of PPAK, an irreversible thrombin inhibitor,[24] with various GPIIb/IIIa receptor antagonists including XL111, XM648, and YY751. For these comparisons, we have evaluated a number of different platelet functions including thrombin-induced platelet aggregation, mobilization of intracellular calcium, fibrinogen binding, serotonin and PAI-1 release. PAI-1 and serotonin secretion were studied because of possible role in physiological and pathological thromboembolism and in vasospastic episodes post-thrombolysis.[2,26] Consequently, the effects of a thrombin inhibitor on platelet PAI-1 and serotonin release may have clinical relevance. Our results revealed that as expected, IIb/IIIa antagonists prevented aggregation against all platelet agonists in contrast to the thrombin inhibitors which blocked only thrombin-induced aggregation. However, thrombin-induced mobilization of intracellular calcium, serotonin release and PAI-1 release, while blocked by thrombin inhibitors, were not affected by GPIIb/IIIa antagonists. These results suggest that platelet secretion is from GPIIb/IIIa receptor binding. Furthermore, these data suggest thrombin inhibitors may have potential for clinical use in combination with GPIIb/IIIa receptor antagonists.

The effects of the direct thrombin inhibitor, and 7E3 the platelet GPIIb/IIIa antagonist, on the prevention of thrombosis and rethrombosis after coronary thrombolysis in a chronic canine model has been recently documented. In contrast to the GPIIb/IIIa antagonist tested in this model which was very effective in preventing thrombosis and rethrombosis, the direct thrombin inhibitor provided little improvement in the incidence of reocclusion and mortality in the same model (Table 6). In that study, animals treated with hirudin exhibited rebound reocclusion after the discontinuation of the infusion.[33,34] Theroux et al.[35] recently reported similar phenomenon with heparin therapy where a significant number of patients experienced reactivation of unstable angina episodes. Furthermore, Gold et al.[36] reported evidences of rebound coagulation with direct thrombin inhibitor in unstable angina. Additionally, oscillation flow or cyclic flow variations occurred in the hirudin treated animals but not in the 7E3 treated group. These cyclic flow variations which occur clinically post-tPA may contribute to myocardial damage, persistent reocclusion, arrythmic events. Cyclic flow variations are intensified under stressful conditions associated with catecholamine release leading to platelet hyperreactivity.[2] Thus, suppressing cyclic flow variations with thrombolysis regimens may be as important as the achievements of a vessel wall patency for successful clinical outcome. In the Folts' model, different platelet GPIIb/IIIa antagonists demonstrated maximal efficacy different platelet GPIIb/IIIa antagonists demonstrated maximal efficacy in inhibiting ex vivo platelet aggregation, totally preventing CFR and maintaining coronary flow and coronary arterial patency.[37,38] These studies suggest the potential of GPIIb/IIIa antagonists in unstable angina. In contrast, aspirin (0.5 - 5.0 mg/kg, IV) was shown to be marginally effective in this model and ineffective against epinephrine re-induced CFR.[38] Hence, it is anticipated that GPIIb/IIIa antagonists might be more effective than aspirin in unstable angina, since platelet activation could be re-induced following elevation of catecholamines, rupture of atherosclerotic plaques, or progression of atherosclerotic lesions. Furthermore, maximal efficacy against epinephrine re-induced CFR was demonstrated with other GPIIb/IIIa antagonists such as SK&F 106760.[38]

Table 6. Comparative Antithrombotic Effects of Thrombin Inhibitor Versus Platelet GPIIb/IIIa Antagonists.[33,34,37,38,42,43]

Antithrombotic Profiles	Thrombin Inhibitors Hirudin	GPIIb/IIIa 7E3	Antagonists SK&F
1. Unstable angina-type model (Folts)			
• Prevention or reduction of CFR	+	+	+
• Prevention or reduction of epinephrine-reinduced CFR	-	+	+
2. Chronic coronary artery electrolytic thrombosis model +r-tPA			
• Improved artery patency/survival	+	+	ND
• Suppression of cyclic flow variations	-	+	ND
• Rebound reocclusion	+	-	ND

CFR = cyclic flow reduction. The models used in the above table were in dogs. ND = Non-determined.

These results suggest greater efficacy for GPIIb/IIIa antagonists as compared to aspirin. Furthermore, it has been also reported that inhibition of thrombin activity alone does not decrease the reocclusion rate following r-tPA thrombolysis in coronary artery thrombosis model in dogs.[39] It is likely that in most circumstances where platelet activation occurs, thrombin is not the sole mediator involved. In man, thromboxane A2 appears particularly important because inhibition of its formation by aspirin has been shown to be effective in a wide variety of thromboembolic disorders.[1,6,9] In experimental models of thrombosis, combined use of thrombin inhibitors with thromboxane A2 inhibitors or aspirin has been shown to be more effective than either alone.[40,41]

Finally, specific inhibitors of thrombin or platelet GPIIb/IIIa receptors are being evaluated in patients with unstable angina, post-angioplasty or thrombolysis. Initial results appear to be promising. However, in the case of short-acting thrombin inhibitors, a three-fold increase in thrombin-antithrombin complexes after discontinuation of the treatment has been shown.[36] This raises the possibility of a rebound thrombosis following drug withdrawal.

Since the platelet GPIIb/IIIa receptors represent the final common pathway for platelet aggregation, a potent and specific GPIIb/IIIa antagonist may prove to be highly effective antithrombotic strategy.[42] However, the combination of other antiplatelet agents such as a safe and effective thrombin inhibitor or aspirin with a platelet GPIIb/IIIa antagonist may prove to be the ultimate antithrombotic strategy in humans because of the lack of platelet anti-secretory efficacy of GPIIb/IIIa antagonists which could be compensated with other antiplatelet approaches. On the other hand, the lack of effect of a specific GPIIb/IIIa antagonist on platelet functions other than platelet aggregation might be an important safety aspect for the maintenance of the hemostatic balance.

REFERENCES

1. P.D. Hirsh, L.D. Hillis, W.B. Campbell, B.G. Firth, J.T. Willerson, Release of prostaglandins and thromboxane into the coronary circulation in patients with ischemic heart disease, *N. Engl. J. Med.* 304:685-691 (1981).
2. M.D. Rubenstein, R.T. Wall, D.S. Bain, D.C. Harrison, Platelet activation in clinical coronary artery disease and spasm, *Am Heart J* 102:363-367 (1981).
3. T.C. Smitherman, M. Milam, J. Woo, J.T. Willerson, E.P. Frenkel, Elevated beta thromboglobulin in peripheral venous blood of patients with acute myocardial ischemia: direct evidence for enhanced platelet reactivity in vivo, *Am J Cardiol* 48:395-402 (1981).
4. D.J. Fitzgerald, L. Roy, F. Catella, G.A. Fitzgerald, Platelet activation in unstable coronary disease, *N Eng J Med* 315:983-989 (1986)
5. V. Fuster, P.M. Steele, J.H. Chesebro, Role of platelets and thrombosis in coronary atherosclerotic disease and sudden death, *J Am Coll Cardiol* 5:175B-184B (1985).
6. J.T. Willerson, P. Golino, J. Eidt, W.B. Campbell, L.M. Buja, Specific platelet mediators and unstable coronary artery lesions. Experimental evidence and potential clinical implications, *Circulation* 198-205 (1989).
7. M.J. Davies, A.C. Thomas, Plaque fissuring-the cause of acute myocardial infarction, sudden ischemic death, and cresendo angina, *Br Heart J* 53: 363-373 (1985).
8. C.W. Hamm, W. Bliefield, W. Kupper, R.L. Lorenz, P.C. Weber, W. Wober, Biochemical evidence of platelet activation in patients with persistent unstable angina, *J Am Coll Cardiol* 10: 998-1004 (1987).
9. B. Ashby, J.L. Daniel, J.B. Smith, Mechanisms of platelet activation and inhibition. Platelet in Health & Disease, 4 (1):1-26 (1990).
10. E.K.O. Kruithof, C. Tran-Thang, F. Bachman, Studies on the release of a plasminogen activator inhibitor by human platelets. *Thromb Haemostas* 55: 201 - 205 (1986).
11. R. Pytele, M.S. Pierschbacher, M.H. Ginsberg, E.F. Plow, E. Ruoslahti, Platelet membrane glycoprotein IIb/IIIa: member of a family of RGD specific adhesion receptors, *Science* 231: 1559-1562 (1986).
12. D.R. Philips, I.F. Charo, R.M. Scarborough, GPIIb/IIIa: The responsive integrin. *Cell* 65:359-362 (1991).
13. S.E. D'Souza, M.H. Ginsberg, T.A. Burke, E.F. Plow, The Ligand Binding Site of the Platelet Integrin Receptor GPIIb-IIIa Is Proximal to the Second Calcium Binding Domain of Its a Subunit. *J Biol Chem* 265:3440-3446 (1990).
14. J.F. Eidt, P. Allison, S. Noble, J. Ashton, P. Golino, J. McNatt, L.M. Buja, J.T. Willerson, Thrombin is an important mediator of platelet aggregation in stenosed canine coronary arteries with endothelial cell injury, *J. Clin. Invest.* 84:18-27 (1989).
15. E. Saltiel, A. Ward, Ticlopidine. A review of its pharmacodynamic and pharmacokinetic properties and therapeutic efficacy in platelet-dependent disease states, *Drugs* 34:222-262 (1987).
16. P. Golino, J.H. Ashton, P. Glas-Greenwalt, J. McNatt, L. M. Buja, J.T. Willerson, Mediation of reocclusion by thromboxane A2 and serotonin after thrombolysis with t-PA in a canine preparation of coronary thrombosis, *Circulation* 77:678-684 (1988).
17. J.M. Walenga, R. Pifarra, J. Fareed, Recombinant hirudin as an antithrombotic agent, *Drugs of The Future* 15(3):267-270 (1990).
18. L.A. Harker, and M. Gent, Antiplatelet agents in the management of thrombotic disorders, *in:* Hemostasis and Thrombosis: Basic Principles and Clinical Practice. Colman RW, Hirsh J, Marder VJ, Salzman EW, eds. 3rd edn. Philadelphia: JP Lippincott, 1993; 1506-1513.
19. A.B. Kelly, U.M. Marzec, W. Krupski, et al., Hirudin interruption of heparin-resistant arterial thrombus formation in baboons, *Blood* 77: 1006-1012 (1991).
20. Markwardt F, Hirudin as an inhibitor of thrombin, *Methods Enzymol* , 69:924-932 (1970).
21. M.D. Talbot, Biology of recombinant hirudin (CGP 39393): a new prospect in the treatment of thrombosis, *Semin Thromb Haemostas* , 15:293-301 (1989).
22. C. Kettner, L. Mersinger, R. Knabb, The selective inhibition of thrombin by peptides of boroarginine, *J Biol Chem* 265:18289-18297 (1990).
23. A.T. Chiu, S.A. Mousa, L.J. Pease, W.A. Roscoe, J.M. Bozarth, T.M. Reilly, R.D. Smith, P.B.M.W.M. Timmermans, Inhibition of the thrombin-platelet reactions by DUP 714, *Biochem Biophys Res Commn* 179:1500-1508 (1991).
24. S.R. Hanson, L.A. Harker, Interruption of acute platelet-dependent thrombosis by the synthetic antithrombin D-phenyalanyl-L-propyl-L- arginylchloromethyl ketone. *Proc Natl Acad Sci USA* , 85: 3184-3188 (1988).

25. I. Jang, H.K. Gold, A.A. Ziskind, R.C. Leinbach, J.T. Fallon, D. Collen, Prevention of platelet-rich arterial thrombosis by selective thrombin inhibition, *Circulation* 81:219-225.
26. T.M. Reilly, R.M. Knabb, S.M. Hassell, J.M. Bozarth, M.S. Forsythe, M.C. Mayo, A.L. Racanelli, S.A. Mousa, Effect of thrombin inhibitors on platelet functions: Comparative analysis of DUP 714 and hirudin, *Blood Coagulation and Fibrinolysis* 3: 513-517 (1992).
27. N.S. Cook, G. Kottirsch, H-G. Zerwes, Platelet glycoprotein IIb/IIIa antagonists, *Drugs of the Future* 19:135-159 (1994).
28. J.S. Bennet, G.O. Vilaire, Exposure of platelet fibrinogen receptors by ADP and epinephrine, *J Clin Invest* 64: 1393 -1401 (1979).
29. S.A. Mousa, J. Bozarth, M. Forsythe, P. Tsao, L. Pease, T. Reilly, Role of platelet GPIIb/IIIa receptors in the modulation of PAI-1, *Life Sciences* 54:1155-1162 (1994).
30. C.K. Derian, P.A. Friedman, Effect of ticlopidine ex vivo on platelet intracellular calcium mobilization, *Thromb Res* 50: 65 - 76 (1988).
31. S.O. Sage and T.J. Rink. Kinetic differences between thrombin-induced and ADP-induced calcium influx and release from internal stores in fura-2 loaded human platelets. *Biochem Biophys Res Commn* 136:1124-1129 (1986).
32. K.A. Ault, M.H. Rinder, J.G. Mitchell, C.S. Rinder, C.T. Lambrew, R.S. Hillman, Correlated measurement of platelet release and aggregation in whole blood. *Cytometry* 10: 448-455 (1989).
33. W.E. Rote Rote, D-X Mu, Bates ER, et al: Prevention of rethrombosis after coronary thrombolysis in a chronic canine model . I. adjunctive therapy with monoclonal antibody 7E3-F(ab')2, J Cardiovas Pharmacol 1993; 23:194 -202.
34. W.E. Rote Rote, D-X Mu, E.R. Bates, et al: Prevention of rethrombosis after coronary thrombolysis in a chronic canine model . II. adjunctive therapy with r-Hirudin, *J Cardiovas Pharmacol* 23:203 -211 (1993).
35. P. Theroux, D. Waters, J. Lam, M. Junean, J. McCans, Reactivation of unstable angina after discontinuation of heparin, 327:141-145 (1992).
36. H.K. Gold, F.W. Torres, H.D. Garabedian, W. Werner, I. Jang, et al., Evidence of a rebound coagulation phenomenon after cessation of a 4-hour infusion of a specific thrombin inhibitor in patient with unstable angina. *J. Am. Coll. Cardiol.* 21:1039-1047 (1993).
37. S. Mousa, J. Bozarth, M. Forsythe, S. Jackson, A. Leamy, M. Diemer, R. Kapil, R. Knabb, M. Mayo, S. Pierce, W. De Grado, M. Thoolen, T. Reilly, Antiplatelet, antithrombotic efficacy of DMP 728, a novel platelet GPIIb/IIIa receptor antagonist. *Circulation* 89: 3-12 (1994).
38. A.J. Nichols, R.R. Ruffolo, W.F. Huffman, G. Poste, J. Samanen, Development of GPIIb/IIIa antagonists as antithrombotic drugs. Trends in Pharmacological Sciences 13(11):413-417 (1992).
39. F.A. Nicolini, G. Rios, P. Lee, K. Kottke-Marchant, E.J. Topol, Inhibition of thrombin activity alone does not decrease the reocclusion rate following r t-PA-induced thrombolysis in canine coronary artery thrombosis, *Throm Heomost* 69:992 (1993).
40. R.J. Clarke, G. Mayo, G.A. Fitzgerald, D.J. Fitzgerald, Combined administration of aspirin and a specific thrombin inhibitor in man, *Circulation* 83:1510-1518, 1991.
41. T. Yasuda, H.K. Gold, H. Yaoita, R.C. Leinbach, et al., Comparison effects of aspirin, a synthetic thrombin inhibitor and a monoclonal antiplatelet glycoprotein IIb/IIIa antibody on coronary artery reperfusion, reocclusion and bleeding with recombinant tissue-type plasminogen activator in a canine preparation. *J. Am. Coll. Cardiol.* 16:714-722, 1990.
42. B.S. Coller, J.D. Folts, L.E. Scudder, S.R. Smith, Antithrombotic effect of a monoclonal antibody to the platelet glycoprotein IIb/IIIa receptor in an experimental animal model, *Blood* 68: 783-786, 1986.
43. J.D. Folts, Experimental arterial platelet thrombosis, platelet inhibitors, and their possible clinical relevance, *Cardiovascular Rev Rep* 3 : 370-382, 1982.

BIOSYNTHESIS OF DOCOSANOIDS BY HUMAN PLATELET: CARDIOVASCULAR PROPERTIES

John W. Karanian, Hee Yong Kim and Norman Salem, Jr.

Laboratory of Membrane Biochemistry and Biophysics, DICBR, NIAAA, NIH, 12501 Washington Avenue, Rockville, MD 20852

INTRODUCTION

Arachidonic acid (20:4n6) and eicosapentaenic acid (20:5n3) are lipoxygenated to the well known 20-carbon hydroxylated eicosanoids, 12-OH-20:4n6 (12-HETE) and 12-OH-20:5n3 (12-HEPE), respectively. A 13-OH-18:2n6 is also produced from linoleic acid (18:2n6). However, very little is known about the biological activity of hydroxylated 22-carbon docosanoids on cardiovascular cell function and the effects of these products on enzymatic activity. The metabolism of docosahexaenoic acid (22:6n3) to hydroxylated derivatives has been observed in several mammalian tissues and cells including basophils [1], macrophages[2], retina[3], liver[4], and brain[5,6]. In the human platelet, an 11-, 14- and 17-hydroxy derivative are formed *in vitro* from exogenously administered 22:6n3[7,8,9]. Similarly, the human platelet metabolizes docosapentaenoic acid (22:5n3) to a pair of isomeric 11- and 14-hydroxy fatty acids[10] and the 22:5n6 to a 14-hydroxy fatty acid[11]. Metabolism of the essential fatty acids linoleic (18:2n6) and linolenic acid (18:3n3) to the hain polyunsaturated fatty acids and their lipoxygenation to the various hydroxylated fatty acids is shown in Figure 1.

Whether these metabolites are produced from endogenous substrate under more physiological conditions[12] and are of physiological significance in the human is still an open question[13]. However, the hydroxylated metabolites of 22:5n3 and 22:6n3 are biologically active in a number of cell preparations[14-17]. It is proposed that the hydroxylated metabolites may be physiologically active in platelet-vascular smooth muscle and other cell interactions [17,18]. A report by Sinzinger[19] on the occurrence of myocardial infarct in young patients with platelet 12-lipoxygenase deficiency points towards a significant involvement of hydroxylated derivatives in the regulation of platelet function. These lipoxygenase metabolites may interfere with TXA2/PGH2 receptors[20]. In addition, docosanoids are known to alter the endogenous production of eicosanoids[21]. These effects may directly result in changes in cell function. Moreover, nutritional modification of the precursor fatty acyl pool with n-3 or n-6 dietary supplementation alters the eicosanoid-docosanoid profile which may result in altered receptor and cell function[22,23].

The biosynthesis and preparation of these 22-carbon hydroxylated derivatives is necessary since these derivatives are not commercially available. In this paper we present a reliable purification and quantification method used to characterize the metabolism and production of the hydroxylated derivatives of 22:5n3, 22:6n3, 22:5n5 and 22:5n6 from mammalian platelets. Their biological properties in platelet and vascular smooth muscle cell function will be discussed.

Cardiovascular Disease 2
Edited by L.L. Gallo, Plenum Press, New York, 1995

MATERIALS AND METHODS

Preparation of Washed Platelet Suspensions

Venous blood was obtained from human, monkey, dog and rabbit and truncal blood was obtained from guinea pig and rat and collected into vials containing acidic citrate-dextrose, 5:1 v/v (86 mM sodium citrate, 65 mM citric acid and 111 mM dextrose). After centrifugation for 20 min at 250xg, platelet-rich plasma was removed and immediately centrifuged at 1000xg for 10 minutes. The platelet pellet was then resuspended in 0.9% NaCl (containing 0.01 M Na_2EDTA) and recentrifuged at 650xg for 10 minutes. The pellet was resuspended in Tris-HCl (0.2 M, pH 7.5, 0.01 M Na_2EDTA) and adjusted to 0.1-1.6 x 10^9 cells/ml. The platelet count was verified using an Electrozone /Celloscope (Particle Data Inc) or a hemocytometer. In some cases, platelets were collected from human donors by platelet pheresis[24]. The leukocyte/platelet ratio of the preparation was less than 1/1000 in all experiments as verified by a complete blood count after staining.

Incubation of Platelets

Platelet suspensions (0.1-1.6x10^9 cells/ml) were incubated for up to 120 min with shaking at 37°C in the presence of 0.25 uCi of 1-[^{14}C]-22:6n3 (4.4 uM, 56.9 mCi/mmol, NEN) that was brought up to 10 uM with respect to the fatty acid concentration by the addition of the non-labeled compound (5.6 uM, Nu Check Prep). Alternatively, for preparative purposes, 100 uM of the non-radiolabeled 22:5n3, 22:6n3, 22:5n5 and 22:5n6 were incubated with 1.5 x 10^9 cells/ml. The 17-OH-22:6n3 was prepared by soybean lipoxygenase treatment of 22:6n3 followed by reduction of the hydroperoxide intermediate with $NaBH_4$[25]. The fatty acids 22:5n5 and 22:6n3 were obtained from the Check

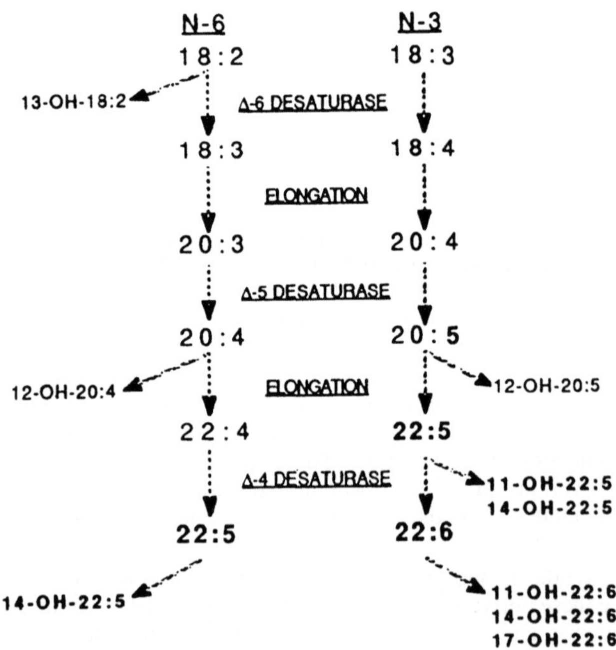

Figure 1. Essential fatty acid metabolism of cis-linoleic (n-6) and alpha-linolenic acid (n-3) to long chain polyunsaturated fatty acids and their lipoxygenation to hydroxylated metabolites

(Elysian, MN). 22:5n5 was chemically synthesized by chain elongation of the methyl end of 20:5n3. The 22:5n3 was obtained from Nippon Suisan Kaisha Ltd (Tokyo, Japan). The 22:5n6 was chemically synthesized (Deva Chemical) and purified by RP-HPLC in our laboratories.

In order to determine whether product formation was enzyme-dependent, the enzyme inhibitors 5,8,11,14-eicosatetraynoic acid (ETYA, 1.0 uM Biomol), indomethacin (0.5 mM, Sigma), aspirin (5.0 mM, Sigma), NDGA (50 uM, Biomol), baicalein (10 uM, Biomol), caffeic acid (20 uM, Biomol), esculetin (10 uM, Biomol), piroprost (50 uM, UpJohn) and SKF 525-A (50 uM, Biomol) were preincubated with the platelet suspension for 10 min prior to the addition of the fatty acid substrate. A control was established by boiling the platelet suspension for 30 min prior to assay. All platelet reactions were terminated immediately following the incubation period by acidification to pH 3.2 with formic acid in the presence of the antioxidant BHT (50 ug/ml).

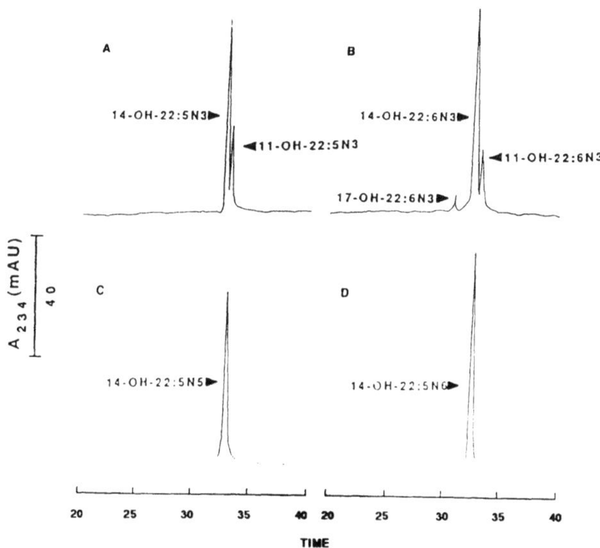

Figure 2. Reversed phase UV chromatogram of the hydroxylated fatty acids (2% of total sample) formed by incubating 100uM 22:5n3 (A), 22:6n3 (B), 22:5n5 (C) and 22:5n6 (D) with human platelets (1.6 x 10^9 cells/ml).

Extraction and Purification

The fatty acid metabolites were extracted twice with equal volumes of dichloromethane. The extracts were evaporated under nitrogen, dissolved in methanol and separated by reversed phase chromatography using an HP-1090 HPLC system. A 5 u Axxiom-ODS column (4.6mm x 25 cm) was used with a mobile phase initially consisting of 0.1 M aqueous ammonium acetate, pH 7.0, and methanol (40/60) and then ramped linearly over a 50 min period to 15/85. The metabolites of interest eluted within this period. The flow rate was 0.8 ml/min. The eluant was passed through a diode array UV detector.

To further purify the monohydroxy fatty acid fractions produced in the preparative experiment and to verify their overall recovery, each fraction was rechromatographed by reversed phase HPLC. The recovery of hydroxy fatty acid in this case was 80% or greater. The purified hydroxy fatty acids were quantified using UV absorbance at 235 nm assuming the molar extinction coefficient (E=30,000)[26]. The metabolites were stored in methanol, under nitrogen at -70°C.

Statistics

All the results were evaluated using analysis of variance and Students t-test for unpaired samples. A P value of <0.05 was considered statistically significant.

RESULTS

Chemical Characterization

In Figure 2, typical reversed phase UV chromatograms are shown for the preparations incubated with 100 uM 22:6n3, 22:5n3, 22:5n5 and 22:5n6 for 20 min. The chromatograms show the hydroxy fatty acids of interest eluting bewteen 25 and 35 min. Maximum absorbance occurred at 234 nm for these compounds, which is consistent with a conjugated diene structure.

Gas chromatography-mass spectrometric analysis was performed on the methyl ester-trimethylsilyl ether derivatives of the 22-carbon metabolites produced by human platelet. Electron-ionization mass spectra of these derivatives exhibited fragment peaks at m/z 321 and 281 which are characteristic for the 14-OH and 11-OH 22:6n3 positional isomers, respectively. The spectra of the isolated isomers were similar to those originally published for human platelets[7,10,11]. The structure of 17-OH-22:6n3 produced from platelet lipoxygenase was also confirmed by GC/MS and contained the expected fragment ion at m/z 361[9].

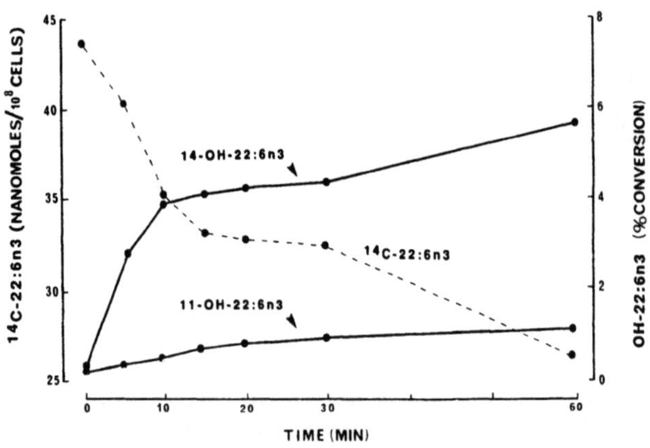

Figure 3. Effect of incubation time on production of 11- and 14-OH-22:6n3 from ^{14}C-22:6n3.

Species Comparison and Localization of OH-22:6n3

All mammalian platelets tested produced 14-OH-22:6n3, but to varying extents (Table 1). The rank order of 14-OH-22:6n3 yield from the various platelet suspensions incubated with 10 uM ^{14}C-22:6n3 was human > rat > monkey > dog > guinea pig > rabbit. An appreciable amount of 11-OH-22:6n3 was detected only in the human platelet incubate. By pelleting the human platelet suspension (1000xg for 10 min) immediately following the incubation period, it was possible to determine the percentage of total radioactive OH-22:6n3 in the platelet compared to that in the supernatant: 54±5% of recoverable OH-22:6n3 was present in the platelet pellet extract, whereas the remaining 46±6% was present in the supernatant (n=4).

Enzyme Inhibitor Effects on OH-22:6n3 Production by Human Platelet

The effect of various enzyme inhibitors on OH-22:6n3 production was determined and is summarized in Table 2. The boiled control showed no 22:6n3 conversion to OH-

22:6n3. The cyclooxygenase inhibitors tested (indomethacin and aspirin) had no effect on OH-22:6n3 production, whereas a specific 12-lipoxygenase inhibitor, baicalein[26], completely blocked OH-22:6n3 production.

Parameters Influencing OH-22:6n3 Production

The yield of the hydroxylated derivatives produced from washed human platelets incubated with 10 uM ^{14}C-22:6n3 for various periods up to 60 min is shown in Figure 3. The production of OH-22:6n3 reached a plateau in 20 min; therefore, a 20 min incubation period was used for all subsequent experiments. The yield of 14-OH-22:6n3 was directly related to the concentration of 22:6n3 over the 1-100 uM range. In the presence of 100 uM 22:6n3, 1.2 and 4.3 nanomoles/10^8 cells of the 11-OH- and 14-OH-22:6n3 were produced, respectively The 11-OH-22:6n3 was not detectable following incubation of the platelet suspension with 1 uM 22:6n3.

Table 1. Comparison of 14-OH-22:6n3 production in mammalian platelets, *in vitro.*

	NANOMOLES/10^8CELLS/20 MIN[a]	N[b]
HUMAN	1.56±0.10	8
RAT	0.64±0.08[c]	8
MONKEY	0.54±0.10[c]	4
DOG	0.38±0.04[d]	4
GUINEA PIG	0.32±0.04[d]	4
RABBIT	0.17±0.04[d]	4

[a]Platelets were incubated with 10 uM ^{14}C-22:6n3.
[b]N denotes the number of determinations.
[c]Denotes significance compared to the human platelet at $p<0.01$, [d] at $p<0.001$.

OH-22:6n3 production was directly related to platelet number; in the presence of 10 uM 22:6n3, incubation with 1.6, 0.8, 0.4, 0.2 and 0.1x10^8 cells/ml resulted in conversion percentages of 23, 14, 8, 6 and 3, respectively. In addition, OH-22:6n3 production was inversely related to the age of the platelet preparation. Lipoxygenase activity was reduced by 18, 32, 49 and 90% on 1, 2, 4 and 6 days of age after blood withdrawal when the cells were maintained as whole blood at 4°C.

Substrate Specificity

Platelet suspensions (1.6x10^9 cells/ml) were incubated with 100 uM of either 22:6n3, 22:5n3, 22:5n5 or 22:5n6 for a 20 min period prior to extraction and purification of the respective hydroxylated derivatives. The rank order of hydroxy fatty acid production was 22:6n3 ≥ 22:5n3 > 22:5n6 > 22:5n5; production in picomoles/min/ml of platelet suspension was 260±29, 250±31, 195±21 and 115±14, respectively.

DISCUSSION

Hydroxylated fatty acids were enzymatically produced in platelets from 22:6n3, 22:5n3, 22:5n5 and 22:5n6. As previously reported, human platelets are capable of hydroxylating 22:5n6 at position 14[11] 22:5n3 at positions 11 and 14[10] and 22:6n3 at positions 11, 14[7] and 17[8]. The inhibition of OH-22:6n3 production by specific lipoxygenase inhibitors and the stereochemical purity of these positional isomers[8] are evidence of enzymatic production. The structure of these hydroxy compounds was confirmed by UV and GC/MS analysis. The hydroxylated 22:6n3 was biosynthesized in all mammalian species tested. Production of these derivatives was dependent upon the substrate concentration, the cell number and the age of the cells after withdrawal. At least one-half of the OH-22:6n3 biosynthesized by the human platelet were found in the platelet supernatant. If maximal product yield is required, it would therefore be necessary to extract both the cells and the supernatant after incubation.

Very little is known about the biological activity of these 22-carbon derivatives on cell function and the effects of these products on receptor and enzymatic activity. It has been reported[15-17] that submicromolar concentrations of these hydroxy derivatives specifically antagonize both the vascular smooth muscle contractile effects of a thromboxane-mimetic, U46619, and its platelet aggregating effect. In addition, OH-22:6n3 has been shown to block thromboxane-induced decreases in cerebral blood flow in the rat (unpublished results). These functional changes may be attributed to antagonism at the level of the vascular smooth muscle cell receptor or the platelet thromboxane receptor. The hydroxy fatty acids specifically interact with the thromboxane receptor in both platelet[20] and smooth muscle. These properties may be directly related to aspects of structural anology

Table 2. Inhibitor profile of human platelet 22-carbon hydroxy-docosahexaenoic acid (22:6n3) production.

INHIBITOR[a]	CONCENTRATION	% INHIBITION	
		14-OH	11-OH
CYCLOOXYGENASE			
INDOMETHACIN	0.5 mM	0	0
ASPIRIN	5.0 mM	0	0
LIPOXYGENASE			
ETYA	1.0 uM	100	100
BAICALEIN	10 uM	100	100
NDGA	50 uM	88	92
ESCULETIN	10 uM	55	62
PIROPROST	50 uM	0	0
CAFFEIC ACID	20 uM	0	0
P-450			
SKF 525-A	50 uM	0	0

[a]Platelet suspension was preincubated with the inhibitor for 10 minutes prior to the addition of 10 uM ^{14}C- docosahexaenoic acid.

with thromboxane which may determine their ability to decrease specific binding of thromboxane to its receptor. Analysis of binding parameters indicates these derivatives induce a marked decrease in the affinity of the receptor for thromboxane with a mild change in the number of receptor sites. A direct correlation has been demonstrated between the inhibition of thromboxane receptor binding and inhibition of thromboxane-induced platelet aggregation[13].

In addition, the OH-22:6n3 has been shown to be capable of contracting the guinea pig lung parenchymal preparation[21]. Moreover, the 22-carbon n-3 fatty acids have been reported to increase leukotriene production in airway smooth muscle and neutrophils[21]. A variety of stimuli, which may now include these hydroxylated fatty acids, are known to promote the endogenous production of lipoxygenase products which may directly alter cell function[17,28]. The 22 carbon n-3 hydroxy fatty acids are the most potent, especially in the platelet, in comparison to the n-6 hydroxy fatty acids and their parent fatty acids. The chain length (22-C>20-C), double bond position (N-3>N-6), isomer position (14-OH>11-OH>17-OH) and chirality (S>R) are significant determinants of their structure activity relationship in platelet and smooth muscle cells. These are the first studies which identify any differences in the biological functions of these hydroxylated derivatives.

Dietary fish oil has been reported to increase the production of these n-3 hydroxy fatty acids and was associated with a decrease in thromboxane-induced platelet aggregation and vascular smooth muscle contraction[18]. A 5-fold increase in the total lipid content of 22:6n3 was observed following fish oil feeding of rats and correlated with a 4-fold increase in OH-22:6n3 production, to a level of approximately 340 picomoles per ml (0.34 uM) of platelet suspension. These data suggest that OH-22:6n3 may reach physiologically significant levels (>0.25 uM) with regards to platelet-vascular smooth muscle cell interactions involving thromboxane. Considering their relatively high potency as thromboxane antagonists they may contribute in part to the previously reported inhibitory effect of docosahexaenoic acid on arachidonic acid or thromboxane-induced vascular contraction[29] and platelet aggregation[30]. Changes in cellular function, such as platelet aggregation and smooth muscle contraction, may be regulated by the collective action of the various positional isomers produced from a number of polyunsaturated fatty acids. Nutritional modification of the hydroxy fatty acid profile in platelet may eventually offer clinical applications in the control of thromboxane-mediated responses related to thrombosis and hemostasis.

The biological significance of 22-carbon lipoxygenation requires considerable elucidation. However, the 22-carbon hydroxylated derivatives, produced by platelets, may play a role in cell function. Methods for the efficient biosynthesis and purification of stereochemically pure 22-carbon lipoxygenase derivatives from human platelets, as described here, will be necessary in order to continue the investigation of their biological effects.

REFERENCES

1. E.J.Corey, C. Shih, and J.R.Cashman, Docosahexanoic acid is a strong inhibitor of prostoglandin but not leukotriene biosynthesis. Proc Natl Acad Sci USA. 80: 3581 (1988)..
2. M. Rigaud. Glass capillary-gas chromotography-mass spectrometry analysis of hydoxy and hydroxy-epoxy polyunsaturated fatty acids. Prostaglandins. 27:358 (1984).
3. N.C. Bazan, D.L. Birkle, and T.S. Reddy, Docosahexanoic acid (22:6n3) is metabolized to lipoxygenase reaction products in the retina. Biochem Biophys Res Commun. 125: 741 (1984).
4. M. Van Rollins, R.C. Baker, H.W. Sprecher, and R.C. Murphy, Oxidation of docosahexanoic acid by rat liver microsomes, J Biol Chem. 259: 5776 (1984).
5. T. Shingu and N.Salem Jr, *in*:: Prostaglandins and Lipid Metabolism , T Waldron Jr and N Hughes, eds., pp. 103-108, Plenum Press, New York (1987).
6. T. Shingu, J.W. Karanian, H.Y. Kim, J.A. Yergey, and N. Salem Jr., Discovery of novel brain lipoxygenase products formed from docosahexaenoic acid (22:6n3), Adv Alcohol Subst Abuse. 7: 235 (1988).

7. M.I. Alvedano and H. Sprecher, Synthesis of hydroxy fatty acids from 4,7,10,13,16,19-[1-14C] docosahexanoic acid by human platelets, J Biol Chem. 258: 9339 (1983).
8. H.Y.Kim, J.W.Karanian,T. Shingu, and N. Salem Jr.,. Stereochemical analysis of hydroxylated docosahexaenoates produced by human platelets and rat brain homogenate, Prostaglandins. 40:473 (1990).
9. H.Y. Kim, J.W. Karanian, and N. Salem Jr., Formation of 15-LO product from docosahexaenoic acid by human platelets, Prostaglandins. 40:539 (1990).
10. M.M. Careaga and H. Sprecher, Synthesis of two hydroxy fatty acids from 7,10, 13, 16, 19-docosapentaenoic acid by human platelets, J Biol Chem. 258:14413 (1984).
11. M.M. Milks and H. Sprecher, Metabolism of 4,7,10,13,16-docosapentaenoic acid by human platelet cyclooxygenase and lipoxygenase, Biochim Biophys acta, 835:29 (9185).
12.S. Fisher, C.V. Schacky,W. Siess, T. Strasser, and P.C. Weber, Uptake, release, and metabolism of docosahexanoic acid (DHA, C22: 6w3) in human platelets and neutrophils, Biochem Biophys Res Commun. 120: 907 (1984).
13. J.W. Karanian and N. Salem Jr., Biological role of hydroxylated fatty acids in platelet and vascular smooth muscle function, *in* : "Advances in Polyunsaturated Fatty Acids," T. Yasugi, H. Nakamura, and M. Soma, eds., Vol. 1/1025, pp.169-172, Excerpta Medica, Amsterdam (1993)
14. J.W. Karanian, H.Y. Kim, and N. Salem Jr., Physiological functions of hydroxy docosahexaenoic acid. *in* : "Essential Fatty Acids and Eicosanoids," A. Sinclair and B. Gibson, eds., pp. 142-144, American Oil Chemists Society, Champagne, Il.(1993).
15. M. Croset, A. Sala, C. Folco, and M. Lagarde, Inhibition by lipoxygenase products of TXA2-like responses of platelets and vascular smooth muscle, Biochem Pharmacol. 37: 1275 (1988).
16. J.W. Karanian, H.Y. Kim,T. Shingu, J.A. Yergey, A. Yoffe, and N. Salem Jr., Smooth muscle effects of hydroxylated docosahexaenoates produced from human platelets, Biochim Biophys Acta. 47: S79 (1988).
17. J.W. Karanian, H.Y. Kim, and N. Salem Jr., Inhibitory effects of n-6 and n-3 hydroxy-fatty acids on thromboxane (U46619)-induced smooth muscle contraction, J Pharmacol Exp Ther. in press (1994).
18. J.W. Karanian and N. Salem Jr., Hydroxylated 22-carbon fatty acids in platelet and vascular smooth muscle function:interference with TXA2/PGH2 receptors, Agents and Actions. in press (1994).
19. H. Sinzinger, J. Kaliman, and J. O'Grady, Platelet lipoxygenase defect (Wien-Penzing defect) in two patients with myocardial infarction, Am J Hematol. 36:202 (1991).
20. D.E. Mais, D.L. Saussy, D.E.Magee, and N.L. Bowling, Interactions of 5-HETE, 15-HETE and 5,12-diHETE at the human platelet thromboxane A2/prostaglandin H2 receptor, Eicosanoids. 3:121 (1993).
21. J.W. Karanian, H.Y. Kim, J.A. Yergey, and N. Salem Jr., Lipoxygenase stimulating effects of hydroxylated docosahexaenoates produced by human platelets, Prostaglandins Leukotrienes and Essential Fatty Acids. 50:271 (1994).
22. J.W. Karanian and N. Salem Jr., Interactions of alcohol and prostanoids in the vascular system: Implications for cardiovascular disease, *in:* "Cardiovascular Disease," L. Gallo, ed., Plenum Press, New York, 299 (1987).
23. Salem, N.Jr., Omega-3 fatty acids: Molecular and biochemical aspects, *in:* "New protective roles for selected nutrients," C Spiller and J Scala, eds., pp. 213-332, Alan R Liss Inc, London (1989).
24. J.A. Lopez and M. Hausz, Therapeutic apheresis, Amer J Nurs. 82:1572 (1982).
25. H.Y. Kim. and N. Salem Jr., Preparation and the structural determination of hydroperoxy derivatives of docosahexaenoic acid and other polyunsaturates by thermospray LC/MS, Prostaglandins. 37:105 (1989).
26. M. Hamberg and B. Samuelsson, On the specificity of the oxygenation of unsaturated fatty acids catalyzed by soybean lipoxidase, J Biol Chem. 242:5329 (1967).
27. K. Sekiya and H. 0kuda, Selective inhibition of platelet lipoxygenase by baialein, Biochem. Biophys. Res. Comm. 105:1090 (1982).

28. J.W. Karanian, H.Y. Kim, T. Shingu, J. Yergey, and N. Salem Jr., Cardiovascular properties of hydroxylated docosahexaenoates, *in:* "Prostaglandins in Clinical Research: Cardiovascular System", K. Schror and H. Sinzinger, eds., pp. 511-515, Alan R. Liss, Inc. London (1988).
29. J. Talesnik, Arachidonic acid induced coronary reactions and their inhibition by docosahexaenoic acid, Can J Physiol Pharmacol. 64:77 (1986).
30. G.H.R.Rao, E. Radha, and J.G. White, Effect of docosahexanoic acid (DHA) on arachidonic acid metabolism and platelet function, Biochem Biophys Res Commun. 17:549 (1983).

THE HEMOSTATIC RESPONSE TO ARTERIAL INJURY IN VIVO: BEHAVIOR OF PROTHROMBIN AT THE DE-ENDOTHELIALISED AORTA WALL

M.W.C. Hatton, S.M.R. Southward, S.D. Serebrin
M. Kulczycky and M.A. Blajchman

Department of Pathology
McMaster University Health Sciences Centre
Hamilton, ON
Canada, L8N 3Z5

INTRODUCTION

This chapter reviews the behaviors of certain key hemostatic proteins, fibrinogen, antithrombin-III (ATIII) and thrombin, after administering a balloon de-endothelialising injury to the rabbit aorta in vivo. Also, the behavior of prothrombin after balloon injury is described in relation to fibrinogen and ATIII.

HEMOSTASIS AND VESSEL WALL REPAIR

The property of blood to gel in a controlled manner at the site of vascular injury serves several short-term purposes: the prevention of blood loss from the circulation; the prevention of blood entering the immediate extravascular space; the maintenance of the fluidity of the blood and continued blood flow. These functions of the hemostatic pathways are presumed to occur spontaneously when injurious stimuli trigger blood coagulation pathways. In the longer term, two other important functions of the blood coagulation process are evident: 1. to expedite vascular wall repair; 2. to induce replacement of those blood components depleted during hemostatic activity.

When the vascular endothelium is adversely stimulated (or blatantly injured), the perturbed cells (endothelial, smooth muscle) react to the stimulus by expressing a specific profile of gene products presumably to alleviate the conditions of a changed environment. Thus, specific cell adhesion molecules[1] and extracellular components[2] may be expressed transiently after an acute injury, or continuously during a chronic injury.[3]

BALLOON CATHETER INJURY OF THE RABBIT AORTA IN VIVO

The balloon de-endothelialised rabbit aorta is a valuable model for measuring vascular wall uptake, turnover and distribution of blood proteins and platelets[4,5] during early and late hemostasis following injury.[3] Such results provide a picture of the events that appear to take place immediately after de-endothelialisation of the endothelium in vivo:

1. The freshly-exposed subendothelium becomes saturated with fibrinogen (15 - 25 pmol/cm^2 of intima-media) in less than 5 min after injury;[6] Platelets saturate the de-endothelialised aorta (max. >40000/mm^2) in a similarly short time interval after balloon injury.[5]

2. Plasminogen saturates the subendothelium (8 - 10 pmol/cm^2) at ca. 20 min after balloon injury.[9] ATIII takes longer to saturate the surface (30 - 60 min; ca. 4 pmol/cm^2),[6,8] although ATIII-β, the minor glycoform,[10] is taken up 6 times faster than ATIII-α.[8]

3. From turnover studies in vivo,[6] bound fibrin(ogen) is replaced rapidly on the ballooned intima-media although at a declining rate over a 10 day interval after injury. This compares with the slowly declining generation of thrombin (measured as releasable thrombin activity) by the damaged tissue over the same time period. The profile of thrombin generation correlates well with ATIII turnover over the 10d period.[6] Thus, as thrombin production within the intima-media decreases over 10d after injury, uptake of fibrinogen and ATIII also decline.

4. From studies of the aorta which has been de-endothelialised in vitro, quantities of intima-media bound thrombin over a certain range (35 - 260 fmol/cm^2) correlate closely with the quantities of fibrinogen adsorbed.[7] Fibrinogen binding is not influenced by bound thrombin at a concentration less than 35 fmol/cm^2. We propose that this value is equivalent to the concentration of endogenous active thrombin inhibitors.

5. From a study of thrombin binding to the ballooned aorta in vitro,[11] at a concentration which is less than 30 fmol/cm^2 of intima-media, approximately 50% of bound thrombin is associated with dermatan sulfate (DS) chains within the extracellular matrix; the remaining thrombin binding sites are not sensitive to chondroitinase ABC. Having bound to DS, thrombin becomes associated rapidly and irreversibly with an unknown component (endogenous thrombin inhibitors ?).

6. Thrombin bound by the ballooned aorta in vitro at a concentration of <30 fmol/cm^2 is not displaced by either hirudin or heparin at high concentration.[12] In contrast, much of thrombin adsorbed by fibrin in vitro is complexed, and subsequently displaced from fibrin, by hirudin.[13] We hypothesize that thrombin produced at the site of injury is attracted by various components which occupy the wound site, including the DS chains of dermatan sulfate proteoglycans of the ECM (from which point thrombin probably attracts fibrinogen[14] or is inactivated by inhibitors[15]) and polymerised fibrin. Thrombin which is not bound by cell surfaces and extracellular components of the subendothelium is presumably free to leak into the circulation, if the production rate allows.

7. Recombinant hirudin has been used in vivo to measure directly the thrombin concentration at the site of balloon catheter injury to the rabbit aorta.[16] Using ^{125}I-r-hirudin injected i.v., the quantity of hirudin adsorbed by the exposed subendothelium

at 90 min after balloon injury of the aorta was equivalent to 26 (±18) fmol/cm^2, compared to ca. 3 fmol/cm^2 for uninjured aortas. As ^{125}I-r-hirudin reacts with and displaces thrombin bound to fibrin,[13] the quantity of subendothelium-bound r-hirudin should reflect the concentration of active thrombin which is not associated with fibrin in the extracellular matrix at the time of maximum thrombin production after injury.

A quantitative study of the behavior of prothrombin at the de-endothelialised aorta is essential information which has been missing from this mosaic. In the present report, we have studied the metabolism of rabbit prothrombin (as ^{131}I-prothrombin) in the circulation of healthy rabbits. Having obtained plasma clearance results which were comparable with published reports, the behavior of circulating ^{131}I-prothrombin was investigated relative to ^{125}I-fibrinogen or ^{125}I-ATIII at various times after a single balloon injury to the aorta. With this information of the metabolism of prothrombin, we believe that the dynamics of the production and fate of thrombin at a site of vascular injury are understood more clearly.

METHODS

Purification of Proteins from Rabbit Plasma

Prothrombin. The procedure used[17] was adapted from previous reports[18,19] The entire process was carried out at 4°C in the presence of suitable protease inhibitors. In brief, crude prothrombin was isolated from rabbit ACD plasma by $BaCl_2$ precipitation. Prothrombin-containing concentrate was chromatographed, first, on DEAE-Sephacel and then purified further on dextran-sulfate-Sepharose. Purified prothrombin was stored in 20 mM Na citrate-200 mM NaCl, pH 6.0, containing 5μM each of dansyl-glu-gly-arg chloromethyl ketone and D-phe-pro-arg chloromethyl ketone (PPACK), at -80°C.

Purity was evaluated by electrophoresis, using both Na dodecyl sulfate-polyacrylamide gel electrophoresis (SDS-PAGE)[20] and non-denaturing PAGE[21] techniques at pH 8.4. After blotting to Immobilon (Millipore), blots were visualised by reacting first with anti-rabbit prothrombin (raised in a laying hen; IgG was extracted from egg yolks[22]) and then with alkaline phosphatase-linked anti-chicken IgG (Zymed) as described by Blake et al.[23] Prothrombin was activated to thrombin by adding human factor Xa, thromboplastin and a trace of human plasma (as a source of factor V). Using a plasma clotting assay, a specific activity of 1.3IU/μg prothrombin was calculated.[17]

ATIII. The purification from rabbit plasma has been described before.[8] Only glycoform α, which makes up 90% of total plasma ATIII,[8,10] was used for the experiments described below. On SDS-PAGE, one band (M_r, approximately 60kDa) was observed for ATIII-α.

Fibrinogen. Fibrinogen was isolated from rabbit plasma as described previously[6] by adapting two published procedures.[24,25] By SDS-PAGE, the fibrinogen peak contained only a single high MW band which hardly entered the 7.5% acrylamide gel; after reduction by β-mercaptoethanol, fibrinogen appeared entirely as α, β and γ chains which migrated as described previously.[6,7] Clottability by thrombin was 95 - 98%.

Albumin. Rabbit albumin was purified from plasma by $(NH_4)_2SO_4$ precipitation followed by chromatography on CM-Sephadex as described previously.[26] On SDS-PAGE, albumin appeared as a single band (M_r, approximately 65kDa).

Radiolabeling of Proteins

A procedure using Iodo-Gen-coated glass vials was used.[27] For all proteins, 100 - 200 μg of protein (in 200μl 0.1M Na phosphate, pH 7.4) was reacted with 1 mCi of ^{125}I (or ^{131}I) in a flat-bottomed glass vial, coated with 5μg Iodo-Gen (Pierce Co.) to a height of 3mm, containing 200 μl 0.2M Na phosphate, pH 7.4. ϵ-Aminohexanoic acid (ϵ-AHA) was added to the reaction mixture (final concentration: 25mM) containing prothrombin to prevent aggregation of ^{131}I-prothrombin. After reaction (3 min), each labeled protein was dialysed (4 x 250ml) for 20h at 4°C, ^{125}I-fibrinogen, ^{125}I-albumin and ^{125}I-ATIII against 0.14M NaCl buffered by 0.01M Na phosphate, pH 7.45. For ^{131}I-prothrombin, 2.5 mM ϵ-AHA was included in the dialysate solution, and 200 nmol of PPACK was added to the dialysing protein at each change of dialysate (to suppress thrombin formation). All radiolabeled proteins were stored at 4°C and were only used for experimental purposes within 4d of iodination.

All preparations of radiolabeled proteins were assessed by SDS-PAGE and non-denaturing PAGE (see above) relative to the respective unlabeled protein, followed by autoradiography using Kodak X-AR5 film.

In Vivo Studies

(i) Plasma Clearance of ^{131}I-Prothrombin and ^{125}I-Albumin

The procedure was similar to that used in this laboratory for other plasma proteins.[28] In brief, rabbits were injected intravenously (ear vein) with a mixed dose containing ^{125}I-albumin and ^{131}I-prothrombin in sterile saline. Blood samples (1 ml) were taken at prescribed intervals into 0.25 ml ACD[29] and processed to determine the proportion of protein-bound radioactivity. From the corrected protein-bound radioactivity values, a plasma clearance curve was drawn for each radiolabeled protein.

Calculation of the plasma curve components was undertaken by the 'curve-peeling' technique and by using a three-compartment model (i.e. compartment 1: intravascular space; compartment 2: non-circulating vascular wall; compartment 3: extravascular space) described by Carlson <u>et al.</u>[30] viz:

$$*A_p = C_1e^{-a_1t} + C_2e^{-a_2t} + C_3e^{-a_3t}$$

where $*A_p$ is the fraction of injected radiolabeled protein remaining in the circulation after t days, C_1, C_2 and C_3 are the fractional constants for compartments 1, 2 and 3, and a_1, a_2 and a_3 are respective rate constants of exchange between those compartments. From the respective 'C' and 'a' values, the fractional catabolic rates of each protein were calculated for the plasma (j_3), plasma plus non-circulating vascular wall ($j_{3.5}$) compartments, and for the total body (j_T). The 'j' values were then used to calculate the distribution of each protein among the plasma (A_p), non-circulating vascular wall (A_w) and extravascular (A_e) compartments as described.[30]

(ii) Uptake of ^{131}I-prothrombin by the aorta wall in vivo

The uptake of ^{131}I-prothrombin and ^{125}I-fibrinogen by the luminal surface of the thoracic aorta wall was compared to the uptake of these proteins by the freshly-injured (i.e. balloon de-endothelialised) aorta surface. The rabbit was anesthetized (Na pentobarbital; 35 mg/kg) and a femoral artery and a carotid artery were exposed. A cannula was placed in the carotid artery for exsanguination later. The two radiolabeled proteins were injected separately into an ear vein, ^{125}I-fibrinogen followed by ^{131}I-prothrombin, each in 1 ml sterile saline. At 10 min after injection, a balloon catheter was passed twice into the aorta, via a femoral artery, to de-endothelialise the aorta wall.[31] For the sham-operated controls, a femoral artery was exposed and ligated, but not penetrated. The radiolabeled proteins were allowed to circulate for 5, 10, 15, 30 or 60 min. The rabbit was then rapidly exsanguinated. The aorta was excised and cut into eight 1-cm long segments. Radioactivity measurements were made on the separated vessel layers; for undamaged aortas, these were endothelium, intima-media, and adventitia, and for balloon de-endothelialized aortas, platelet monolayer, intima-media and adventitia.[9,31] Radioactivity bound by the area-corrected aorta surface (i.e. cpm/cm^2) was related to the quantity of radioactivity in the circulation (i.e. cpm/ml of blood) at the time of euthanization.

Uptake of ^{131}I-prothrombin by the de-endothelialised rabbit aorta was also measured at various times (up to 69d) after balloon injury. Separate doses of ^{131}I-prothrombin and either ^{125}I-ATIII or ^{125}I-fibrinogen were injected intravenously into rabbits which had received previously a balloon injury to the aorta. After a circulation time of 10 min, a blood sample was taken and the rabbit exsanguinated through a carotid cannula. The aorta was excised and processed as described above.

RESULTS

Properties of Rabbit Prothrombin and ^{131}I-Prothrombin

Prothrombin migrated as a single band when electrophoresed using a non-denaturing PAGE method at pH 8.4; ^{131}I-prothrombin, which had been iodinated in the presence of ϵ-AHA, migrated in a similar position as the native protein (Fig. 1). In the absence of ϵ-AHA, the iodination reaction caused prothrombin to aggregate; aggregated prothrombin did not enter the stacking gel (4% polyacrylamide, w/v) when non-denaturing conditions for PAGE were used. Aggregated prothrombin was entirely de-aggregated in the presence of SDS.

Using SDS-PAGE, ^{131}I-prothrombin co-migrated with unlabeled prothrombin. After activation by factor Xa in the presence of Ca^{++} and thromboplastin alone, or with factor V (as diluted human plasma), the electrophoretic profile of ^{131}I-prothrombin products obtained by autoradiography was qualitatively similar to that obtained for unlabeled prothrombin by immunoblotting. From these results in the non-denaturing and in the denaturing PAGE systems, it appeared that radiolabeled prothrombin behaved as the unlabeled prothrombin.

Clearance of ^{131}I-Prothrombin and ^{125}I-Albumin from the Rabbit Circulation

The clearance of ^{131}I-prothrombin from the circulation was compared with that are shown in Fig. 2. Prothrombin was cleared quickly from the circulation (j_3=2.00 d^{-1}) and was catabolised rapidly (j_T=0.41 d^{-1}) compared with albumin (j_3=0.32 d^{-1}; j_T=0.13

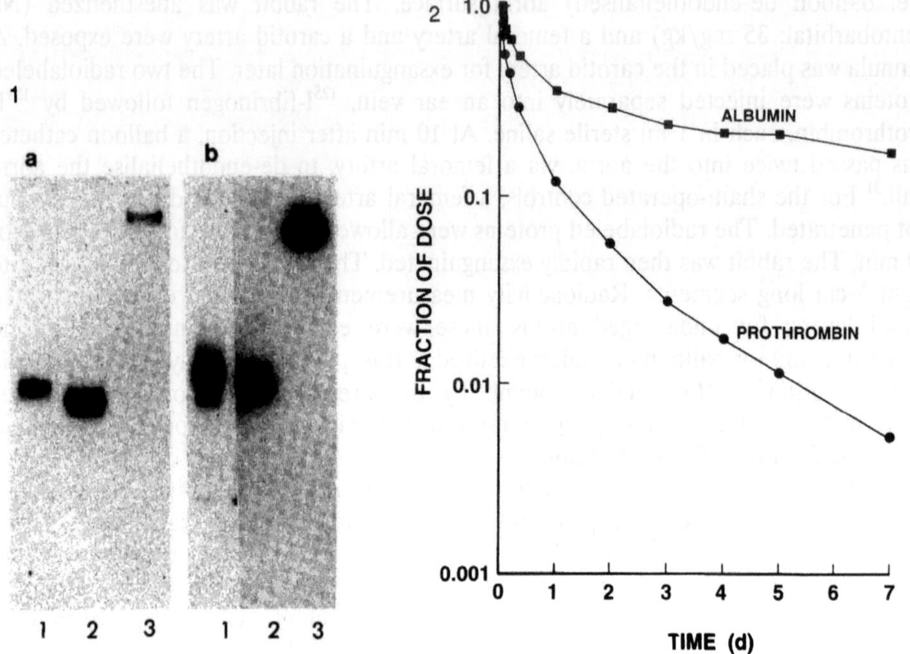

Figure 1. Polyacrylamide gel electrophoresis of unlabeled and radiolabeled rabbit proteins; non-denaturing conditions were used. a. Gel stained with Coomassie Brilliant Blue; b. autoradiograph of same gel. Loads: lane 1, prothrombin; lane 2, ATIII; lane 3, fibrinogen.

Figure 2. Plasma clearance curves of ^{125}I-albumin and ^{131}I-prothrombin injected intravenously into a 2.6 kg rabbit. The curves represent protein-bound radioactivities.

d^{-1}). The values determined for j_T are equivalent to half-lives ($T_{½}$) of 1.70d and 5.5d for prothrombin and albumin respectively.

From the measurements of compartmental distribution, 21% of total body prothrombin was contained within the vascular compartment (A_p), 24% was associated with the blood vessel wall (A_w) and the remaining 56% was located in the extravascular space (A_e). The respective values measured for albumin, 40%, 17% and 43%, were similar to those reported previously.[26]

Behavior of Prothrombin, ATIII and Fibrinogen at the Aorta Wall in vivo

By using ^{125}I-fibrinogen and ^{131}I-prothrombin as plasma markers, uptake of fibrinogen and prothrombin by the uninjured intima-media of sham-operated rabbits ranged from 0.58 - 1.9 pmol/cm^2 and 0.18 - 1.0 pmol/cm^2, respectively, during a circulation time of 5 - 60 min.

After de-endothelialising the aorta with a balloon catheter, fibrinogen was rapidly taken up by the exposed intima-media, saturating (20.0$\pm$8.7 pmol/cm^2) the exposed surface within 5 min after injury. The concentration of bound fibrinogen had decreased slightly at 1h after injury. Prothrombin was taken up more slowly than fibrinogen by the exposed intima-media; saturation (5.3$\pm$1.9 pmol/cm^2) was complete by 15 min and changed little up to 1h after balloon injury. By contrast, ATIII, as ^{125} I-ATIII-α, was taken up more slowly than prothrombin; saturation (3.9-4.6 pmol/cm^2) of the intima-

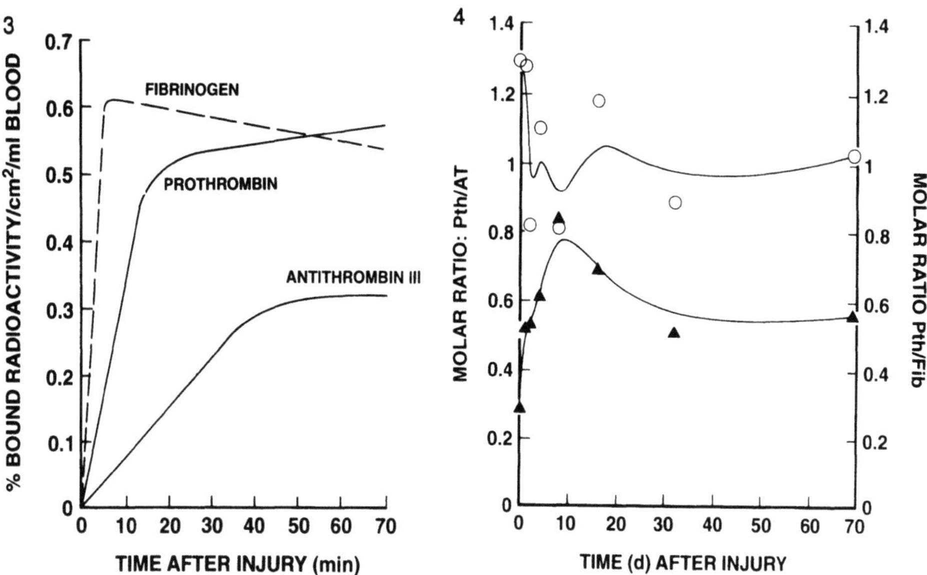

Figure 3. Curves depicting the mean uptakes of prothrombin, fibrinogen and ATIII by the balloon-de-endothelialised intima-media of the rabbit thoracic aorta in vivo. Radiolabeled proteins were injected 10 min before balloon injury (0 min), and were allowed to circulate for up to 70 min after injury. The quantity of intima-media bound protein was calculated from the radioactivity content of the aorta intima-media (eight segments, each corrected to 1 cm^2) expressed as a percentage of the radioactivity content of 1 ml of blood at exsanguination. Each curve represents the mean data from 10 - 15 aortas.

Figure 4. Relative demand for (a) prothrombin and fibrinogen, and (b) prothrombin and ATIII by the thoracic aorta intima-media at various times after balloon de-endothelialisation. Either pair of radiolabeled proteins was injected into a rabbit whose aorta had been ballooned previously. At 10 min after injection, a blood sample was taken and the animal quickly exsanguinated. The quantities of intima-media bound protein were determined as described in the text. Results are given as molar ratios of bound proteins; prothrombin:ATIII, O----O; prothrombin:fibrinogen, ▲-----▲. Data points are results from single aortas.

media was reached at 30-60 min after balloon injury.[6,8] A composite graph showing the relative behaviors of these proteins is given in Fig. 3.

The demand for prothrombin, ATIII and fibrinogen by the ballooned aorta surface was measured in rabbits which had received previously a balloon injury to the aorta. The injected proteins, viz. ^{131}I-prothrombin with either ^{125}I-fibrinogen or ^{125}I-ATIII-α, were allowed to circulate for 10 min before a rapid exsanguination of the rabbit was commenced. Fig. 4 shows the changes in molar ratios of these proteins bound to the intima-media during the healing process over a 69 day period. At all times after injury, the concentration of bound prothrombin was matched closely by bound ATIII, the molar ratio, prothrombin:ATIII, decreasing to unity with time. In contrast, the molar ratio of bound prothrombin:bound fibrinogen increased with time after injury.

DISCUSSION

Purified rabbit prothrombin, iodinated in the presence of ε-AHA, displayed several properties in vitro which were similar to those of unlabeled prothrombin. These results encouraged us to study the behavior of ^{131}I-prothrombin in vivo particularly since no study of the metabolism of rabbit prothrombin in the rabbit has been reported.

From the total-body fractional catabolic rate of ^{131}I-prothrombin (j_T), a half life, $T_{½}$, was calculated as 1.70 (±0.17)d. This compares well with the clearance of bovine

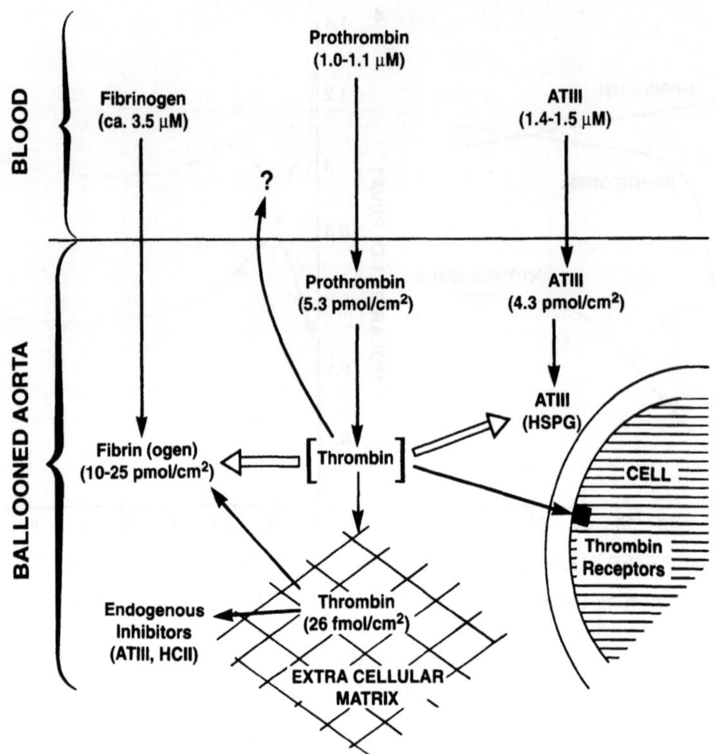

Figure 5. Summary of the behaviors of prothrombin, fibrinogen and ATIII at 1h after balloon injury to the rabbit thoracic aorta. HSPG and HCII refer to heparan sulfate proteoglycan and heparin cofactor-II respectively. Proteins (in 'blood') are given as their approximate concentrations in rabbit blood, other concentrations as per cm^2 of intima-media.

^{125}I-prothrombin in calves ($T_{½}$, 1.77d)[32] and of <u>bovine</u> ^{125}I-prothrombin in rabbits ($T_{½}$, 1.66d)[33] but differs from that of human prothrombin in humans ($T_{½}$, 2.8d[34]; 2.29d[35]). Respecting the ratio of extravascular:intravascular prothrombin (i.e. $A_e/(A_p + A_w)$, a value of 1.22 is calculated for rabbit prothrombin. This value compares with 0.56[34] and 0.90 ($\pm$0.20)[35] for human prothrombin in humans, and 1.81 for <u>bovine</u> prothrombin in rabbits.[33]

^{131}I-Prothrombin was therefore used as an indicator of the demand for plasma prothrombin by the rabbit aorta wall during the first 1h after a balloon catheter injury <u>in vivo</u>. The plasma level of prothrombin in the healthy rabbit[17] was found to be 90 ($\pm$3)% of the level (2μM) in human plasma[36]; i.e. 1.8 μM. From the data in Fig. 3, uptake by the de-endothelialised aorta wall amounted to 5.3 pmol of prothrombin/cm^2 of intima-media at 15 min after balloon catheter injury. From earlier measurements,[6,8] approximately 2.0 pmol of ATIII was taken up by 1 cm^2 of de-endothelialised rabbit aorta during 15 min after balloon injury. Thus, at 15 min after balloon injury, the rate of replacement of intima-media-bound ATIII was considerably less than that of prothrombin. During this early stage after injury, a relative excess of thrombin appears to be produced to encourage and consolidate fibrin production at the wound site. To admit a large quantity of antithrombin to this focus of hemostatic activity might inactivate thrombin too rapidly, leading to excessive blood loss from the circulation. The factors (or binding sites) within the subendothelium which control prothrombin and ATIII uptake are not understood although previous reports suggest that heparan sulfate proteoglycan located in the smooth muscle cell basement membrane[37,38] or extracellular matrix[39] may bind ATIII.

The behavior of prothrombin relative to fibrinogen and ATIII at the ballooned aorta wall at 1h after a balloon de-endothelialising injury is summarized in Fig. 5. In addition to fibrinogen and prothrombin, the de-endothelialised aorta was saturated with ATIII, equivalent to approximately 4.3 pmol/cm^2.[6,8] At present, no information is available concerning the concentration of other thrombin inhibitors, particularly heparin cofactor-II, within the aorta wall after balloon injury. The concentration of endogenous ATIII is less than that of endogenous prothrombin, viz. 5.3 pmol/cm^2. We predict that, at 1h after balloon injury, the flux of prothrombin and its conversion to thrombin within the site of injury is controlled by the presence of adequate amounts of endogenous inhibitors. Thus, the concentration of active thrombin within the intima-media would be relatively small. A recent attempt to measure thrombin contained within the rabbit aorta de-endothelialised in vivo (measured by using ^{131}I-r-hirudin injected i.v.) reported 26 ($\pm$18) fmol of thrombin/cm^2 of intima-media at 1 - 1.5h after injury.[15] We consider that this small concentration of thrombin reflects the effective control of thrombin production by antithrombins at the site of injury. Of thrombin that is generated, perhaps only a relatively small portion is actively involved in fibrinogen recruitment and conversion into fibrin.[6,7]

In an earlier report,[6] we showed that, after balloon injury in vivo, the excised thoracic aorta wall released a measurable quantity of thrombin from the damaged luminal surface in vitro. The amount released was dependent upon the time interval between balloon injury of the aorta and the death (by exsanguination) of the rabbit. Briefly, thrombin activity rose to a maximum at 1-2h after balloon injury and then decreased over the subsequent 4h. Thrombin activity continued to fall over the ensuing 10d after injury but, even at 10d, the level was significantly greater than that released by control (uninjured) aortas. In comparison, prothrombin uptake by the ballooned aorta was maximal by 15 min (Fig. 3). Clearly, a better understanding of the relationship between prothrombin uptake in vivo and production of releasable thrombin by the injured aorta surface (as measured in vitro) is required.

The ratio of intima-media-bound prothrombin:ATIII decreased during 69 d after balloon injury (Fig. 4). As the ratio tends towards unity, we assume that the capacity for thrombin inactivation within the site of injury increases with time. If this is the case, one would expect the turnover of fibrinogen to decrease with time. In effect, fibrinogen turnover decreased markedly over the 69 d interval after injury, as shown by the increase in intima-media-bound prothrombin:fibrinogen ratio.

In conclusion, our findings are consistent with the hypothesis that a delicate balance of pro- and anti-thrombotic forces has evolved in the aorta wall, with the prothrombotic forces having an advantage over the antithrombins during the early stage after vessel injury. This mechanism may have evolved to secure a fibrin breach. However, once this has happened, the antithrombin forces assume a more active role to control thrombin during wound healing.

REFERENCES

1. M.P. Bevilacqua, J.S. Pober, D.L. Mendrick, R.S. Cotran, and M.A. Gimbrone. Identification of an inducible endothelial-leukocyte adhesion molecule. Proc. Natl. Acad. Sci. USA 84: 9238 (1987).
2. M. Richardson and M.W.C. Hatton. Transient morphological and biochemical alterations of arterial proteoglycan during early wound healing. Exp. Mol. Pathol. 58: 77 (1993).
3. A.E. Koch, J.C. Burrows, G.K. Haines, T.M. Carlos, J.M. Harlan, and S.J. Leibovich. Immunolocalization of endothelial and leukocyte adhesion molecules in human rheumatoid and osteoarthritic synovial tissues. Lab. Invest. 64: 313 (1991).

4. H.R. Baumgartner, M.B. Stemerman, and T.H. Spaet. Adhesion of blood platelets to subendothelial surface: Distinct from adhesion to collagen. Experientia 27: 283 (1971).
5. H.M. Groves, R.L. Kinlough-Rathbone, M. Richardson, S. Moore, and J.F. Mustard. Platelet interaction with damaged rabbit aorta. Lab. Invest. 40: 194 (1979).
6. M.W.C. Hatton, S.L. Moar, and M. Richardson. De-endothelialisation in vivo initiates a thrombogenic reaction at the rabbit aorta surface. Correlation of uptake of fibrinogen and antithrombin-III with thrombin generation by the exposed subendothelium. Am. J. Pathol. 135:499 (1989).
7. M.W.C. Hatton, S.L. Moar, and M. Richardson. Enhanced binding of fibrinogen by the subendothelium after treatment of the rabbit aorta with thrombin. J. Lab. Clin. Med. 115: 356 (1991).
8. M.R. Witmer and M.W.C. Hatton. Antithrombin-III β associates more readily than antithrombinIII α with uninjured and de-endothelialized aortic wall in vitro and in vivo. Arterioscler. Thromb. 11: 530 (1991).
9. M.W.C. Hatton, S.L. Moar, and M. Richardson. Behavior of plasminogen at the luminal surface of the normal and de-endothelialized rabbit aorta in vivo and in vitro. Blood 71: 1260 (1988).
10. T.H. Carlson and A.C. Atencio. Isolation and partial characterisation of two distinct types of antithrombin III from rabbit. Thromb. Res. 27: 23 (1982).
11. M.W.C. Hatton. Evidence for thrombin binding to dermatan sulfate sites in the rabbit aorta subendothelium in vitro. Blood. Coag. Fibrinol. 4: 927 (1993).
12. M.W.C. Hatton and S.L. Moar. Comparison of the effects of heparin and hirudin on thrombin binding to the normal and the de-endothelialized rabbit aorta in vitro. Thromb. Haemostas. 66: 208 (1991).
13. F. Rubens, B. Ross-Ouellet, C. Dennie, G. Coates, R.L. Kinlough-Rathbone, and M.W.C. Hatton. Evaluation of ^{131}I-recombinant hirudin as an agent for the detection of pulmonary emboli in the rabbit. Thromb. Haemostas. In Press (1994).
14. R. Bar-Shavit, A. Eldor, and I. Vlodavsky. Binding of thrombin to subendothelial extracellular matrix. Protection and expression of functional properties. J. Clin. Invest. 84: 1096 (1989).
15. M.W.C. Hatton, L.R. Berry, and E. Regoeczi. Inhibition of thrombin by antithrombin III in the presence of certain glycosaminoglycans found in the mammalian aorta. Thromb. Res. 13: 655 (1978).
16. M.W.C. Hatton and B. Ross-Ouellet. Radiolabeled r-hirudin as a measure of thrombin activity at, or within, the rabbit aorta wall in vitro and in vivo. Thromb. Haemostas. 71: 499 (1994).
17. M.W.C. Hatton, S.M.R. Southward, S.D. Serebrin, M. Kulczycky, and M.A. Blajchman. Catabolism of rabbit prothrombin in rabbits: Behavior at the aorta wall before and after a balloon de-endothelialising injury in vivo. Submitted to Arterioscleros. Thromb. (1994).
18. S.P. Bajaj and K.G. Mann. Simultaneous purification of bovine prothrombin and factor X: Activation of prothrombin by trypsin-activated factor X. J. Biol. Chem. 248: 7729 (1973).
19. G.J. Modi, M.A. Blajchman and F.A. Ofosu. The isolation of prothrombin, factor IX and factor X from factor IX concentrates. Thromb. Res. 36: 537 (1984).
20. U.K. Laemmli. Cleavage of structural proteins during the assembly of the head of bacteriophage T4. Nature 227: 680 (1970).
21. J.T. Clarke. Simplified "disc" (polyacrylamide gel) electrophoresis. Anal. NY Acad. Sci. 121: 428 (1964).
22. A. Polson, T. Coetzer, J. Kruger, E. von Maltzahn, and K.J. van der Merwe. Improvements to the isolation of IgY from the yolk of eggs laid by immunized hens. Immunol. Invest. 14: 323 (1985).
23. M.S. Blake, K.H. Johnston, G.J. Russell-Jones, and E.C. Gotschlich. A rapid sensitive method for detection of alkaline-phosphatase-conjugated anti-antibody on western blots. Anal. Biochem. 136: 175 (1984).
24. W. Straughn and R.H. Wagner. A simple method for preparing fibrinogen. Thromb. Diath. Haemorrh. 16: 198 (1967).
25. J.S. Lawrie, J. Ross and G.C. Kemp. Purification of fibrinogen and the separation of its degradation products in the presence of calcium ions. Biochem. Soc. Trans. 7: 693 (1979).
26. M.W.C. Hatton, M. Richardson, and P.D. Winocour. On glucose transport and non-enzymic glycation of proteins in vivo. J. Theor. Biol. 161: 481 (1993).
27. E. Regoeczi. "Iodine-Labeled Plasma Proteins," CRC Press, Boca Raton, (1984), Vol 1, pp.49-56.
28. M.R. Witmer, S.J. Hadcock, S.L. Peltier, P.D. Winocour, M. Richardson, and M.W.C. Hatton. Altered levels of antithrombin III and fibrinogen in the aortic wall of the alloxan-induced diabetic rabbit: Evidence of a prothrombotic state. J. Lab. Clin. Med. 119: 221 (1992).

29. R.H. Aster and J.H. Jandl. Platelet sequestration in man. I. Methods. J. Clin. Invest. 43: 843 (1964).
30. T.H. Carlson, A.C. Atencio, and T.L. Simon. In vivo behavior of radioiodinated rabbit antithrombin III. Demonstration of a non-circulating vascular compartment. J. Clin. Invest. 74: 191 (1984).
31. M.W.C. Hatton and S.L. Moar. Uptake and catabolism of ^{125}I-thrombin by the rabbit thoracic aorta in vitro: Permeability of the endothelium, intima-media and adventitial layers. Thromb. Haemostas. 52: 105 (1984).
32. Y. Takeda. Studies of the effects of heparin, Coumadin, and vitamin K on prothrombin metabolism and distribution in calves with the use of ^{125}I-prothrombin. J. Lab. Clin. Med. 75: 355 (1970).
33. K.A. Mitropoulos and M.P. Esnouf. Turnover of factor X and prothrombin in rabbits fed on a standard or cholesterol-supplemented diet. Biochem. J. 244: 263 (1987).
34. S.S. Shapiro and J. Martinez. Human prothrombin metabolism in normal man and in hypocoagulable subjects. J. Clin. Invest. 48: 1292 (1969).
35. Y. Takeda. Studies of the metabolism and distribution of prothrombin in healthy men with homologous ^{125}I-prothrombin. Thromb. Diath. Haemorr. 27: 472 (1972).
36. F.C. McDuffie, C. Giffin, R. Niedringhaus, K.G. Mann, C.A. Owen, E.J.W. Bowie, J. Peterson, G. Clark, and G.G. Hunder. Prothrombin, thrombin and prothrombin fragments in plasma of normal individuals and of patients with laboratory evidence of disseminated intravascular coagulation. Thromb. Res. 16: 759 (1979).
37. M.W.C. Hatton, S.L. Moar, and M. Richardson. On the interaction of rabbit antithrombin-III with the luminal surface of the normal and de-endothelialised rabbit thoracic aorta in vitro. Blood 67: 878 (1986).
38. M.W.C. Hatton, S.L. Moar, and M. Richardson. Evidence that rabbit antithrombin-III binds to proteoheparan sulfate at the subendothelium of the rabbit aorta in vitro. Blood Vessels 25: 12 (1988).
39. A.I. de Agostini, S.C. Watkins, H.S. Slayter, H. Youssouffian, and R.D. Rosenberg. Localization of anticoagulantly active heparan sulfate proteoglycans in vascular endothelium: Antithrombin binding on culturd endothelial cells and perfused rat aorta. J. Cell Biol. 111: 1293 (1990).

PLASMA COAGULATION FACTORS IN EMERGENCY ROOM PATIENTS WITH ACUTE CHEST PAIN AND SUBSEQUENT HOSPITALIZATION: MYOCARDIAL INFARCTION, CORONARY ARTERY DISEASE, AND HYPERTENSION

C.F. Saladino, V. Misra, N. Sathish, R. Fox, S.E. Feffer and E.A. Jonas

Department of Medicine, Nassau County Medical Center, East Meadow, NY 11554

INTRODUCTION

It is well known that the vascular pathology leading to a myocardial infarction involves both the long term formation of an atherosclerotic plaque and a thrombus, which often leads to final closure of a coronary vessel. Numerous cells types, including the platelet, contribute to the development of an atherosclerotic lesion. Of considerable importance is the observation that various cardiovascular risk factor conditions, such as hypertension (HTN) and hyperlipidemia, alter the physiology of the platelets and, therefore, increase their activity level. For example, we have previously demonstrated that hypertriglyceridemia-producing infusions of parenteral lipid emulsions into rats induce 1) myofibroelastic changes indicative of early aortic atherogenesis[1] and 2) significant platelet hyperaggregability.[2-4] Regarding hypertension, in general, the literature indicates an enhanced risk of thromboembolic disease in hypertensives, due to an increase in platelet function and a decrease in fibrinolytic activity. For example, Hoffmann et al.[5] have shown that intraplatelet levels of cAMP are significantly lower in an untreated essential hypertension group, compared to that of healthy control subjects. Gleerup et al.[6] have shown that the threshold for irreversible platelet aggregation is reduced in hypertensives, while tPA activity is significantly lower vs. that of normal patients.

Thus, it stands to reason that if platelet activity is enhanced in coronary artery disease (CAD) and its accompanying risk factor conditions, such as hypertension, then it is likely that changes could also be occurring in the hemostatic system of which the coagulation cascade proteins are a critical part. Several studies have begun to address this question. For example, it has been reported that fibrinogen is an independent risk indicator for coronary heart disease and is enhanced by, but can still independent of, plasma LDL cholesterol levels.[7] No changes in factor VIIc were noted in this study. However, others[8] have indicated that factor VIIc is important in CAD, although this could be related primarily to plasma lipid levels.[9, 10], Anderson[11] has reported high levels of fibrinogen, factor VIIc, and plasminogen activator inhibitor to be independent risk factors for CAD or recurrent myocardial infarction, the later two proteins being associated with plasma lipid levels. Jansson et al.[12] have demonstrated high concentrations of von Willebrand factor as an index of increased risk for reinfarction and mortality in survivors of myocardial infarction. Finally, Irie et al.[13] have suggested that coronary vasospasm associated with myocardial ischemia might induce stasis of the blood, resulting in fibrinogen-fibrin conversion in the coronary vessels.

These studies certainly point to an enhanced coagulation cascade in CAD. However, data from other studies conflict with some of the above-cited reports. For example, in a study by Bounameaux et al.,[14] it is concluded that myocardial ischemia, silent or symptomatic, is not associated with an activation of plasma coagulation or fibrinolysis that can be distinguished from excercise-induced thrombin generation. Further, data from a study by Kovacs et al.[15] using CAD patients suggest that elevated fibrinogen is not a causative factor for either CAD or arterial thrombosis. Of additional importance is the modulating role of medications,[16] lipid levels, smoking[17], physical activity, and psychosocial factors[18] in determining the levels of plasma coagulation proteins, such as fibrinogen.

Thus, the data are often conflicting. Further, only a very small number of studies have investigated more than a few of the many coagulation proteins in CAD and MI, with little information available on the relationship of these coagulation cascade factors to hypertension. The present study, therefore, was designed to determine the levels of eighteen different coagulation cascade proteins within the setting of MI, CAD, and hypertension.

MATERIALS AND METHODS

Patients. Thirty two patients presenting to the emergency room (ER) with chest pain and requiring hospitalization were accepted into the study after informed consent was obtained, as required by the Institutional Review Board of the Nassau County Medical Center. The study included 17 male (mean age 60) and 15 female (mean age 59) subjects. The patients were divided into three groups: a) those with documented myocardial infarction (MI) (N=6), b) those without MI, but with documented coronary artery disease (CAD)(N=12), and c) those without MI or CAD (N=14). MI was established by cardiac serum enzymes and/or electrocardiography. CAD was determined by positive graded exercise stress test and/or coronary angiography. Patients with a history of hypertension (HTN) were also evaluated as a separate subgroup (N=16, 9 males and 7 females). HTN was defined as a systolic pressure $\geq$ 160mm Hg and a diastolic pressure $\geq$ 90mm Hg.

Hemostatic Determinations. Fifteen ml of venous blood was obtained from each patient and mixed with sodium citrate to a final concentration of 0.38% of the anticoagulant. The plasma was removed after centrifugation at 1000 x g for 20 minutes, aliquoted, and frozen at -70°C. The following coagulation proteins were measured: a) chronometric tests - prothrombin time, activated partial thromboplastin time, fibrinogen, factors V, VII, VIII, and X, functional proteins C and S, b) chromogenic tests - antithrombin III, plasminogen, alpha-2-antiplasmin, plasminogen activator inhibitor, c) enzyme-linked immunoassay - tissue plasminogen activator, d-dimers, von Willebrand antigen, and thrombomodulin (American Bioproducts, Parsippany, NJ).[19-26] Chronometric testing was carried out using an ST-4 (American Bioproducts Co.) or 16S Profiler (Ortho Diagnostics, Raritan, NJ). All chromogenic tests were performed manually, aliquoted to microtiter plates, and then read on a SLT Microtiter Plate Reader (SLT Lab Instruments, Hillsborough, NC). Elisa assays were also read on an SLT Microtiter Plate Reader.

Statistics. Standard t-tests and ANOVA were determined using a Kwikstat 3.3 data analysis and graphics program (TexaSoft, Cedar Hill, TX). Error bars in the figures represnt the standard error of the mean.

RESULTS

The patients were divided into three groups to distinguish a) those with documented MI, b) those without MI but with doumented CAD, and c) those without MI or CAD. Out of the eighteen coagulation proteins measured, only protein S was significantly altered in the three patient groups. Specifically, Figure 1 shows that protein S is greater $p \leq 0.03$ and elevated above normal in the patients with MI, compared to that of the other two groups, which were statistically the same. Thus, we were not able to detect any other changes in the coagulation cascade within or between any of the other patient groups.

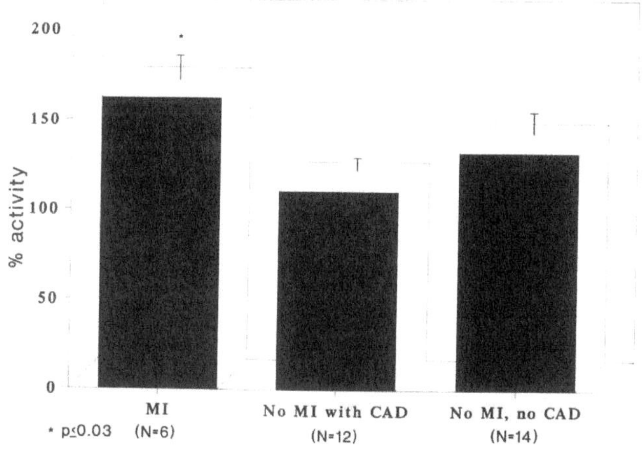

Figure 1. Percent activity of protein S in the three patient groups. Those with MI show significantly greater levels of protein S than the other two groups, which have normal levels.

Out of the total of thirty two patients in the present study, 16 were determined to have hypertension, as defined in the Materials and Methods. As shown in Figure 2, when compared to patients without HTN, hypertensives showed significantly elevated levels of the following three coagulation cascade proteins: fibrinogen (385 vs. 285mg/dl) (p≤p0.002), d-dimers (360 610%) (p≤0.04), and von Willebrand factor (160 vs. 110%) (p≤0.02). It should be noted, that thrombomodulin was also higher in HTN patients, (266 vs. 145%, respectively), although the difference in the means was not significant.

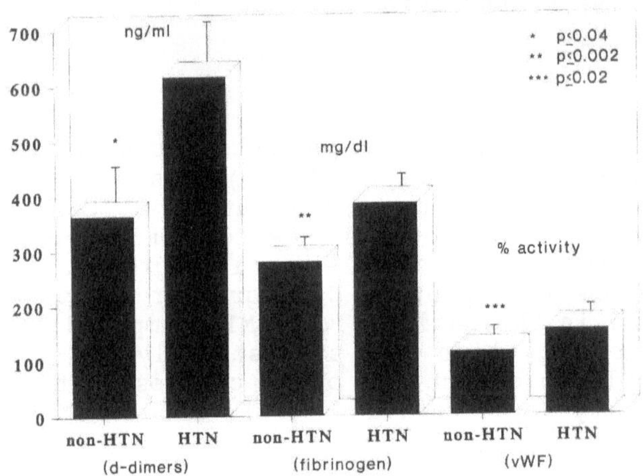

Figure 2. Plasma levels of d-dimers, fibrinogen, and von Willebrand factor in HTN vs. non-HTN patients.

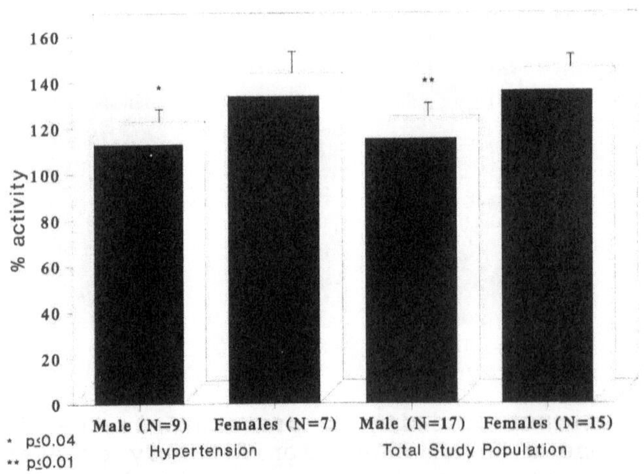

Figure 3. Plasminogen activity in male vs. females in HTN patients and in the total study population.

Of the 16 patients with HTN, 9 were males and 7 were females. An interesting observation is shown in Figure 3. Mean plasminogen levels were higher in female patients with HTN vs. females without HTN (135 vs. 108%, respectively) ($p \leq 0.04$). In addition, plasminogen was also higher in the total female study population than in the total male study population (138 vs. 110%) ($p \leq 0.01$).

DISCUSSION

The purpose of this study was to gain new insight into the role of the coagulation cascade within the setting of MI, CAD, and HTN. Eighteen different coagulation proteins were evaluated in thirty two ER patients presenting with chest pain and subsequently requiring hospitalization. In addition, a subset of 16 patients with hypertension was studied, and the important variable of gender was also considered.

The only change noted in patients with MI is an elevation of plasma protein S. Unfortunately, it is difficult to explain the meaning of this, in view of the lack of change in the levels of other related endothelial proteins which are part of the hemostatic/anticoagulation system. It might be noted, however, that protein S is not only synthesized by the endothelium,[27] but it is also localized in and released by the platelet.[28] Perhaps our observed increase in protein S is related to expected enhanced platelet activity during the thrombosis phase of MI development. Yet, until a greater undertstanding of the complete role of protein S in hemostasis (and possibly atherogenesis) is achieved, we can only speculate as to the meaning of our observation. In this regard, larger numbers of MI patients would have to be studied than were utilized in the present study.

There have been various reports in the literature that plasma fibrinogen is elevated in MI patients and might even constitute an independent risk factor for CAD.[7,11] However, we do not observe this in the present study, which was small in patient number but did utilize individuals selected in a randomized fashion. Rather, our results agree with studies such as that of Kovacs et al.,[15] which conclude that fibrinogen is not a causative factor for CAD. However, it is interesting that we do observe a significant elevation of plasma fibrinogen in patients with HTN. This increase in fibrinogen might result from increased thrombin formation with subsequent increased fibrin degradation. This idea is in good agreement with the fact that d-dimers are a product of fibrin degradation, and, as expected, we do observe a significant increase in plasma d-dimers in HTN patients. Thus, the elevated levels of fibrinogen and d-dimers could together indicate an ongoing or compensating thrombotic process in HTN patients, where fibrinogen production and degradation are part of the hemostatic response.

We also observe in the plasma of HTN patients a significant increase in von Willebrand Factor (vWF). vWF is localized in the vascular endothelium, acts as a carrier protein for factor VIII, and causes platelets to adhere to subendothelial tissue.[29] The increased level of plasma vWF might indicate endothelial damage resulting from the enhanced shear forces that would be expected to occur with elevated blood pressures. This endothelial damage would then result in release of the vWF. In fact, we also observe higher (although) observe higher (although not significantly) levels of thrombomodulin, an endothelium-localized protein which acts as a cofactor for thrombin-catalyzed activation of anticoagulant protein C[30] and which inhibits thrombin activity on fibrinogen, factor V, and platelets.[31]

Finally, we observe a significant difference in plasma plasminogen levels of female patients vs. males, both in the total study population and in the HTN patients. Perhaps this indicates that females are afforded better protection against thrombosis via fibrinolysis than their male counter-parts. However, the clinical significance of this can only be speculative at this time, especially in view of the fact that the mean ages of the males and females in our study population are 60 and 59 years, respectively. At these ages, the beneficial effect of the premenopausal female being a negative risk factor for CAD would be minimized. Thus, estogen levels might not be the critical factor in explaining our observed differences in plasma plasminogen levels between male and female patients, whether or not they have HTN.

In conclusion, it appears that whereas we can detect few changes in the coagulation cascade system in patients with MI or CAD, our study subpopulation of patients with HTN reveals significant changes in the levels of several coagulation proteins. These changes indicate possible endothelial damage, as well as a shift in hemostasis toward a prothrombotic state in patients with hypertension.

REFERENCES

1. C.F. Saladino, R.A. Klein, and E.A. Jonas, Induction of early atherosclerosis in rats using parenterally-administered lipid emulsions, *Artery.* 14:304-315 (1987).
2. C.F. Saladino, R.L.Fox, Q. Yeh, F. Karpowicz, S.E. Feffer, and E.A. Jonas, Platelet aggregability in rats with early atherosclerotic changes induced by parenterally-administered lipid emulsions, *Atherosclerosis.* 66:19-28 (1987).
3. C.F. Saladino, S. Sulimovici, A. Jacob, R.L. Fox, M.S. Roginsky, and E.A. Jonas, Altered cAMO production in platelet membranes from rats receiving atherogenic doses of parenteral lipid emulsions. *Biochem. Archiv.* 6:275-286 (1990).
4. C.F. Saladino, C. Kosacolsky-Singer, R.L. Fox, V. Nethala, S.E. Feffer, and E.A. Jonas, The effect of parenteral lipid emulsion-induced hyperlipidemia on prostaglandin E1 modulation of platelet function, *Artery.* 20:303-313 (1993).
5. G. Hoffmann, B.O. Gobel, U. Harbrecht, H. Vetter, and R. Dusing, Platelet cAMP and cGMP in essential hypertension, *Am. J. Hypertens.* 5:847-850 (1992).
6. K. Winther, G. Gleerup, and T. Hedner, Enhamced risk of thromboemolic disease in hypertension from platelet hyperfunction and decreased fibrinolytic activity: has antihypertensive therapy any influence? *J. Cardiovasc. Pharmacol.* 19 Suppl 3:S21-S24 (1992).
7. J. Heinrich, L. Balleisen, H. Schulte, G. Assmann, J. van de Loo, Fibrinogen and factor VII in the prediction of coronary risk, *Arterioscler. Thromb.* 14:54-59 (1994).

8. T.W. Meade, S. Mellows, M. Brozovic, G.J. Miller, R.R. Chakrabarti, W.R.S. North, A.P. Haines, Y. Stirling, J.D. Imeson, and S.G. Thompson, Haemostatic function and ischaemic heart disease: principal results of the Northwick Park Heart Study, ***Lancet.*** ii:533 (1986).
9. G.J. Miller, J.K. Cruickshank, L.J. Ellis, R.L. Thompson, H.C. Wilkes, Y. Stirling, K.A. Mitropoulos, J.V. Allison, T.E. Fox, and A.O. Walker,Fat consumption and factor VII coagulant activity in middle-aged men, ***Atheroscler.*** 78: 19-24 (1989).
10. A. Silveira, F. Karpe, M. Blomback, G. Steiner, G. Walldius, and A. Hamsten, Activation of coagulation factor VII during alimentary lipemia, ***Arterioscler. Thromb.*** 14:60-69 (1994).
11. P. Andersen, Hypercoagulability and reduced fibrinolysis in hyperlipidemia: relationship to the metabolic cardiovascular syndrome, ***J. Cardiovasc. Pharmacol.*** 20 Suppl 8:S29-S31 (1992).
12. J.H. Jansson, T.K. Nilsson, and O. Johnson, von Willebrand factor in plasma: a novel risk factor for recurrent myocardial infarction and death, ***Brit. Heart J.*** 66:351-355 (1991).
13. T. Irie, T. Imaizumi, T. Matuguchi, S. Koyanagi, H. Kanaide, A. Takeshita, and M. Nakamura, Increased fibrinopeptide A during anginal attacks in patients with variant angina, ***J. Am. Col. Cardiol.*** 14:589-94 (1989).
14. H. Bounamaeux, A. Righetti, P. de Moerloose, O. Bongard, and G. Reber, Effects of exercise tests on plasma markers of an activation of coagulation and/or fibrinolysis in patients with symptomatic or silent myocardial ischemia, ***Thromb. Res.*** 65:27-32 (1992).
15. I.B. Kovacs, C.P. Ratnatunga, C.D. Ridler, P. Gorog, S.J. Edmondson, and G.M. Rees, Significance of plasma fibrinogen in coronary arterial disease: marker or causative factor for arterial thrombosis? ***Intl. J. Cardiol.*** 35:57-64 (1992).
16. J. Mayer, T. Eller, P. Brauer, E.M. Solleder, R.M. Schafer, F. Keller, and K. Kochsiek, Effects of long term treatment with lovastatin on the clotting system and blood platelets, ***Ann. Haematol.*** 64:196-201 (1992).
17. W.B. Kannel, R.B. D-Agostino, A.J. Belanger, Fibrinogen, cigarette smoking, and risk of cardiovascular disease: Insights from the Framingham Study, ***Am. Heart J.*** 113: 1006-1010 (1987).
18. H.L.J. Markowe, M.G. Marmot, M.J. Shipley, C.J. Bulpitt, T.W. Meade, Y. Stirling, M.V. Vickers, and M.V. Semmence, A possible link between social class and coronary heart disease, ***Brit. Med. J.*** 291:1312-1314 (1985).
19. D.A. Triplett and C.S. Harms, "Procedures for the Laboratory," Am. Soc. Clin. Pathol, Chicago (1982).
20. O.R. Odegard, M. Lie, and U. Abildgaard, Heparin cofactor activity measured with amidolytic method, ***Thromb. Res.*** 6:287-294 (1975)
21. C. Soria, J. Soria, and M. Samama, Dosage du plasminogene a laide d'un substrat chromogene tripeptideque, ***Pathol. Biol. Fr.*** 24:275-725 (1976).
22. J.L. Martinoli and K. Stocker, Functional protein C assay using protac, a novel protein C activator, ***Thromb. Res.*** 43: 253-264 (1986).

23. P.C. Comp, D. Doray, D. Patton, and C.T. Esmon, An abnormal plasma distribution of protein S occurs in functional protein S deficiency, ***Blood.*** 67:504-508 (1986).
24. D.B. Rylatt, A.S. Blake, L.E. Cottis et al., An immunoassay for human d-dimer using monoclonal antibodies, ***Thromb. Res.*** 31:767-778 (1983).
25. A. Bartlett, K.M, Dormandy, C.M. Hawkey, P.Stableforth, and A. Voller, Factor VIII related antigen: measurement by enzyme immunoassay, ***Brit. Med. J.*** 1:994-996 (1976).
26. M.C. Alessi, M.C. Boffa, C. Menart, M.F. Aillaud, G. Parrot, and I. Juhan-Vague, Venous occlusion does not induce the release of thrombomodulin from endothelial cells in patients with thromboembolic disease, ***Thromb. Haemostas.*** 68:483-484 (1992).
27. D.S. Fair, R.A. Marlar, and E.G. Levin, Human endothelial cells synthesize protein S, ***Blood.*** 67:1168-1171 (1986).
28. H.P. Schwartz, M.J. Heeb, and J.G. GRiffin, Human platelet protein S, ***Thromb. Haemostas.*** 54:57 Abstr. (1985).
29. K.J. Gasapard, Alterations in blood coagulation and hemostasis, *in:* "Pathophysiology," C.M. Porth, ed., J.B. Lippincott Co., Philadelphia (1986).
30. N.L. Esmon, Thrombomodulin, *Prog.* ***Hemostas. Thromb.*** 9:29-55 (1989).
31. M.C. Bourin, E. Lundgren-Akerlund, and U. Lindahl, Isolation and characterization of the glycosaminoglycan component of rabbit thrombomodulin proteoglycan, ***J. Biol. Chem.*** 265:1524-1531 (1990).

GENETIC FACTORS CONTRIBUTING TO ATHEROSCLEROSIS: FROM HUMANS TO MICE AND BACK AGAIN

Craig H. Warden and Aldons J. Lusis

Departments of Medicine, Division of Cardiology
UCLA
Los Angeles, CA 90024-1679
310-206-0133
FAX 310-794-7345
e-mail: warden@biovx1.biology.ucla.edu

INTRODUCTION

Genetic methods to analyze atherosclerosis in mice and in humans have recently been surveyed by other reviews and will not be repeated here [1-3]. This review will provide examples of the synergy gained by simultaneous studies of atherosclerosis in mice and humans. This synergy arises largely because analysis of atherosclerosis in humans is greatly complicated by environmental and genetic heterogeneity [4], problems that do not exist in mouse models. There are, however, some significant differences between the mouse model of atherosclerosis and human atherosclerosis [5]. Another potential advantage of coordinated studies of atherosclerosis in mice and humans is that systematic linkage mapping in humans can detect novel genetic loci underlying atherosclerosis, but methods are not currently available to clone novel loci for complex multifactorial disease. In contrast, methods are available to clone novel loci underlying complex diseases in mouse models [6-8]. Furthermore, a wide variety of methods are available that can be used to test hypotheses in mouse models about the roles of specific genes in the atherosclerosis disease process -- construction of transgenics and targeted mutations [9,10]. Thus, studies of atherosclerosis in mice and humans produce a synergism that provides a deeper understanding of the mechanisms, causes and treatments for atherosclerosis.

An abbreviated outline of the approaches to analysis of complex diseases is given in Figure 1. The numbers shown correspond to the sections of this review:

1. Complete linkage maps in animal models and humans.
2. Congenic mouse strains are a powerful resource for studies of atherosclerosis
3. Identifying candidate genes from chromosomal location
4. Testing hypotheses by genetic modifications in animals.
5. Testing whether genes identified in animal models also underlie human complex disease.
6. Back to mice.

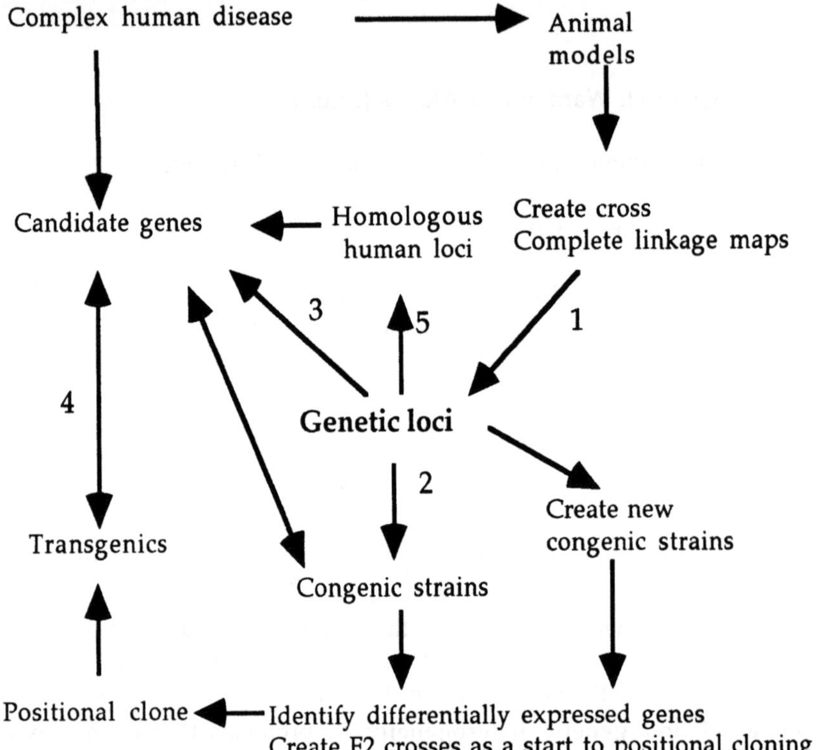

Figure 1. Outline of the steps involved in the analysis of common complex diseases. Numbers in the figure correspond to the numbers of each step in this review.

(1) COMPLETE LINKAGE MAPS

A method has recently been developed for mapping of quantitative trait loci (QTLs) in animal models [11,12]. QTL mapping is a general method that does not require any previous knowledge about the underlying physiology of the disease being studied. QTL mapping involves four steps: (1) Two different inbred strains are crossed to produce F2 or backcross progeny; (2) The progeny are individually genotyped for markers which span the genome at 10 - 20 cM intervals; (3) Each of the progeny is phenotyped for the traits of interest; (4) QTLs are located by a statistical approach, such as that provided by Mapmaker/QTL [13].

Example of mapping complex traits in a mouse model: apoA-II

The principals of QTL mapping can be illustrated with a mouse backcross that has been analyzed for genes underlying obesity and atherosclerosis. A backcross was performed by crossing F1 females, resulting from the cross of female C57BL/6J with male *M. spretus* mice, with male C57BL/6J mice. We have designated these backcross progeny as BSB [14]. Linkage of these genes with the quantitative traits has been determined by analysis of variance (ANOVA) and by LOD score analysis with the MAPMAKER/QTL program [15].

We have measured plasma apolipoproteinA-II (apoA-II) levels in the BSB mice. A peak LOD score of 4.0 was revealed on mouse chromosome 1 centered around the Apoa2 gene locus. The 90% confidence interval for this QTL spans approximately 10 centimorgans and thus includes many diverse genes. However, the co-incidence of the QTL for plasma apoA-II levels and the *Apoa2* gene, suggests that the *Apoa2* gene controls plasma apoA-II levels. This hypothesis will be tested later in congenic mice, transgenic mice and in humans.

Complete linkage maps in humans

Methods to rapidly detect genes underlying complex traits with systematic linkage maps in humans have recently been developed [4,13]. Detection of loci underlying atherosclerosis with systematic linkage maps requires that hundreds of markers by typed in each member of dozens of families. Simple sequence length polymorphism (SSLP) polymerase chain reaction (PCR) markers are ideal for this purpose. There are currently more than 1000 SSLP PCR markers available for human linkage mapping [16,17]. These markers can be typed rapidly at low cost and are highly polymorphic. While these tools will greatly aid in the detection of genes underlying atherosclerosis in humans with systematic linkage maps, several problems remain.. For instance: genes underlying atherosclerosis may not be detected due to genetic and environmental heterogeneity, incomplete penetrance, gene interactions, and because there may be more underlying genes than can be handled by current methods (approximately five or six) [4].

(2) CONGENIC MOUSE STRAINS

Congenic mouse strains provide a rich resource for the rapid identification of the genes underlying complex disease. A congenic mouse strain is genetically identical to a background strain, except for a small chromosomal region derived from a donor strain. A congenic strain is created by placing a gene from various genetic sources onto a standard inbred-strain background by a regimen of crossing and selection [18,19]. Thus, comparison of phenotypes in background strains with those in congenic strains allows study of the effects of single genes derived from the donor strain, isolated from the effects of other donor strain genes [20]. If the donor DNA of a congenic strain and a QTL cover the same genetic locus, and if the congenic differs from the background strain for plasma cholesterol levels, then this will provide strong confirming evidence that the QTL is real. Studies of congenic strains are also useful because they can be used to investigate the biochemistry and physiology of single loci underlying atherosclerosis, whether or not the underlying genes are known. Finally, congenic strains can be used as tools for positional cloning because only one gene will control a trait in a cross of the congenic strain with its background strain.

Congenic mouse strains have been useful in the study of atherosclerosis. The *Ath-1* gene for atherosclerosis susceptibility has been located on distal mouse chromosome 1, near the apo A-II (*Apoa2*) gene locus [5]. C57BL/6J mice are susceptible to fatty streak development on a high fat diet, while BALB/cJ mice do not develop fatty streaks while on the same diet. The B6.C.H-25^{c} congenic strain uses C57BL/6J as the background strain with BALB/cJ as the donor strain for H-25. The apoA-II gene of B6.C.H-25^{c} is also derived from the resistant BALB/cJ.. The congenic mice were susceptible to development of atherosclerosis when placed on the high fat diet and thus the *Apoa2* gene cannot be *Ath-1* [20]. BALB/cJ mice have apoA-II levels twice those found in C57BL/6J mice [21] The B6.C.H-25^{c} congenic mice also showed a two-fold increase of plasma apoA-II compared to the C57BL/6J background strain [22]. These results are consistent with the QTL for apoA-II found in the BSB backcross.

Congenic mouse strains provide a general resource for studies of atherosclerosis. There are at least 38 known mouse histocompatibility genes with at least one congenic strain per histocompatibility antigen. These congenics cover approximately 50 % of the mouse genome.

(3) CANDIDATE GENES

A candidate gene is one that might, because of its properties, underlie a disease [23,24]. Identification of loci for complex traits, either by QTL mapping or by study of congenic strains can be used to guide the identification of candidate genes that should be studied in greater detail. Genes underlying many traits and diseases have been identified by this coincidence mapping approach [25].

(4) TESTING HYPOTHESES BY GENETIC MODIFICATIONS IN ANIMALS

As shown above, a QTL for serum apoA-II levels was demonstrated at the locus surrounding the mouse *Apoa2* gene. It was then demonstrated that B6.C.H-25^{c} congenic mice have increased apoA-II levels compared to C57BL/6J. While these studies suggest that the *Apoa2* gene controls plasma apoA-II levels, we next used transgenic mice to test the hypothesis that the

Apoa2 gene controls plasma apoA-II levels. We generated mice that are transgenic for the mouse *Apoa2* gene [22]. These studies revealed that overexpression of mouse apoA-II elevated HDL-cholesterol concentrations and increased atherosclerotic lesion development [10]. Thus, both the composition and amount of HDL appear to be important determinants of atherosclerosis. These results also suggest that the *Apoa2* gene underlies the QTL previously observed in the BSB cross.

(5) TESTING WHETHER GENES CONTRIBUTE TO HUMAN DISEASE

Extension of mouse mapping results to humans is possible because comparative gene mapping has demonstrated that linked homologous genes are found in 101 conserved autosomal segments [26]. Extension of mouse mapping data to humans is also aided by comparing positions of expressed genes that have been mapped in both species [27]. Mouse (*Apoa2*) and human (APOA2) genes are both found on chromosome 1, in a large conserved linkage group. To test whether variations of the APOA2 gene influence plasma lipid metabolism in humans, we studied 306 individuals in 25 families enriched for coronary artery disease (CAD). The segregation of the APOA2 gene was followed using an informative simple sequence repeat in the second intron of the gene [28]. Robust sib-pair linkage analysis was performed on members of these families using the SAGE linkage programs. The results show linkage between the human APOA2 gene and a gene controlling plasma apoA-II levels ($p=0.01$). Plasma apoA-II levels were also significantly correlated with plasma free fatty acid (FFA) levels ($p=0.007$). Moreover, the APOA2 gene exhibited significant linkage with a gene controlling FFA levels ($p=0.025$) [29].

(6) BACK TO THE MOUSE

We have also identified two loci controlling plasma cholesterol levels in BSB mice. These two loci do not contain any candidate genes for plasma cholesterol synthesis or turnover, and thus likely represent novel genes [15]. These loci are good candidates for studies with congenic strains to start positional cloning of the underlying genes (Figure 1). The methods already demonstrated for analysis of apoA-II suggest that these studies can identify genes that are relevant in mice and in humans.

CONCLUSIONS

Our work with mice strongly suggests that complete linkage maps are very likely to identify genes underlying complex traits. One of the biggest advantages of the systematic linkage approach is that one can narrow the list of candidate genes for a disease to those that are present in the QTLs. Furthermore, systematic linkage maps may also identify novel genes or loci that are important in the etiology of a disease.

Mouse models do not have to directly recapitulate human disease to be useful, since studies in mice may serve to identify pathways and to test mechanisms [30]. For instance, in mice genes underlying QTLs can be isolated on congenic backgrounds and these congenic strains then used as tools to aid in positional cloning. This contrasts with the situation in human families where positional cloning of genes for complex traits is not currently possible. Thus, in mouse studies we can identify novel loci by QTL mapping, the regions containing the QTL genes can be narrowed by the use of congenics, and these can be used as a tool for identifying the underlying genes. Finally, our results show the synergy gained by studying atherosclerosis in both mice and in humans, with studies in the mouse acting to complement difficulties and problems that naturally occur in human studies.

REFERENCES

1. A.J. Lusis, L.W. Castellani and J. Fisler. Fitting pieces from studies of animal models into the puzzle of atherosclerosis, *Curr.Opin.Lipidol.* 3:143-150 (1992).

2. C.H. Warden, A. Daluiski and A.J. Lusis. Identification of new genes contributing to atherosclerosis: The mapping of genes contributing to complex disorders in animal models. In: Monographs in Human Genetics: Molecular Genetics of Coronary Artery Disease, A.J. Lusis, J.I. Rotter and R.S. Sparkes. ed., Karger, Basel: (1992), p. 419-441.

3. J.L. Breslow. The genetic basis of lipoprotein disorders. Introduction and overview, *J.Intern.Med.* 231:627-631 (1992).

4. E.S. Lander and D. Botstein. Mapping complex genetic traits in humans: new methods using a complete RFLP linkage map, *Cold Spring Harbor Symposia on Quantitative Biology* LI:49-62 (1986).

5. B. Paigen, D. Mitchell, K. Reue, A. Morrow, A.J. Lusis and R.C. LeBoeuf. Ath-1, a gene determining atherosclerosis susceptibility and high density lipoprotein levels in mice, *Proc.Natl.Acad.Sci.U.S.A.* 84:3763-3767 (1987).

6. S. Hirotsune, H. Shibata, Y. Okazaki, H. Sugino, H. Imoto, N. Sasaki, K. Hirose, H. Okuizumi, M. Muramatsu, C. Plass, V.M. Chapman, S. Tamatsukuri, C. Miyamato, Y. Furuichi and Y. Hayashizaki. Molecular cloning of polymorphic markers on RLGS gel using the spot target cloning method, *Biochem.Biophys.Res.Commun.* 194:1406-1412 (1993).

7. I. Hatada, Y. Hayashizaki, S. Hirotsune, H. Komatsubara and T. Mukai. A genomic scanning method for higher organisms using restriction sites as landmarks, *Proc.Natl.Acad.Sci.,USA* 88:9523-9527 (1991).

8. N.A. Lisitsyn, J.A. Segre, K. Kusumi, N.M. Lisitsyn, J.H. Nadeau, W.N. Frankel, M.H. Wigler and E.S. Lander. Direct isolation of polymorphic markers linked to a trait by genetically directed representational difference analysis, *Nature Genet.* 6:57-63 (1994).

9. J.L. Breslow. Transgenic mouse models of lipoprotein metabolism and atherosclerosis, *Proc.Natl.Acad.Sci.USA* 90:8314-8318 (1993).

10. C.H. Warden, C.C. Hedrick, J.-H. Qiao, L.W. Castellani and A.J. Lusis. Atherosclerosis in transgenic mice overexpressing apolipoprotein A-II, *Science* 261:469-472 (1993).

11. E.S. Lander, P. Green, J. Abrahamson, A. Barlow, M.J. Daly, S.E. Lincoln and L. Newburg. MAPMAKER: an interactive computer package for constructing primary genetic linkage maps of experimental and natural populations, *Genomics.* 1:174-181 (1987).

12. A.H. Paterson, S. Damon, J.D. Hewitt, D. Zamir, H.D. Rabinowitch, S.E. Lincoln, E.S. Lander and S.D. Tanksley. Mendelian factors underlying quantitative traits in tomato: Comparison across species, generations, and environments, *Genetics* 127:181-197 (1991).

13. E.S. Lander and D. Botstein. Mapping Mendelian factors underlying quantitative traits using RFLP linkage maps, *Genetics* 121:185-199 (1989).

14. C.H. Warden, M. Mehrabian, K. He, M.-Y. Yoon, A. Diep, Y.-R. Xia, P.-Z. Wen, K.L. Svenson, R.S. Sparkes and A.J. Lusis. Linkage mapping of 40 randomly isolated liver cDNA clones in the mouse, *Genomics* 18:295-307 (1993).

15. C.H. Warden, J.S. Fisler, M.J. Pace, K.L. Svenson and A.J. Lusis. Coincidence of genetic loci for plasma cholesterol levels and obesity in a multifactorial mouse model, *J.Clin.Invest.* 92:773-779 (1993).

16. J. Weissenbach, G. Gyapay, C. Dib, A. Vignal, J. Morissette, P. Millasseau, G. Vaysseix and M. Lathrop. A second-generation linkage map of the human genome, *Nature* 359:794-801 (1992).

17. T.J. Hudson, M. Engelstein, M.K. Lee, E.C. Ho, M.J. Rubenfield, C.P. Adams, D.E. Houseman and N.C. Dracopoli. Isolation and chromosomal assignment of 100 highly informative human simple sequence repeat polymorphisms, *Genomics* 13:622-629 (1992).

18. R.J. Graff and G.D. Snell. Histocompatibility genes of mice: VIII. The alleles of the *H-1* locus, *Transplantation* 6:598-617 (1968).

19. D.W. Bailey. Genetics of histocompatibility in mice: I. New loci and congenic lines, *Immunogenet.* 2:249-256 (1975).

20. M. Mehrabian, J.-H. Qiao, R. Hyman, D. Ruddle, C. Laughton and A.J. Lusis. Influence of the ApoA-II gene locus on HDL levels and fatty streak development in mice, *Arterioscler.Thromb.* 13:1-10 (1993).

21. M.H. Doolittle, R.C. LeBoeuf, C.H. Warden, L.M. Bee and A.J. Lusis. A polymorphism affecting apolipoprotein A-II translational efficiency determines high density lipoprotein size and composition, *J.Biol.Chem.* 265:16380-16388 (1990).

22. C.C. Hedrick, L.W. Castellani, C.H. Warden, D.L. Puppione and A.J. Lusis. Influence of mouse apolipoprotein A-II on plasma lipoproteins in transgenic mice, *J.Biol.Chem.* 268:20676-20682 (1993).

23. A.J. Lusis. Genetic factors affecting blood lipoproteins: the candidate gene approach, *J.Lipid.Res.* 29:397-429 (1988).

24. M. Mehrabian and A.J. Lusis. Genetic markers for studies of atherosclerosis and related risk factors. In: Monographs in Human Genetics: Molecular Genetics of Coronary Artery Disease, A.J. Lusis, J.I. Rotter and R.S. Sparkes. ed., Karger, Basel: (1992), p. 363-418.

25. A.J. Lusis. Molecular genetics of coronary artery disease: Candidate genes and processes in atherosclerosis. Introduction, *Monogr.Hum.Genet.* 14:XV-XVII (1992).

26. N.G. Copeland, N.A. Jenkins, D.J. Gilbert, J.T. Eppig, L.J. Maltais, J.C. Miller, W.F. Dietrich, A. Weaver, S.E. Lincoln, R.G. Steen, L.D. Stein, J.H. Nadeau and E.S. Lander. A genetic linkage map of the mouse: Current applications and future prospects, *Science* 262:57-66 (1993).

27. S.J. O'Brien, J.E. Womack, L.A. Lyons, K.J. Moore, N.A. Jenkins and N.G. Copeland. Anchored reference loci for comparative genome mapping in mammals, *Nature Genet.* 3:103-112 (1993).

28. J.L. Weber and P.E. May. Abundant class of human DNA polymorphisms which can be typed using the polymerase chain reaction, *Am.J.Hum.Genet* 44:388-396 (1989).

29. C.H. Warden, A. Daluiski, X. Bu, D.A. Purcell-Huynh, C. De Meester, B.-H. Shieh, D.L. Puppione, R.M. Gray, G.M. Reaven, Y.-D.I. Chen, J.I. Rotter and A.J. Lusis. Evidence for linkage of the apolipoprotein A-II locus to plasma apolipoprotein A-II and free fatty acid levels in mice and humans, *Proc.Natl.Acad.Sci.USA* 90:10886-10890 (1993).

30. N. Risch, S. Ghosh and J.A. Todd. Statistical evaluation of multiple-locus linkage data in experimental species and its relevance to human studies: Application to nonobese diabetic (NOD) mouse and human insulin-dependent diabetes mellitus (IDDM), *Am.J.Hum.Genet.* 53:702-714 (1993).

GENETIC MANIPULATION OF LIPOPROTEIN RECEPTORS: IMPLICATIONS FOR LIPID METABOLISM AND ATHEROSCLEROSIS

Thomas E. Willnow, Shun Ishibashi, and Joachim Herz

Department of Molecular Genetics
University of Texas Southwestern Medical Center
Dallas, TX 75235

INTRODUCTION

The metabolism of plasma lipoproteins is a complex process which involves the interplay of a number of apoproteins, enzymes and cell surface receptors. The primary sites of synthesis for lipoproteins are the gut and the liver which secrete large, triglyceride-rich particles into the lymph ducts (chylomicrons) or directly into the venous circulation (very low density lipoproteins, VLDL). In the capillaries of the peripheral tissues the triglycerides are hydrolyzed by lipoprotein lipase (LPL) and as a consequence the particles shrink in size and become chylomicron remnants or low density lipoproteins (LDL). Eventually, these lipoproteins are taken up into the liver by cell surface receptors which recognize specific apoproteins present on the lipoprotein particle[1].

Chylomicron remnants, the carriers of dietary cholesterol, are taken up into the hepatocytes by at least two independent receptor systems[2]. Binding of the remnants to these receptors is dependent upon apolipoprotein (apo) E. Two endocytic receptors that are known to be expressed in the liver and that bind apoE with high affinity are the low density lipoprotein receptor (LDLR)[3] and the LDL receptor-related protein (LRP)[4-6]. Genetic defects of the LDL receptor are the underlying cause for the inherited disease 'Familial Hypercholesterolemia' (FH). Affected patients have elevated plasma cholesterol levels because they accumulate LDL in their circulation. This, in turn, results in an increased risk for early coronary heart disease[3].

A role for LRP in remnant metabolism has been proposed, because a) LDLR-deficient ($LDLR^{-/-}$) patients and animals do not accumulate chylomicron remnants in their circulation[2,7,8], and b) LRP binds apoE-enriched lipoprotein particles as well as LPL[9]. Hepatic LRP efficiently removes circulating ligands from the bloodstream[10].

In an effort to better characterize the physiological functions of LDLR and LRP we have experimentally altered their activity in animal model systems. This was accomplished by two different approaches: Gene transfer using adenovirus vectors and gene disruption by homologous recombination in embryonic stem cells.

LDL RECEPTOR-DEFICIENT (LDLR$^{-/-}$) MICE

Mice lacking functional LDL receptors (LDLR$^{-/-}$) were generated by inserting a neo expression cassette into exon 4 of the murine LDL receptor gene in embryonic stem cells[8]. Chimeric mice which transmitted the disrupted allele through their germ line were generated using established procedures and the resulting heterozygous offspring were bred to generate homozygous animals. LDLR$^{-/-}$ mice are fertile and appear superficially normal. In contrast to wild type animals which do not have significant levels of circulating LDL, LDLR$^{-/-}$ mice accumulate large amounts of this lipoprotein in their bloodstream. However, the total plasma cholesterol levels of these mice (~250 mg/dl) do not reach the levels of human FH-homozygotes (~600 - 1000 mg/dl).

In human patients FH strongly predisposes to premature atherosclerosis. LDLR$^{-/-}$ mice, on the other hand, do not develop significant atherosclerotic lesions on a normal mouse chow diet which is relatively poor in cholesterol. Such lesions spontaneously form in apolipoprotein E (apoE) deficient mice[11,12]. However, if the LDLR$^{-/-}$ animals are fed a cholesterol-rich diet, they also develop pronounced atherosclerotic lesions over the duration of a few months[13].

LDLR$^{-/-}$ MICE: A MODEL FOR *IN VIVO* GENE TRANSFER AND GENE THERAPY

FH and other metabolic genetic diseases that manifest themselves in the liver could in principle be causally cured by transferring functional copies of the defective gene into the somatic cells of the affected tissue, in this case the hepatic parenchymal cells. This requires an efficient gene transfer system which is capable of transferring the therapeutic gene into a majority of the liver cells. There, under ideal circumstances, the gene should be integrated into the genome of the hepatocytes leading to a permanent correction of the disease phenotype.

A vector which satisfies the first of these requirements, efficient gene transfer into a high percentage of the hepatocytes, is the human adenovirus. This virus, which normally infects the respiratory epithelium, has nevertheless a broad tissue tropism. In fact, after intravenous injection of recombinant adenovirus into the systemic circulation of laboratory animals (mice and rabbits) adenovirus predominantly targets the liver hepatocytes[14,15].

We have exploited these obvious advantages of adenovirus to transfer the human LDLR cDNA driven by a strong viral promoter into the liver of the FH-mice. This resulted in the complete correction of the hypercholesterolemic phenotype[8].

Albeit the reversal of the gene defect was only transient, it nevertheless demonstrates the principal feasibility of the exogenous supplementation of dysfunctional genes in the liver. If ways can be found to prolong the expression of the exogenous gene, by readministration of the gene or by integrating it into the genome of the host cell, gene therapy of a variety of metabolic disorders would become a reality.

Because of the ease by which the therapeutic effect of gene transfer measures can be assessed (simple determination of the plasma cholesterol level and resistance to diet-induced atherosclerosis), $LDLR^{-/-}$ mice are in many ways an ideal model system on which future gene therapy strategies could be tested.

THE ROLE OF LRP IN CHYLOMICRON REMNANT CLEARANCE

Because of its close structural similarity to the LDL receptor, LRP has originally been proposed to function as a chylomicron remnant receptor. The existence of such a receptor which could function as a backup for the LDL receptor had been postulated on the grounds that FH patients and rabbits lacking functional LDL receptors accumulate only LDL, but not chylomicron remnants, in their circulation[2,7,8]. In contrast, apoE-deficiency results in the accumulation of mainly chylomicron remnants but not of LDL[11,12]. Therefore, the remnant receptor would be expected to be specific for apoE and unable to recognize apoB100, the sole apoprotein in LDL.

This binding specificity agrees with that of LRP which only binds apoE, but not apoB100, *in vitro* and in cultured cells. However, it has been difficult to demonstrate this function of LRP *in vivo*. An animal model for an LRP gene defect has been generated. Unfortunately, LRP-deficient embryos die *in utero* early during development thus precluding an analysis of remnant metabolism in adult LRP-deficient mice. The lethal phenotype of the LRP gene defect is most likely due to the multifunctional nature of this receptor. Besides its postulated function in lipoprotein metabolism, LRP also functions as a receptor for α_2-macroglobulin and for a variety of other proteases and protease inhibitors[16-18]. The importance of regulated protease activity in the developing organism is well established and a disturbance of this system may be incompatible with normal development.

To circumvent this problem of the early embryonic lethality we made use of a receptor-associated protein (RAP) which was first noticed during the purification of LRP on α_2-macroglobulin affinity columns[17,18]. This protein of approximately 39 kDa was subsequently found to block the binding of all known ligands to this multifunctional receptor. The physiological function of RAP is still unclear but it has been proposed that RAP could act as a fast-acting modulator of LRP activity in response to an unknown extracellular signal[16]. Alternatively, RAP might be a chaperone required for the proper folding of LRP in the endoplasmic reticulum. In the liver, LRP is the only protein which binds RAP with high affinity. The affinity of the LDL receptor for RAP is between 100 and 1000 times lower[19,20].

Because of its potent inhibitory properties and its nearly selective action on LRP, we have overexpressed RAP in the livers of mice[21]. To accomplish this we have again used the adenovirus system which allows high expression of foreign genes in virtually all hepatic parenchymal cells. Overexpression of RAP from the strong cytomegalovirus promoter resulted in the release of RAP from the endoplasmic reticulum where it usually resides and in secretion of the protein into the circulation of the mice. There, it accumulated in concentrations approaching 0.5 mg/ml plasma. As expected from previous *in vitro* studies, RAP completely blocked hepatic LRP function in the infected mice. This was shown by measuring the turnover of ^{125}I-labeled α_2-macroglobulin which was completely abolished in these animals. The clearance of an unrelated ligand, asialofetuin, which is also taken up into the hepatocytes, but by an unrelated receptor, was not affected. RAP overexpression resulted in a mild elevation of plasma cholesterol levels in wild type mice which contain two functional LDLR alleles. The cholesterol was mainly contained in chylomicron remnants, because only lipoprotein particles containing apoE and apoB48 (a diagnostic marker for these lipoproteins) accumulated in these mice. In contrast, LDLR-deficient mice which had been injected with the RAP virus accumulated large amounts of chylomicron remnants in their circulation. These findings add further support to the hypothesis that both, LDLR and LRP, are involved in the uptake of dietary cholesterol into the liver.

CONCLUDING REMARKS

Gene disruption and gene transfer are powerful and complementing tools to elucidate the function of genes which function in complex biological systems such as the metabolism of plasma lipoproteins. In our experiments we have used these methods to dissect the two receptor system which mediates the uptake of dietary lipoproteins into the liver. The increasing number of genetically altered animals lacking or overexpressing genes intrinsically involved in lipid metabolism or in the formation of atherosclerotic lesions will allow us in the future to dissect the physiological mechanisms in which these gene functions *in vivo*. Ultimately, this might allow better treatment or prevention of cardiovascular disease.

REFERENCES

1. Havel, R.J. and Kane, J.P. (1989). Structure and Metabolism of Plasma Lipoproteins. In The Metabolic Basis of Inherited Disease. C.R. Scriver, A.L. Beaudet, W.S. Sly, and D. Valle, eds. (New York: McGraw-Hill), pp. 1129-1138.

2. Kita, T., Goldstein, J.L., Brown, M.S., Watanabe, Y., Hornick, C.A., and Havel, R.J. (1982). Hepatic uptake of chylomicron remnants in WHHL rabbits: A mechanism genetically distinct from the low density lipoprotein receptor. Proc. Natl. Acad. Sci. USA *79*, 3623-3627.

3. Goldstein, J.L. and Brown, M.S. (1989). Familial Hypercholesterolemia. In The Metabolic Basis of Inherited Disease. C.R. Scriver, A.L. Beaudet, W.S. Sly, and D. Valle, eds. (New York, N.Y.: McGraw-Hill Publishing Co.), pp. 1215-1250.

4. Herz, J., Hamann, U., Rogne, S., Myklebost, O., Gausepohl, H., and Stanley, K.K. (1988). Surface location and high affinity for calcium of a 500-kd liver

membrane protein closely related to the LDL-receptor suggest a physiological role as lipoprotein receptor. EMBO J. *7*, 4119-4127.

5. Beisiegel, U., Weber, W., Ihrke, G., Herz, J., and Stanley, K.K. (1989). The LDL receptor-related protein, LRP, is an apolipoprotein E-binding protein. Nature *341*, 162-164.

6. Kowal, R.C., Herz, J., Goldstein, J.L., Esser, V., and Brown, M.S. (1989). Low density lipoprotein receptor-related protein mediates uptake of cholesteryl esters derived from apoprotein E-enriched lipoproteins. Proc. Natl. Acad. Sci. USA *86*, 5810-5814.

7. Rubinsztein, D.C., Cohen, J.C., Berger, G.M., van der Westhuyzen, D.R., Coetzee, G.A., and Gevers, W. (1990). Chylomicron remnant clearance from the plasma is normal in familial hypercholesterolemic homozygotes with defined receptor defects. J. Clin. Invest. *86*, 1306-1312.

8. Ishibashi, S., Brown, M.S., Goldstein, J.L., Gerard, R.D., Hammer, R.E., and Herz, J. (1993). Hypercholesterolemia In LDL Receptor Knockout Mice And Its Reversal By Adenovirus-Mediated Gene Delivery. J. Clin. Invest. *92*, 883-893.

9. Beisiegel, U., Weber, W., and Bengtsson-Olivecrona, G. (1991). Lipoprotein lipase enhances the binding of chylomicrons to low density lipoprotein receptor-related protein. Proc. Natl. Acad. Sci. USA *88*, 8342-8346.

10. Herz, J., Kowal, R.C., Ho, Y.K., Brown, M.S., and Goldstein, J.L. (1990). Low density lipoprotein receptor-related protein mediates endocytosis of monoclonal antibodies in cultured cells and rabbit liver. J. Biol. Chem. *265*, 21355-21362.

11. Plump, A.S., Smith, J.D., Hayek, T., Aalto-Setälä, K., Walsh, A., Verstuyft, J.G., Rubin, E.M., and Breslow, J.L. (1992). Severe Hypercholesterolemia and Atherosclerosis in Apolipoprotein E-Deficient Mice Created by Homologous Recombination in ES Cells. Cell *71*, 343-353.

12. Zhang, S.H., Reddick, R.L., Piedrahita, J.A., and Maeda, N. (1992). Spontaneous hypercholesterolemia and arterial lesions in mice lacking apolipoprotein E. Science *258*, 468-471.

13. Ishibashi, S., Goldstein, J.L., Brown, M.S., Herz, J., and Burns, D.K. (1994). Massive Xanthomatosis and Atherosclerosis in Cholesterol-Fed LDL Receptor-Negative Mice. J. Clin. Invest. *93*, 1885-1893.

14. Stratford-Perricaudet, L.D., Makeh, I., Perricaudet, M., and Briand, P. (1992). Widespread Long-term Gene Transfer to Mouse Skeletal Muscle and Heart. J. Clin. Invest. *90*, 626-630.

15. Herz, J. and Gerard, R.D. (1993). Adenovirus-mediated transfer of low density

lipoprotein receptor gene acutely accelerates cholesterol clearance in normal mice. Proc. Natl. Acad. Sci. USA *90*, 2812-2816.

16. Herz, J. (1993). The LDL-receptor-related protein - portrait of a multifunctional receptor. Current Opinion in Lipidology *4*, 107-113.

17. Strickland, D.K., Ashcom, J.D., Williams, S., Burgess, W.H., Migliorini, M., and Argraves, W.S. (1990). Sequence identity between the a_2-macroglobulin receptor and low density lipoprotein receptor-related protein suggests that this molecule is a multifunctional receptor. J. Biol. Chem. *265*, 17401-17404.

18. Kristensen, T., Moestrup, S.K., Gliemann, J., Bendtsen, L., Sand, O., and Sottrup-Jensen, L. (1990). Evidence that the newly cloned low-density-lipoprotein receptor related protein (LRP) is the a2-macroglobulin receptor. FEBS Lett. *276*, 151-155.

19. Mokuno, H., Brady, S., Kotite, L., Herz, J., and Havel, R.J. (1994). Effect of 39-kDa Receptor-associated Protein on the Hepatic Uptake and Endocytosis of Chylomicron Remnants and Low Density Lipoproteins in the Rat. J. Biol. Chem. *269*, 13238-13243.

20. Goldstein, J.L., Herz, J., and Brown, M.S. Unpublished observations.

21. Willnow, T.E., Sheng, Z., Ishibashi, S., and Herz, J. (1994). Inhibition of Hepatic Chylomicron Remnant Uptake by Gene Transfer of A Receptor Antagonist. Science *264*, 1471-1474.

GENETIC MODELS OF VASCULAR DISEASE

Elizabeth G. Nabel[1], David Gordon[3], Diane P. Carr[1], Takeshi Ohno[1], Zhiyong Yang[4], Hong San[1], and Gary J. Nabel[1,2,4]

Departments of Internal Medicine[1]
Biological Chemistry[2] and Pathology[3]
Howard Hughes Medical Institute[4]
University of Michigan Medical Center
Ann Arbor, MI 48109-0688

INTRODUCTION

Direct gene transfer is the introduction of recombinant genes into host cells of a recipient organism, and represents a new approach to the study of gene expression and regulation in vivo. Our laboratory has had an interest in using direct gene transfer to investigate the pathobiology of vascular diseases and to develop potential new treatments for these diseases.

Early studies in our laboratory were focused on developing methods for direct gene transfer of recombinant genes into focal segments of arteries in vivo (1,2). Over the past five years, different viral and nonviral vectors have been used by others and us to perform vascular gene transfer. Each vector has its own advantages and limitations with regard to transfection efficiency, integration, stability of gene expression, and safety profile. In addition, modifications are being made to these vectors by many investigators in order to improve transfection efficiency and minimize toxicity within the host.

GROWTH FACTOR GENE EXPRESSION

A major recent interest of our laboratory has been to investigate the mechanisms of cellular proliferation following the expression of growth factor genes in porcine arteries in vivo. In particular, we have examined three recombinant growth factor genes, PDGF B gene (3), a secreted form of FGF-1 (4), and a secreted active form of TGF-β 1 (5). While data from human atherosclerosis specimens and gene targeting studies have suggested an important role for these growth factors in human vascular disease, the effect of these growth factors in arteries in vivo have been poorly understood. An advantage of gene transfer is that a recombinant gene can be directly transfected into a focal arterial segment where its expression and effects on arterial function can be evaluated. Therefore, we used direct gene transfer to investigate the pathobiology of these growth factor genes in vivo. In these experiments, the recombinant gene was transfected into pig arteries

Cardiovascular Disease 2
Edited by L.L. Gallo, Plenum Press, New York, 1995

using cationic liposomes and biological expression was analyzed twenty-one days later. Control experiments were performed by transfecting a reporter gene into porcine arteries. Briefly, arteries transfected with PDGF BB demonstrated severe intimal thickening, characterized by increased cellularity and smooth muscle cell proliferation in contrast to control arteries transduced with a reporter gene. Intimal thickening was also observed in FGF-1 transduced arteries; however, in FGF-1 transduced vessels with expanded intima, capillaries were present within the intima. This intimal angiogenesis was not observed in vessels transduced with other growth factor genes, including PDGF and TGF-β. This data supports a role for FGF-1 in simulating vascular angiogenesis in vivo. Porcine arteries transduced with TGF-β 1 demonstrated increased extracellular matrix synthesis compared with PDGF B or *lac*Z transduced arteries. These studies suggested that expression of PDGF, FGF-1 and TGF-β 1 genes have distinct effects on arterial morphology, including different effects smooth muscle cell proliferation, angiogenesis, and extracellular matrix synthesis. These studies also suggested that selective targeting of a single growth factor may be insufficient to reduce intimal thlckening following arterial injury.

MODIFICATION OF SMOOTH MUSCLE CELL PROLIFERATION

An alternate approach to inhibit smooth muscle cell proliferation in vivo includes local delivery of antiproliferative agents. Since accumulation of vascular smooth muscle cells constitutes a major feature of vascular proliferative disorders, we hypothesized that local delivery of antiproliferative agent during the peak of smooth muscle cell division might limit intimal hyperplasia. One approach to the selective elimination of dividing cells is to express a recombinant herpes virus thymidine kinase gene (HSV-tk) in smooth muscle cells. HSV-tk converts the nucleoside analog ganciclovir into a phosphorylated form in transduced cells, and the subsequent incorporation of phosphorylated ganciclovir into cellular DNA induces cell death. We evaluated the feasibility of using this approach to limit smooth muscle cell proliferation in porcine arteries by infecting injured porcine arteries with an adenoviral vector expressing the HSV thymidine kinase gene. Previous studies in our laboratory suggested that infection of injured porcine arteries with an adenoviral vector encoding the reporter gene, alkaline phosphatase, resulting in gene expression in smooth muscle cells in the intima and lumenal region of the media, appropriate targets for studies of the HSV-tk gene. We have introduced adenoviral vectors encoding HSV-tk into balloon injured porcine arteries. Introduction of the tk gene into intimal smooth muscle cells in vivo rendered them sensitive to treatment with ganciclovir. A decrease in intimal to medial area ratios was observed in HSV-tk infected arteries transfected with ganciclovir compared with HSV-tk infected arteries treated with saline and arteries infected with a control vector treated with ganciclovir saline. These studies suggest that direct gene transfer of a HSV-tk gene into a local vessel segment allows sufficient conversion of ganciclovir at the site of pathology to limit cellular proliferation in vivo. The ability to inhibit vascular cell proliferation in response to different stimuli could provide further insight into the cellular pathogenesis of vascular diseases.

REFERENCES

1. E.G. Nabel, G. Plautz, F.M. Boyce, J.C. Stanley, et al, Recombinant gene expression in vivo within endothelial cells of the arterial wall, Science 244:1342 (1989).

2. E.G. Nabel, G. Plautz, and G.J. Nabel. Site-specific gene expression in vivo by direct gene transfer into the arterial wall, Science 249:1285 (1990).
3. E.G. Nabel, Z. Yang, S. Liptay, et al: Recombinant platelet-derived growth factor B gene expression in porcine arteries induces intimal hyperplasia in vivo, J. Clin. Invest. 91:1822 (1993).
4. E.G. Nabel, Z. Yang, G. Plautz, et al: Recombinant fibroblast growth factor- 1 promotes intimal hyperplasia and angiogenesis in arteries in vivo, Nature 362:844 (1993).
5. E.G. Nabel, L. Shum, V.J. Pompili, et al: Direct gene transfer of transforming growth factor β1 into arteries stimulates fibrocellular hyperplasia, Proc. Natl. Acad. Sci. U. S. A. 90:10759 (1993).

GENERATION AND CHARACTERIZATION OF APOLIPOPROTEIN C1-DEFICIENT MICE

Marten H. Hofker[1], Janine H. van Ree[1,2], Walther J.A.A. van den Broek[3], Jan M.A. van Deursen[3], Hans van der Boom[2], Rune R. Frants[1], Bé Wieringa[3], and Louis M. Havekes[2]

[1]MGC-Department of Human Genetics, Leiden University, Leiden
[2]PG-TNO, Gaubius Laboratory, Leiden
[3]Department of Cell Biology and Histology,
Nijmegen University, Nijmegen
The Netherlands

INTRODUCTION

Genetic defects in humans have greatly improved our insight in the function of genes controlling the lipoprotein metabolism. A good example is the apolipoprotein (APO) E gene, for which both dominant and recessive mutations have been observed. These findings have clearly established an important role for apoE in lipoprotein remnant metabolism[1]. In contrast, mutations within the APOC1 gene have not been observed hampering insight in the *in vivo* function of apoC1.

The apoC1 protein is a component of chylomicrons, very low density lipoproteins (VLDL) and high density lipoproteins (HDL). *In vitro* studies have shown that APOC1 can activate the enzyme lecithin:cholesterol acyltransferase (LCAT)[2]. In addition, apoC1 can inhibit apoE mediated binding of apoE-enriched β-VLDL to the low density lipoprotein (LDL) receptor and the LDL receptor related protein (LRP)[3]. The aim of our study is to complement these data with *in vivo* studies by generating a mouse model carrying a null-allele for the endogenous *Apoc1* gene.

Cardiovascular Disease 2
Edited by L.L. Gallo, Plenum Press, New York, 1995

The human APOC1 gene, together with APOE and APOC2 genes are located in a cluster on chromosome 19. The structure of the APOE-C1-C2 gene cluster (figure 1) is conserved between man and mouse[4]. The individual mouse genes share extensive sequence homology with their human counterparts. Detailed analysis of the mouse gene cluster has lead to the discovery of an additional conserved gene located between *Apoc1* and *Apoc2*. This gene has been named *Apoc2* Linked Gene (*Acl*)[5] and has an unknown function. The high degree of similarity between the mouse and human loci underscores

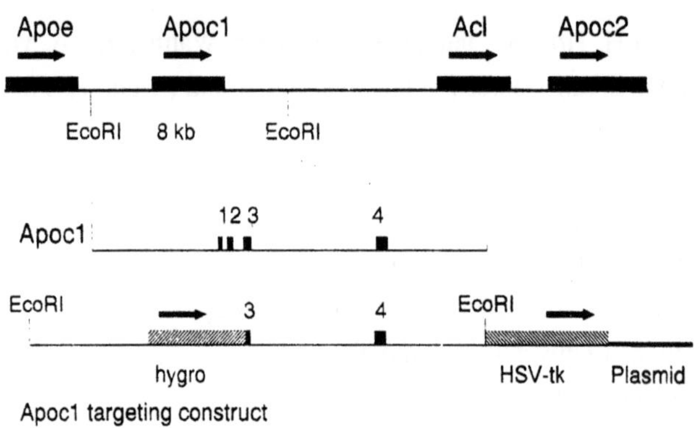

Figure 1. The mouse *Apoe-c1-Acl-c2* gene cluster is schematically represented by the top bar, genes are solid boxes, and arrows indicate the direction of transcription. The middle bar represents the mouse *Apoc1* gene, small solid boxes are the exons. At the bottom, the targeting construct is shown. Striped boxes indicate the hygromycin resistance gene and the Herpes simplex thymidine kinase gene (HSV-TK).

the feasibility of making a valid mouse model, which would mimic human apoC1 deficiency.

DISRUPTION OF *Apoc1* BY GENE TARGETING

The mouse *Apoc1* gene is 3.3 kb in size. The 409 bp mRNA is encoded by 4 exons[6]. The procedures used for gene targeting of *Apoc1* were essentially the same as previously used to disrupt the *Apoe* gene[7]. A replacement type vector was based on an 8 kb *EcoRI* fragment containing *Apoc1*, and was designed to delete exons 1, 2 and a part of exon 3 (figure 1). After transfection of mouse embryonic stem cells (ES), hygromycin resistant colonies were tested for proper homologous recombination of the targeting vector with the endogenous *Apoc1* gene. Hybridizations were carried out using a probe that was located just outside the *EcoRI* fragment. Clones in which the targeting vector has integrated via homologous recombination show a distinct 9.3 kb band, and clones with randomly integrated vectors yield a native 8.0 kb band (figure 2A). For *Apoc1* a frequency of homologous recombination of 1:5 was found. Importantly, one of our clones gave rise to highly chimeric mice, which transmitted the *Apoc1*-null allele through the germline. Figure 2B shows the result of DNA typing of a cross between two

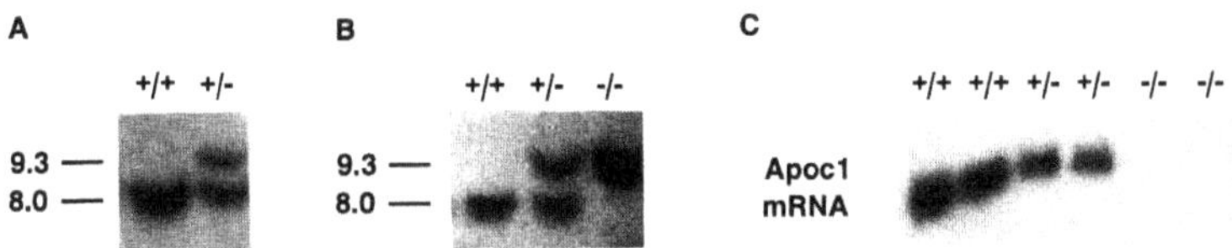

Figure 2. Panels A and B show Southern blots containing *HindIII* DNA of ES cells and mice tailtips, respectively. Normal alleles (+) are 8.0 kb in size; alleles generated by homologous recombination (-) have a size of 9.3 kb. Panel C shows a Northern blot containing liver mRNA of control (+/+), heterozygous (+/-), and homozygous (-/-) apoC1 deficient mice, hybridized with a mouse *Apoc1* cDNA probe.

heterozygous mice, producing mice homozygous for the *Apoc1*-null allele, which are healthy and do not exhibit overt abnormalities.

PHENOTYPE OF *Apoc1*-null MICE

Apoc1 mRNA is absent in the livers of these mice, and present at a reduced level in heterozygous mice (figure 2C). Antibodies specific for mouse apoC1 do not detect any apoC1 protein in homozygous mice. Hence the targeted *Apoc1* gene is completely silenced. For each genotype 9 - 10 mice were given 3 different diets for three weeks. These diets were: I) regular chow, II) a sucrose based diet containing 0.25% cholesterol and 15% saturated fat (HFC), III) a similar diet with 1% cholesterol and 0.5% cholate (HFC0.5%)[8]. The serum lipid levels are depicted in table 1. Absence of apoC1 leads to a small increase of serum triglyceride levels in mice kept on both the chow and HFC diet.

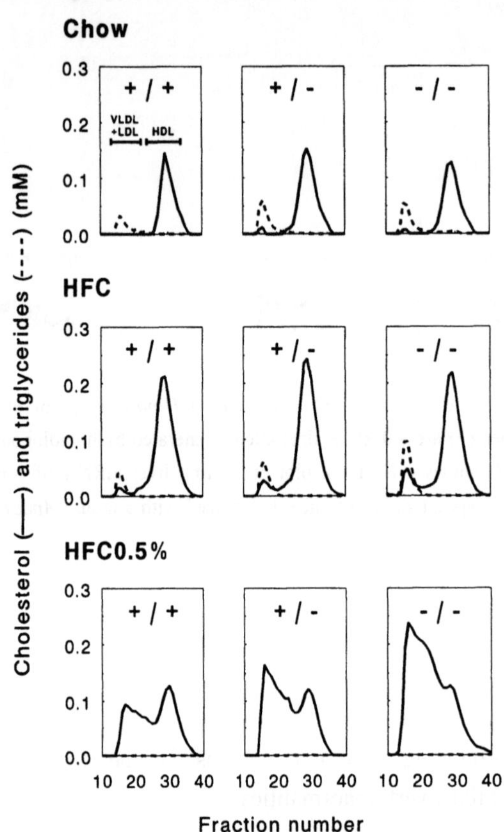

Figure 3. Lipoprotein profiles of control (+/+), heterozygous (+/-), and homozygous apoC1 deficient (-/-) mice on chow, HFC and HFC0.5% diets. Fractions containing VLDL/LDL and HDL are 14-23 and 24-33, respectively, as shown in the left upper panel.

Table 1. Serum cholesterol and triglyceride levels in apoC1 deficient mice.

Apoc1-genotype	N	Diet	Serum lipids (mean ± S.D.)	
			TC	TG
+/+	10	chow	3.0 ± 0.6	0.2 ± 0.1
+/-	10	chow	3.0 ± 0.4	0.4 ± 0.1
-/-	10	chow	2.4 ± 0.3	0.4 ± 0.1
+/+	10	HFC	3.9 ± 1.0	0.6 ± 0.4
+/-	10	HFC	3.9 ± 0.6	0.7 ± 0.3
-/-	9	HFC	4.0 ± 0.5	1.1 ± 0.4
+/+	10	HFC0.5%	5.1 ± 1.6	N.D.
+/-	10	HFC0.5%	6.7 ± 1.8	N.D.
-/-	9	HFC0.5%	10.7 ± 3.3	N.D.

S.D., standard deviation; N, number of animals analyzed; TC, total cholesterol; TG, triglycerides. +/+, *Apoc1* alleles unaffected; +/-, heterozygous for the *Apoc1*-null allele; -/-, homozygous for the *Apoc1*-null allele; N.D., not detectable.

Total serum cholesterol levels are unaffected. Lipoprotein fractions were analyzed by fast protein liquid chromatography (FPLC) of the pooled serum. Figure 3 shows that the accumulation of triglycerides takes place in the VLDL/LDL sized fractions. Although on the HFC diet the serum cholesterol values were not changed, the distribution of cholesterol has shifted from the HDL fraction to the VLDL/LDL fraction. A more pronounced phenotype was observed in mice kept on the HFC0.5% diet. The cholesterol values of apoC1 deficient mice are two times higher than those found in the controls (table 1). This increase is due to the accumulation of VLDL/LDL sized particles (figure 3).

CONCLUSIONS

The apoC1-deficient mice show, on a regular diet and on a moderate high fat diet (HFC), small changes in the amount and composition of the VLDL. A severe hypercholesterolemic (HFC0.5%) diet leads to a marked accumulation of cholesterol in the VLDL/LDL fraction. However, the phenotype associated with apoC1 deficiency is rather subtle, and only becomes prominent during the administration of a severe diet. Therefore, it is posssible that apoC1 shares functions with other apolipoproteins. It would be interesting to test this hypothesis by breeding these *Apoc1* knock out mice with mice deficient for other apolipoproteins. If certain functions are shared between different lipoproteins, then the combination of two knock-out mutations in one mouse should unmask these functional redundancies

REFERENCES

1. R.W. Mahley, T.L. Innerarity, S.C. Rall, Jr, K.H. Weisgraber, and J.M. Taylor, Apolipoprotein E: genetic variants provide insight into its structure and function, *Curr. Opin. Lipid.* 1:87-95 (1990).
2. A.K. Soutar, C.W. Garner, H.N. Bakker, J.T. Sparrow, R.L Jackson, A.M. Gotto, and L.C. Smith, Effect of the human plasma apolipoproteins and phosphatidylcholine acyl donor on the activity of lecithin:acyltransferase, *Biochem.* 14:3057-3064 (1975).
3. K.H. Weisgraber, R.W. Mahley, R.C. Kowall, J. Herz, J.L. Goldstein, and M.S. Brown, *J. Biol. Chem.* 265:22453-22459 (1990).
4. M.J.V. Hoffer, M.H. Hofker, M.M. van Eck, L.M. Havekes, and R.R. Frants, Evolutionary conservation of the mouse apolipoprotein E - C1 - C2 genecluster; structure and genetic variability in inbred mice. *Genomics*, 15:62-67 (1993).
5. M.M. van Eck MM, M.J.V Hoffer, L.M. Havekes, R.R. Frants, and M.H. Hofker, Detection of a new gene within the mouse apolipoprotein *e-c1-c2* gene cluster, *Genomics* (in press).
6. M.J.V. Hoffer, M.M. van Eck, L.M. Havekes, M.H. Hofker, and R.R. Frants, The mouse apolipoprotein C1 gene: structure and expression. *Genomics*, 18:37-42 (1993).
7. J.H. van Ree, W.J.A.A. van den Broek, V.E.H. Dahlmans, P.H.E. Groot, M. Vidgeon-Hart, R.R. Frants, B. Wieringa, L.M. Havekes, and M.H. Hofker, Diet-induced hypercholesterolemia and atherosclerosis in heterozygous apolipoprotein E-deficient mice, *Atherosclerosis* (in press).
8. B. van Vlijmen, A.M.J.M. van den Maagdenberg, M. Gijbels, H. van der Boom, R.R Frants, M.H. Hofker, and L.M. Havekes, Diet induced hyperlipipoproteinemia and atherosclerosis in transgenic APOE3-Leiden mice *J. Clin. Invest.* 93: 1403-1410 (1994).

STUDIES OF APOLIPOPROTEIN B AND LIPOPROTEIN(a) IN TRANSGENIC MICE

Edward M. Rubin

Human Genome Center
Lawrence Berkeley Laboratory
University of California
Berkeley, California 94720

SUMMARY

Two of the many differences in lipoprotein metabolism between mice and humans are a predominance of high density lipoproteins (HDL) in mice compared with high levels of low density lipoproteins (LDL) in humans, and the lack of either apolipoprotein(a) (apo[a]) or lipoprotein(a) (Lp[a]) in the mouse. In order to develop a strain of mice with higher levels of LDL and with Lp(a), a P1 phagemid library was screened for a human apoB genomic clone. A clone containing the entire human apoB gene was isolated and used to create multiple lines of mice expressing high levels of human apoB. The lipoproteins of the human apoB transgenic animals differed from control mice due to the presence of significant amounts of low-density lipoprotein particles containing human apoB[1]. The human apoB transgenic mice were crossed with a previously described line of transgenic mice expressing human apo(a) cDNA[2]. The resulting combined transgenic animals expressing both human apoB and human apo(a) produced an Lp(a) particle sharing many properties with Lp(a) present in the plasma of humans. The present series of transgenic animals will serve as useful substrates to address issues relevant to LDL and Lp(a) that are difficult to approach in either humans or naturally occurring animal models.

STUDIES OF APO(a) AND ATHEROGENESIS IN MICE

The mouse has been used as an experimental model for lipoprotein research for only a short time. Sophisticated genetic analysis of the species has recently been used to develop mouse models having the ability to address hypotheses of lipoprotein metabolism not possible in other systems. The genetic technologies available for mice, including transgenesis and gene targeting, have enabled the creation of animals to serve as substrates providing insights into lipoprotein metabolism not available in other organisms[3, 4].

The low levels of LDL and the absence of apo(a) and Lp(a) in the mouse has limited its use for analysis of this aspect of human lipoprotein metabolism. Studies of apo(a) and Lp(a) in general have been limited to only a few organisms: old world monkeys, humans, and the European hedgehog, thereby restricting the ability to study the biology of this molecule, including factors determining plasma level and atherogenic properties. This

Cardiovascular Disease 2
Edited by L.L. Gallo, Plenum Press, New York, 1995

deficit has partially been overcome with the creation of transgenic mice that overexpress a human apo(a) cDNA driven by the transferrin promoter[5]. Characterization of the apo(a) transgenic mice revealed that human apo(a) in these animals existed in the lipid-free plasma fraction. In humans, however, the vast majority of apo(a) is associated with LDL via a covalent linkage between apo(a) and apoB of LDL[6-8]. Analysis of human apo(a) transgenic mice revealed an inability of the human apo(a) to interact with murine apoB[2, 9]. Infusion of human LDL into the apo(a) transgenic mice did result in the production of an Lp(a) particle *in vivo*, suggesting that production of human apoB-containing LDL in mice would be necessary to produce an Lp(a) particle[2].

Despite the failure to form Lp(a) particles *in vivo*, analysis of the apo(a) transgenic mice has provided insights into *in vivo* properties of apo(a). Although apo(a) is normally lacking in mice, the apo(a) transgenics had a significantly increased risk of diet-induced atherosclerosis compared to non-transgenic control animals[9]. Since both the apo(a) transgenic and the non-transgenic control animals had nearly identical lipoprotein profiles, these results suggested that the presence of the apo(a) protein, although not associated with LDL, may be directly linked to the atherogenic properties attributed to Lp(a). Furthermore, support for this comes from analysis of the fatty streak lesions which developed in the apo(a) transgenic mice[9]. These lesions contained human apo(a), suggesting a causative relationship between this protein and formation of the fatty streak lesion. This finding was most unexpected, due to the fact that the majority of apo(a) in these animals was free in the plasma, not associated with apoB and LDL.

Although the above studies with the apo(a) transgenic mice provided interesting insights into potential properties of apo(a), the absence of Lp(a) particles in the plasma of these mice clearly question its relevance to studies of apo(a)'s properties in humans.

CREATION OF APOB TRANSGENIC MICE

Due to the importance of apoB in determining LDL structure, as well as its interactions with apo(a), a significant effort has gone into creating transgenic mice expressing human apoB. The apoB cDNA is approximately 14 Kb and the gene is greater than 40 Kb. A variety of mini gene constructions have either failed to express apoB or express it at extremely low levels. The difficulty in expressing apoB cDNAs suggest that flanking or intronic genomic DNA sequences, absent in the apoB mini-gene constructs, may be necessary for expression of this gene. The availability of P1 phagemid libraries containing the human genome have enabled two groups to independently isolate P1 phagemid clones containing the entire human apoB gene, plus significant amounts of 5' and 3' flanking DNA[1, 10]. Comparing multiple founder lines of human apoB transgenic mice revealed plasma levels of apoB in these animals to be roughly proportional to the copy member of the apoB transgene, ranging from 1.6 mg/dl to 71 mg/dl[1]. A surprising feature of these animals, despite greater than 15 KB of 5' and 3' flanking DNA, was discovering that the human transgene expressed exclusively in the liver and lacked intestinal expression, a tissue where both human and murine apoB is normally expressed. This finding suggests that intestinal apoB expression from the 75 Kb genomic clone may require DNA elements located distal to the flanking regions included in the apoB fragment. Similarly, the apoAI gene has demonstrated tissue-specific elements necessary for expression located distal to the coding sequences of apoAI in transgenic mice[11].

Although mice have extremely low concentrations of LDL cholesterol, even when fed a diet high in fat and cholesterol, transgenic animals expressing human apoB have increased levels of an LDL-like particle. This triglyceride-rich particle was found in the same density range where human LDL is normally found. The fact that this LDL particle was more triglyceride-rich rather than human LDL, may reflect other basic differences in lipoprotein metabolism between mice and humans. Mice do not produce cholesteryl ester transfer protein (CETP), involved in the exchange of triglycerides and cholesterol between HDL and LDL. Crosses between the human apoB transgenic and human CETP transgenic mice are presently underway to investigate whether the lack of CETP is responsible for the triglyceride-rich LDL particle in apoB transgenic animals.

PRODUCTION OF MICE PRODUCING LP(a)

Mice expressing human apoB were bred with animals expressing human apo(a) resulting in animals expressing both transgenes – apo(a)/apoB transgenics. Characterization of apo(a) in these animals demonstrated that it was now present primarily in the plasma lipid fraction. The lipid-associated apo(a) in the mice was found in a density range similar to that of plasma human Lp(a). The association between human apo(a) and human apoB in the apo(a)/apoB transgenic mice is covalent, again resembling the interaction of these molecules in humans.

FUTURE DIRECTIONS

The availability of transgenic mice producing a human-like Lp(a) particle provides a unique substrate to derive insights into the properties of Lp(a) as well as its role in atherogenesis. Approaches for manipulating these substrates are presently available to investigate the particular amino acid sequences involved in the association between apo(a) and apoB *in vivo*, as well as sequences that may play a role in atherogenic properties. In addition, the analysis of atherogenesis development in apo(a) transgenic mice versus apo(a)/apoB transgenic mice producing Lp(a) will provide further insights into the role of these molecules in atherogenesis.

A distinct drawback to the present Lp(a) mouse is that the apo(a) transgene is a cDNA construct linked not to the apo(a) promoter, but rather to the transferrin promoter. Thus, it will be difficult to derive information regarding the regulation of expression of apo(a) and the determination of Lp(a) plasma levels. A transgenic mouse expressing a genomic clone of apo(a), including extensive 5' and 3' DNA would provide a much more useful model for studying a variety of factors relevant to apo(a) and the determination of its plasma levels. The difficulty in creating such a mouse is related to the fact that apo(a), located on chromosome 6, is an extremely large gene of between 120 and 220 Kb. The availability of yeast artificial chromosome (YAC) libraries of the human genome, containing human DNA inserts of up to 1000 Kb, as well as recent success at introducing YACs into mice, provide an approach presently being pursued for creating such animals in the future.

ACKNOWLEDGMENTS

This work was supported by National Institutes of Health Grants to E.R., PPG HL18574, and a grant funded by the National Dairy Promotion and Research Board and administered in cooperation with the National Dairy Council. E.R. is an American Heart Association Established Investigator. Research was conducted at the Lawrence Berkeley Laboratory (Dept. of Energy Contract DE-AC0376SF00098), University of California, Berkeley.

REFERENCES

1. M.J. Callow, L.J. Stoltzfus, R.M. Lawn, & E.M Rubin, Expression of human apolipoprotein B and assembly of lipoprotein(a) in transgenic mice, *Proc. Nat'l Acad. Sci. USA* **92**:21130-21136 (1994).
2. G. Chiesa, Expression of human apolipoprotein B100 in transgenic mice, *J Bio Chem* **268**:23747-23750 (1993).
3. E.M. Rubin, & D.J. Smith, Getting to the heart of a polygenic disorder: atherosclerosis in mice, *Trend Gen* (submitted, 1994).
4. J.L. Breslow, Transgenic mouse models of lipoprotein metabolism and atherosclerosis, *Proc Natl Acad Sci, USA* **90**:8314-8318 (1993).

5. G. Chiesa, *et al.*, Reconstitution of lipoprotein (a) by infusion of human low density lipoprotein into transgenic mice expressing human apolipoprotein (a), *J Biol Chem* **267**:24369-24374 (1992).
6. J.W. Gaubatz, C. Heideman, A.M.J. Gotto, J.D. Morrisett, G.H. Dahlen, Human plasma lipoprotein(a); structural properties, *J Bio Chem* **258**:4582-4589 (1983).
7. G. Utermann, W. Weber, Protein composition of Lp(a) lipoprotein from human plasma, *FEBS Let* **154**:357-361 (1983).
8. J.J. Guevara *et al.*, Proposed mechanisms for binding of apo(a) kringle type 9 to apo B-100 in human lipoprotein(a), *Biophys J* **64**:686-700 (1993).
9. R.M. Lawn, *et al.*, Atherogenesis in transgenic mice expressing human apolipoprotein (a). *Nature,* **360**:670-672 (1992).
10. M.F. Linton, *et al.*, Transgenic mice expressing high plasma concentrations of human apolipoprotein B100 and lipoprotein(a), *J Clin Invest* **92**:3029-3037 (1993).
11. A. Walsh, *et al.*, Intestinal expression of the human apoA-I gene in transgenic mice is controlled by a DNA region 3' to the gene in the promoter of the adjacent convergently transcribed apoC-III gene, *J. Lipid. Res.* **34**:617-623 (1993).

ISOLATION OF NOVEL GENES REGULATED BY DIETARY CHOLESTEROL BY A PCR-BASED SUBTRACTION LIBRARY

Alan T. Remaley, U. Kurt Schumacher,
H. Bryan Brewer, Jr., and Jeffrey M. Hoeg

National Institutes of Health
National Heart, Lung and Blood Institute
Bethesda, MD 20892

INTRODUCTION

Cholesterol, an integral part of cellular membranes, is maintained within a narrow concentration range by several mechanisms. Part of cholesterol regulation is mediated by cholesterol-induced changes in the expression of genes that encode for proteins that are involved in either the biosynthesis or catabolism of cholesterol[1]. In addition, cholesterol affects the activity of numerous membrane-bound proteins[2]. Excess cellular cholesterol has been proposed, in part, to account for the pathogenesis of atherosclerosis through its ability to indirectly modify the activity of cellular proteins. The pathophysiologically relevant changes in proteins induced by excess cholesterol and whether these changes secondarily lead to homeostatic alterations in gene expression has not been thoroughly investigated.

In order, to fully understand the molecular basis of atherosclerosis, more information is needed on the effect of cholesterol on both genes that are directly regulated by cholesterol, as well as those genes that are secondarily affected by cholesterol. We describe an PCR-based subtraction library approach[5], which should be widely applicable to several models of atherosclerosis, for investigating in a nontargeted manner the overall effect of cholesterol on gene expression.

SUBTRACTION LIBRARY

Several methods exist for isolating differentially expressed genes[6,7,8], but subtraction libraries have proven to be one of the most sensitive. The principles of a subtraction library are shown in Fig. 1. Two sources of mRNA are needed to make a subtraction library. One source, usually referred to as the tracer or treated mRNA, contains the mRNAs that are differentially expressed. In the example shown in Fig. 1, the differentially expressed mRNAs (solid line) are up regulated by the treatment. The other source of mRNA is the driver or control mRNA and shares all of the same mRNA species (striped line) as the tracer, except for the differentially expressed mRNAs. The goal of a subtraction library is to

Cardiovascular Disease 2
Edited by L.L. Gallo, Plenum Press, New York, 1995

subtract out from the tracer mRNA pool the shared driver mRNA sequences (striped line) to yield only the differentially expressed mRNAs (solid line). This is achieved by first converting the mRNA to cDNA by reverse transcriptase. The double stranded tracer and driver cDNA are then heated to produce single strands, and the tracer is mixed with a 10-fold excess of driver cDNA and allowed to reanneal. Because the driver cDNA is in excess, when the tracer cDNA reanneals, it forms mostly heteroduplexes of the tracer and driver cDNA. Because the differentially expressed gene is not expressed in the driver mRNA pool, the differentially expressed tracer cDNA either remains as a single strand or rehybridizes with the complementary strand within the tracer population. The heteroduplexes can be subtracted or removed away from the differentially expressed cDNA by several methods. In Fig. 1, this is conveniently done by first biotinylating (solid circles) the driver cDNA before mixing it with the tracer cDNA[9]. After the tracer and the biotinylated driver cDNA reanneal and form heteroduplexes, avidin is added and the DNA-biotin-avidin complexes that form can be removed by phenol extraction[9].

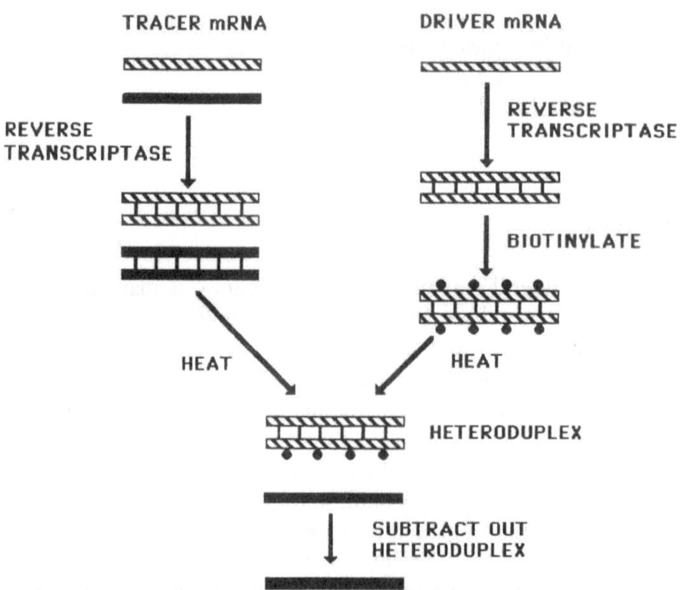

Figure 1. Diagram of a Standard Subtraction Library. The solid line indicates differentially expressed mRNAs. The striped line indicates non-differentially expressed mRNAs shared between the tracer and driver pool.

In practice, one problem with a standard subtraction library is that the subtraction process is never completely efficient, and because of the small amount of cDNA that is left after the first subtraction, the subtraction step can not be readily repeated. This problem has recently been solved by combining PCR with subtraction libraries[5], as shown in Fig. 2. The tracer and driver mRNA are converted to cDNA as before but the cDNA is then cut with a restriction enzyme to convert the cDNA into fragments of approximately 500 base pairs, which can be readily PCR amplified. Oligonucleotide linkers are then attached to the cDNA fragments, and the linkers are then later used as a source of primers for PCR amplification. The driver cDNA is then subtracted out from the tracer cDNA, as previously described. At this point, when one would normally stop with a standard subtraction library, the subtracted cDNA can then be PCR amplified by using the linkers as primers, thus generating a source of cDNA for additional subtractions. The subtraction step followed by PCR amplification can be used as many times as necessary to completely remove the shared cDNA fragments between the tracer and driver pools. After the final subtraction, the subtracted cDNA is ligated into a bacterial plasmid and used to transform *E.coli*, in order to produce a bacterial cDNA library. The library can then be screened for the differentially expressed genes by standard methods, involving nitrocellulose lifts and probing with the final subtracted ^{32}P-labelled cDNA pool. An additional advantage of this technique is that genes isolated from the initial screen can be biotinylated and subtracted out of the library, in order to screen for additional differentially expressed genes.

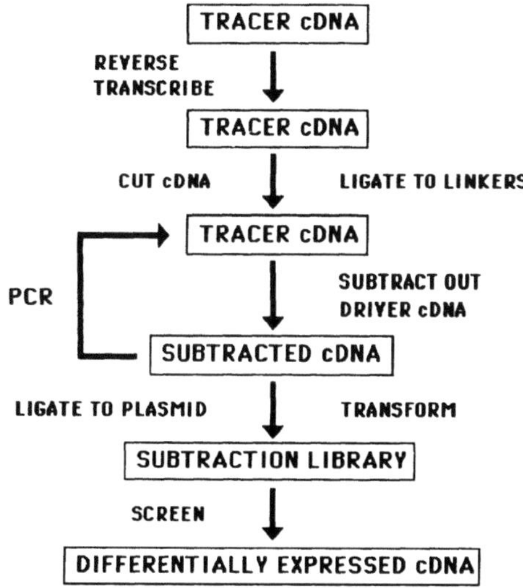

Figure 2. Diagram of a PCR-based Subtraction Library.

DIFFERENTIAL GENE EXPRESSION IN CHOLESTEROL-FED RABBIT LIVER

In the following section, we describe the use of a PCR-based subtraction library for isolating differentially expressed genes in cholesterol-fed rabbit liver. New Zealand White rabbits were fed either a standard rabbit chow or standard chow supplemented with 1.0 % cholesterol. After 4 weeks on the diet, the plasma cholesterol of the cholesterol-fed rabbits increased from a mean of 35 mg/dL to 1100 mg/dL. Poly (A) mRNA was isolated from the chow-fed and cholesterol-fed rabbit liver, converted into cDNA, and used to prepare two PCR-based subtraction libraries, as described in Fig. 2. Cholesterol-up-regulated (CUR) genes were isolated from a library in which the cholesterol-fed rabbits were used as the source of tracer mRNA and the chow-fed rabbits as the driver mRNA. Cholesterol-down-regulated (CDR) genes were isolated from a library in which the chow-fed rabbits were used as the source of tracer mRNA and the cholesterol-fed rabbits as the driver mRNA.

A convenient method for determining whether the subtraction process is complete is to perform cross hybridization analysis by Southern blots. If the subtraction is complete, the CUR pool of cDNA should not share any cDNA transcripts with the CDR pool, because the same gene can not be both up regulated and down regulated at the same time. In Fig. 3, we show that the cDNA from the CUR pool self hybridized but did not significantly cross hybridize with the CDR pool, indicating that the subtraction process was complete. The converse was also true, as shown in Fig. 3. This was achieved after six rounds of subtraction and PCR amplification.

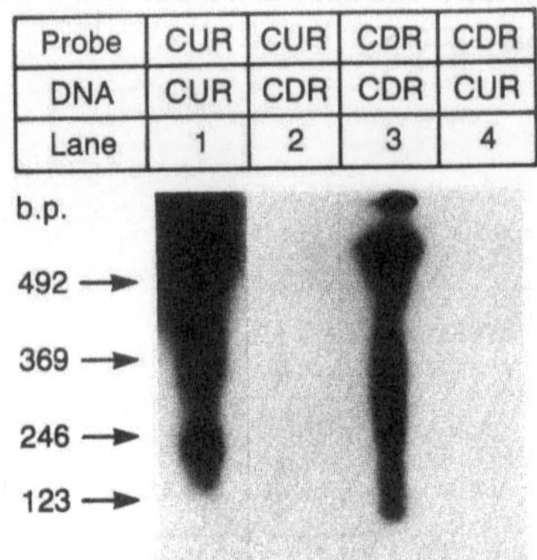

Figure 3. Cross-hybridization Analysis. cDNA from either the CUR or CDR pool, as indicated, was electrophoresed and blotted onto nitrocellulose and probed with either ^{32}-P labelled CUR or CDR cDNA, as indicated.

After the final subtraction, the subtracted cDNA was ligated into a prokaryotic plasmid and used to prepare a bacterial cDNA library. The two libraries were then screened with either the CUR or CDR cDNA pools. Numerous positive clones were obtained for each library, and the positive clones were re-screened by dot blot analysis, as shown for the CUR genes in Fig. 4. Plasmids were purified from each positive clone, using standard techniques, and were spotted onto nitrocellulose filters and probed with the CUR and CDR cDNA pools. As can be seen in Fig. 4, most of the CUR clones that were positive in the primary screen were also positive in the secondary screen. In addition, none of the CUR clones hybridized with the CDR probe, thus showing the specificity of the method.

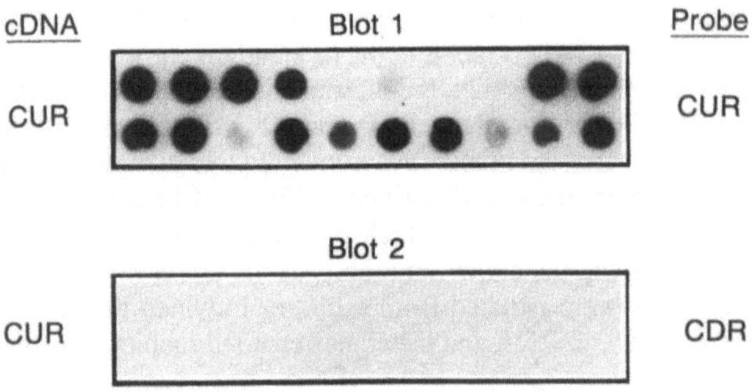

Figure 4. Secondary Screen of Differentially Expressed Genes. Plasmids from the positive clones from the primary screen of the CUR library were spotted onto nitrocellulose and probed with ^{32}-P labelled cDNA from either the CUR or CDR pool, as indicated.

To confirm that the isolated cDNA fragments responded to the cholesterol feeding as predicted, a Northern blot analysis was performed. An example of a Northern blot for both an up regulated and down regulated cDNA is shown in Fig. 5. Each lane contains approximately the same amount of RNA, as shown by the actin blot in panel C. Lanes 1, 2, 3, and 4 contain total hepatic RNA from rabbits that were fed the cholesterol-rich diet for 0, 1, 2, and 4 weeks, respectively. Panel A is an example of a CDR gene that showed a significant decrease in expression after 4 weeks on the cholesterol-rich diet. In contrast, the CUR gene in panel B was not significantly expressed before cholesterol feeding but was up regulated after 1 week on the cholesterol-rich diet. A total of 13 non-redundant differentially expressed cDNA fragments were isolated and confirmed by Northern blot analysis. Nine of the cDNA fragments were up regulated and 4 down regulated by cholesterol feeding.

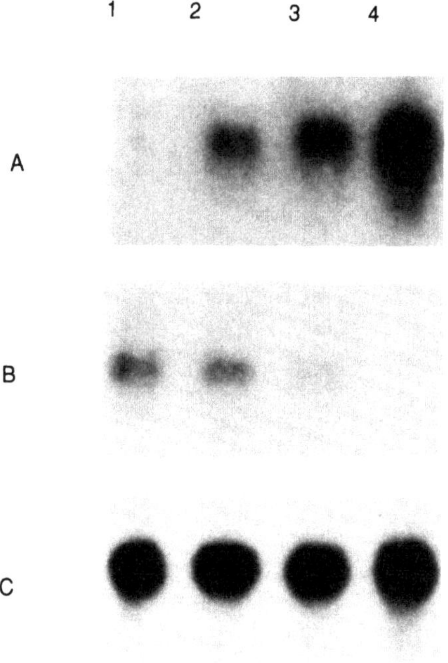

Figure 5. Northern Blot Analysis of Differentially Expressed Genes. cDNA inserts from either a CDR clone (panel A), a CUR clone (panel B), or beta-actin (panel C) was used to probe total hepatic RNA from rabbits fed a cholesterol-rich diet for 0 weeks (lane 1), 1 week (lane 2), 2 weeks (lane 3), or 4 weeks (lane 4).

Once differentially expressed cDNA fragments are isolated, several more steps can be performed to fully characterize the regulated genes. After the secondary screen, not all of the positive clones may contain unique cDNA fragments; the same cDNA fragment may be present in multiple clones. To exclude redundant cDNA fragments, the cDNA inserts from each positive clone can be isolated and sized by agarose gel electrophoresis. Those fragments that are identical in size can be studied by cross hybridization analysis on dot blots to determine if they represent the same cDNA fragment. In order to identify the remaining cDNA fragments, DNA sequence homology searches[10] can be performed. The cDNA fragments not identified by homology searches can be analyzed for functional motifs[11], in order to gain insight into their possible function. The cDNA fragments can also be used to isolate the complete cDNA by screening a full length cDNA library. Ultimately, any isolated cDNA can be used for modifying the expression of the gene in either cell culture by transfection studies or in transgenic animals for exploring the functional role of the differentially expressed gene.

SUMMARY

We describe a PCR-based subtraction library approach for investigating the overall effect of cholesterol on gene expression. We isolated by PCR-based subtraction libraries several differentially expressed genes from cholesterol-fed rabbit liver. These genes may be directly involved in cholesterol metabolism or may be secondarily affected by the high intracellular concentration of cholesterol in the cholesterol-fed rabbit liver. Future studies will be aimed at exploring the functional role of the isolated genes and utilizing PCR-based subtraction libraries for studying gene expression in other model systems of atherosclerosis.

REFERENCES

1. H. Rudney, and S.R. Panini, Cholesterol biosynthesis, Curr. Opin. Lipidol. 4: 230 (1993).
2. P.L. Yeagle, Modulation of membrane function by cholesterol, Biochimie 73:1303 (1991).
3. D. Papahadjopoulos, Cholesterol and cell membrane function: ahypothesis concerning the etiology of atherosclerosis, J. Theor. Biol. 43:329 (1974).
4. R.L. Jackson, and A. M. Gotto, Hypothesis concerning membrane structure, cholesterol, and atherosclerosis, Athero. Rev. 1:1 (1976).
5. Z. Wang, and D.D. Brown, A gene expression screen, Proc. Natl. Acad. Sci. USA 88:11505 (1991).
6. S.M. Hanash, J. R. Strahler, J.V. Neel, N. Hailat, R. Melhem, D. Klim, X.X. Zhu, D. Wagner, D.A. Gage , and J.T. Watson, Highly resolving two-dimensional gels for protein sequencing, Proc. Natl. Acad. Sci.USA 88:5709 (1991).
7. S. J. Lemke, T. Burke, G.B. Bodor, and R.E. Moore, SPOT: an improved differential screening protocol that allows the detection of marginally induced mRNAs, Biotech. 14:415 (1993).
8. P. Liang, and A.B. Pardee, Differential display of eukaryotic messenger RNA by means of the polymerase chain reaction, Science 257: 967 (1992).
9. H.L. Sive, and T. St. John, A simple subtractive hybridization technique employing photoactivatable biotin and phenol extraction, Nucleic Acids Res. 16:10937 (1988).
10. D. Benson, D.J. Lipman, and J. Ostell, GenBank, Nucleic Acids Res. 1:2963 (1993).
11. A. Bairoch, The PROSITE dictionary of sites and patterns in proteins, its current status, Nucleic Acids Res. 21:3097 (1993).

CHOLESTEROL AND MORTALITY:

WHAT CAN META-ANALYSIS TELL US?

David J. Gordon

National Heart, Lung, and Blood Institute
Division of Heart and Vascular Diseases
Bethesda, Maryland, U.S.A.

INTRODUCTION

Replication of experimental results has long been one of the cardinal principles of scientific endeavor. Yet because of their size, duration, and cost, this principle has been difficult to apply to clinical trials, where therapeutic recommendations affecting thousands of patients must often be inferred from a relatively small number of studies of varying power and quality, using disparate interventions and study designs, carried out over many years or even decades. Traditionally, it has been the province of "the experts," often in constituted panels, to sift and weigh the evidence obtained from clinical trials and to derive standards for clinical practice. More recently, the statistical technique of meta-analysis has been applied to groups of related clinical trials to bring more objectivity and quantitative rigor to this process.

Whatever its successes elsewhere, the application of meta-analysis to cholesterol-lowering trials has not brought objectivity and consensus[1-8]. Although the etiologic role of cholesterol in coronary atherogenesis is well established and the effectiveness of cholesterol-lowering therapy in reducing the incidence of myocardial infarction (MI) and other clinical sequelae of atherosclerotic coronary heart disease (CHD) has been demonstrated repeatedly in randomized clinical trials using a wide variety of cholesterol-lowering treatments, the beneficial effect of these treatments on CHD mortality in these trials has been offset by increased mortality from causes other than CHD, and overall mortality has not been reduced. While meta-analysts have all described this phenomenon more or less, their interpretations and recommendations have diverged widely. Indeed, disputes concerning which trials are included or excluded and the emphasis given to various aspects of the results have tended to overshadow the general quantitative agreement among meta-analyses. For example, Holme[1] reported odds ratios of 1.037 for total mortality ($Z = 1.28$) and 0.905 for CHD incidence ($Z = -3.69$), while Ravnskov[2] reported odds ratios of 1.02 for total mortality ($Z = 0.79$), 0.94 for CHD death ($Z = -1.90$), and 0.88 for nonfatal MI ($Z = -2.74$). These results are essentially in close agreement. Yet, Holme writes, "This study shows that cholesterol reduction is effective in lowering CHD incidence, but cholesterol reduction must be at least 8-9% to be

Cardiovascular Disease 2
Edited by L.L. Gallo, Plenum Press, New York, 1995

effective in lowering total mortality", while Ravnskov writes, "Lowering serum cholesterol concentrations does not reduce mortality and is unlikely to prevent coronary heart disease. Claims of the opposite are based on preferential citation of supportive trials." While preferential citation of supportive trials may be a genuine potential source of bias, it is clearly has little to do with the minor differences between these two particular meta-analyses.

While meta-analysis has often been viewed as a way in effect to combine the results of many similar small-to-medium-sized trials of limited statistical power into a single large trial with abundant power, meta-analysts have disagreed as to which of the nearly 50 completed randomized cholesterol-lowering trials employing more than 60 treatment regimens are most relevant and as to how to interpret the heterogeneity of their designs and results. The purpose of this paper is to outline some of the important issues and to illustrate these issues with a fresh meta-analysis of the cholesterol-lowering trials.

INCLUSIONS AND EXCLUSIONS

Which cholesterol-lowering trials should be included in a meta-analysis? Three types of inclusion/exclusion criteria are commonly considered -- those pertaining to study design (e.g., randomization, duration, blinding, co-interventions), those pertaining to the quality of the study's implementation (e.g., compliance, data quality, analysis by intent to treat), and those pertaining to study outcomes (e.g., toxicity of drug regimen). The least controversial exclusions are those based on principles of sound study design, since these can be agreed upon a priori without reference to specific subject matter and are least subject to be influenced by whether or not a particular trial supports the meta-analyst's point of view. Exclusions based on study quality are more prone to bias, since they require particular knowledge of the post-randomization course each study, yet are often unavoidable. For example, it makes little sense to include a large study that used an intervention that failed to lower cholesterol to address the hypothesis is that lowering cholesterol reduces mortality and morbidity (though some have done so). However, exclusions based on actual study end points, however rational they might seem, should be avoided, since they inevitably introduce the biases of the meta-analyst. It is a good idea to ask the following questions before excluding a trial from consideration: (1) Would an excluded negative trial have been included if the trial's outcome were positive? (2) Would an included positive trial have been excluded if the trial's outcome were negative? If either is answered affirmatively, the objectivity of the analysis is in doubt.

A meta-analysis of 22 treatment regimens from 18 trials meeting the following criteria was performed: (1) randomized, results reported by intent-to-treat; (2) relevant to hypothesis that cholesterol lowering is beneficial -- i.e., a meaningful cholesterol reduction (≥ 4%) was attained, there was no systematic co-intervention on important non-lipid risk factors, and the study population involved appropriate candidates for cholesterol-lowering treatment; (3) at least 3 years planned duration; and (4) reported before November, 1993.

These criteria exclude trials that intervened on multiple risk factors while lowering cholesterol by 1-2% and trials in which cholesterol lowering was a side effect of a drug given for another purpose (e.g., tamoxiphen in a breast cancer treatment trial, in which >90% of deaths are due to breast cancer). However, they do not automatically exclude trials that used hormones (dextrothyroxine and estrogen) to lower cholesterol.

The 22 qualifying trials are listed, classified by their cholesterol-lowering modality, in Table 1. The total number of person-years of follow-up accrued by patients randomized to active treatment are given for each treatment modality. Nearly 60% of the 86,660 person-years of follow-up in the active treatment groups of these trials was contributed by the participants treated with the fibric acid derivatives clofibrate (5 trials) and gemfibrozil (2 trials) -- more than triple any other treatment modality[9-17].

Cholestyramine resin[18-20] (mainly the Lipid Research Clinics Coronary Primary Prevention Trial (LRC-CPPT)[18]) contributed 14,491 person-years, and niacin[9-10,15,21] (mainly the Coronary Drug Project (CDP)[9-10]) contributed 8,760 person-years. The remaining follow-up came from 6 small to mid-sized diet trials[20,22-26], the Program on the Surgical Control of Hypercholesterolemia (POSCH) partial ileal bypass surgery trial[27], and three trials that used pharmacologic doses of d-thyroxine[21,28] or estrogen hormones[29] to lower cholesterol.

Table 1. Randomized trials of cholesterol lowering.

Fibrates (50,333 person-yr)	**Diet (6356 person-yr)**
CDP[9-10]	*LA VA[22]
*Helsinki[11]	MRC Low Fat[23]
Helsinki Ancillary[12]	MRC Soya[24]
Newcastle[13]	Oslo[25]
Scottish Physicians[14]	STARS[20]
**Stockholm[15]	Sydney[26]
*WHO[16-17]	
Resins (14,491 person-yr)	**Surgery (4084 person-yr)**
*LRC-CPPT[18]	POSCH[27]
NHLBI Type II[19]	
STARS[20]	
Niacin (8760 person-yr)	**Hormones (4031 person-yr)**
CDP[9-10]	CDP (d-thyroxine)[28]
**Stockholm[15]	USVA Drug Lipid (d-thyroxine)[21]
USVA Drug Lipid[21]	Oliver et al (estrogen)[29]

* Primary prevention trial
** Stockholm study participants were treated with both clofibrate and niacin

RESULTS

Meta-analyses of mortality due to CHD, non-CHD causes, and all causes combined and of the combined incidence of CHD death and nonfatal MI (CHD incidence) were performed for the trials listed in Table 1, using the Mantel-Haenszel procedure as modified by Yusuf et al.[30] The results of the overall meta-analysis are shown in Table 2. The mean 10% cholesterol reduction in these trials was associated with a mean 17% decrease ($P<0.001$) in the combined incidence of fatal and nonfatal myocardial infarction. However, the 9% decrease ($P=0.03$) in CHD mortality associated with active treatment in these trials was offset by a 24% increase ($P=0.001$) in non-CHD mortality; all-causes mortality was not reduced (relative risk = 1.01).

Table 2. Results of meta-analysis.

#Trials	22
#Treated	15,487
Person-Years	86,660
Mean Cholesterol Reduction	10%
Reduction in:*	
Total Mortality	+1.0%
CHD Mortality	**-9.1%**
Non-CHD Mortality	**+23.8%**
CHD Incidence	**-16.8%**

* Statistically significant reductions ($P<0.05$) in boldface type.

However, subgroup analyses of these trials indicated that the impact of cholesterol lowering on all-causes mortality was not uniform among the different categories of trials. For example, mortality was reduced by 3.4% in the 18 secondary prevention trials, but increased by 12.6% in the 4 primary prevention trials (Table 3). This difference was due entirely to the higher proportion of deaths attributed to CHD in the secondary trials. Davey Smith[5] has taken a more general approach to essentially the same phenomenon by stratifying the trials by baseline CHD risk, as indicated by the CHD mortality rate in the each trial's control group, and similarly finds that the all-cause mortality results are most favorable in the trials incorporating subjects at the highest initial risk.

Table 3. Results of meta-analysis -- primary *vs.* secondary prevention.

Type of Trial	Primary	Secondary
#Trials	4	18
#Treated	9,713	6,134
Person-Years	55,589	31,071
Mean Cholesterol Reduction	9%	11%
Reduction in:*		
Total Mortality	+12.6%	-3.4%
CHD Mortality	-7.1%	**-9.5%**
Non-CHD Mortality	**+26.0%**	**+21.3%**
CHD Incidence	**-23.2%**	**-14.2%**

* Statistically significant reductions (P<0.05) in boldface type.

As might be expected, the impact of cholesterol lowering on mortality was also strongly related to the degree by which the mean cholesterol level was reduced by treatment (Table 4). When trials in which cholesterol reductions of more or less than the median (12%) were compared, all-cause mortality was reduced by 19.7% (P=0.02) in the 11 trials in which the mean cholesterol level was lowered by at least 12%, but increased by 9.6% (P=0.03) in the 11 trials with lesser reductions in mean cholesterol level. Similar effects were observed for CHD incidence and mortality. These results are in basic agreement with regression analyses performed by several other meta-analysts.[1,6,7] It is also worth noting that the increase in non-CHD mortality seen in Table 2 was attributable mainly to the trials with the least cholesterol lowering. This absence of a dose-response effect argues against the hypothesis that cholesterol lowering *per se* increases non-CHD mortality rate. It should also be noted that this meta-analysis did not include any trials that used HMG CoA reductase inhibitors ("statins"), which typically produce total and LDL cholesterol reductions comparable to those achieved (along with a quite favorable clinical outcome) in the POSCH trial[29]. Although several small statin trials were reported before the November 1993 cutoff, most with favorable end point trends, none met the minimum 3-year duration requirement of this meta-analysis.

Table 4. Results of meta-analysis by degree of cholesterol lowering.

Degree of Cholesterol Lowering	> 12%	< 12%
#Trials	11	11
#Treated	2,399	6,134
Person-Years	12,974	72,770
Mean Cholesterol Reduction	15%	9%
Reduction in:*		
Total Mortality	**-19.7%**	**+9.6%**
CHD Mortality	**-27.1%**	-1.5%
Non-CHD Mortality	+10.8%	**+29.7%**
CHD Incidence	**-31.0%**	**-11.3%**

* Statistically significant reductions (P<0.05) in boldface type.

The dose response relationship for the intensity of cholesterol lowering can also be demonstrated for its duration. Law et al, analyzing unpublished year-by-year event rates and adjusting for degree of cholesterol lowering, reported that CHD incidence rates were reduced by only 7% in Years 1-2, 22% in Years 3-5 (in trials with more than 2 years follow-up), and 25% thereafter (in trials with more than 5 years follow-up).[7-8] This effect (handled in the present analysis by excluding trials of less than 3 years duration) may be another important confounder of more inclusive meta-analyses, which treat 1-year and 8-year trials as equivalent.

The results of the cholesterol-lowering trials also vary considerably according to treatment modality (Table 5). In the trials that used surgery or diet to lower cholesterol level, there was no increase in non-CHD mortality to offset the observed decrease in CHD mortality. The resins and niacin produced proportionate decreases in CHD and increases in non-CHD mortality, with small net decreases in all-cause mortality. In the seven fibrate trials, the increase in non-CHD mortality outweighed the decrease in CHD mortality, and the net effect on all-cause mortality was unfavorable. In the three trials that used hormones to lower cholesterol, both CHD and non-CHD mortality increased with treatment.

Table 5. Results of meta-analysis by treatment modality.

Modality	Total Mortality	CHD Mortality	Non-CHD Mortality	CHD Incidence
Surgery	-24%	-30%	-7%	**-43%**
Resin	-11%	-32%	+33%	**-21%**
Diet	-6%	-21%	+0%	**-24%**
Niacin	-4%	-7%	+8%	**-17%**
Fibrates	+3%	-8%	**+32%**	**-18%**
Hormones	+18%	+5%	+77%	+7%

* Statistically significant reductions (P<0.05) in boldface type.

The 10 trials that used fibrates or hormones were then compared to the 12 trials that used other cholesterol-lowering modalities (Table 6). In the former group of trials, the significant 42% increase in non-CHD mortality more than offset the 4% decrease in CHD mortality, and all-cause mortality increased by 8%. However, in the latter group of trials, the significant 15% reduction in CHD mortality was not offset by the nonsignificant 6% increase in non-CHD mortality, and total mortality decreased by 7%. Note that CHD incidence was significantly reduced by treatment in both groups of trials.

Table 6. Results of meta-analysis by degree of cholesterol lowering.

Fibrate/Hormone	No	Yes
#Trials	12	10
#Treated	4,877	10,970
Person-Years	32,295	54,365
Mean Cholesterol Reduction	11%	9%
Reduction in:*		
Total Mortality	-6.9%	+8.3%
CHD Mortality	**-14.7%**	-3.9%
CHD Mortality	+6.4%	**+41.7%**
Non-CHD Mortality	**-22.1%**	**-12.0%**

* Statistically significant reductions (P<0.05) in boldface type.

CONCLUSION

The following two points should not be lost in the considerable controversy generated by the meta-analysis of clinical trials of cholesterol lowering: (1) No published comprehensive meta-analysis of cholesterol-lowering trials has failed to show a significant overall reduction in CHD incidence. (2) No published comprehensive meta-analysis of cholesterol-lowering trials has succeeded in showing a significant overall reduction in all-causes mortality. While some meta-analysts have emphasized one of these findings and/or minimized the other, their objective results, like those presented here, are in fundamental agreement in these two regards, despite substantive differences in trial selection and methodology. The results presented here futher suggest that the failure of past cholesterol-lowering trials, considered in aggregate, to demonstrate a net reduction in all-cause mortality may be due at least in part to the insufficiently elevated mean CHD risk of the participants they recruited, the limited cholesterol-lowering efficacy and duration, and/or the significant adverse non-CHD effects of the particular drugs that they have employed. If this interpretation is correct, trials now in progress using more potent (and hopefully safe) cholesterol-lowering drugs in high-risk patients can be expected to provide a truer and more powerful test of the potential impact of cholesterol lowering on mortality.

REFERENCES

1. Holme I. An analysis of randomized trials evaluating the effect of cholesterol reduction on total mortality and coronary heart disease incidence. Circulation 1990; 82:1916-1924.
2. Ravnskov U. Cholesterol lowering trials in coronary heart disease: frequency of citation and outcome. BMJ 1992; 305:15-19.
3. Muldoon MF, Manuck SB, Mathews KA. Lowering cholesterol concentrations and mortality: A quantitative review of primary prevention trials. Br Med J 1990; 301:309-314.
4. Rossouw JE, Lewis B, Rifkind BM. The value of lowering cholesterol after myocardial infarction. N Engl J Med 1990; 323:1112-1119.
5. Davey Smith G, Song F, Sheldon TA. Cholesterol lowering and mortality and mortality: the importance of considering initial level of risk. BMJ 1993; 306: 1367-1373.
6. Holme I. Relation of coronary heart disease incidence and total mortality to plasma cholesterol

reduction in randomised trials: use of meta-analysis. Br Heart J 1993; 69 (suppl):S42-S47.
7. Law MR, Wald NJ, Thompson SG. By how much and how quickly does reduction in serum cholesterol concentration lower risk of ischaemic heart disease. Br Med J 1994; 308:367-373.
8. Law MR, Thompson SG, Wald NJ. Assessing possible hazards of reducing serum cholesterol. Br Med J 1994; 308:373-379.
9. Coronary Drug Project Research Group. The Coronary Drug Project: Clofibrate and niacin in coronary heart disease. JAMA 231:360-381, 1975.
10. Canner PL, Berge KG, Wenger NK, Stamler J, Friedman L, Prineas RJ, Friedewald W. Fifteen year mortality in Coronary Drug Project patients: long-term benefit with niacin. J Am Coll Cardiol 8:1245-55, 1986.
11. Frick MH, Elo O, Heinonen O, Heinsalmi P, Helo P, Huttunen JK, Kaitaniemi P, Koskinen P, Manninen V, Mäenpää H, Mälkönen M, Mänttäri M, Norola S, Pasternack A, Pikkarainen J, Romo M, Sjöblom T, Nikkilä E. Helsinki Heart Study: Primary-prevention trial with gemfibrozil in middle-aged men with dyslipidemia. Safety of treatment, changes in risk factors, and incidence of coronary heart disease. N Engl J Med 1987; 317:1237-1245.
12. Frick MH, Heinonen OP, Huttunen JK, Koskinen P, Manttari M, Manninen V. Efficacy of gemfibrozil in dyslipidemic subjects with suspected heart disease. An ancillary study in the Helsinki Heart Study frame population. Ann Med 1993; 25:41-45.
13. Group of Physicians of the Newcastle Upon Tyne Region. Trial of clofibrate in the treatment of ischaemic heart disease. Five year study. BMJ 1971; 4:767-775.
14. Research Committee of the Scottish Society of Physicians. Ischaemic heart disease: a secondary prevention trial using clofibrate. BMJ 1971; 4:775-784.
15. Carlson LA, Rosenhamer G. Reduction of mortality in the Stockholm ischaemic heart disease secondary prevention study by combined treatment with clofibrate and nicotinic acid. Acta Med Scand 1988; 223:405-418.
16. Report from the Committee of Principal Investigators. A cooperative trial in the primary prevention of ischaemic heart disease using clofibrate. Br Heart J 1978; 40:1069-1103.
17. Heady JA, Morris JN, Oliver MF. WHO clofibrate/cholesterol trial: clarifications. Lancet 1992; 340:1405-1406.
18. Lipid Research Clinics Program. The Lipid Research Clinics Coronary Primary Prevention Trial Results I. Reduction in incidence of coronary heart disease. JAMA 251:351-364, 1984.
19. Levy RI, Brensike JF, Epstein SE, Kelsey SF, Passamani ER, Richardson JM, Loh IK, Stone NJ, Aldrich RF, Battaglini JW, Moriarty DJ, Fisher ML, Friedman L, Friedewald W, and Detre KM. The influences of changes in lipid values induced by cholestyramine and diet on progression of coronary artery disease: Results of the NHLBI Type II Coronary Intervention Study. Circulation 69:325-337, 1984.
20. Watts GF, Lewis B, Brunt JNH, Lewis ES, Coltart DJ, Smith LDR, Mann JI, Swan AV. Effects on coronary artey disease of lipid-lowering diet, or diet plus cholestyramine, in the St. Thomas' Atherosclerosis Regression Study (STARS). Lancet 1992; 339: 563-569.
21. Schock HK. The U.S. Veterans Administration cardiology drug-lipid study: an interim report. Adv Exp Med Biol 1968; 405-420.
22. Dayton S, Pearce ML, Hashimoto S, Dixon WJ, Tomiyasu U. A controlled clinical trial of a diet high in unsaturated fat in preventing complications of atherosclerosis. Circulation 39 and 40:Suppl. 2, 1969.
23. Research Committee. Low-fat diet in myocardial infarction: a controlled trial. Lancet 1965; 2:501-504.
24. Research Committee to the Medical Research Council. Controlled trial of soybean oil in myocardial infarction. Lancet 1968; 2:693-700.
25. Leren P. The effect of plasma cholesterol lowering diet in male survivors of myocardial infarction. A controlled clinical trial. Acta Med Scand 1966; Suppl 466: 1-92.
26. Woodhill JM, Palmer AJ, Leelerthaepin B, McGilchrist C, Blacket RB. Low fat, low cholesterol diet in secondary prevention of coronary heart disease. Adv Exp Med Biol 1978; 109:317-330.
27. Buchwald H, Varco RL, Matts JP, Long JM, Fitch LL, Campbell GS, Pearce MB, Yellin AE, Edmiston A, Smink RD, Sawin HS, Campos CT, Hansen BJ, Tuna N, Karnegis J, Sanmarco ME, Amplatz K, Castaneda-Zunida WR, Hunter DW, Bissett JK, Weber FJ, Stevenson JW, Leon AS, Chalmers TC, and the POSCH Group. Effect of partial ileal bypass surgery on morbidity and mortality from coronary heart disease in patients with hypercholesterolemia. Report of the Program on the Surgical Control of Hypercholesterolemia (POSCH). N Engl J Med 1990; 323:946-955.
28. Coronary Drug Project. Findings leading to further modifications of its protocol with respect to

dextrothyroxine. JAMA 1972; 220: 996-1008.

29. Oliver MF, Boyd GS. Influence of reduction of serum lipids on prognosis of coronary heart disease. A five-year study using oestrogen. Lancet 1961; 2: 499-505.

30. Yusuf S, Peto R, Collins R, Sleight P. Beta blockade during and after myocardial infarction: an overview of randomized trials. Prog Cardiovasc Dis 1985; 27:335-371.

SCREENING FOR HIGH BLOOD CHOLESTEROL: A RISKY BUSINESS

Stephen B. Hulley

Department of Epidemiology and Biostatistics
University of California, San Francisco
San Francisco, CA 94143

INTRODUCTION

The National Cholesterol Education Program (NCEP) has recommended blood cholesterol testing for all adults, followed by evaluation and treatment of those with high levels(1). Those with total cholesterol exceeding specified cutpoints should then undergo lipoprotein testing and, if the LDL-cholesterol also exceeds cutpoints, become candidates for intensive intervention with diet and lipid-lowering drugs. The cutpoints for intensive intervention are risk-specific, i.e., lower for those who already have coronary disease or several other risk factors in order to favor more vigorous treatment in those at higher short term risk.

Another set of guidelines, produced by the Toronto Working Group for the province of Ontario(2), differs from the NCEP chiefly in not recommending universal cholesterol screening for young adults—men under age 35 and women before menopause. This concern with age is a logical extension of the NCEP premise that those at lowest risk have the least to gain from preventive interventions.

This report discusses the role of relatively short term (up to one decade) risk status in considering the harms and benefits of screening. The goal is to provide guidance for deciding such issues as whether or not to screen young adults for blood cholesterol.

BENEFITS OF SCREENING ARE RELATED TO THE MAGNITUDE OF CHD RISK, BUT HARMS ARE NOT

The benefit from cholesterol intervention—the reduction in a person's likelihood of a CHD event and death—depends on that individual's probability of having a CHD event. Thus, people at high short-term risk of CHD have the most to gain from modifying blood cholesterol levels. On the other hand, the potential harm from cholesterol intervention tends to be unrelated to the risk of CHD. This means that for patients who already have coronary disease, the reduction in subsequent CHD mortality from treating high blood cholesterol is likely to outweigh any adverse effects of the treatment. On the other hand, for patients at low risk of CHD there is a greater chance that adverse effects could predominate.

POTENTIAL BENEFITS AND HARMS—THE RANDOMIZED CONTROLLED TRIALS

Clinical trials have established the effectiveness of lowering blood cholesterol in preventing

Cardiovascular Disease 2
Edited by L.L. Gallo, Plenum Press, New York, 1995

CHD events (chiefly non-fatal myocardial infarctions), but have not had the power to detect significant effects on mortality. Recently, meta-analyses that combine the results of the primary prevention trials (thereby increasing the power to detect an effect on mortality) have found that the expected decrease in CHD death rates is more than offset by unexpected statistically significant increases in death rates from injuries and cancer (3-6). The excess in non-CHD deaths cannot be due to competing mortality (the fact that those who do not die of CHD will necessarily die of something else) because competing mortality can only have a trivial effect in cohorts with a large majority of people still alive. It is unlikely to be due to problems in conducting the meta-analyses, because several independent investigators have obtained essentially the same result. It could be due to chance, but the p values make this unlikely. There remains the most straightforward explanation for randomized trial results—that one or more of the cholesterol-lowering interventions may cause non-CHD deaths.

One strategy for understanding the increased non-CHD mortality in the primary prevention trials is to examine the trials of secondary prevention in patients who already have CHD. A meta-analysis of this second set of studies found no evidence for an increase in non-cardiovascular mortality, although the wide confidence intervals do not exclude a true increase in the population (4, 7-9). There was a decrease in CHD deaths, and because of the very high risk of CHD death relative to other causes in patients with coronary disease, a decrease in total mortality.

Meta-analyses of secondary prevention trials have less favorable total mortality findings if they include studies of dextro-thyroxine and high dose estrogen, drugs no longer in use for lowering blood lipids because clinical trials revealed very serious adverse effects including death(10, 11). The unexpected adverse experience with these two drugs underscores the wisdom of a cautious approach to the use of any lipid-lowering drug in healthy people. There are recent reports from the Helsinki trials of increased mortality with gemfibrozil(12, 13), and little is known about the long-term effects of the most widely used cholesterol-lowering drugs, the HMG CoA reductase inhibitors. These drugs have received FDA approval even though full scale trials with mortality endpoints are not yet available and a preliminary trial of lovastatin (not included in the meta-analysis) showed a non-significant increase in total mortality in the treated groups ($p = .08$)(14).

POTENTIAL BENEFITS AND HARMS—OTHER LINES OF EVIDENCE

The evidence that high blood cholesterol is a reversible cause of CHD is convincing, based on a large set of epidemiologic, basic science, and animal evidence. Interventions that lower blood cholesterol may also have other beneficial effects—epidemiologic studies suggest that fat-controlled diets may be associated with lower rates of cancers of the breast, colon and prostate.

However, there is a considerable body of animal evidence indicating that cholesterol lowering treatments may be harmful. These include a randomized trial in which monkeys fed a "prudent" diet low in fat and cholesterol exhibited significantly more aggressive behavior(15), and another trial in which monkeys given clofibrate (a fibric acid derivative related to gemfibrozil) had a 2 to 5 fold increase in mortality (cause unspecified)(16). Most ominous is the significant increase in cancer among rodents noted in the product information for lovastatin, pravastatin, symvastatin, cholestyramine, clofibrate and gemfibrozil; this increase occurs with doses that are similar to those used in humans(16).

There are also disconcerting epidemiologic findings: a recent NIH conference on the pooled observational findings found no association between high blood cholesterol and cardiovascular death in women, and revealed highly significant associations between low levels of blood cholesterol (below 160 mg/dl) and deaths from cancer, injuries, respiratory disease, gastro-intestinal disease, and other non-CHD causes(17). Although we do not know which, if any, of these associations represent cause and effect, there is speculation on mechanisms by which lowering blood cholesterol may alter cellular function and increase proclivity to injuries or cancer(18-20). The potential harms also include the general problems with screening, such as the psychological effects of labelling(21).

CLINICAL IMPLICATIONS FOR HIGH RISK PATIENTS

How should this evidence alter our efforts to screen for and treat high blood cholesterol? The answer depends on the level of CHD risk in the population under consideration(7). People at high risk of CHD have more to gain from modifying blood cholesterol levels than do those at average risk. The presence of coronary artery disease is a factor associated with very high risk of subsequent CHD events and death; recent evidence indicates that the presence of substantial disease in other arteries, particularly the carotids and the femorals, also conveys very high risk of subsequent CHD. Other well known risk factors include male sex, blood pressure, smoking, low HDL-cholesterol, diabetes, and sedentary life style; an individual who has several of these present is at substantially higher risk, and thus has a high potential benefit-to-harm ratio.

CLINICAL IMPLICATIONS FOR LOW RISK ADULTS

The strongest determinant of CHD is age, accounting for a 500-fold difference in risk between men aged 22 and 62 (Table 2). The CHD death rate among young adults is so low that the short term benefits of CHD prevention in this age group are miniscule. Non-CHD deaths (predominantly those due to injuries) are 100 times more common for both men and women in the youngest age group, so even a small percentage increase in non-CHD deaths would outweigh the tiny projected reduction in CHD deaths.

It is widely believed that efforts to lower blood cholesterol that are begun at an early age and continued for decades are more effective in preventing CHD deaths than efforts begun in middle age. However, there is no direct evidence that this is true. More importantly, the magnitude of any additional reduction in the risk of CHD that comes from beginning cholesterol-lowering intervention at a younger age must be small. Clinical trials of treatment begun in middle age have consistently shown a 2% reduction in CHD event rates for every 1% reduction in cholesterol within about two years of beginning therapy, two thirds of the maximum attainable risk reduction projected from the cholesterol-CHD risk relationships in observational studies(22-24).

Whether earlier onset of treatment would allow an individual to achieve more than two-thirds of the projected benefit years later is not known. This possibility might lead some to begin monitoring their blood cholesterol levels for possible treatment at an earlier age than others. However, screening and treatment can be witheld at least until age 35 in men and 45 in women, the cutpoints recommended for beginning universal screening in the Toronto Working Group Guidelines. These ages identify the lower risk half of our adult population (roughly 80 million men and women), and are conservative since the annual CHD death rate does not reach 1/1000 in either gender until a dozen years later. An exception would be those few young adults who have known CHD, or several other CHD risk factors, or a first degree relative with familial hypercholesterolemia; these individuals are at higher short term risk and thus more likely to benefit from cholesterol screening.

There are two other reasons for not screening young adults. The first is the concern with ethics. Particularly when considering an intervention for people who are in good health, we need a scientific basis for being confident that the harms will not exceed the benefits. Any adverse effects of treating young adults will accumulate over decades before CHD begins to be common. Until we have compelling reasons to dismiss the increase in non-CHD mortality that has been observed in primary prevention trials, it is imprudent to proceed as though the adverse effects of treatment are negligible(25-27).

The second consideration is cost-effectiveness. Because the short term CHD risk of the average young adult is extremely low, drug treatment to lower high blood cholesterol in this population is an inordinately expensive means of prolonging life; the estimated $1 to 10 million per year of life is 100 to 1000 times the cost of secondary CHD prevention, or of many other medical treatments(28, 29). Dietary recommendations for individual patients is considerably less expensive than drug treatment. However, clinical trials in outpatients reveal dietary intervention to be surprisingly ineffective, the usual step I fat-controlled diet producing a sustained decrease in blood cholesterol level of 5% or less(30, 31). More intensive diets produce larger reductions in blood cholesterol, but cost more for adequate supervision and are less acceptable to many patients.

CLINICAL IMPLICATIONS FOR CHILDREN

The same reasons for the policy of rejecting blood cholesterol screening in young adults apply even more strongly to screening of children(32, 33), whether or not there is a family history of lipid disorders or early CHD. Any reduction in CHD deaths from screening and treating blood cholesterol in children is very small in size, temporally remote, unproven, and un-needed in light of the evidence that most of the risk associated with high blood cholesterol is reversible with intervention begun in middle age. The harmful effects of cholesterol screening and treatment in children are better established than the benefits, and potentially serious. Such a program would be expensive, cause malnutrition in some, have the adverse consequences of labelling, and might cause increased injury or cancer mortality rates. Regardless of the family history of the children, cholesterol screening and treatment is inappropriate until we have strong evidence from clinical trials in children that the benefits outweigh the harms.

CONCLUSIONS

Screening for high blood cholesterol makes sense only if the result of the test will affect clinical decisions, particularly the decision to prescribe cholesterol-lowering diets or drugs. And this intervention is justified only if the potential benefits clearly outweigh the possible harms. It seems clear that the policy of screening and treating high blood cholesterol is appropriate for high risk populations, particularly patients with prior evidence of CHD or other symptomatic arterial disease, and also middle aged or older men with several other risk factors. Cholesterol screening and treatment in young adults should be limited to rare individuals with known coronary artery disease or other unusual factors that place them at high short term risk of CHD. For the rest, it has unfavorable benefit-to-harm and cost-effectiveness ratios, and it is also not necessary—most CHD events associated with high blood cholesterol in this population will not occur for decades and can be prevented by treatment that is not begun until middle age. Cholesterol screening before age 20 should be avoided entirely.

REFERENCES

1. Expert Panel. Summary of the second report of the National Cholesterol Education Program (NCEP) Expert Panel on Detection, Evaluation, and Treatment of High Blood Cholesterol in Adults. JAMA 1993;269:3015-3023.

2. Toronto Working Group. Asymptomatic Hypercholesterolemia: A Clinical Policy Review. J Clin Epidemiol 1990;43:1029-1121.

3. Davey Smith G, Pekkanen J. Should there be a moratorium on the use of cholesterol lowering drugs? BMJ 1992;304:431-434.

4. Law MR, Thompson SG, Wald NJ. Assessing possible hazards of reducing serum cholesterol. BMJ 1994;308:373-379.

5. Muldoon M, Manuck S, Matthews K. Lowering cholesterol concentrations and mortality: a quantitative review of primary prevention trials. BMJ 1990;301:309-314.

6. Newman TB, Browner WS, Hulley SB. Childhood cholesterol screening: contraindicated. JAMA 1992;267:100-101.

7. Davey Smith G, Song F, Sheldon TA. Cholesterol lowering and mortality: the importance of considering initial level of risk. BMJ 1993;306:1367-1373.

8. Rossouw J, Lewis B, Rifkind B. The value of lowering cholesterol after myocardial infarction. N Engl J Med 1990;323:1112-1119.

9. Rossouw JE, Canner PL, Hulley SB. Deaths from injury, violence, and suicide in secondary prevention trials of cholesterol lowering. N Engl J Med 1991;325:1813 (letter).

10. Coronary Drug Project Research Group. The CDP: Findings leading to further modifications of its protocol with respect to dextrothyroxine. JAMA 1972;220:996-1008.

11. Coronary Drug Project Research Group. The CDP: Initial findings leading to modifications of its research protocol. JAMA 1970;214:1303-1313.

12. Huttunen JK, Heinonen OP, Manninen V, et al. The Helsinki Hear Study: an 8.5-year safety and mortality follow-up. J Intern Med 1994;235:31-39.

13. Newman TB. Possibly disappointing results of treatment with gemfibrozil. N Engl J Med 1993;328:139-140.

14. Bradford RH, Shear CL, Chremos AN, et al. Expanded Clinical Evaluation of Lovastatin (EXCEL) Study results. Arch Intern Med 1991;151:43-49.

15. Kaplan J, Manuck S, Shively C. The effects of fat and cholesterol on social behavior in monkeys. Psychosom Med 1991;53:634-642.

16. Duffy MA. Physicians' Desk Refence, 48th Edition.Montvale, N.J.: Medical Economics Data, 1994.

17. Jacobs D, Blackburn H, Higgins M, etal. Report of the Conference on Low Blood Cholesterol: Mortalty Associations. Circulation 1992;86:1046-1060.

18. Jacobs DR. Why is low blood cholesterol associated with risk of non-theroscerotic disease death? Ann Rev Pub Hlth 1993;14:95-114.

19. Engelberg H. Low serum cholesterol and suicide. Lancet 1992;339:727-8.

20. Mason RP, Herbette LG, Silverman DI. Can altering serum cholesterol affect neurologic function? J Mol Cell Cardiol 1991;23:1339-1342.

21. Brett AS. Psychologic effects of the diagnosis and treatment of hypercholesteroloemia: lessons from case studies. Am J Med 1991;91:642-647.

22. Davis C, Rifkind B, Brenner H, Gordon D. A single cholesterol measurement underestimates the risk of CHD. JAMA 1990;264:3044-3046.

23. LRC Program. The LRC Coronary Primary Prevention Trial results, II: the relationship of reduction in incidence of CHD to cholesterol lowering. JAMA 1984;251:365-74.

24. Tyroler H. Review of lipid-lowering clinical trials in relation to observational epidemiologic studies. Circulation 1987;76:515-522.

25. Oliver M. Might treatment of hypercholesterolaemia increase non-cardiac mortality? Lancet 1991;337:1529-1531.

26. Sackett DL, Haynes RB, Guyatt GH, Tugwell P. In: Clinical Epidemiology: A Basic Science for Clinical Medicine. Boston: Little, Brown, 1991: 163-167.

27. Eddy DM. Practice policies: Where do they come from? JAMA 1990;263:1265-1268.

28. Hulley SB, Newman TB, Grady D, Garber AM, Baron RB, Browner WS. Should we be measuring blood cholesterol levels in young adults? JAMA 1993;269:1416-1419.

29. Goldman L, Weinstein M, Goldman P, Williams L. Cost-effectiveness of HMG-CoA reductase inhibition for primary and secondary prevention of CHD. JAMA 1991;265:1145-1151.

30. Hunninghake DB, Stein EA, Dujovne CA, et al. The efficacy of intensive dietary therapy alone or combined with lovastatin in outpatients with hypercholesterolemia. N Engl J Med 1993;328:1213-1219.

31. Ramsay LE, Yeo WW, Jackson PR. Dietary reduction of serum cholesterol concentration: time to think again. BMJ 1991;303:953-957.

32. Newman TB, Garber AM, Holtzman NA, Hulley SB. Problems with the report of the Expert Panel of Blood Cholesterol Levels in Children and Adolescents. Arch Peds 1994; in press.

33. Hulley SB, Newman TB. Position Statement: Cholesterol screening in children is not indicated, even with positive family history. J Am Coll Nutr 1992;11:20S-22S.

CHOLESTEROL LOWERING, LOW CHOLESTEROL, AND MORTALITY

John C. LaRosa

Dean for Research
The George Washington University Medical Center
Washington, DC 20037

INTRODUCTION

Lowering of circulating cholesterol levels can prevent, arrest, and even reverse coronary atherosclerosis. Moreover, cholesterol lowering in patients with established coronary disease appears to be beneficial.

Meta-analyses of secondary prevention trials[1-3] indicate a slight but statistically insignificant decline in total mortality. A decline in cardiovascular and coronary heart disease mortality is achieved in these early studies by even modest cholesterol-lowering interventions. Meta-analyses of these secondary prevention trials do *not* demonstrate an excess mortality from non-cardiovascular causes, including cancer and violent death. The same is true of meta-analyses that *combine* primary and secondary trials.[4] Such an approach is reasonable if atherosclerosis is considered to be a continuous process with clinical manifestations only late in the course of the disease. Arguments about primary prevention do not necessarily apply to individuals with established coronary disease.

ISSUES RELATED TO PRIMARY PREVENTION

Both individual and meta-analyses of primary prevention studies have been reported.[5-11] Taken individually, these studies have demonstrated favorable effects in preventing myocardial infarction but have not demonstrated favorable effects on total mortality.[5] Even in meta-analyses of these studies, benefits on all-cause mortality cannot be demonstrated. The slight decline demonstrable in coronary mortality is offset by increased mortality from cancer and from non-cardiovascular, non-cancer causes, including accidents, suicides, and homicides.[5] These findings have led to speculation that either cholesterol lowering itself or the drugs used in these studies, including the bile acid sequestrants, clofibrate, and gemfibrozil, might themselves be promoters of non-cardiovascular mortality.

CANCER MORTALITY

The apparent increase in cancer mortality seen in meta-analyses of primary prevention

Cardiovascular Disease 2
Edited by L.L. Gallo, Plenum Press, New York, 1995

trials is largely the consequence of World Health Organization clofibrate study results.[8,12] An excess mortality from cancer, mostly of the gastrointestinal and biliary tracts, occurred in patients taking clofibrate during the course of this study. After 8 years, observations in the surviving subjects no longer taking study medication did not show the effect to be continued.[12] Dropouts from this study were not included in the published mortality analyses. Re-analysis of these data, in which dropouts are included, diminishes, but does not eliminate, the difference.[13] However, the Coronary Drug Project, utilizing clofibrate,[14] and the Helsinki Study, utilizing gemfibrozil, demonstrated no evidence of excess cancer mortality.[11]

ACCIDENTS, SUICIDES, AND HOMICIDES

In meta-analyses of selected primary prevention trials, there is an excess mortality from accidents, suicides, and homicides.[5] This finding is so far unexplained. The relationship of these mortal events to cholesterol lowering, however, is dubious. In a case-by-case audit of cases of violent death in the Lipid Research Clinic Coronary Primary Prevention Trial and the Helsinki Study, no relationship between the degree of cholesterol lowering and violent death could be demonstrated.[15] That is, there was no dose-response relationship. Indeed, the majority of suicide victims were not taking the prescribed medication at the time of their deaths.

In another analysis, Jacobs et al[16] and Yusef et al[17] attempted to relate non-cardiovascular, non-cancer mortality in intervention studies to the "strength" of the intervention (strength equalled the degree of cholesterol lowering multiplied by the duration). Again, no "dose-response" relationship could be demonstrated.

In some clinical observational studies, individuals who were prone to suicide, violence, or homicidal behavior (as *perpetrator*, not as victim) were observed to have low cholesterol levels.[18-20] These studies have not systematically examined individuals with other mental disorders, however, and by themselves, prove little.

There is some evidence from animal studies that animals put on low cholesterol diets exhibit more aggressive behavior.[21] However, a recent report in 305 men and women indicates that a cholesterol-lowering diet resulted in both lower cholesterols and reductions in depression and aggressive hostility.[22] It has been hypothesized that low cholesterol levels are associated, through a complex hypothetical mechanism involving changes in membrane cholesterol, with low levels of brain serotonin and, therefore, with changes in mood.[23] There is little direct evidence, however, to support this hypothesis.

Whatever the nature of these relationships, it should be pointed out that in current meta-analyses of primary prevention trials, there is no overall *excess* mortality in the treated groups. That is, at worst, all-cause mortality rates are unaffected and morbidity is favorably affected. Therefore, the case can be made that cholesterol lowering in primary prevention should be undertaken, even if only to lower the rate of morbid events.

RELEVANCE OF SPONTANEOUSLY OCCURRING LOW CHOLESTEROL LEVELS

Some population studies indicate that, particularly in men, the curve relating cholesterol to mortality is U-shaped, with higher mortality rates at both ends of the cholesterol distribution.[16]

A number of apparently disparate diseases are associated with low cholesterol levels. These range from violent deaths to a variety of cancer deaths (including hematological cancers), and also deaths from hemorrhagic stroke, chronic obstructive pulmonary disease, and "benign" cirrhotic liver disease. It is difficult to imagine that low cholesterol, in fact, *causes* all of these diseases. On the contrary, there is considerable evidence that these diseases themselves often result in a low cholesterol level (an effect-cause relationship).

This is particularly true in hematological cancers,[24] in cirrhosis of the liver, and in chronic hepatitis.[25]

The possibility that low cholesterol-mortality relationships are of the effect-cause variety, is buttressed by observations that, in a study of a statistically representative sample of the U.S. population, low cholesterol is most strongly associated with non-cardiovascular mortality in people over 70 years of age.[26] Other studies have indicated that the relationship is stronger among people of low socioeconomic status and among those with other evidence of declining health (i.e., weight loss, lack of exercise).[26,27] Thus, a large portion of this relationship occurs in individuals at risk of declining health, whose disease results in, rather than is caused by, low cholesterol.

It is also possible (although by no means proven) that many of these associations are a result of "confounding" variables. For example, it is a plausible hypothesis that in the Japanese, whose cholesterols are low and whose rates of hemorrhagic stroke are high,[28] the real culprit is a diet low in saturated fat but high in salt and fish, the former predisposing to hypertension, the latter (through fish oil effects) predisposing to decreased platelet adhesiveness and increased cerebral hemorrhage.

These relationships at the lower end of the cholesterol spectrum are poorly understood. Contrary to relationships between high cholesterol levels and atherosclerosis, cause and effect has not been established. It is imprudent to state, as one investigator did, that those with low blood cholesterol levels ought to try to raise them.[29] It is important, moreover, not to equate observations made in prevention trials where cholesterol has, in general, been lowered only 5-to-15%, to spontaneously occurring, much lower cholesterol levels in population studies. Cholesterol levels of 160 mg/dL (4.1 mmol/L) seen in the lowest quartile of populations have not been achieved in large clinical trials.

IMPLICATIONS FOR PUBLIC HEALTH INTERVENTIONS

It has been argued that if low cholesterol is causally associated with non-cardiovascular mortality, population strategies to lower the mean cholesterol level in the population by dietary change might increase the risk of non-cardiovascular mortality. Indeed, even the most vocal champions of cholesterol lowering do not claim that cholesterol lowering will provide immortality. It is certain that individuals who do not die of coronary atherosclerosis will die of some other cause. It is further possible that some of the dietary changes resulting in a low cholesterol level also bring about unfavorable metabolic changes that predispose individuals to other diseases.

There is danger, however, in applying poorly understood observations to public health policy. Theoretical concerns about the potential dangers of exposing a population to lower cholesterol levels should be tempered by remembering the adverse effects of our current diet and current cholesterol levels, leading to high mortality rates from coronary atherosclerosis (still the leading cause of death in Western countries), and perhaps increasing the risk of common cancers.[30]

As we go about sorting out these relationships, it is perhaps instructive to note that the Japanese smoke more cigarettes per capita than are consumed in the United States, have higher average blood pressures, and just as much diabetes. They still, however, have lower cholesterol levels, lower mortality rates, and a longer life expectancy.[28]

REFERENCES

1. J.E.Rossouw, B. Lewis, and B.M. Rifkind, The value of lowering cholesterol after myocardial infarction, *N Engl J Med.*323:1112(1990).
2. J.E.Rossouw, P.L.Canner, and S.B.Hulley, Deaths from injury, violence, and suicide in secondary prevention trials of cholesterol lowering, *N Engl J Med.* 325:1813(1991).

3. J.S.Silberberg and D.A.Henry, The benefit of reducing cholesterol levels: the need to distinguish primary from secondary prevention,*Med J Aust.* 155:665(1991).
4. S.MacMahon, Lowering cholesterol: effects on trauma death, cancer death and total mortality, *Aust N Z J Med.* 22:580(1992).
5. M.F.Muldoon, S.B.Manuck, and K.A.Matthews, Lowering cholesterol concentrations and mortality: a quantitative review of primary prevention trials, *BMJ.*301:309(1990).
6. S.Dayton, M.L.Pearce, S.Hashimoto, W.J.Dixon, and U.Tomiyasu,A controlled clinical trial of a diet high in unsaturated fat in preventing complications of atherosclerosis, *Circulation.* 40(suppl.II)II-I(1969).
7. I.D.Frantz, E.A.Dawson, P.L.Ashman and et al, Test of effect of lipid lowering by diet on cardiovascular risk, *Arteriosclerosis.*9:129(1989).
8. Committee of Principal Investigators, A co-operative trial in the primary prevention of ischaemic heart disease using clofibrate, *Br Heart J.* 40:1069(1978).
9. A.E.Dorr, K.Gundersen, J.C.Schneider, T.W.Spencer, and W.B.Martin, Colestipol hydrochloride in hypercholesterolaemic patients: effect on serum cholesterol and mortality, *J Chronic Dis.*31:5(1978).
10. Lipid Research Clinics Program, the Lipid Research Clinics coronary primary prevention trial results, *JAMA.* 251:351(1984).
11. M.H.Frick, O.Elo, K.Haapa, and et al, Helsinki Heart Study: primary-prevention trial with gemfibrozil in middle-aged men with dyslipidemia, *N Engl J Med.* 317:1237(1987).
12. Committee of Principal Investigators, WHO cooperative trial on primary prevention of ischaemic heart disease with clofibrate to lower serum cholesterol: final mortality follow-up, *Lancet.* ii:600(1984).
13. J.A.Heady, J.N.Morris, and M.F.Oliver, WHO clofibrate/cholesterol trial: clarifications, *Lancet.* 340:1405(1992).
14. The Coronary Drug Project Research Group, Clofibrate and niacin in coronary heart disease, *JAMA.* 231:360(1975).
15. D.A.Wysowski and T.P.Gross, Deaths due to accidents and violence in two recent trials of cholesterol-lowering drugs, *Arch Intern Med.*150:2169(1990).
16. D.Jacobs, H.Blackburn, M. Higgins, and et al, Report of the conference on low blood cholesterol: mortality associations, *Circulation.* 86:1046(1992).
17. S.Yusef, J.Wittes, and L.Friedman, Overview of results of randomized clinical trials in heart disease. II. Unstable angina,heart failure, primary prevention with aspirin, and risk factor modification, *JAMA.* 260:2259(1988).
18. M.Virkkunen, Serum cholesterol levels in homicidal offenders, *Neuropsychobiology.*10:65(1983).
19. M.Virkkunen, Serum cholesterol in antisocial personality, *Neuropsychobiology.* 5:27(1979).
20. M.Virkkunen, J. DeJong, J.Bartko, and M. innoila, Psychobiological concomitants of history of suicide attempts among violent offenders and impulsive fire setters, *Arch Gen Psychiatry.*46:604(1989).
21. J.R.Kaplan, S.B.Manuck, and C.Shively, The effects of fat and cholesterol on social behavior in monkeys, *Psychosom Med.* 53:634(1991).
22. G.Weidner, S.L.Connor, J.F.Hollis, and W.E.Connor, Improvements in hostility and depression in relation to dietary change and cholesterol lowering: the Family Heart Study, *Ann Intern Med.*117:820(1992).

23. H.Engelberg, Low serum cholesterol and suicide, *Lancet.* 339:727(1992).
24. D.Budd and H.Ginsberg, Hypocholesterolemia and acute myelogenous leukemia: association between disease activity and plasma low-density lipoprotein cholesterol concentrations,*Cancer.* 58:1361(1986).
25. Z.Chen, R.Peto, R.Collins, S.MacMahon, J.Lu, and W.Li, Serum cholesterol and coronary heart disease in populations with low cholesterol concentrations, *BMJ.*303:276(1991).
26. T.Harris, J.J.Feldman, J.C.Kleinman, W.H.Ettinger, Jr., D.M.Makuc, and A.G.Schatzkin, The low cholesterol-mortality association in a national cohort, *J Clin Epidemiol.* 45:595(1992).
27. G.D.Smith, M.F.Shipley, M.G.Marmot, and G.Rose, Plasma cholesterol concentration and mortality: the Whitehall Study, *JAMA.* 267:70(1992).
28. Y.Goto and E.H.Moriguchi, Diet and ischemic heart disease in Japan, *Atherosclerosis Rev.* 21:21(1990).
29. G.Kolata, Cholesterol's new image: high is bad; so is low, *New York Times.*(August 11, 1992).
30. Public Health Service, *The Surgeon General's Report on Nutrition and Health.*DHHS(PHS)Pub.No.88-50210, U.S.Government Printing Office, Washington, DC (1988).

BAVARIAN CHOLESTEROL SCREENING PROJECT (BCSP)

Peter Schwandt, Werner O. Richter, Andreas C. Sönnichsen

Medical Department II
Klinikum Grosshadern of the University
Marchioninistr. 15
81366 Munich, FRG

INTRODUCTION

In Germany more than 50 percent of the deaths were caused by cardiovascular disease (more females than males) followed by all cancer mortality (which was about 1/2 of the cardiovascular death rate) in 1993. From these more than 440,000 cardiovascular deaths 105,736 were caused by stroke and 89,049 by myocardial infarction (9.92 % of all cause mortality) with considerable regional differences.

Thus there were several reasons to start a risk factor screening program in Bavaria (about 11 Million inhabitants) in the South of Germany:

- to increase the awareness of the silent risk "hypercholesterolemia"
- to detect severe hypercholesterolemia for immediate evaluation and treatment by the family doctor
- to find out patients with borderline and slightly elevated cholesterol concentrations which mostly can be normalized by dietary and life style changes.

SUBJECTS AND METHOD

Cholesterol was measured enzymatically by the dry chemistry method with the REFLOTRON (Boehringer Mannheim, Germany) with a high accurracy (± 3 %) compared to automated analysis (r = 0.98). The team of medical students and technicians was well trained and controlled. Internal quality control was carried out at the beginning and after every 50 measurements. External quality control was performed twice a year with INSTAND (Düsseldorf, Germany).

From all participants (104,934 women, 86,765 men) a history was taken including age, bodyweight, height, cigarette smoking, hypertension, diabetes mellitus, drug consumption, previous cholesterol measurement, known hyperlipoproteinemia, myocardial infarction and stroke in the participant and his family.

Cardiovascular Disease 2
Edited by L.L. Gallo, Plenum Press, New York, 1995

The participants were advised according to their cholesterol levels as shown in table 1.

Table 1. BCSP - Advice according to cholesterol level

- **< 200 mg/dl without risk factors:** repeat within 5 years
- **< 200 mg/dl with ≥ 2 risk factors:** repeat and add HDL cholesterol and triglycerides within 1 year
- **200-250 mg/dl without risk factors:** full lipid profile within 3 months
- **200-250 mg/dl with ≥ 2 risk factors:** life style modification, contact physician within 3 months
- **> 250 mg/dl:** contact your physician immediately

The age distribution of the population screened from 1989-1993 is shown in figure 1.

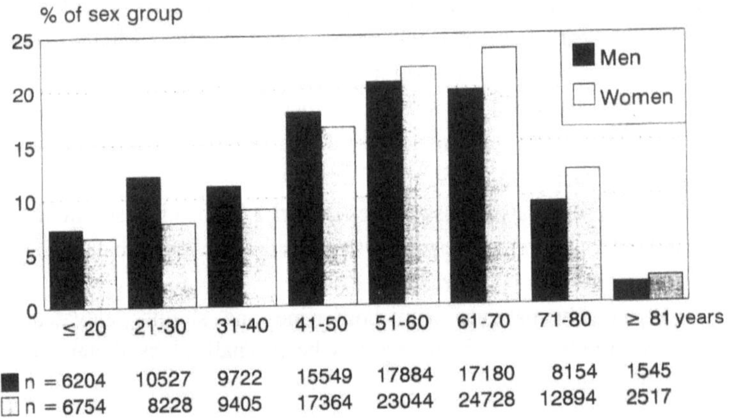

Figure 1. BCSP - Age distribution of participants (n = 191,699)

RESULTS

In table 2a and table 2b the plasma cholesterol levels for women and men are given in comparison with the NHANES III data [1]. The mean cholesterol concentration in Bavaria is considerably higher than in the USA for women (243 mg/dl compared to 207 mg/dl) as well as for men (231 mg/dl compared to 205 mg/dl).

Table 2a. NHANES III - Plasma cholesterol levels (mg/dl) in the US population, 1988-1991 [1]

Men		Selected percentile			
Age (y)	n	Mean	5th	50th	95th
20-34	1 186	189	134	186	260
35-44	653	207	144	205	269
45-54	508	218	152	215	283
55-64	535	221	154	221	285
65-74	557	218	157	214	286
≥ 75	514	205	145	202	275
total	**3 953**	**205**	**143**	**201**	**276**

Women		Selected percentile			
Age (y)	n	Mean	5th	50th	95th
20-34	1 177	185	134	182	254
35-44	709	195	142	193	254
45-54	464	217	158	212	297
55-64	503	237	168	228	323
65-74	493	234	168	232	308
≥ 75	539	230	163	227	316
total	**3 885**	**207**	**143**	**202**	**287**

Table 2b. BCSP - Plasma cholesterol levels (mg/dl) in the Bavarian population, 1989-1993

Men		Selected percentile			
Age (y)	n	Mean	5th	50th	95th
20-34	20 327	186	118	181	271
35-44	10 839	231	154	229	318
45-54	18 471	241	167	239	324
55-64	17 509	244	170	242	327
65-74	13 287	242	168	240	322
≥ 74	6 332	232	157	320	312
total	**81 383**	**231**	**149**	**230**	**318**

Women		Selected percentile			
Age (y)	n	Mean	5th	50th	95th
20-34	12 288	199	138	195	273
35-44	11 578	215	153	211	290
45-54	21 609	238	169	235	320
55-64	23 322	260	189	257	343
65-74	20 403	263	190	260	345
≥ 75	9 952	257	180	255	341
total	**99152**	**243**	**163**	**240**	**331**

32 % of the males and 40 % of the females participating in BCSP had cholesterol concentrations above 250 mg/dl (table 3). By combining elevated cholesterol levels and the other risk factors, the BCSP demonstrated that a large portion of the polulation of nearly 200,000 people are at on risk for cardiovascular disease.

Table 3. BCSP - Risk profile (%) of 191,699 participants

Risk factors	Men (n = 86,765)	Women (n = 104,934)
Total cholesterol		
< 200 mg/dl	31.3	23.3
200-250 mg/dl	36.9	36.8
251-300 mg/dl	23.5	27.9
> 300 mg/dl	8.3	12.0
Cigarette smoking	20.7	12.3
Hypertension (> 160 / > 95 mmHg)	17.1	19.5
Obesity (BMI > 27.5 kg/m^2)	20.2	16.4
Diabetes mellitus	4.0	4.0
Myocardial infarction	7.4	3.9
Stroke	1.4	1.2
Familiy history of myocardial infarction	25.3	33.5

Only 23.6 % of the women and 18.7 % of the men knew their cholesterol values which had been measured previously. As shown in table 4 only 4,853 participants received lipid-lowering drugs. Furthermore, most of them obviously were not treated efficiently.

Table 4. BCSP - Cholesterol levels in men and women on treatment with lipid-lowering drugs

Cholesterol (mg/dl)	Men (n = 1,885)	Women (n = 2,968)
< 200	11.4 %	5.3 %
200 - 250	38.0 %	34.0 %
251 - 300	34.4 %	39.6 %
> 300	16.2 %	21.2 %

DISCUSSION

These data - which at least in part are representative for Germany - explain the only slight decrease in deaths from coronary heart disease in the western part of this country and the increase in the eastern part during the same observation period (figure 2) [2]. Much has to be done in Germany to spread the knowledge about the effects of primary and secondary prevention, which led to a dramatic reduction in coronary heart disease in some other countries.

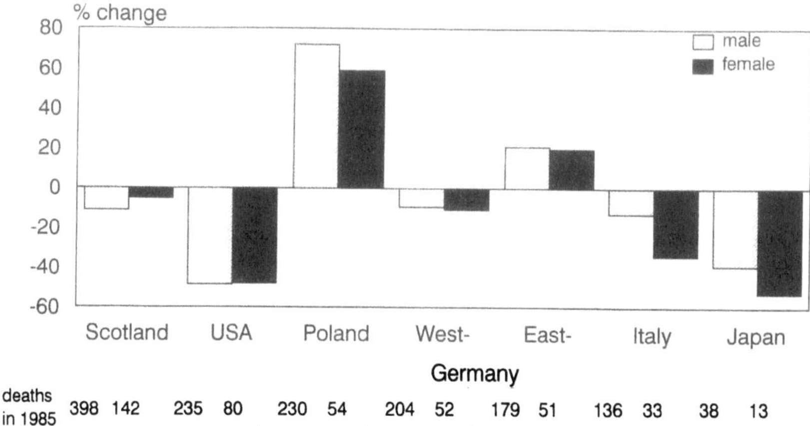

Figure 2. Age-Standardized Mortality per 100,000 Population aged 30-69 Years in 1985 and Percentage Change (1970-1985) from Ischaemic Heart Disease

REFERENCES

1 Report of the Expert Panel of the National Cholesterol Education Program: Second report of the Expert Panel on Detection, Evaluation, and Treatment of High Blood Cholesterol in Adults (Adult Treatment Panel II). Circulation, 89:1329-1444 (1994)

2 Uemura K., Pisa Z.: Trends in Cardiovascular Disease Mortality in Industrializes Countries since 1950. Wld. hlth. statist. quart. 41:155-178 (1988)

INDEX

Accidents, cholesterol lowering and, 347, 348
Acidic fibroblast growth factor, *see* Fibroblast growth factor 1
Acquired immunodeficiency syndrome (AIDS), 179
Actin, 147
Acute chest pain, 291–296
Acylation stimulatory protein (ASP), 69
Acyl coenzyme A:Acylcholesterol acyltransference (ACAT), 58, 75, 76
Adenosine 5'-diphosphate (ADP)
  platelet effects of, 255, 256, 257, 262, 263
  P-selectin phosphorylation and, 194
Adhesion molecules, 159–167, *see also* specific types
  in atherosclerosis, 165
  in chronic periaortitis, 166
  in normal aorta, 164
Adipsin, 69
Albumin, 281, 282, 283–284
Alcohol, 99–100
Alkaline phosphatase anti-alkaline phosphatase (APAAP), 162
Alzheimer's disease, 57, 64–65
Angina, *see* Stable angina; Unstable angina
Angiogenesis, 133–140
Antioxidants, 2
  vascular cell adhesion molecule-1 and, 121–125
α-2-antiplasmin, 293
Anti-thrombin III (AT-III)
  acute chest pain and, 293
  de-endothelialised aorta wall and, 279, 280, 281, 282, 283, 284–285, 286–287
  tissue factor and, 236, 240
Aorta
  adhesion molecules in, 159–167
  hemostatic response to de-endothelialised, 279–287
APOA2 (human) gene, 303
Apoa2 (mouse) gene, 302–303
APOB gene, 69
APOC1 (human) gene, 318
Apoc1 (mouse) gene, 317–322, 318–319
APOC2 (human) gene, 318
Apoc2 (mouse) gene, 318
Apoc2 Linked Gene (*Acl*), 318
APOE gene, 318
Apolipoprotein(a) (apo(a)), 45–46, 323–324
Apolipoprotein A-1 (ApoA-1), 35, 67
  cholesterol efflux and, 89, 90–92, 93
  deficiency of, 36
  metabolism of, 31
  particles containing in atherosclerosis, 97–101
Apolipoprotein A-2 (ApoA-2), 35, 97
  genetic effects on level of, 301, 302–303
  metabolism of, 31
Apolipoprotein B (ApoB), 57, 323–325
  cholesterol efflux and, 92
  low density lipoprotein receptor-related protein and, 227
  metabolism of, 31, 32
Apolipoprotein B-48 (ApoB-48), 310
Apolipoprotein B-100 (ApoB-100), 46, 223, 309
Apolipoprotein C-1 (ApoC-1), 317–322
Apolipoprotein C-2 (ApoC-2), 32
Apolipoprotein C-3 (ApoC-3), 58
Apolipoprotein E (ApoE), 57–65, 317
  chylomicrons and, 307–308, 309, 310
  lipoprotein metabolism and, 59–60
  low density lipoprotein receptor-related protein and, 223, 227, 228
  metabolism of, 32
  paradoxes of, 60–65
  very low density lipoprotein receptor and, 224–225
Arachidonic acid, 255, 269
Arginine-Glycine-Aspartate, *see* RGD
Artery wall, lipid oxidation in, 41–42
Aspirin, 256, 264, 265, 271, 273
Atherosclerosis, 1
  adhesion molecules in, 165
  antioxidants and vascular cell adhesion molecule-1 expression, 121–125
  apolipoprotein(a) and, 323–324
  apolipoprotein A-I-containing particles and, 97–101
  gelatinase 92kDa expression in, 11–16
  genetic factors in, 299–304
  genetic manipulation of lipoprotein receptors and, 307–310
  lipoproteins in, 31–37

Atherosclerosis (*cont.*)
matrix metalloproteinases and, 20, 22, 25, 26
plasminogen activator 2 and, 214
platelet activation and, 256
platelet-derived growth factor potentiation in, 129–131
Ath-1 gene, 302
Autocrine hypothesis, 129
5-Azacytidine, 131

Baicalein, 271, 273
Balloon catheter injuries, 279–287
Basic fibroblast growth factor, *see* Fibroblast growth factor 2
Basic protein I (BP I), 69–72
Basic protein II (BP II), 69–72
Basic protein III (BP III), 69
Bavarian Cholesterol Screening Project (BCSP), 353–357
bcl-2 gene, 130
Beta-adrenergic receptor kinase 1 (BARK1), 252
Beta-adrenergic receptor kinase 2 (BARK2), 252
Bile acid sequestrants, 347
Breast cancer, 238, 334

C:8, 70–71
C5a, 154
Caffeic acid, 271
Calcium
L-selectin and, 173
in Tangier fibroblasts, 80, 84, 85–86
thrombin inhibitor vs. GPIIb/IIIa effect on, 255, 258, 261–262
Cancer, *see also* specific types
cholesterol lowering and, 342, 347–348, 348
thromboembolic disease and, 238
vascular endothelial growth factor and, 137–138
Candidate genes, 302
γ-Carboxyglutamic acid, 239
CCAIT, 1
cdc2 protein kinase, 111
Cellular proliferation
fibroblast growth factor 1-induced genes and, 111
genetic modification of in smooth muscle, 314
c-fos gene, 110, 111, 146
Children, cholesterol lowering in, 344
Chloramphenicol acyl transferase (CAT), 130
Cholesterol, 2, 3, 67
Bavarian Cholesterol Screening Project on, 353–357
efflux from cultured cells, 89–94
genes regulated by dietary, 327–332
high density lipoprotein receptor removal of cellular, 75–78
high density lipoprotein $_3$ removal of cellular. 79–86
meta-analysis on, 333–338, 342, 347–348
removal from Tangier fibroblasts, 79–86
spontaneously low levels of, 348–349
Cholesterol-down-regulated (CDR) genes, 329–331
Cholesterol esters, 2
Cholesterol ester transfer protein (CETP), 32, 59, 324
Cholesterol lowering
in children, 344
mortality and, 347–349
risks of, 341–344
Cholesterol monohydrate crystals, 2, 3–4
Cholesterol-up-regulated (CUR) genes, 329–331
Cholestyramine, 2, 335, 342
Chronic obstructive pulmonary disease, 348
Chronic periaortitis, 166
Chylomicrons, 307–308
apolipoprotein C1 and, 317
apolipoprotein E and, 57
low density lipoprotein receptor-related protein and, 309–310
metabolism of, 31, 32
Cirrhosis of the liver, 348, 349
CIV 22, 161
c-Jun gene, 110, 111
CLAS, 1
Clofibrate, 334, 342, 347, 348
c-myb gene, 110, 111
c-myc gene, 110, 111, 146
Coagulation factors, 291–296, *see also* specific types
Cobalt chloride, 135
Collagen, 3
platelet effects of, 255, 256, 257
P-selectin phosphorylation and, 194
restenosis and, 21, 22, 23
wound healing and, 20
Collagen IV, 23, 161
Collagenase, 20–21, 21, 25
Colon carcinoma, 137, 138, 238
Connexin-43, 42
Coronary Angioplasty versus excisional Atherectomy Trial (CAVEAT), 12
Coronary artery disease (CAD), *see also* Premature coronary artery disease
apolipoprotein A-I-containing particles and, 99
coagulation factors and, 291–296
Coronary Drug Project (CDP), 2, 335, 348
Coronary heart disease (CHD)
cholesterol lowering and, 341–344
meta-analysis of cholesterol levels in, 333–338
c-Ras gene, 111
Culprit lesions, 4, 5, 15
Cysteamine, 59–60

3D4, 161
De-endothelialised aorta wall, 279–287
Delayed-early genes, 110, 113
Dermatan sulfate, 280
Developmental maturation, 145–149
Dextrothyroxine, 334
Diabetes, 67, 343
hyperapobetalipoproteinemia and, 68
L-selectin and, 185–186
triglyceride-rich lipoprotein metabolism and, 49–53

Diabetic retinopathy, 138–139
Diacylglycerol
  apolipoprotein A-I-containing particles and, 98
  in Tangier fibroblasts, 81, 84, 86
Diapedesis, 153, 154
Diet, 1, 67, 99–100
Dietary cholesterol, 3, 327–332
Differential display, 109–115
D-dimers, 293, 294, 295
Dimyristoyl phosphatidylcholine (DMPC), 91–92
Disseminated intravascular coagulation (DIC), 238
DMP 728, 257, 258, 259
Docosahexaenoic acid, 269
Docosanoids, 269–275
Docosapentaenoic acid, 269, 270–271
Double-stranded RNA, 123
DUP 714, 255, 257, 259–260, 264
Dysbetalipoproteinemia, *see* Hyperlipidemia type III
Dyslipoproteinemias, 99

7E3, 264
Early erythroid progenitor cells (BFU-E), 174
Early growth response gene-1, 110
EDRF (nitric oxide), 3
Eicosapentaenoic acid, 269
5,8,11,14-Eicosatetraynoic acid (ETYA), 271
ELAM-1, 42, 153
Elastase, 20, 22
Elastin, 3, 20–21, 22
EMBII, 161
Endothelial cells
  lipid oxidation and, 41
  vascular endothelial growth factor binding to, 136
Endothelium
  consequences of neutrophil adherence to, 153–156
  as integrator of pathophysiologic stimuli, 105–106
Epidermal growth factor (EGF), 21
Epidermal growth factor (EGF) domain
  L-selectin, 173–174, 179, 185, 193–194
  P-selectin, 191, 193–194, 200
Epinephrine, 194, 256, 257
EPR-1, 236
Esculetin, 271
E-selectin, 106, 122, 123, 124, 162, 185, 187, 192
  in atherosclerosis, 165
  in chronic periaortitis, 166
  inflammation and, 121
  L-selectin and, 177, 178
  neutrophil adherence and, 153, 154, 155, 156
  in normal aorta, 164
Estrogen, 62–64, 296, 334, 335
Extracellular matrix (ECM)
  matrix metalloproteinases and, 19–20, 21–26
  vascular endothelial growth factor and, 135
  wound healing and, 19–20
Extracellular staining pattern, 14, 15

Factor II, 256
Factor V
  acute chest pain and and, 293, 296
  de-endothelialised aorta wall and, 283
Factor VII
  acute chest pain and, 293
  tissue factor and, 235, 236, 240
Factor VIIa, *see* Tissue factor-factor VIIa complex
Factor VIIc, 291–292
Factor VIII
  acute chest pain and, 293, 295
  vascular endothelial growth factor and, 136
Factor IX, 239, 240
Factor IXα, 236
Factor IXa, 236, 241
Factor X, 256
  acute chest pain and, 293
  tissue factor and, 236, 239, 240
Factor Xa
  de-endothelialised aorta wall and, 283
  tissue factor and, 236, 241
Familial combined hyperlipidemia (FCHL)
  apolipoprotein E in, 61–62, 64
  hyperapobetalipoproteinemia and, 67, 68–69
Familial dyslipidemic hypertension, 69
Familial hypercholesterolemia (FH), 67, 307, 308, 309
Familial hypertriglyceridemia (FHT), 67
Familial hypertriglyceridemia-hypoalpha-lipoproteinemia syndrome, 36
Familial hypo-alphalipoproteinemia, 36
FATS, 1, 5
Fenofibrate, 100
F-31/H19, 147–149
Fibrin, 192, 211
Fibrinogen
  acute chest pain and, 291–292, 293, 294, 296
  de-endothelialised aorta wall and, 279, 280, 281, 282–283, 284–285, 287
  thrombin inhibitor vs. GPIIb/IIIa effect on, 255, 257–258, 259–260
Fibrinolysis, 206–207
Fibroblast growth factor (FGF), 21
Fibroblast growth factor 1 (FGF-1), 109–115
Fibroblast growth factor 1 (FGF-1) gene, 313–314
Fibroblast growth factor 2 (FGF-2), 109, 134, 135, 139
Fibronectin, 19–20, 42
Fish oil, 275, 349
Flk-1/KDR tyrosine kinase, 136–137
Flt-1 tyrosine kinase, 136–137
FR-1 gene, 110, 112, 113–114
FR-2 gene, 113
FR-3 gene, 112
FR-12 gene, 114, !15
Fructose, 50–51
Fucose, 155

Gelatinase, 92kDa
  plaque rupture and, 11–16
  restenosis and, 20–21, 23

Gemfibrozil, 334, 342, 347, 348
Gender
  coagulation factors and, 295, 296
  coronary heart disease and, 343
  premature coronary artery disease and, 67
Genes
  atherosclerosis and, 299–304
  candidate, 302
  developmental expression in smooth muscle cells, 145–149
  dietary cholesterol regulation of, 327–332
  fibroblast growth factor 1-inducible, 109–115
  growth factor, 313–314
Gene targeting, 318–319
Genetic manipulation
  of lipoprotein receptors, 307–310
  of vascular disease, 313–314
Genistein, 71–72
Glucose, 50–51
GlyCAM-1, 186, 187
Glyceraldehyde-3-phosphate dehydrogenase, 110
Glycoprotein 330 (gp330), 224, 225
Glycosyl-phosphatidylinositol (GPI), 218
GMP-140, 154, 199, 249
GPIIb/IIIa antagonists, 255–265
Granulocyte-macrophage colony stimulating factor, 154
Growth factor gene expression, 313–314

H-7, 70
HBP, 77
HD37, 161
Heidelberg Study, 1
Helsinki Study, 342, 348
Hematological cancer, 348, 349
Hemorrhagic stroke, 348, 349
Hemostasis, 279–287
Heparin, 201, 256, 264
Heparinase, 227
Heparin sulfate, 21
Hepatic triglyceride lipase (HTGL), 58, 64
Hepatitis, 349
12-HEPE, 269
Herpes simplex virus, 238
Herpes virus thymidine kinase (HSV-tk) gene, 314
12-HETE, 269
High density lipoprotein (HDL), 67, 303, 343
  apolipoprotein C-1 and, 317
  apolipoprotein E and, 57, 58, 59–60, 62
  cholesterol efflux and, 92–93
  hyperapobetalipoproteinemia and, 69
  low density lipoprotein oxidation and, 41
  metabolism of, 31, 32
  molecular heterogeneity of, 34–37
  types of apolipoprotein A-I particles in, 97–101
High density lipoprotein $_2$ (HDL_2), 92
High density lipoprotein $_3$ (HDL_3), 92, 93
  apolipoprotein A-I-containing particles and, 98
  in Tangier fibroblasts and, 79–86
High density lipoprotein (HDL) receptors, 75–78
Hirudin, 255, 256–257, 259, 260, 264
  de-endothelialised aorta wall and, 280–281
Histamine, 154
HLA DR, 161
Homicide, cholesterol lowering and, 347, 348
Hydrogen peroxide, 123, 154
Hydroxymethylglutaryl CoA reductase (HMG-CoA) inhibitors, 58, 100, 336, 342
Hyperalphalipoproteinemia, 99
Hyperapobetalipoproteinemia (hyperapoB), 67–72
Hyperinsulinemia, 49–52
Hyperlipidemia type III, 61, 62
Hyperlipoproteinemia type III, 34
Hypertension, 67
  coagulation factors and, 291–296
  hyperapobetalipoproteinemia and, 68
  in Japan, 349
Hypertriglyceridemia
  diabetes and, 49–53
  hyperapobetalipoproteinemia and, 68
Hypoxia, 135

ICAM-1, *see* Intracellular cell adhesion molecule-1
Immediate-early genes, 110, 113
Indomethacin, 271, 273
Inflammation, 121–122
Initiation complex, 235, 239, 242
Inositol biphosphate ($InsP_2$), 80–81, 85–86
Inositol monophosphate ($InsP_1$), 80–81
Inositol triphosphate ($InsP_3$), 80–81, 85–86
Insulin, 50, 51–52
Insulinemia, *see* Hyperinsulinemia
Insulin-like growth factor (IGF), 21
Insulin-like growth factor-II (IGF-II), 147
Integrelin, 257
β_1-Integrins, 178–179
β_2-Integrins
  leukocyte rolling and, 179
  neutrophil adherence and, 153, 154, 156
Interferon-γ, 154
Interleukin-1 (IL-1), 42, 105, 154
Interleukin-1β (IL-1β), 122, 123, 211
Interleukin-4 (IL-4), 154
Interleukin-6 (IL-6), 42
Interleukin-8 (IL-8), 154, 155
Intermediate density lipoprotein (IDL)
  apolipoprotein E in, 57, 58–59, 61–64
  metabolism of, 32
  very low density lipoprotein receptor and, 225
Intimal fibrous proliferation (IFP) response, 21, 22
Intracellular cell adhesion molecule-1 (ICAM-1), 106, 122, 123, 124, 162, 163, 173
  in atherosclerosis, 165
  in chronic periaortitis, 166
  inflammation and, 121
  neutrophil adherence and, 153
  in normal aorta, 164
Intracellular cell adhesion molecule-2 (ICAM-2), 153, 173
Intracellular cell adhesion molecule-3 (ICAM-3), 173

Intracellular staining pattern, 14, 15
Intracranial tumors, 137
Ischemia/reperfusion injury, 155–156

Japan, 349
JC70, 161

4KB128, 161

LAM-1, 154
LECAM-1, 154
Lecithin:cholesterol acyltransferase (LCAT), 317
Lectin domain
  L-selectin, 173, 185
  P-selectin, 191, 193, 199–203
Leucine zipper, 111, 112
Leukemia, 179
Leukocyte adherence, 153–156
Leukocyte adhesion deficiency (LAD) type-1, 155, 156
Leukocyte adhesion deficiency (LAD) type-2, 155, 156
Leukocyte rolling, 153, 177–179
Leukocyte sticking, 153
Leukotriene B4, 154
Leumedin, 41–42
LFA-1, 153, 173, 174
LFA-3, 174
Ligand binding
  in P-selectin, 192–193
  in thrombin receptor, 250–251
Limb ischemia, 139
Linoleic acid, 269
Linolenic acid, 269
Linoleyl hydroperoxide (13-HPODE), 124
Lipid Research Clinics Coronary Primary Prevention Trial (LRC-CPPT), 1, 335, 348
Lipids
  genetic manipulation of lipoprotein receptors and, 307–310
  homeostasis in oxidation of, 41–42
  lowering of and atherosclerosis progression, 1
  lowering of and prevention of clinical events, 1–2
  as triggering risk factor, 1–6
Lipid transfer zone, 59
Lipopolysaccharide (LPS)
  antioxidants and, 123
  neutrophil adherence and, 154
  plasminogen activator 2 and, 211, 213
  tissue factor and, 236, 238, 242
Lipoprotein
  apolipoprotein E and, 59–60
  in atherosclerosis, 31–37
  major atherogenic/anti-atherogenic, 32–33
  metabolism of, 31–32, 59–60
  triglyceride-rich in diabetes, 49–53
Lipoprotein(a) (Lp(a)), 2, 32, 323–325
  lysine binding polymorphism of, 45–46
  properties of, 34
Lipoprotein lipase, 226–227, 228–229
Lipoprotein receptors, 307–310
Lovastatin, 1, 342
Low density lipoprotein (LDL), 32–33
  acute chest pain and, 291
  apolipoprotein A-I-containing particles and, 98
  apolipoprotein C-1 and, 317, 321
  apolipoprotein E in, 57, 60–62, 63–64
  cholesterol efflux and, 92, 93
  homeostasis of oxidation in artery wall, 41–42
  hyperapobetalipoproteinemia, 67–69
  lipoprotein(a) and, 323–324
  lowering of, 1–2, 341
  metabolism of, 32
  plaque disruption and, 5
  plasminogen activator 2 increase and, 211–214
  in Tangier fibroblasts, 84
  vascular cell adhesion molecule-1 and, 124–125
Low density lipoprotein (LDL) hypothesis, 122
Low density lipoprotein receptor (LDLR)
  apolipoprotein E and, 57
  family organization in, 223–225
  genetic manipulation of, 307–310
  high density lipoprotein receptor and, 77
Low density lipoprotein receptor-related protein (LRP), 223–230
  apolipoprotein C1 and, 317
  genetic manipulation of, 307–310
  lipoprotein lipase regulation and, 226–227, 228–229
  lipoprotein metabolism and, 32
  proteinase regulation and, 225–226
  urokinase receptor and, 218, 219
LpA-I, 35–37, 98–101
  cholesterol efflux and, 92, 93
  isolation and composition of, 97–98
  quantitative determination of, 98–99
LpA-I:A-II, 35–37, 98–101
  isolation and composition of, 97–98
  quantitative determination of, 98–99
LpA-I:A-IV, 35
LpA-I:E, 35
LpA-II, 92
LpE, 35
LPLC, 228, 229, 230
L-selectin, 173–180, 185–187, 193–194
  expression of, 174
  function of, 175
  lymphocyte homing and, 175–177, 185
  neutrophil adherence and, 154, 155
  as P-selectin ligand, 192, 193
  regulation of function, 179–180
  structure of, 173–174
LTC4/LTD4, 154
Lung tumors, 238
Lymphocyte homing, 175–177, 185
LYP-20, 199, 200–201, 201
Lysine binding, 45–46
Lysophosphatidylcholine (lyso-PC), 124

Mac-1, 153, 173
MadCAM-1, *see* Mucosal addressin cell adhesion molecule 1

MARS, 1
Matrix metalloproteinases (MMPs), 19–26
activators of, 24–26
inhibitors of, 24–26
plaque rupture and, 11, 15, 16
restenosis and, 20–26
wound healing and, 19–20
MECA-79, 176, 177, 189–190
MEL-14 antibody, 177
Meta-analysis, 333–338, 342, 347–348
Metalloproteinases, *see* Matrix metalloproteinases
Methyl DNA binding protein-2 (MeCP-2), 131
Microdomain HDL receptor model, 76–77
Mitogenic signal transduction, 109–110, 114
MK383, 257
Mo-1, 153, 173
Monoclonal antibodies, *see also* specific antibodies
to adhesion molecules, 161
to P-selectin, 200–201
Monocyte chemotactic protein-1 (MCP-1), 41
Monocytes, plasminogen activator 2 in, 211–214
Mortality
cholesterol lowering and, 342, 347–349
meta-analysis on cholesterol in, 333–338
Mucosal addressin cell adhesion molecule 1 (Mad-CAM-1), 153, 176
Myocardial infarction, 335
cholesterol and, 333
coagulation factors and, 291–296
platelet activation and, 256

N-acetylcysteine (NAC), 123
National Cholesterol Education Program (NCEP), 341
NDGA, 271
Neutrophils
adherence of, 153–156
P-selectin mediation in platelet interactions, 199–203
NFκB system, 105
lipid oxidation and, 42
tissue factor and, 236
vascular cell adhesion molecule-1 regulation through, 123
NHLBI, 1
Niacin, 1, 2, 335
Nicotinic acid, 100
Nitric oxide (EDRF), 3

Obesity, 67, 68, 301
OH-22:6n3, 272–273, 274–275
Ornithine decarboxylase, 110
Oxygen radicals, 123

p53, 111
p150/95, 173
PADGEM, 154, 199
Pancreas tumors, 238
PECAM-1, 154
Periaortitis, 166
Phagocytes, 153
Phe-Pro-Arg-chloromethyl ketone, *see* PPACK
Phosphatidyl inositol 4,5-biphosphate (PIP_2), 72
Phosphofructokinase, 110
Phospholipase C γ-1 (PLCγ-1), 72
Phospholipids, 2
Piroprost, 271
Plaque rupture, 1–6
determinants of, 4–5
gelatinase 92kDa expression following, 11–16
prevention of, 5–6
Plasmin, 21
Plasminogen
acute chest pain and and, 293, 294, 296
lysine binding by, 45
wound healing and, 19–20
Plasminogen activator inhibitor, 292, 293
Plasminogen activator inhibitor 1 (PAI-1)
low density lipoprotein receptor-related protein and, 225, 227
low density lipoprotein stimulation and, 211, 212, 214
thrombin inhibitor vs. GPIIb/IIIa effect on, 255, 256, 258, 260, 264
urokinase receptor and, 218, 219, 220, 225
vascular endothelial growth factor and, 134
vascular fibrinolysis regulation by, 206–207
Plasminogen activator inhibitor 2 (PAI-2)
low density lipoprotein stimulation and, 211–214
urokinase receptor and, 218, 219
Plasmodium falciparum, 238
Platelet-activating factor (PAF), 154, 155
Platelet-derived growth factor (PDGF), 21, 106
A-chain of, 129–131, 135
B-chain of, 129–130, 135
mechanisms of potentiation in atherosclerosis, 129–131
vascular endothelial growth factor homology to, 135, 137
Platelet-derived growth factor B (PDGF B) gene, 313–314
Platelets
docosanoid biosynthesis by, 269–275
P-selectin mediation in neutrophil interactions, 199–203
thrombin inhibitor vs. GPIIb/IIIa antagonist effect on, 255–265
Pleiotrophin (PTN), 131
Poly(I:C) (PIC), 123
Polymerase chain reaction (PCR), 301
Polymerase chain reaction (PCR)-based subtraction library, 327–332
Polyunsaturated fat, 99
PPACK, 255, 256, 259, 260, 264, 282
Pravastatin, 100, 342
preβ-HDL, 35
Premature coronary artery disease (CAD), 97
hyperapobetalipoproteinemia and, 67–72
Procollagen I, 22
Procollagen III, 22
Procollagenase, 21

Program on the Surgical Control of Hyper-cholesterolemia (POSCH), 1, 335, 336
Proliferating cell nuclear antigen, 111
Proliferin, 110
Protease cascades, 236–237
Proteinase, 19–26, *see also* specific types
  low density lipoprotein receptor-related protein regulation of, 225–226
  restenosis and, 20–26
  wound healing and, 19–20
Protein b, 293
Protein C, 293, 296
Protein kinase C
  apolipoprotein A-I-containing particles and, 98
  hyperapobetalipoproteinemia and, 70–72
  impaired activation in Tangier fibroblasts, 79–86
  thrombin receptor and, 252
  tissue factor and, 237
Protein S, 293, 295
Protein tyrosine kinase, 111, 112
Proteoglycans, 20, 22
Prothrombin, 279, 281, 282, 283–284, 285–287
P-selectin, 185, 187, 191–195
  ligand of, 192–193
  L-selectin compared with, 174, 175, 177–178
  neutrophil adherence and, 153, 154, 155, 156
  neutrophil-platelet interactions and, 199–203
  phosphorylation of, 194–195
  platelet binding and, 255, 259, 262–263
  structure-function relationships in, 193–194
  thrombosis and, 191–192
P-Selectin Glycoprotein Ligand-1 (PSGL-1), 154, 192–193
Pyrrolidine dithiocarbamate (PDTC), 123, 124, 125

Quantitative trait loci (QTL) mapping, 300–301, 302, 303–304

R4/23, 161
Receptor-associated protein (RAP), 225, 226, 227, 228, 309–310
Redox-sensitive signals, 123–125
Renal cell carcinoma, 137
Resin, 1, 335
Restenosis, 19–26
  growth and cellular response of, 20–21
  proteinases and, 20–26
  vascular endothelial growth factor and, 139–140
Retinopathy, 138–139
RGD, 255, 257
  cyclic, 255, 257
  linear, 255, 259, 260
Rheumatoid arthritis, 138
Rickettsia conorii, 238
Ristocetin, 257
Rolling, *see* Leukocyte rolling
Rosetting inhibition, 200–201

SCRIP, 1
Second messenger pathways, 67–72
Septic shock, 238
Serotonin
  cholesterol lowering and, 348
  thrombin inhibitor vs. GPIIb/IIIa effect on, 255, 256, 258, 260–261, 264
SFLLR, 194
Shear stress response element (SSRE), 105–106
Shed L-selectin (sL-selectin), 179
Sialyl Lewis x (sLex), 187, 192, 193
Simple sequence length polmorphism (SSLP), 301
Simvastatin, 100
SKF 525-A, 271
SK&F 106760, 264
Smoking, 67, 343
Smooth muscle cells
  developmentally associated gene expression in, 145–149
  genetic modification of proliferation, 314
  intimal fibrous proliferation response and, 21
  lipid oxidation and, 41
Somatomedin C, 21
Src homology 2 domains, 111, 112
Stable angina, 12, 14, 15
STARS, 1
Sticking, *see* Leukocyte sticking
Strokes, 256, 348, 349
Stromelysin, 11, 15
Suicide, cholesterol lowering and, 347, 348
Symvastatin, 342
Syndrome X, 69
Synthetic peptides, 89, 90–91, 93

T3–10, 161
TAL.1B5, 161
Tamoxiphen, 334
Tangier disease, 36
  protein kinase C impairment in, 79–86
Tethered ligand hypothesis, 250
TF8–5G9, 240, 242
TFPI, 236, 238, 239, 240
Thrombin, 256, 259–262
  acute chest pain and, 295
  de-endothelialised aorta wall and, 280
  neutrophil adherence and, 154
  P-selectin phosphorylation and, 194
Thrombin inhibitors, 255–265
Thrombin receptors, 249–252
  shut off mechanism of, 251–252
  structure of, 249–250
Thrombomodulin, 293, 294, 296
Thrombosis
  plaque rupture and, 11
  P-selectin and, 191–192
  tissue factor and, 235, 238–239
Thrombospondin, 146
Thromboxane, 263, 265
  docosanoids and, 275
  platelet effects of, 256
d-Thyroxine, 335
Ticlopidine, 256
Tissue factor (TF), 211, 235–242
  cellular expression of, 237–239

Tissue factor (TF) (*cont.*)
function of, 239–241
protease cascades and, 236–237
selective molecular intervention in vivo, 241–242
structure of, 239
thrombosis and, 236, 238–239
Tissue factor (TF)-factor VIIa complex, 235, 236, 237, 239, 242
binding of, 240–241
catalytic function of, 240
Tissue inhibitors of metalloproteinase (TIMPs), 19, 20, 24–26
Tissue-type plasminogen activator (tPA), 211, 291
acute chest pain and, 293
low density lipoprotein receptor-related protein and, 225
restenosis and, 25–26
vascular endothelial growth factor and, 134
vascular fibrinolysis regulation by, 206
TK gene, 130
Toronto Working Group, 341
Transforming growth factorβ (TGF-β)
matrix metalloproteinases and, 21, 23
plasminogen activator 1 and, 211
urokinase receptor and, 217
Transforming growth factorβ1 (TGF-β1) gene, 313–314
Triggering risk factors, 1–6
Triglyceride-rich lipoprotein metabolism, 49–53
Triglycerides
apolipoprotein C1 and, 319
cholesterol efflux and, 92
Trousseau's syndrome, 238
Trypsin, 21
Tu102, 161
Tumor angiogenesis, 137–138
Tumor necrosis factor (TNF), 105
neutrophil adherence and, 154, 155
Tumor necrosis factorα (TNF-α)
plasminogen activator 2 increase and, 211
vascular cell adhesion molecule-1 and, 122, 123, 124
Tunicamycin, 213

U46619, 274
UCHL1, 161
Unstable angina
plaque rupture and, 12, 14, 15
platelet activation and, 256
thrombin receptor and, 252
Urokinase receptor (uPAR)
low density lipoprotein receptor-related protein and, 225–226, 227
structure, function and regulation of, 217–220
Urokinase-type plasminogen activator (uPA)
low density lipoprotein receptor-related protein and, 225–226, 227
low density lipoprotein stimulation and, 211, 212, 214
restenosis and, 21, 25
vascular endothelial growth factor and, 134
Urokinase-type plasminogen activator (uPA) (*cont.*)
vascular fibrinolysis regulation by, 206
US-SCOR, 1

Vascular cell adhesion molecule-1 (VCAM-1), 106, 162, 163, 173
antioxidants and, 121–125
in atherosclerosis, 165
in chronic periaortitis, 166
inflammation and, 121–122
leukocyte rolling and, 178
neutrophil adherence and, 153, 154
in normal aorta, 164
Vascular disease, genetic manipulation of, 313–314
Vascular endothelial growth factor (VEGF)
angiogenesis regulation by, 133–140
biological properties of, 134
isoforms of, 135
receptors of, 136–137
structural and genetic properties of, 134–135
Vascular endothelium, 105–106
Vascular fibrinolysis, 206–207
Vascular permeability factor (VPF), 133, 134
VCAM-1, *see* Vascular cell adhesion molecule-1
Vegetarian diet, 1, 3–4
VEGF, *see* Vascular endothelial growth factor
Very late activation antigen-4 (VLA-4), 122, 173
leukocyte rolling and, 178
L-selectin and, 174
neutrophil adherence and, 153, 154, 155
Very low density lipoprotein (VLDL), 307
apolipoprotein C-1 and, 317, 321
apolipoprotein E and, 57, 58–59, 60–65
hyperapobetalipoproteinemia and, 67, 68
insulin and, 51–52
metabolism of, 31, 32
β-Very low density lipoprotein (VLDL), 32, 225
apolipoprotein C-1 and, 317
apolipoprotein E and, 60, 62
properties of, 34
Very low density lipoprotein (VLDL) receptor, 223–224
Vitamin C, 41
VLA-4, *see* Very late activation antigen-4
von Willebrand Factor (vWF), 161, 293, 294, 295–296
v-sis gene, 129

Wilms' tumor, 130
Wound healing
proteinases and, 19–20, 21
vascular endothelial growth factor and, 137
WT1, 130–131

Xanthoma tuberosum, *see* Hyperlipidemia type III
XL086, 255, 257, 259, 262
XL111, 255, 257, 259, 264
XM648, 255, 257, 259, 264

Yeast artificial chromosome (YAC), 325
YY751, 255, 257, 259, 262–263, 264

Zinc finger, 111, 112

MIX
Papier aus verantwortungsvollen Quellen
Paper from responsible sources
FSC® C105338

If you have any concerns about our products,
you can contact us on
ProductSafety@springernature.com

In case Publisher is established outside the EU,
the EU authorized representative is:
Springer Nature Customer Service Center GmbH
Europaplatz 3, 69115 Heidelberg, Germany

Printed by Libri Plureos GmbH
in Hamburg, Germany